Performance, Technology and Application of High Performance Marine Vessels Volume One

Performance, Technology and Application of High Performance Marine Vessels Volume One

Edited by

Liang Yun, Raju Datla and Xinfa Yang

Cambridge Scholars Publishing

Performance, Technology and Application of High Performance Marine Vessels Volume One

Edited by Liang Yun, Raju Datla and Xinfa Yang

This book first published 2018

Cambridge Scholars Publishing

Lady Stephenson Library, Newcastle upon Tyne, NE6 2PA, UK

British Library Cataloguing in Publication Data
A catalogue record for this book is available from the British Library

ISBN (10): 1-5275-0356-9
ISBN (13): 978-1-5275-0356-4

CONTENTS

Preface by the Editors-in-Chief

Liang Yun Raju Datla Xinfa Yang

Liang Yun

Marine Research & Design Institute of China (MARIC), Shanghai, China

Raju Datla

Stevens Institute of Technology, New Jersey, USA

Xinfa Yang

Shanghai Association of the Shipbuilding Industry, Shanghai, China

The search for greater speed—humanity has a tendency always to want to go faster—has been enabled through increasingly efficient and lightweight power plants such as high speed diesel and gas-turbine engines, and lighter and stronger structural materials (aluminum alloy, GRP, titanium alloy), which have enabled the designers of fast boats, hydrofoils (HYC), and air cushion craft (ACC) and wing in ground effect craft (WIG) to develop performance close to the physical limits of speed on a seaway.

Over the last 30 years or so, a revolution in electronics has given us the possibility for automated stabilization of motion that was simply not possible before, together with big strides in power plant efficiency, not to mention satellite navigation. These have been important enablers of comfort at higher speeds and high-speed vessel development.

Over the last century, and particularly heavily over the last 50 years, designers, scientists, and various organizations, including commercial, military, and governmental, have dedicated resources to find ways in which combinations of hull geometries, hydrofoils, and static or dynamic air cushions can be used to deliver high-speed vessels that can perform very challenging missions. This work continues, still strongly driven by military objectives, and increasingly now by energy efficiency and environmental impact rather than simply the mission envelope defined by speed, payload, environment, and range.

A series of new variations of high-speed craft, or high-performance marine vessels (HPMV) have been developed in the last half century around the world, including improved high-speed monohull (HSMH) and planing monohull craft (PMC) from the 1940s, hydrofoils (HYC) from the 1950s, air cushion vehicles (ACV) and surface effect ships (SES) from the 1960s, small water plane area twin-hull craft (SWATH) from the 1970s, high-speed catamarans (HSCAT) from the 1980s, wave-piercing craft (WPC) from the 1990s, high-speed trimarans (HSTRI) in the first decade of the twenty-first century, and wing in ground effect craft (WIG) from the 1970s to the present. These various concepts and hybrids form an interacting group of vehicle concepts.

In this book we refer to this family of craft as *high-performance marine vessels* (HPMV), as the vessels not only are built for high speed but also may have other attributes, such as amphibious capability (air cushion vehicles) or extreme seaworthiness (SWATH). Specialists from some countries refer to such craft simply as high-speed craft (HSC); however, as use of the term *high-performance marine vessels* (HPMV) is now more common, we will use that description and acronym in this book.

Due to the increasing development of HPMV around the world, the International High-Performance Marine Vessels (HPMV) Conference—organized by The Royal Institute of Naval Architects, UK (RINA), Shanghai Association of Shipbuilding Industry (SASI) of China, Ship Design Academy Committee of CSNAME, and China Boat Industry and Trade Association (CBITA)—has since 1997 been held annually eighteen times in Shanghai, China. In that time we have collected about 500 papers contributed by hundreds of authors from various countries and regions around the world.

We know that many people have a strong interest in this technology; however, while many HPMV are in operation in different parts of the world, until now there has not been a publication giving an overview and discussing the performance, design, hydrodynamics, technology, and applications of various HPMV, as well as the differences and special features of them, and the approach to selection taken in various cases for both civil and military applications. Therefore, we have utilized the hundreds of papers presented by worldwide specialists at these conferences held over the past dozen years to create the present book—*Performance, Technology, and Applications of High-Performance Marine Vessels*—for readers around the world. Since the contents are so comprehensive and large, the book has been divided into two volumes, concerning the whole HPMV family. The book is ordered as follows:

Volume One

Chapter 1: High-Speed Monohull Craft (HSMH)
Chapter 1 covers the performance, technology, and application of conventional high-speed displaced monohull craft; deep-V monohull craft; planing hulls with and without steps, "M" craft, and other transit operation mode monohulls; axe bow shape monohulls; and other appendages and installations for improving the performance of high-speed monohulls, as well as their development evolution and prospect, and so on.

Chapter 2: Multihull Craft
Chapter 2 covers the evolution of trimaran craft around the world, the theoretical and experimental investigation of wave-making, wave-making interference between trihulls, and its drag as well as sea-keeping quality, and maneuverability of trimarans and pentamarans; the practical design of both trimaran ferries and main hull form and side hull form arrangement and investigation; and evolution and performance, technology of slice craft, and so on.

Chapter 3: Hydrofoil Craft (HYC)
Chapter 3 covers the development, research, and design of various types of hydrofoil craft, such as surface piercing HYC, automatic-control deep-submersible HYC, single and twin foils as well as triple-foil HYC, and foil-assisted catamarans; the theoretical and experimental investigation

and design of hydrofoils, practical applications of hydrofoil craft, and so on.

Chapter 4: Air Cavity Craft (ACC)
Chapter 4 covers the basic principles of low-speed air-lubrication craft and high-speed air cavity craft and their classifications; the evolution of and prospects for ACC; the research, design, production, and application of ACC in Russia, Ukraine, and Norway; experimental investigation of drag-reduction technology on both high-speed and non-high-speed craft through the use of air lubrication technology and sea-keeping of ACC, and so on.

Chapter 5: Wing in Ground Effect Craft (WIG)
Chapter 5 covers the development, research, design, and production of classic WIGs, power-augmented-ram wing in ground effect (PARWIG), dynamic air cushion wing in ground effect craft (DACWIG), and amphibious WIG (AWIG) in countries including Russia, Germany, China, and Japan; theoretical and experimental investigation of WIG craft on longitudinal stability, automatic control conception, and take-off ability improvement of the crafts, and so on.

Volume Two

Chapter 6: Air Cushion Vehicles (ACV), Surface Effect Crafts (SES), and Amphibious Seaplanes (ASP)
Chapter 6 covers the evolution and development of ACV, SES, and ASP; the research and design of the ACV and its skirt system as well as automatic control system for improving motion of ACV; concept design of large-type SES; theoretical and experimental investigation of speed performance, stability, impact of loads on amphibious seaplanes during take-off, and so on.

Chapter 7: Small Waterplane Area Twin-Hull Craft (SWATH), and Semi Small Waterplane Area Twin-Hull Craft (SEMI-SWATH)
Chapter 7 covers the development, research, performance, technology, and application of SWATH in China and around the world; experimental investigation of SWATH and power-performance compared with slice craft; the research and experimental investigation of the sea-keeping quality of SWATH; research and design of Semi-SWATH "X" craft for Littoral warfare requirement, and experimental investigation of Semi-SWATH crafts for the Taiwan Strait, and so on.

Chapter 8: Wave-Piercing Craft (WPC) and conventional catamarans
Chapter 8 covers the evolution and development of conventional catamarans and WPC; theoretical and experimental investigation of the powering performance and sea-keeping quality of conventional catamaran, and WPC; investigation, design, and optimization of hull lines of WPC; design of large-type WPC ferry crafts for civil applications in Taiwan Strait, and so on.

Chapter 9: Other Hybrid High-Performance Marine Vessels
Chapter 9 covers development, research, design, performance, technology, and application of various hybrid HPMV, such as planing catamarans (PCAT), supercritical planing twin-hull craft (SPTH), hydrofoil submersible SWATH (HYSWAS), high-speed trimarans with anti-wave-making hydrofoil craft, and so on.

Chapter 10: Material and Hull Structure on High-Performance Marine Vessels
Chapter 10 covers the research and design of special characteristic steel and GRP materials that apply to HPMVs; the research and design of aluminum hull structures of ACVs using extruded panel structures; theoretical and experimental investigations on wave-induced loads, dynamic response, and hull-structure analysis of SWATH, HSMH, and also HSCAT. It also considers mechanical impedance analysis methods applied to hull design of HPMVs, and hydro-elastic analysis of SWATH hull structure in waves.

Chapter 11: Propulsion and Power Plant on High-Performance Marine Vessels
Chapter 11 covers CFD and the dynamic simulation of water inlets and water pipelines of water jet propulsion systems; water jet propulsion technology and its application to HPMVs; study of two-phase-flow water-jet propulsion; the latest development of surface-piercing propellers at the China Ship Scientific Research Center (CSSRC), the introduction of air fan propulsion on ACVs; simulation study of intelligent control water jet propulsion system of WPC, and on the lift and propulsion power installation of the control system of an ACV.

Chapter 12: General survey
Chapter 12 covers the general description of HPMVs; classification, development, performance, technology, and application of HPMVs; test methods of HPMVs; some HPMVs and technical themes that are not

covered in other chapters, such as some additional appendages for improving performance of HPMV, and so on.

The possible readers of this book

The book is written particularly for the following readers:

- Students and teachers in naval architecture and marine engineering, aviation engineering, mechanical engineering, and other concerned faculties of universities and institutes
- Staff members, technicians, engineers, researchers of marine transportation units, shipping companies, shipyards, and ship and aviation research institutes, as well as other concerned units for both civil and naval organizations
- Everyone who is interested in HPMVs in both military and civil applications

The editorial committee of the book is composed as follows:

Liang Yun	Marine Research & Design Institute of China (MARIC)	Editor-in-Chief, and Editor for chapters 4 and 5
Raju Datla	Stevens Institute of Technology, New Jersey, USA	Editor-in-Chief
Xinfa Yang	Shanghai Association of the Shipbuilding Industry, Shanghai, China	Editor-in-Chief
Zushun Dong	Naval University of Engineering, Wuhan, China	Editor for Chapters 1, 8, and 9
Qijun Ni	China Ship Scientific Research Center (CSSRC), Wuxi, China	Editor for Chapters 3 and 7
Yunbo Li	Harbin Engineering University, Harbin, China	Editor for Chapter 2
Debo Huang	Harbin Engineering University, Harbin, China	Editor for Chapter 2
Xiaoping Li	Marine Design & Research Institute of China (MARIC)	Editor for Chapters 6 and 11
Guoyu Chen	Shanghai Marine Steel Institute, Shanghai, China	Editor for Chapter 10
Ning Ma	Shanghai Jiao Tong University, Shanghai, China	Editor for Chapter 12
Qikang Liang	Marine Design & Research Institute of China (MARIC)	English Language Editor
Jiang Yuqing Zhao Lan Sherry Li Shao Jie Lv	China Boat Industry & Trade Association (CBITA) Shanghai Association of Shipbuilding Industry, Shanghai, China	Secretaries of the Editorial Committee

Acknowledgements

The editors of this book would like to express their sincere thanks to: Mr. Huang Pingtao, President of China Society Naval Architecture & Marine Engineering (CSNAME); Mr. Trevor Blakeley, CEO of the Royal Institution of Naval Architects (RINA); Mr. Zhou Zhenbai, President of Shanghai Association of Shipbuilding Industry, (SASIC) of China, for their warn help during the editing of these books, and also Xianhu Wu, Fellow, Institute of Naval Equipment, China, Youjing Zhang, Fellow, China Ship Scientific Research Center (CSSRC), Hongye Zhuang, Fellow, CSSRC, and Zhenyu Huang, Fellow, CSSRC, for their English language editing of this book.

List of authors of the papers presented at the Annual China HPMV Conference included in the books

No.	First name	Family name	Affiliation	Country	Section
1	Zushun	Dong	Naval University of Engineering	China	1.1, 1.10, 1.11, 1.17
2	Yoshida	Yasushi	Wessex Institute of Technology	UK	1.2
3	Dong Jin	Kim	Korea Institute of Ocean Science and Technology	Korea	1.3
4	Sun Young	Kim	Korea Institute of Ocean Science and Technology	Korea	1.3
5	Seong Hwan	Kim	FAR EAST Ship Design and Engineering Co.,Ltd.	Korea	1.3
6	Jeong Hwa	Seo	Seoul National University	Korea	1.3
7	Shin Hyung	Rhee	Seoul National University	Korea	1.3
8	Lixin	Xu	American Bureau of Shipping	USA	1.4
9	Yung S.	Shin	American Bureau of Shipping	USA	1.4
10	Xiaohong	Shen	China Ship Development and Design Center	China	1.5, 1.6
11	Wugang	Huang	China Ship Development and Design Center	China	1.6
12	Qingxu	Wang	China Ship Development and Design Center	China	1.7
13	Lei	Yue	China Ship Development and Design Center	China	1.7
14	Chengxia	Zhang	Hudong Shipbuilding Group, Shanghai	China	1.8
15	L.	Koh	University of Newcastle upon Tyne	UK	1.9
16	C.-E.	Janson	Chalmers University of Technology	Sweden	1.9
17	M.	Atlar	University of Newcastle upon Tyne	UK	1.9
18	L.	Larsson	Chalmers University of Technology	Sweden	1.9
19	E.	Mesbahi	University of Newcastle upon Tyne	UK	1.9
20	C.	Abt	FRIENDSHIP-Systems GmbH	Germany	1.9
21	Song	Wei	Naval University of Engineering	China	1.10
22	Qing	Zeng	Naval University of Engineering	China	1.10, 1.11
23	Xiaopeng	Gao	Naval University of Engineering	China	1.10, 1.11, 1.17
24	Qingwei	Wu	Naval University of Engineering	China	1.11, 1.17
25	Marcos	Salas	University Austral of Chile	Chile	1.12
26	José	Rosas	University Austral of Chile	Chile	1.12
27	Andrea	Mancini	Insean—Italian Ship Model Basin	Italy	1.13

28	Alessandro	Moriconi	Insean—Italian Ship Model Basin	Italy	1.13
29	Shuai	Yang	China Ship Development and Design Center	China	1.14
30	Wei	Xie	China Ship Development and Design Center	China	1.14
31	Huimin	Li	China Ship Development and Design Center	China	1.14
32	Shuzheng	Sun	Harbin Engineering University	China	1.15, 1.16
33	Quanming	Miao	China Ship Scientific Research Center	China	1.15
34	Xiaodong	Zhao	Harbin Engineering University	China	1.15, 1.16
35	Jide	Li	Harbin Engineering University	China	1.15, 1.16
36	Jun	Shi	Shenzhen Hispeed Boat Tech & Dev Co. Ltd	China	1.17
37	Yunbo	Li	Harbin Engineering University	China	2.1, 2.2
38	Dobo	Huang	Harbin Engineering University	China	2.2
39	Wenyang	Duan	Harbin Engineering University	China	2.2
40	Guangli	Zhou	Harbin Engineering University	China	2.2
41	Kaijia	Han	Harbin Engineering University	China	2.2
42	Jianghua	Li	Harbin Engineering University	China	2.2
43	Yuming	Sun	Naval University of Engineering	China	2.3
44	Xiaoping	Lu	Naval University of Engineering	China	2.3, 2.4, 2.5
45	Zhong	Wang	Naval University of Engineering	China	2.3
46	Yucun	Pan	Naval University of Engineering	China	2.3
47	Jinlin	Zhan	Naval University of Engineering	China	2.5
48	Shulong	Zhang	Naval University of Engineering	China	2.5
49	Yang	Chen	Naval University of Engineering	China	2.5
50	Zhirong	Shen	Shanghai Jiao Tong University	China	2.6
51	Juan	Ma	Shanghai Jiao Tong University	China	2.6
52	Decheng	Wan	Shanghai Jiao Tong University	China	2.6
53	Lei	Gao	China Ship Scientific Research Center	China	2.7
54	Liang	Chang	China Ship Scientific Research Center	China	2.7
55	Gengyao	Qiu	China Ship Scientific Research Center	China	2.7
56	Zhi	Zong	Dalian University of Technology	China	2.8
57	Jingbei	Jia	Dalian Ocean University	China	2.8

58	Mei	Tan	CIMC Ship & Offshore Research Institute	China	2.9
59	Fei	Xiong	CIMC Ship & Offshore Research Institute	China	2.9
60	Zhongfu	Wang	CIMC Ship & Offshore Research Institute	China	2.9
61	N. A.	Armstrong	Austal Ships	Australia	2.10
62	V.	Moretti	Austal Ships	Australia	2.10
63	E.	Begovic	University of Naples “Federico II”	Italy	2.11
64	C.	Bertorello	University of Naples “Federico II”	Italy	2.11
65	S.	Caldarella	University of Naples “Federico II”	Italy	2.11
66	P.	Cassella	University of Naples “Federico II”	Italy	2.11
67	John	Kecsmar	Ad Hoc Marine Designs	Japan	2.12
68	Nigel	Warren	Ad Hoc Marine Designs	England	2.12
69	John	Moore	Lockheed Marine	USA	2.12
70	Qijun	Ni	China Ship Scientific Research Center	China	3.1, 4.5, 4.7, 4.9, 4.10
71	Daxiong	Zhang	China Ship Scientific Research Center	China	3.2.1, 3.3.2
72	Günther	Migeotte	CAE-Marine, Mechanical Engineering Laboratories, Stellenbosch University	South Africa	3.3.1
73	Nikolai	Kornev	University of Rostock	Germany	3.3.1
74	Mo	Gui	Director of Port Division, Headquarter of Shanghai Deepwater Port Construction	China	3.3.2
75	Gennady A.	Pavlov	Joint Stock Company FEODOSIA SHIPBUILDING COMPANY “MORYE”	Ukraine/Russia	3.3.2, 4.2, 4.3
76	Chengjun	Gao	China Ship Scientific Research Center	China	3.3.3
77	Haigang	Cheng	China Ship Scientific Research Center	China	3.3.3
78	Guoping	Pan	China Ship Scientific Research Center	China	3.3.3, 3.3.4
79	Faming	Zhao	China Ship Scientific Research Center	China	3.3.3, 3.3.4
80	Dong	Guo	China Ship Scientific Research Center	China	3.3.4, 3.5.1
81	Quanshui	Huang	China Ship Scientific Research Center	China	3.3.4
82	Wei	Wu	Naval University of Engineering	China	3.4.1
83	Ying	Xiong	Naval University of Engineering	China	3.4.1
84	Xiaochuan	Wang	China Ship Scientific Research Center	China	3.5.1

85	Liang	Yun	Flying Dragon (& Engain) Science & Technology LTD (Hong Kong)	China	4.1, 3.3.2, 4.3, 5.1, 5.5
86	U. S.	Tudem	Effect Ships International AS	Norway	4.4
87	T. K.	Livgård	Effect Ships International AS	Norway	4.4
88	Hongxia	Cheng	China Ship Scientific Research Center	China	4.5, 4.7, 4.9
89	Dong	Yang	China Ship Scientific Research Center	China	4.5
90	Shengde	Xing	China Ship Scientific Research Center	China	4.5, 4.7, 4.9
91	Jinqi	Cai	Shanghai Qing Hang Air-filmed Resistance-reduced Ship Co., Ltd	China	4.6
92	Shulong	He	China Ship Scientific Research Center	China	4.7
93	Wencai	Dong	Naval University of Engineering	China	4.8
94	Rixiu	Guo	National Laboratory of Hydrodynamics	China	4.8
95	Xiwu	Liu	National Laboratory of Hydrodynamics	China	4.8
96	Yangchun	Yu	China Ship Scientific Research Center	China	4.9
97	Liyan	Wang	China Ship Scientific Research Center	China	4.10
98	Shengzhong	Li	China Ship Scientific Research Center	China	4.10
99	Yajun	Shi	China Ship Scientific Research Center	China	5.2, 5.7
100	Wei	Zhang	China Ship Scientific Research Center	China	5.2
101	Xiunai	Ye	China Ship Scientific Research Center	China	5.2
102	Enguang	Wang	Aviation Industry Corporation of China Special Vehicle Research Institute	China	5.3
103	Lintang	Chu	Aviation Industry Corporation of China Special Vehicle Research Institute	China	5.3
104	Mingde	Song	Aviation Industry Corporation of China Special Vehicle Research Institute	China	5.3
105	V. V.	Kolganov	Joint Stock Company "Research & Production Complex “TREK”	Russia	5.4
106	V. G.	Sergeev	Joint Stock Company "Research & Production Complex “TREK”	Russia	5.4

107	Hiromichi	Akimoto	Department of Applied Mathematics & Physics, Faculty of Engineering, Tottori University	Japan	5.6
108	Syozo	Kubo	Department of Applied Mathematics & Physics, Faculty of Engineering, Tottori University	Japan	5.6
109	Takahiro	Taketsume	Department of Applied Mathematics & Physics, Faculty of Engineering, Tottori University	Japan	5.6
110	Xinfa	Xu	China Ship Scientific Research Center	China	5.7, 5.8
111	Changhua	Yuan	China Ship Scientific Research Center	China	5.7, 5.8
112	Qian	Zhou	China Ship Scientific Research Center	China	5.8

PREFACE

TREVOR BLAKELEY FRINA FIMAREST, FIMECHE CENG

Chief Executive, Royal Institution of Naval Architects

Over the past twenty years, the high-speed sector of the maritime industry has seen much development in the design, construction, and operation of commercial, military, and recreational marine vessels, as it has responded to the challenge of operators, regulators, and society for greater efficiency and safety and, increasingly, the protection of the environment. Such development has sought to reduce production costs, extend operating envelopes, reduce downtime, and increase comfort, reliability, and safety. In more recent years, the sector has also faced the challenge of society to reduce the environmental impact of operating high-performance marine vessels through reducing emissions.

The high-speed sector has responded successfully to these challenges, and a key to that success has been the innovation that the sector has readily sought and adopted. Essential to that innovation has been research and development. The High-Performance Marine Vessels (HPMV) Conferences have provided a valuable opportunity for individuals from all parts of the sector to exchange and discuss the technical and scientific information that has resulted from that research and development.

As well as providing a record of the technical and scientific information presented at the HPMV Conferences, the book also serves to chart the development of the high-speed sector over 20 years, recording its success in meeting the challenges it has faced. The book is also a testament to the contribution that the HPMV Conferences have made to the successful development of the sector. The Royal Institution of Naval Architects is proud to have been associated with the HPMV Conferences as a co-organizer.

December 2, 2016

PREFACE

GUO DA-CHENG

President of China Association of the National Shipbuilding Industry

Since 1996 the Royal Institution of Naval Architects (RINA), UK, the China Boat Industry and Trade Association, and the Ship Design Academic Committee and Mechanics Academic Committee of the China Society of Naval Architecture and Marine Engineering (CSNAME) have jointly held the China International Annual High-Performance Marine Vessels Conference (HPMV) in Shanghai, China. The exploration and persistence exhibited over the past twenty years have developed the conference into a famous academic event in China's high-performance shipbuilding industry. The HPMV Conference has published 19 proceedings with more than 553 papers from 30 countries and regions across the world. It reflects the continual endeavors and hard work of specialists and organizers in China and around the world who are dedicated to the high-performance shipbuilding industry. The HPMV proceedings have been recognized not only by the Royal Institute of Naval Architects (RINA) but also by domestic universities, research institutes, and the manufacturing industry. The collected papers have been enthusiastically spread and exchanged.

Therefore, Cambridge Scholars Publishing has decided to sponsor the publication of the HPMV Collected Papers. The first volume contains papers that have been selected from among those on many types of high-performance vessels, on new materials, new technologies, new design ideas, new equipment, new structural styles of the HPMV, and so on. This first volume of the Collected Papers will be distributed globally. It not only encourages Chinese professionals engaged in high-performance ship research circles, but also promotes high-performance ship research worldwide. Furthermore, Cambridge Scholars Publishing promotes worldwide academic exchange with charity. We do hope that the publication of the anthology will motivate more extensive academic exchanges and interaction between people around the world. HPMV is a great worldwide event in the field of high-performance ship research and industry that is highly deserving of recognition and celebration and is of far-reaching significance.

As one of the largest and oldest domestic and international academic conferences, the HPMV Conference has set up an effective international academic exchange platform and has provided an opportunity for circulating the most cutting-edge technologic research and development information. It will promote the healthy and sustainable development of China's high-performance shipbuilding industry effectively. My hope is that that HPMV conference will continue to be organized better and better and will attract even more professionals from home and abroad engaged in the field. At the same time, excellent results can be applied to social practice. The HPMV Conference will become a compass for leading high-performance ship research and design in China.

With sincere thanks to Cambridge Scholars Publishing for their understanding, support, and philanthropic action.

December 28, 2016

PREFACE

HUANG PING-TAO

Chairman of China Society of Naval Architecture and Marine Engineering (CSNAME)

With the growth of the Chinese economy and the expansion of the shipbuilding industry, high-performance marine vessels are gradually growing and developing. Last century, in the mid-1990s, hydrofoil crafts appeared on the Pearl River Delta, then hydrofoils and hovercrafts; second-hand hydrofoils purchased from Russia operated in the Chongqing area. After the highways were completed along the Yangtze River, these high-speed ships soon disappeared. Nevertheless, the curtain had been raised on research and construction of high-performance ships in China. Next came high-performance ships, including SWATH, catamarans, wave-piercing catamarans, high-speed crafts with jet propulsion, semi-submersible propeller ships, and deep-V boats, as well as wing in ground effect craft and other high-performance ships with maximum speeds up to 70 kn. These craft were used not only as commercial passenger ships, yachts, and workboats, but also for military applications. Since 1996 the Royal Institute of Naval Architects (RINA), the China Boat Industry and Trade Association, the Shanghai Association of Shipbuilding Industry, the Ship Design Committee of China Society of Naval Architecture and Marine Engineering (CSNAME), and the Mechanics Committee of CSNAME have co-hosted the International Annual Conference on High-

Performance Marine Vessels (HPMV). The HPMV Conference has been held successfully for twenty-one years. The work of the organizing committee is very insightful and fruitful. The HPMV Conference has effectively strengthened exchanges and communications between domestic and overseas technology circles on the research, design, application, and development of high-performance ships.

Here, we would like to thank Prof. Yun Liang. Many years ago, in the late 1950s, when I was a student, Professor Yun researched and developed the hovercraft. He was engaged in the research, design, development, and application of high-performance ships for 60 years. As the organizer of the HPMV Conference he has done a great deal of work, and has obtained the support of Cambridge Scholars Publishing to issue the Collected Papers of the HPMV Conference. This is a highly influential and meaningful event for domestic scientists, engineers, and technicians in the field of high-performance marine vessels. The Collected Papers presents the latest achievements in the research, design, and application of high-performance ships at home and abroad, which is helpful for the innovation of ideas, technology, and application in this field. At same time we also thank the authors; the papers are the crystallization of their many years of labor, their experiences of success and failure, and their extremely hard work.

The publication of the collection is also a powerful push and inspiration for the HPMV Conferences; it will appear as a new starting point, and inspire greater and newer results.

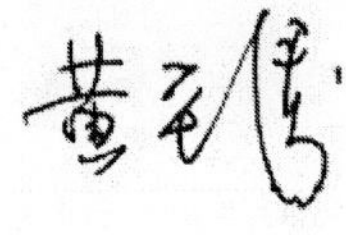

December 28, 2016

PREFACE

ZHANG GUANGQIN 张广钦

Former Vice Minister of the Commission of Science, Technology, and Industry for National Defense of the People's Republic of China

Former Deputy General Manager of China Shipbuilding Industry Corporation

Former President of China Association of National Shipbuilding Industry

The International Annual Conference on High-Performance Marine Vessels, held in Shanghai, China, has collected around 553 papers over the past 21 years. During that time the proceedings of the conference have presented the research, development, and design results of both Chinese and overseas experts on the technology of high-performance marine vessels. The papers gathered from the conference are not only the witnesses of history but also the crystallization of the authors' sweat and wisdom.

Research on high-performance marine vessels is an endless business. I hope that the publication of these Collected Papers will push and spur on the spirit of innovation and creativity in the Chinese shipbuilding industry. At the same time, we firmly believe that the papers will assist the industry's most outstanding talents to study hard, boldly innovate, and realize their new design ideas in the shipbuilding industry. In turn this will lead to more fruitful results and make a great contribution to the prosperity and development of China's and the world's shipbuilding industry by creating an elevated position from which to start.

In publishing this book, Cambridge Scholars Publishing is promoting the academic exchange on high-performance ships globally. Its philanthropy in sponsoring and issuing this book worldwide should be praised. I would like to express my heartfelt thanks to Cambridge Scholars Publishing. Furthermore, this will encourage me to host the HPMV conference better.

Participating in and supporting the HPMV International Annual Conference in Shanghai, China, is my job with joy and happiness.

PREFACE

ZHOU ZHEN-BO

Director of Shanghai Association of Shipbuilding Industry

Co-chairman of Organizing Committee of HPMV International Conference

The Collected Papers of the High-Performance Marine Vessels Conference brings together numerous research results and scientific innovations on high-performance marine vessels, contributed by both domestic and overseas specialists over the past 21 years.

With the rapid development of the Chinese economy it was necessary to meet the requirements of various domestic commercial vessels, working ships, and transport ships running on rivers, lakes, and oceans. Thus, we gathered worldwide specialists' mature experience of high-performance marine vessels and combined this with domestic development and innovation in this field to make the academic forum that became the International Annual HPMV Conference in Shanghai, China. For ship specialists, the conference is not only an important way to improve academic ability and promote professional growth, but also a powerful tool with which to obtain new technical information and achieve new leaps in the development of the R/D idea of HPMV, and a way to renew and transform ideas in the research and design process. The editorial board fully recognizes this and is pleased to see that ship specialists diligently and enthusiastically participate in this meaningful work.

The HPMV Collected Papers presents a wide spectrum of research and design ideas of worldwide scholars, both individual and shared. These papers embody the hard work and wisdom of scholars and specialists in the shipbuilding industry, and many of them are insightful and profound masterpieces. It realizes the future development of the HPMV research field in China and all around the world. Each contributor has studied the basic theory, has carried out experimental investigations and has practiced, and then has researched continuously and repeatedly. Through these processes, they have found extraordinary ideas and innovations in the research and design of HPMV. This is the real wealth—there will be explorations of ideas and designs, developments, and great successes as well as bright expectations.

The book condenses the hard work of scholars and their painstaking study. The HPMV Collected Papers deeply reflects the research and innovative results and hard work of the outstanding specialists in the world. At same time, I would like to express my sincere thanks to the organization committee members of the HPMV conference for their careful preparation of this book. I do hope the HPMV International Annual Conference will be continuously held in Shanghai, China. This is of great significance to technology communication and cooperation between China and the rest of the world in the HPMV field.

周振柏

December 26, 2016

CHAPTER 1

HIGH-SPEED MONOHULL CRAFT

1.1

On the Development of High-Speed Monohull Craft

DONG ZUSHUN

Navy Engineering University, China

1.1.1. Introduction

Abstract

1. Various types of novel high-speed marine vessels have appeared since the 1950s, such as ACV, SES, HYC, HSCAT, WPC, and WIG. Even though such HPMV have different advantages in terms of performance, and have been widely applied in various fields, the most widely used craft (including, in terms of number of craft) among the HPMV community are still high-speed monohulls (HSMH), due to their well-proven and simple technologies and because they are cost-effective, reliable, practicable, and easy to maintain.

2. Even though HSMH craft have been relatively well-proven technically, with many applications around the world, to continuously improve performance, particularly sea-keeping quality, researchers and designers have taken various steps to improve. For example, in the past half century, these have included using a deep-V hull profile with a high dead-rise angle on the hull, using a tunnel planing hull, and using an "M" type of craft.

With respect to semi-planing hulls, or so-called transit planing craft, to improve sea-keeping quality the following have also been widely used: a hull profile with a round bilge and/or mixed hard chine with a round bilge hull profile; a deep-V profile. A newly innovated advanced "X" hull type has also appeared in the United States in recent years.

To sum up, much research and investigation into planing hulls and semi-planing hulls has been carried out worldwide in recent years!

3. In addition, some additional devices or appendages have also been widely installed on such crafts to improve their performance, such as stern wave-press plates, wedge plates, stern interceptors, semi-submersed bows (SSB), and bow hydrofoils, and so on. On the basis of this research and these investigations and innovations, the navigation performance and seaworthiness of HSMV has been improved significantly, especially in terms of navigation performance in waves. This has also been examined in the papers of past proceedings, the findings of which can be outlined as follows.

1.1.2. Planing hull

1. Research and investigation into performance prediction and optimization

Although high-speed planing hulls have been developed over many years, with well-proven technologies and simple structures that have lowered costs and are realizable and practicable, R/D is still continuously undertaken to further improve safety and enhance its transportation efficiency as mentioned above. In particular, some challenges, such as seaworthiness during high-speed planing operation mode in waves, have been realized in the proceedings of the International China HPMV Conference.[1-2-1] This has included presentations on hull profiles, which have significantly influenced the linear or un-linear speed and running attitude equations of planing hulls, and have also included presentations on predicted methods for estimating the drag of HSPH. The author in reference 1-2-1 provided an effective idea and method for quickly calculating the drag of HSMH, without a model towing tank test.

Researchers and designers have stressed the investigation of craft drag for a long time; however, recently they also paid more attention to seaworthiness, and have contributed more papers and research results with integrated consideration of both speed performance and seaworthiness for the hull profile and lines design of HSMH, as shown in reference 1-1-2, where the author used four lines with different features, such as different main hull parameters, via the model towing tank test to compare their drags and sea-keeping quality and guide the hull-lines design effectively.

Ref. 1-1-3 investigated the transverse stability of HSMH with a prismatic planing surface in the case of high-speed running in heeling. With respect to the motion stability of HSPH, researchers used to focus on longitudinal motion stability (i.e., coupled pitching with heaving motion) and the so-called porpoise motion problem but were less concerned about transverse motion. Actually, transverse overturning used to happen, especially at high speeds when turning using a large rudder angle. In the paper, the author established roll-motion equations by considering the influences of rolling angles and rolling angular velocity, and arrived at a solution by analyzing the influences of the main parameters on transverse stability, and developing criteria for judging transverse stability. The analysis is of interest and is useful for craft design to improve transverse stability, even if it only limits hulls with prismatic planing surfaces.

2. R/D of craft types
Other than conventional planing hulls, many papers in the HPMV Proceedings describe the R/D of non-conventional types, such as the multi-step planing hull and M-type planing hull.

Reference 1-1-4 carried out a series of towing tank model tests for planing hulls with double steps. In view of the shortage of prediction methods for speed performance for planing hulls with double steps—and also to widely use such types of craft—the method seems significantly to aid designers in selecting design parameters.

Reference 1-1-5 carried out a series tests for M-type craft with different hull parameters: keeping the hull length and displacement unchanged, and considering different widths of hull and air ducts, the author calculated the effect of hull and air-duct width on drag. The test results demonstrated that increasing the hull beam or air duct is favorable to the drag, and increasing the hull beam also increases the hydrodynamic lift. Increasing duct width leads to an increase in aerodynamic lift; however, excess increased duct width might lead to motion instability, possibly due to flow instability in the air cushion, to which more attention should be paid.

Reference 1-1-6 investigated the calculation method for the drag of M-type craft by means of the numerical simulation method, and compared this with the model test results. This demonstrated that the difference between them can be controlled by 10% at low speed; nevertheless, the calculation couldn't be used for high speeds.

1.1.3. High-speed displaced monohull

High-speed displaced monohull crafts are extremely popular, and are widely and comprehensively used throughout the world. Due to this, even though the HSPMH is characterized by very high speeds, it also suffers from serious craft motion when running in waves, which affects the comfort of passengers and crews. In addition, many craft don't need very high speeds; instead, more equipment needs to be installed to improve living and working conditions, so as to increase the payload and displacement significantly, and decrease the relative speed of the crafts significantly. For example, a craft with a displacement of 80t and a speed of 42kn is F_v=3.32 (Froude number based on craft volume). It is still running at planing mode; however, if the displacement of this craft is increased to 300t at the same speed, 42kn, we get F_v=2.67, which is located in transit operation mode, which may be designed as a high-speed displaced ship type. Such phenomenon had been discovered in the 1960s, when almost all guide missile ships around the world were designed as transit craft with big displacements rather than planing craft with small displacements. To improve the sea-keeping quality of such crafts, the craft profile should transit from a round bilge to a short hard chine, mid hard chine, and long hard chine, respectively, and eventually to a deep-V type.

Reference 1-1-7 carried out an investigation into heeling problems of deep-V craft during turning, and analyzed the difference in hydrodynamics acting on craft between conventional round-bilge craft and deep-V craft; it also established the dynamic equations for deep-V crafts during stable turning. After simplifying the process, the heeling angle of deep-V craft during stable turning could be obtained. The paper also analyzed the effect of main-hull and motion parameters to stable turning, and made a comparison with model-test calculations, with rather a small error margin, so as to apply such methods to the hull configuration design of deep-V craft.

Reference 1-1-8 developed deep-V craft with a lower Froude number (F_n=0.2–0.388), as well as by adding a bulbous bow and improving the stern profile, and via the artificial nervous net method (ANN) and gradient motion (MMA) to optimize the drag performance.

Delft University in the Netherlands and the Daman Shipyard cooperated successfully to research and develop a new type of deep-V craft. Its basic measure was to extend the ship length with an unchanged hull beam. Since the thickness of hull plates is defined according to the requirement for anti-corrosion rather than hull strength, for small ships an increase in craft length does not significantly result in an increase in hull weight; consequently, an increase in length can improve the speed and sea-keeping performance of such craft. Meanwhile, the "axe bow" and the bow keel dropping down of deep-V type craft creates a decrease in the buoyancy variation rate, induced by waves, which is also favorable to sea-keeping quality. After continuous improvement of such types of craft, the "axe bow" type has been applied to ships of medium size, and even of large size. References 1-1-9 and 1-1-10 carried out comparison tests of drag and sea-keeping quality for both conventional deep-V craft and "axe bow" craft, and produced useful results to aid correcting design ideas and principles.

1.1.4. Studies on additional appendages for improving operational performance

Another research area on HSMH concerns additional appendages of the craft to improve operational performance; significant results have been obtained, such as the addition of stern wave depressed plates, stern wedged plates, stern interference plates, bow submerged bodies, stern/bow fins and hydrofoils, and so on.

1. Stern depressed plates and interference plates

Reference 1-1-11 investigated the effect of stern depressed plate size on the drag performance of craft. The author studied six groups of projects focusing on different chords, spans, and installation angles, and carried out tests on drag and trimming angle of a model with half displacement, to compare drag, thus obtaining favorable parameters for stern depressed plates of practical design.

Reference 1-1-12 introduced the action and effect of an interceptor or intruder, and/carried out model tests of three planing and semi-planing hulls. It demonstrated that with the aid of interceptors to adjust the trimming angle, the drag could be reduced significantly in either planing or semi-planing crafts.

Reference 1-1-13 presented the function theory of the interceptor, and also, via analysis of model test results, gave a design method for determining the interference height of the stern interceptor.

2. Application and development of semi-submerged bow (SSB)

Reference 1-1-14 introduced the performance investigation of SSB with a fin installed on a high-speed displaced ship (SSB with fin). The author had carried out research on such a complex body installed on a high-speed displaced ship weighing 400t. The model test and real-ship trial evidenced that the drag of craft in calm water decreased, and sea-keeping quality also improved significantly. However, in the case of big ships, ship drag in calm water was increased, with improvement of sea-keeping quality. Therefore, the author improved the complex body and also added stern plates, to adjust the running attitude, and got a good result—that is, one that improved sea-keeping quality and also did not increase drag significantly in calm water.

Reference 1-1-15 introduced research on operational performance of ships weighing from 400t to 4,000t, with various installed SSB of different profiles and sizes. The research demonstrated that sea-keeping quality could be improved significantly on ships weighing up to 4,000t, but drag in calm water increased significantly.

Reference 1-1-16 described a complex appendage with both SSB with fin and bow fin to further improve sea-keeping quality and simultaneously improve speed performance, which might be useful for development and applications of large axe bow type ships.

References

[1-1-1] Yasushi Yoshida, "The Concise Hull Form Parameters to Estimate a Planing Boat's Resistance," 2003, E-07.

[1-1-2] Dong Jin Kim et al., "Development of High-Speed Planing Hull Forms with Consideration of Resistance and Sea-Keeping Performance," 2013, E-03.

[1-1-3] Lixin Xu & Yung S. Shin, "Simulation Transverse Motion and Dynamic Stability for High Speed Planing Craft," 2001, E-03.

[1-1-4] Shen Xiaohong, "Series Test Research on Linear Form of Double Planing Boat," 2014, E-08.

[1-1-5] Huang Wugang & Shen Xiaohong, "Study on Lines of M-Form Hull," 2014, E-07.

[1-1-6] Wang Qingxu & Yue Lei, "The Hull Characteristics and Numerical Simulation without Grids of M-Form Hull," 2013, E-13.

[1-1-7] Zhang Chengxia, "A Probe into Turning Heel of Deep-V Vessel," 2000, E-06.

[1-1-8] L. Koh, C. E Janson, et al., "Novel Design and Hydrodynamic Optimization of a High Speed Hull Form," 2005, E-01.

[1-1-9] Wei Song, Zeng Qing, et al., "Model Experimental Investigation on Resistance of Axe Bow Ship," 2011, E-24.

[1-1-10] Wu Qingwei, Zeng Qing, et al., "An Axe Bow Ship of Excellent Sea-keeping," 2012, E-06.

[1-1-11] Marcos Salas & Jose Rosas, "Hydrodynamic Analysis of the Performance of Stern Flaps in a Semi-Displacement Hull," 2003, E-06.

[1-1-12] A. Mancini & A. Moriconi, "Intruder: A Device to Improve Hull Performance," 2006, E-07.

[1-1-13] Yang Shuai, Xie Wei, & Li Huimin, "Estimation of Effect and Size on Ship Interceptor," 2014, E-04.

[1-1-14] Sun Shuzheng, Miao Quanming, Zhao Xiaodong, & Li Jide, "Research on Model Test for 1000t Hybrid Monohull with Stern Flap," 2013, E-13.
[1-1-15] Sun Shuzheng, Li Jide, & Zhao Xiaodong, "Investigation of Built-up Appendage Applied on High Speed Round Bilge and Deep-V Hulls for Motion Performance Improvement," 2014, E-03.
[1-1-16] Shi Jun, Wu Qingwei, et al., "Sea-keeping Research on Axe Bow Ship with Appendage," 2014, E-05.

1.2

The Concise Hull-Form Parameters to Estimate a Planing Boat's Resistance (2003)

Yasushi Yoshida

Wessex Institute of Technology, UK

Nomenclature

C_f	Frictional resistance coefficient
C_i	A set of coefficients
C_t	Model's total resistance coefficient. Total resistance is divided by $(\frac{1}{2})\rho V^2 (\sqrt[3]{\nabla})^2$
d_i	A set of coefficients
EHP	Effective horsepower
F	Froude number based on displacement volume. $F=V/\sqrt{g \cdot \sqrt[3]{\nabla}}$
F(**X**)	Presumption such as resistance, EHP, etc.
f_i	A set of coefficients
R_S	Total resistance at speed V_S
S	Wetted surface area
T	Thrust
V	Velocity
X	A set of hull form parameters X_1, X_2, X_3, etc.
Z_i	A set of principal components $Z_1 > Z_2 > Z_3 > Z_4$. (Eigenvalue's order)
ΔD	Non-dimensional draft change at transom. The draft change has been divided by ∇. Positive sign is taken for draft decrease
$\Delta\theta$	Trim angle change. Positive sign is taken for bow up
∇	Displacement volume
ρ	Density

Yoshida's planing boat hull form

b_1	Half width at Ord.10 (stern)
b_2	Half width at Ord.5 (midship)
b_3	Half width at Ord.2
H_3	Keel height at Ord.2
H_4	Keel height at Ord.0 (bow)
h_1	Chine height at Ord.10
h_2	Chine height at Ord.5 (midship)
h_3	Chine height at Ord. 2
L	Length from Ord.0 through 10
ℓ	Transom's overhang
Θ	Initial trim angle
χ_i	A set of non-dimensional hull-form parameters

Keuning et al.' planing boat hull form

$A_p/\sqrt[3]{\nabla}$	Planing area
L	Planing length
LCG	Ratio of distance from stern through center of gravity to planing length
L/B_{px}	Ratio of planing length to width

Tsuchiya's fishing boat hull form

B/T	Ratio of maximum width under water to draft at midship
C_M	Coefficient of cross section at midship
L	Total length
LCB	Distance between midship and center of gravity
t/L	Tangent of trim angle
α_{BS}	Angle of buttock-line's slope
$0.5\alpha_e$	Bow's incident angle into water plane
$0.5\alpha_r$	Stern's angle at water plane
ξ_i	A set of non-dimensional hull-form parameters

Tagano's merchant ship hull form

B	Width at midship
D	Draft
C_m	Coefficient of cross section at midship
L	Total length
LBd/∇	Reciprocal of block coefficient

1.2.1. Introduction

The author presents a concise hull-form parameter set to presume resistance performance and define a planing boat's characteristics. Keuning et al.'s hull form parameters of a planing boat, Tsuchiya's ones of a fishing boat, and Tagano's ones of a merchant ship will be compared with the author's parameters, in order to clarify the characteristics of a planing boat's hull-form parameters. In particular, a set of Keuning et al.'s hull form parameters is helpful in selecting which are effective.

First, the author's selection of a set of hull-form parameters will be presented. The author has selected this set from the data of a series of towing tank tests by using thirteen kinds of 2.5m-length models in the Japan Defense Agency's Towing Tank (called the Meguro Towing Tank), Tokyo.[1-2-1] [1-2-2] Each model's displacement and center of gravity was changed in six ways, which makes 78 data in total. The model's principal dimensions are shown in Fig. 1.2.1. In order to analyze the models' resistance performance, the author first selected the 11 non-dimensional hull-form parameters in Fig. 1.2.2. The resistance performance means a set of C_t, $\Delta\theta$, and Δ D. C_t's denominator selects the dynamic pressure. $\Delta\theta$ and Δ D define a model's floating position. At intervals of 0.5 up to 3.5 in Froude number based on displacement volume, the empirical equations to presume the set will consist of a set of concisely selected parameters.

To obtain the hull forms' characteristics, the principal component analysis of the total data (11 × 78) will be done.

1.2.2. Principal component analysis

The correlation matrix is made from the total hull-form data (11 × 78), which shows all correlation coefficients of each pair of hull-form parameters. Principal component analysis to the matrix gives us principal component axes Z_i, i=1, 2, 3, and 4. Z_1 is the predominant axis, Z_2 next to Z_1, Z_3 next to Z_2, etc., in eigenvalue or data distribution direction. Z_4's eigenvalue is greater than 1. The axes are perpendicular to one another. Each Z_i can be expressed by a linear equation of the 11 hull-form parameters in Fig. 1.2.2. Each hull-form parameter's coefficient is called a factor load, which means the correlation coefficient between the hull-form parameter and the axis Z_i. The coefficient is also the element of Z_i's eigenvector. Taking scores by these axes, we can see the distribution of all data like an ellipse on the rectangular coordinates of Z_1, Z_2, and Z_3. Large coefficients in absolute value on each axis are selected, as shown in Fig. 1.2.2, in order to look at the axis's characteristics, which enable us to make it the standard for discussing hull-form parameters. More than 0.60 in absolute value is selected in Fig. 1.2.2.

Let's summarize the axes' characteristics:

a. Z_1—Judging from the four major factor loads on Z_1 in Fig. 1.2.2, we regard Z_1 as "the axis of chine height and width" or "the axis of cross section." We will call Z_1 the "axis of cross section" from now on.
b. Z_2—Judging from the three major factor loads on Z_2 in Fig. 1.2.2, we regard Z_2 as "the axis of length and transom" or "the axis of planing area." Considering that the theme is a planing boat, we will call Z_2 the "axis of planing area" from now on. Similarly Z_3 and Z_4 are as follows:
c. Z_3—"the axis of width."
d. Z_4—"the axis of initial trim angle."

According to these axes, the author arranges each author's hull form parameters to presume total resistance, EHP, and wave-making resistance and so on, as shown in Fig. 1.2.3.

First, the author's hull-form parameters are discussed.

1.2.3. Yoshida's hull form parameters

1. Boat's total resistance

A planing boat's total resistance R_s at speed V_s is obtained by equation (1.2.1).

$$R_s = [\ C_t - \{C_{fm}S_m / (\sqrt[3]{\nabla m}\ )^2\}]\ (½)\ \rho s\ (\sqrt[3]{\nabla s}\ )^2\ V_s^2 + C_f\ (½)\ \rho s\ S_s\ V_s^2 \quad (1.2.1)$$

In this equation, the "*m*" of each suffix is taken for a model, "*s*" for a boat. The equation is derived from the same residual resistance coefficient as a Froude number between model and boat.

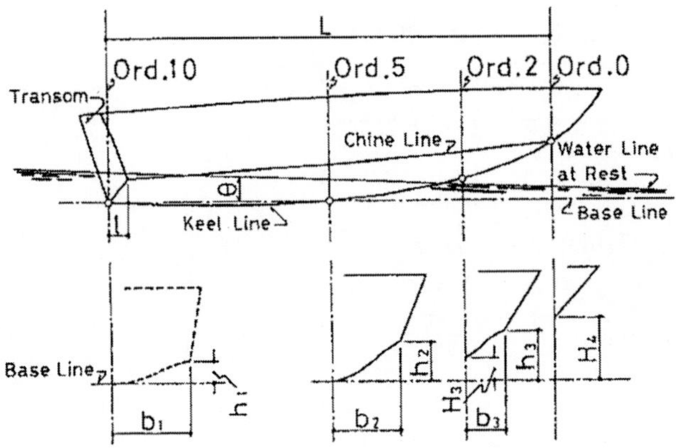

Fig. 1.2.1. Dimensions

	Z_1	Z_2	Z_3	Z_4
$h_1/\nabla^{1/3}\ (=\chi_2)$				
$h_2/\nabla^{1/3}\ (=\chi_3)$	**0.87**			
$h_3/\nabla^{1/3}\ (=\chi_4)$	**0.93**			
$H_3/\nabla^{1/3}\ (=\chi_5)$				
$H_4/\nabla^{1/3}\ (=\chi_6)$	**0.96**			
$L/\nabla^{1/3}\ (=\chi_7)$		**-0.61**		
$b_2/\nabla^{1/3}\ (=\chi_8)$	**0.63**		**-0.65**	
$b_3/\nabla^{1/3}\ (=\chi_9)$			**-0.78**	
$\ell/\nabla^{1/3}\ (=\chi_{10})$		**0.89**		
$\theta\ (=\chi_{11})$				**-0.65**
$b_1/\nabla^{1/3}\ (=\chi_{12})$		**0.80**		

Fig. 1.2.2. Hull-form parameters' major factor loads

<table>
<tr><td></td><td>Yoshida's planing boat</td><td>Keuning et al.'s planing boat</td><td>Tsuchiya's fishing boat</td><td colspan="2">Tagano's merchant ship</td></tr>
<tr><td>F(X)</td><td colspan="2">Total resistance & floating position</td><td>EHP</td><td colspan="2">Wave-making resistance</td></tr>
<tr><td>Z_1</td><td>$b_2/\sqrt[3]{\nabla}$ (χ_8)
$h_2/\sqrt[3]{\nabla}$ (χ_3)
$H_4/\sqrt[3]{\nabla}$ (χ_6)
$h_3/\sqrt[3]{\nabla}$ (χ_4)</td><td></td><td>B/T (ξ_2)
C_M (ξ_3)
$0.5\alpha_e$ (ξ_6)</td><td>B/d
C_m
LBd/ ∇</td><td rowspan="2">Points that are used to form prismatic curve</td></tr>
<tr><td>Z_2</td><td>$L/\sqrt[3]{\nabla}$ (χ_7)
$b_1/\sqrt[3]{\nabla}$ (χ_{12})</td><td>L/Bpx
$Ap/\sqrt[3]{\nabla}$</td><td>$L/\sqrt[3]{\nabla}$ (ξ_1)
$0.5\alpha_r$(ξ7)
α_{BS} (ξ_8)</td><td>LBd/ ∇</td></tr>
<tr><td>Z_3</td><td>$b_2/\sqrt[3]{\nabla}$ (χ_8)
$b_3/\sqrt[3]{\nabla}$ (χ_9)</td><td></td><td></td><td colspan="2"></td></tr>
<tr><td>Z_4</td><td>θ (χ_{11})</td><td>LCG</td><td>LCB (ξ_5)
t/L (ξ_9)</td><td colspan="2"></td></tr>
<tr><td>F</td><td>0.5 ~3.5</td><td>0.75 ~ 3.0</td><td>0.45 ~ 0.80</td><td colspan="2">0.3 ~ 0.65</td></tr>
<tr><td>Terms</td><td>5, 7, 44</td><td>13</td><td>72</td><td colspan="2">Selected-point numbers</td></tr>
<tr><td>Power</td><td>1, 2</td><td>3</td><td>4</td><td colspan="2">7 (coefficients 4 + curve 3)</td></tr>
</table>

Fig. 1.2.3. Authors' hull form parameters

C_t is obtained from equations 1.2.2, 1.2.3, and 1.2.4 in the next section, which result from the principal component analysis.

2. Model's resistance performance

A set of C_t, $\Delta\,\theta$, and Δ D is called resistance performance. C_t is a model's total resistance coefficient presumed by the empirical equation that is made of four non-dimensional hull-form parameters as equation 1.2.2. The parameters are made of length, half width and chine height at midship, and initial trim angle, which have major factor loads on the axes in Fig. 1.2.2. Major factor load means more than 0.60 in the correlation coefficient's absolute value.

$$C_t = d0 + d_1 h_2/\sqrt[3]{\nabla} + d_2 L/\sqrt[3]{\nabla} + d_3 b_2/\sqrt[3]{\nabla} + d_4 \theta \qquad (1.2.2)$$

Also $\Delta\,\theta$ and Δ D, which are necessary to determine the floating position of a model, are expressed by this kind of linear equation.

If we add the equation to two squared hull-form parameters that are made of length and initial trim angle, the quadratic equation (1.2.3) is got.

$$C_t = f_0 + f_1 h_2/\sqrt[3]{\nabla} + f_2 L/\sqrt[3]{\nabla} + f_3 b_2/\sqrt[3]{\nabla} + f_4 \theta + f_5 (L/\sqrt[3]{\nabla})^2 + f6\theta^2 \qquad (1.2.3)$$

If we want to use 11 hull-form parameters to presume resistance performance, two kinds of equations are considered. The first is the linear equation made of 11 hull-form parameters in Fig. 1.2.2. The second one is the linear equations made of principal components as the equation (1.2.4).

$$C_t = C_0 + C_1 Z_1 + C_2 Z_2 + C_3 Z_3 + C_4 Z_4 \qquad (1.2.4)$$

Four principal components have no relation to one another. Thus, the equation (1.2.4) has the highest accuracy among these kinds of equations. As each principal component term has 11 terms, the total number of terms becomes 44. These 11 hull-form parameters are used to obtain a set of optimum hull-form parameters and draw the plan.[1-2-1] [1-2-2]

During the initial design phase, equations 1.2.2 and 1.2.3 were used to contain major hull-form parameters and concise equations. Both equations have reasonable accuracy. As one of the examples, the PT boat's total resistance is presumed as in Fig. 1.2.4. The PT boat is 35m in overall length, 9m in width, 1304kN or 133tonf in displacement, and has three propellers.[1-2-3] The total resistances shown by the equations 1.2.1, 1.2.2, and 1.2.3 have been checked against the total thrusts from the trial test data,[1-2-4] showing that the reasonable estimation has been done. And trim angle changes by the equations 1.2.2 and 1.2.3 are compared with the PT boat's trial test data, showing a lower estimation, as in Fig. 1.2.5. This is due to scale effect.[1-2-5] Thus, the limitation of Yoshida's equations may be about 25m in overall length.

The author adds other authors' hull-form parameters to the author's ones and discusses all the parameters in the next chapter, referring to Fig. 1.2.3.

2.2.4. Discussion

1. Cross section

An against-the-flow planing boat's cross section contributes to the total resistance. The main part of the cross section at midship seems to be shaped by the isosceles triangle having two alternate angles in the deadrise angles on width $2b_2$'s ends in Fig. 1.2.1. Chine height h_2 and half width b_2 at midship determine the deadrise angle. The author has not selected the cross section directly, for two reasons: from low speed to high speed the correlation coefficients at midship between total resistance coefficient and the cross section, between the total resistance coefficient and the chine height, and between the total resistance coefficient and the width are calculated in Table 1.2.1, and have a similar tendency in all speed ranges. The cross section, the chine height and the width have high correlation coefficients with one another, which is the reason they show a similar tendency. If we use a pair of the terms having a high correlation coefficient to each other in these kinds of equations, the sign of the term will be irregular. Each term's sign of the equation 1.2.2 is almost correct because of each other's independent term. For example, increasing the length causes the total resistance to decrease, the length's sign must be minus. All signs have been checked in such a way.

Second, designers face the problem of how to decide chine height and deadrise angle under the given cross section. Thus, including chine height and deadrise angle in the equation is desirable.

Keeping width in constant value, the hull form has a higher chine height (sharp-pointed bottom) and gets less total resistance in the low speed range and more total resistance in the high speed range (see Table 1.2.1).

Seeing the hull form from the side, as in Fig. 1.2.1, we find the height of chine line h_4-h_3-h_2 or bow shape affects the flow or total resistance.

$b_2/\sqrt[3]{\nabla}$, $h_2/\sqrt[3]{\nabla}$, $h_4/\sqrt[3]{\nabla}$ and $h_3/\sqrt[3]{\nabla}$ are included in Z_1.

Tsuchiya has selected B/T and C_M as the cross section factor and $0.5\alpha_e$ against flow to presume the EHP of a fishing boat.[1-2-6]

Tagano has selected B/d and C_m as the cross section factor to presume the wave-making resistance of a merchant ship.[1-2-7]

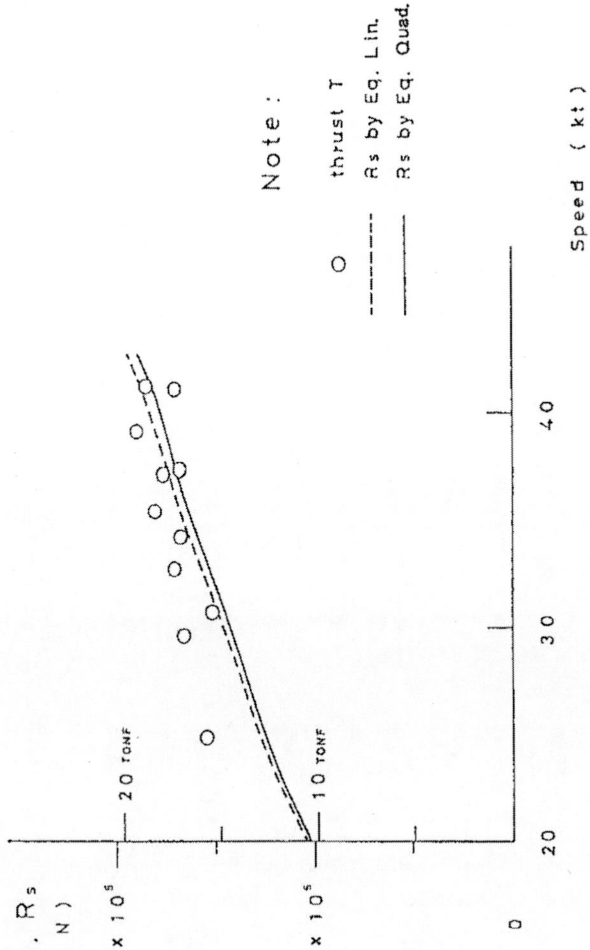

Fig. 1.2.4. Resistance and thrust

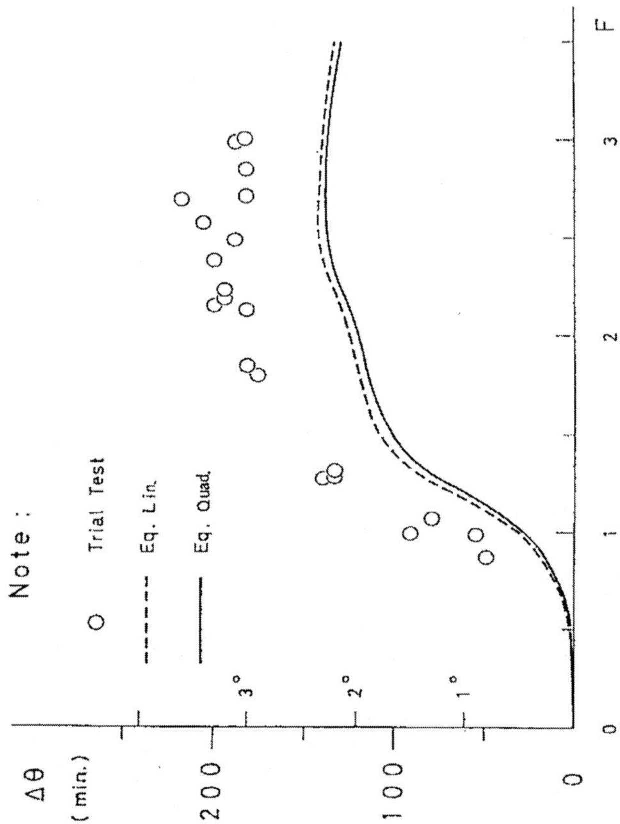

Fig. 1.2.5. Trim angle change

Table 1.2.1. Correlation coefficients between total resistance coefficient and cross section, and so on, at midship

F	0.5	1.0	1.5	2.0	2.5	3.0	3.5
Cross section	-0.68	-0.35	-0.44	-0.19	0.88	0.66	0.00
Chine height	-0.73	-0.35	-0.43	-0.28	0.30	0.57	0.00
Width	-0.32	-0.19	-0.29	-0.00	0.42	0.52	0.00

2. Planing area

If we look at the plane figure of the model in Fig. 1.2.1, we find that lengths L and ℓ affect the wetted length, that is, the total resistance. The area at water plane level is mainly determined by the isosceles triangle having a perpendicular L to the middle of the width 2b1 at transom. The apex corresponds to the incident angle against flow.

$L/\sqrt[3]{\nabla}$, $b_1/\sqrt[3]{\nabla}$, and $\ell/\sqrt[3]{\nabla}$ are included in Z_2.

Keuning et al.[1-2-8] have selected L/B_{px} and $A_p/\sqrt[3]{\nabla}$ to presume resistance performance of a planing boat.

Although the fishing boat does not get the planing condition, Tsuchiya has selected $L/\sqrt[3]{\nabla}$, $0.5\alpha_r$, and α_{BS} by evaluating the area at water plane level to presume the EHP of a fishing boat.

3. Width

The width at Ord. 2 and 5 in Fig. 1.2.1 are important factors in hull form. The author has selected $b_2/\sqrt[3]{\nabla}$ and $b_3/\sqrt[3]{\nabla}$ in Z_3. Tsuchiya has selected B/T to presume the EHP of a fishing boat. Tagano has selected B/L and B/d to presume the wave-making resistance of a merchant ship.

4. Initial trim angle

The location of the center of gravity is connected with the initial trim angle. The author has selected θ in Z_4. Setting the center of gravity toward the transom is desirable to get fine planing. Porpoising will occurred due to the center of gravity near the bow.

Keuning et al. have selected LCG to presume the resistance performance of a planing boat.

Tsuchiya has selected LCB and t/L to presume the EHP of a fishing boat.

5. Block coefficient

Tagano has selected LBd/∇ to presume the wave-making resistance of a merchant ship. LBd/∇ is connected with previous sections 1 through 3.

The author would like to commend Keuning et al.'s research for providing useful information.

1.2.5. Keuning et al.'s hull-form parameter

Keuning et al. have selected hull-form parameters from the consideration of the model tests by Clement and Blount.[1-2-9] L/B_{px}, $A_p/\sqrt[3]{\nabla}$, and LCG are selected. $A_p/\sqrt[3]{\nabla}$ on the water plane will make an approximate isosceles triangle having the perpendicular L/B_{px} to transom, which we may call the planing area. The isosceles triangle on the water plane together with the location of the center of gravity is the essence of discussion on the planing phenomenon. But there are two problems in the view of the designers: (1) how to determine the principal dimension from the planing area; (2) how the change in the deadrise angle should be dealt with.

Finally, the author discusses the number of terms in the equation, each term's power, and the applicable range of the equation and so on, in reference to Fig. 1.2.3.

1.2.6. Equations

1. Yoshida's equation

The term numbers of the author's equations are as follows: equation 1.2.2 has 5 terms; equation 1.2.3 has 7; equation 1.2.4 has 44. In addition to the three kinds of equations, the author has a linear equation composed of 11 hull form parameters expressed by χ_i, i=2, 3, ⊙⊙⊙⊙⊙, 12 (see Fig. 1.2.2).

The power is up to 2.

The author's equations should be used to presume a planing boat's resistance and floating position up to 3.5 in Froude number. The equations are obtained at intervals of 0.5 in Froude number. The wetted surface area at a floating position enables us to calculate the frictional resistance. Except for the clean condition of the transom, the author regards the wetted surface area as the area under the water plane.

The maximum wave-making resistance is achieved at about 1.0 in Froude number. Whenever the speed exceeds 1.0 in Froude number slightly, the clear condition on the transom begins. A semi-planing condition is achieved at about 2.0 in Froude number. A planing condition is achieved at more than 2.5 in Froude number.

2. Keuning's equation
The term numbers of the equation are 13.

The power is up to 3.

Their equation should be used to presume a planing boat's resistance and floating position up to 3.0 in Froude number.

3. Tsuchiya's equation
The term numbers of the equation are 72. The hull form parameters ξ_i, i=1, 2, ◎◎◎◎◎, 9 are used.

The power is up to 4.

His equation should be used to presume a fishing boat's EHP up to 0.8 in Froude number.

The author has converted the length base Froude number into the displacement volume base Froude number, referring to reference 1-2-10.

A fishing boat has the following characteristics: (1) A fishing boat has a complex hull form, and (2) the boat is used at a lower speed range.

4. Tagano's equation
The term numbers equal to the plotted point numbers form the prismatic curve.

The power is 7. Power 4 is used for coefficients and power 3 for curve.

His equation should be used to presume a merchant ship's wave-making resistance up to 0.65 in Froude number.

The Froude number has been converted in the same way as Tsuchiya's case in the previous section.[1-2-10]

A merchant ship has the following characteristics: (1) a merchant ship's main part has a fairly similar cross section, and (2) the ship is used at lower speed range.

1.2.7. Conclusions

Applying the principal component analysis to the models' data, the author has obtained the concise hull-form parameters for estimation of a planing boat's resistance performance. This can be summed up in the following five items:

1. To presume resistance performance it is necessary to have the length, half width and chine height at midship, and the initial trim angle of a planing boat.

2. Comparing hull form parameters of a planing boat with ones of several kinds of ships, the equation for a planing boat can be composed in simpler form than that for other kinds of ships.
3. Considering the scale effect, the limitation of the author's equations may be 25m in the overall length of a planing boat.
4. A planing boat's approximate isosceles triangle area at water plane level together with the location of the center of gravity is important to presume resistance performance.
5. In addition to item 4, a decision on chine heights is necessary.

References

[1-2-1]. A. B. Degtyarev, S. I. Krolenko, & Y. Yoshida, "A Mathematics for Design and Development of Optimum Hull Form Applying to Obtain a Planing Boat's Hull Form Parameter Set with the Minimum Resistance," Proceedings of ISC '98 KSRI, ed. V. E. Spiro & B. P. Polonsky, St Petersburg, Section A, Vol. 1, pp. 207–14, November 1998.

[1-2-2] Y. Yoshida, "Optimum Design of a Planing Boat's Hull Form," *Marine Technology* 3, ed. T. Graczyk, T. Jastrzëbski, & C. A. Brebbia, WIT Press, Southampton, Boston, pp. 3–13, 1999.

[1-2-3] V. Raymond & B. Blackmann, eds., "Jane's Fighting Ships," London, p. 213, 1975–76.

[1-2-4] Y. Yoshida, "Apply Tanaka's Nomogram to the PT Thrust Estimation," Papers of International Conference on High Speed Craft Technology and Operation of RINA, London, November 1–12, 2002.

[1-2-5] Y. Yoshida, "A Scale Effect Check about Resistance Estimation of Planing Boats," Proceedings of 3rd International Euro Conference on High-Performance Marine Vehicles (HIPER '02), ed. V. B. Bergen, pp. 448–56, September 14–17, 2002.

[1-2-6] H. Tsuchiya, "Statistical Analysis to Fishing Boat Hull Forms' Effective Horsepower," *Journal of the Society of Naval Architecture of Japan* 132, pp. 63–80, December 1972.

[1-2-7] H. Tagano, "The Statistical Method of Estimating Hull Forms" "Wave Making Resistance," *Journal of the Kansai Society of Naval Architecture of Japan* 147, pp. 43–52, March 1973.

[1-2-8] J. A. Keuning, J. Gerritsma, & P. F. Van Terwisga, "Resistance Tests of a Series Planing Hull Forms with 30° Deadrise Angle, and a Calculation Model Based on This and Similar Systematic Series," *Int. Shipbuild. Progr.* 40 (424), pp. 333–85, 1993.

[1-2-9] E. P. Clement & D. L. Blount, "Resistance Tests of a Systematic Series of Planing Hull Forms," Transactions of SNAME 71, 1963.
[1-2-10] M. Komatsu, & M. Yamaguchi, "The Propulsive Performance of High Speed Crafts," Proceedings of the Symposium on High-Speed Crafts and the Performance, Society of Naval Architecture of Japan, pp. 75–118, June 1989.

1.3

Development of High-Speed Planing Hull Forms with Consideration of Resistance and Sea-Keeping Performance (2013)

Dong Jin Kim,[1] Sun Young Kim,[1] Seong Hwan Kim,[2] Jeong Hwa Seo,[3] Shin Hyung Rhee[3]

[1] Korea Institute of Ocean Science and Technology
[2] FAR EAST Ship Design and Engineering Co., Ltd.
[3] Seoul National University

Introduction

Planing hulls need to be designed with consideration of various hydrodynamic characteristics, and, if necessary, those hull forms should be modified to improve resistance or sea-keeping performance in the initial design stage. A desired planing vessel in this research is to run at Froude number 3.25 in calm water and operate in rough water above sea state 3—so good resistance and sea-keeping performance are required.

Four hull forms that have almost the same displacement are designed and model tests in calm water and in waves are carried out to analyze their resistance performance and sea-keeping behavior. The running attitudes and resistance are simultaneously measured with various Froude numbers in calm water tests. Heave and pitch motions are measured in regular waves with various wave lengths. And the effects of hull forms on resistance performance and sea-keeping behavior are analyzed.

1.3.1. Hull form design

1. Main particulars of model ships

The main particulars of four model ships are shown in Table 1.3.2. All model ships have warped hull forms, so the deadrise angles at A.P., midship, and F.P. of each model are shown in Table 1.3.1, respectively.

2. Body plans of model ships

Model A has the smallest deadrise angle among the four models. Model B has the largest deadrise angles; in particular, they rapidly increase from the midship to the bow. Models C and D have hull forms that are compromised between models A and B. In comparison with model C, the body cross sections of model D have concave shapes and large climb angles of chines toward the bow.

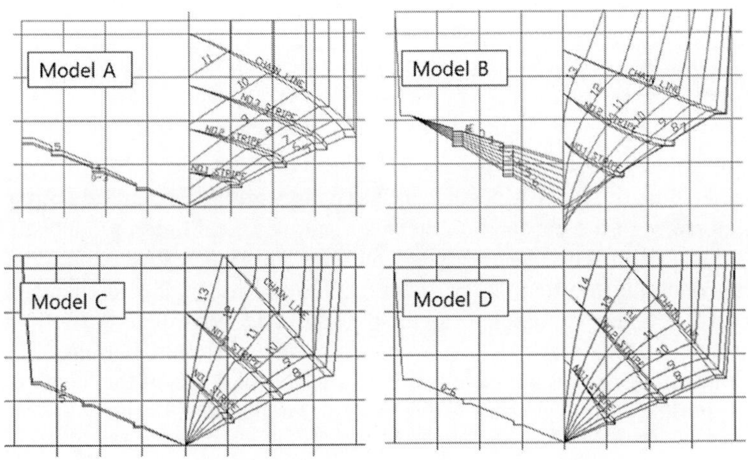

Fig. 1.3.1. Body plans of the model ships

Table 1.3. 2. Main particulars of model ships

Particular	Model A	Model B	Model C	Model D
L_{OA}/m	0.927	1.092	1.123	1.200
L_{WL}/m	0.796	1.026	1.072	1.149
Breadth/m	0.308	0.308	0.292	0.308
Draft/m	0.080	0.087	0.072	0.070
Weight/kgf	9.103	9.467	9.467	9.650
L/B (L_{WL}/chine breadth)	2.586	3.334	3.872	3.930
Deadrise angle at A.P./deg	20	17	22	22
Deadrise angle at midship/deg	23	30	24	22
Deadrise angle at F.P./deg	32	80	75	77

3. Shear plans of model ships
The keel line of model A begins curving up toward the bow at 55% of the L_{WL} forward of the transom; on the other hand, the keel lines of models B, C, and D begin curving up toward the bow at around 80% of the L_{WL} forward of the transom. In particular, the keel line at the forebody of model B is below the baseline, so the bow of model B hardly comes out of the water. Model B is advantageous for wave-piercing motions. The bottoms of all models are extended backward about 3–5% of the length of the waterline—these play a role that is similar to the 0° trim tabs.

4. Half breadth plans of model ships
The length-to-beam ratio of model A is the smallest and the length-to-beam ratio of model D is the largest among the four models. In particular, the projected plane area bounded by chines of the forebody of model D is smaller than that of model C. Model D has smaller trim angles at low speed in comparison with model C, as shown in Fig. 1.3.8.

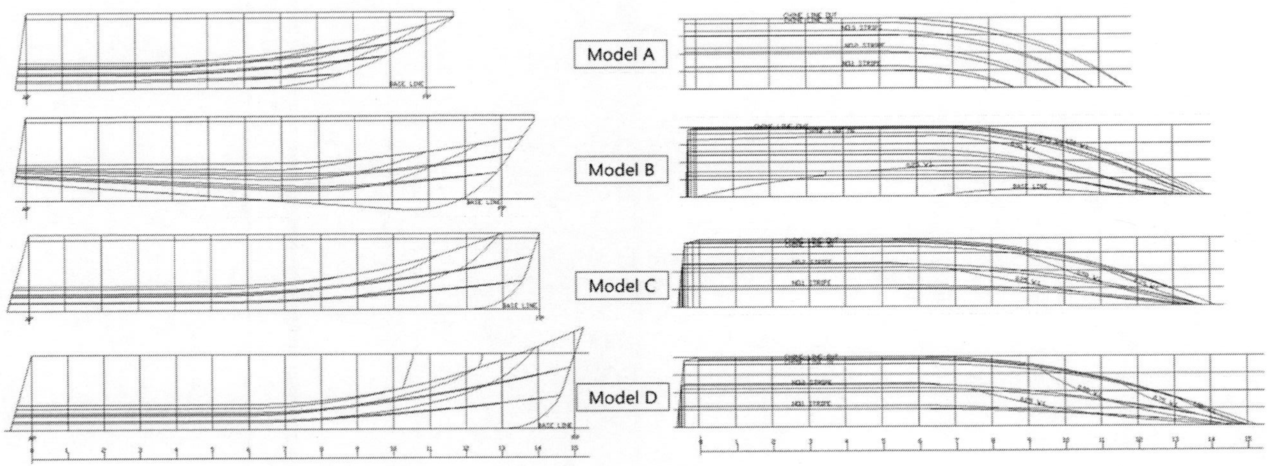

Fig. 1.3.2. Shear plans (left) and half-breadth plans (right) of the model ships

1.3.2. High-speed towing system

Fig. 1.3.3 shows the schematic view of the high-speed towing system at Seoul National University. Fig. 1.3.4 shows the model ship being towed under heave/pitch-free conditions. The heave and pitch resistance of the model and encounter wave elevations are measured in the model tests. The new towing device is developed in order that the model ship can be towed in the inclined thrust direction when the model ship is trimmed while running.

The towing rod is counterbalanced by the counterweight, so there are no effects of its weight on the displacement or the running attitude of a model that is at a steady state in calm water. But in the model tests with waves, the towing rod and the counter weight do influence the motion of a model ship.

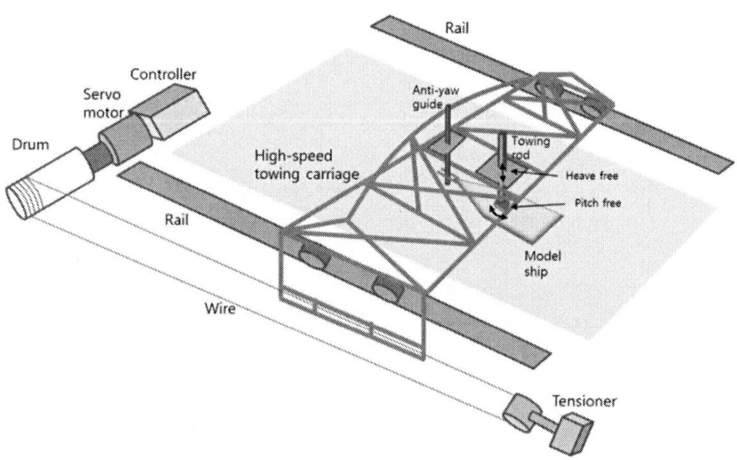

Fig. 1.3.3. Schematic views of a high-speed towing system

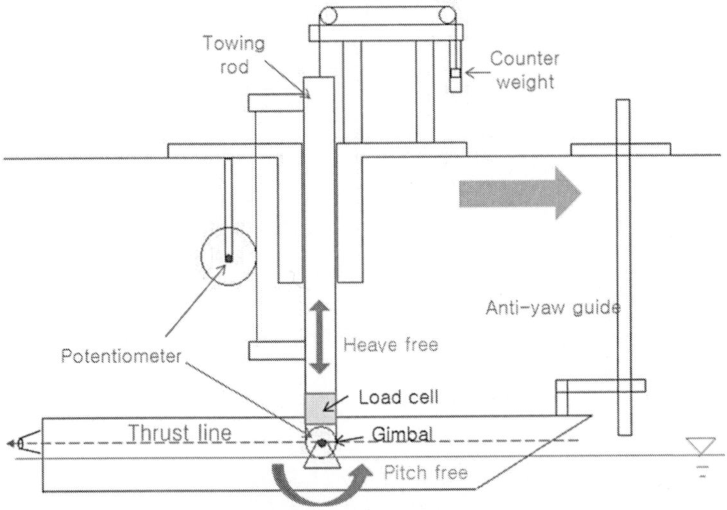

Fig. 1.3.4. Side view of heave/pitch-free towing device

1.3.3. Model tests in calm water

Fig. 1.3.5 shows the running attitudes of four model ships at the maximum Froude number (F_n=3.25).

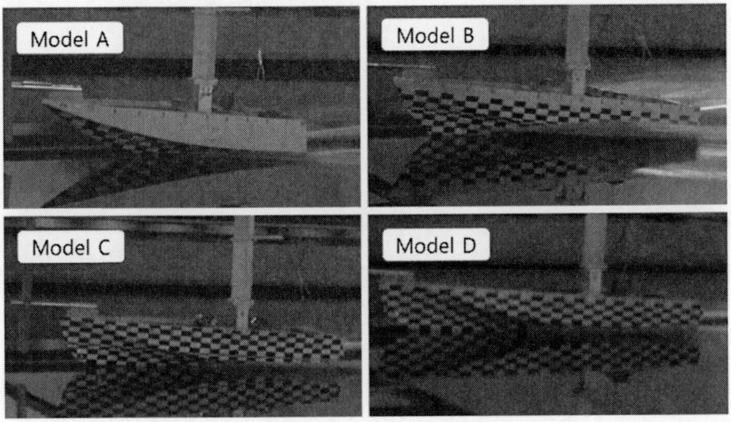

Fig. 1.3.5. Running attitudes of models at F_n.=3.25

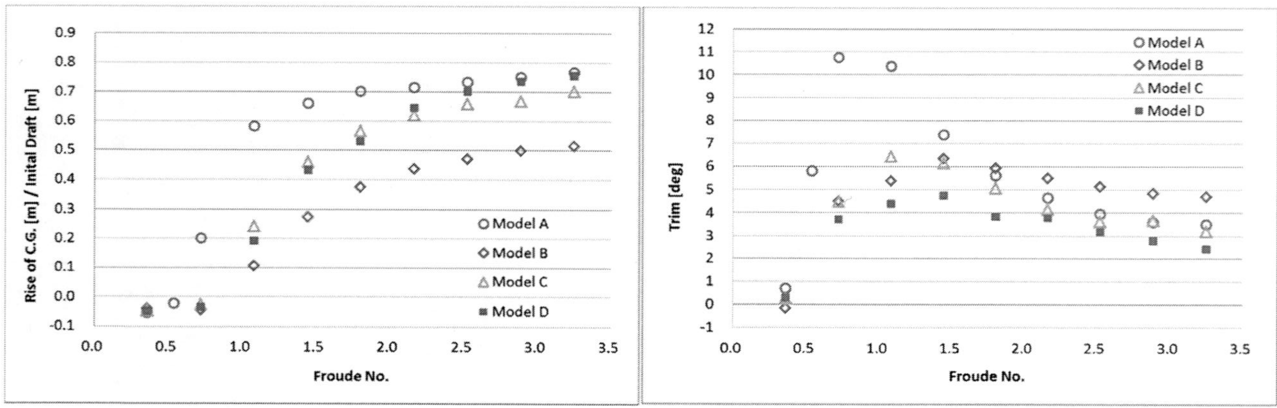

Fig. 1.3.6. Variations of rise of C.G. (left) and trim (right) with F_n

Model B has the largest deadrise angles, so it shows the smallest rise of C.G., the largest trim angle, and the largest resistance at high speed (above Froude no. 2.0).

Model A shows the largest trim angles at low speed (below Froude no. 1.4); this is because model A has the smallest deadrise angles, in particular on the forebody, and has the smallest length-to-beam ratio. Large trim angles create large pressure drag components. Therefore, model A has the largest resistance at low speed, but it shows good resistance performance at high speed due to the highest rise and small trim angle.

The resistance performance of model C is in the middle between that of model A and that of model B. Model D has a larger length-to-beam ratio and has a smaller projected area of the forebody bounded by chines than model C. So model D shows an improved resistance performance in comparison with model C.

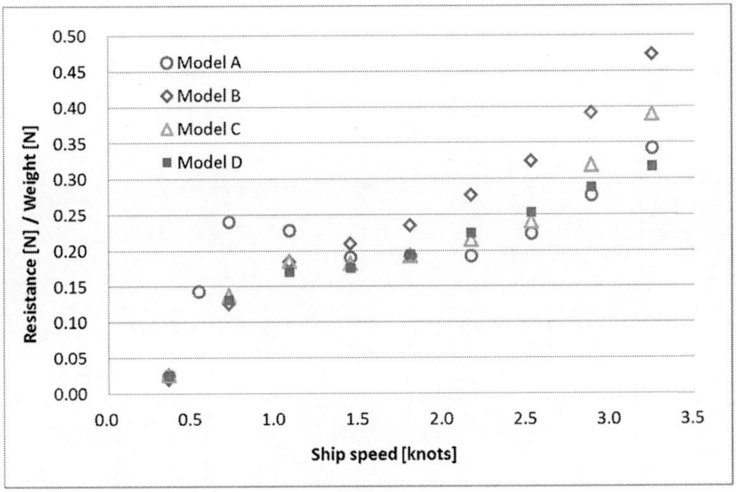

Fig. 1.3.7. Variations of resistance with F_n

1.3.4. Model tests in regular waves

1. A/B/C model (weight of towing rod: 4.4kgf)

Fig. 1.3.8 shows the heave and pitch transfer functions of models A, B, and C at Froude no. 1.44. Model A, which has the smallest deadrise angle, shows very large motion amplitudes in waves. In particular, model A experiences the "fly over" phenomenon around its resonant frequency. On the other hand, model B shows small motion amplitudes, compared with

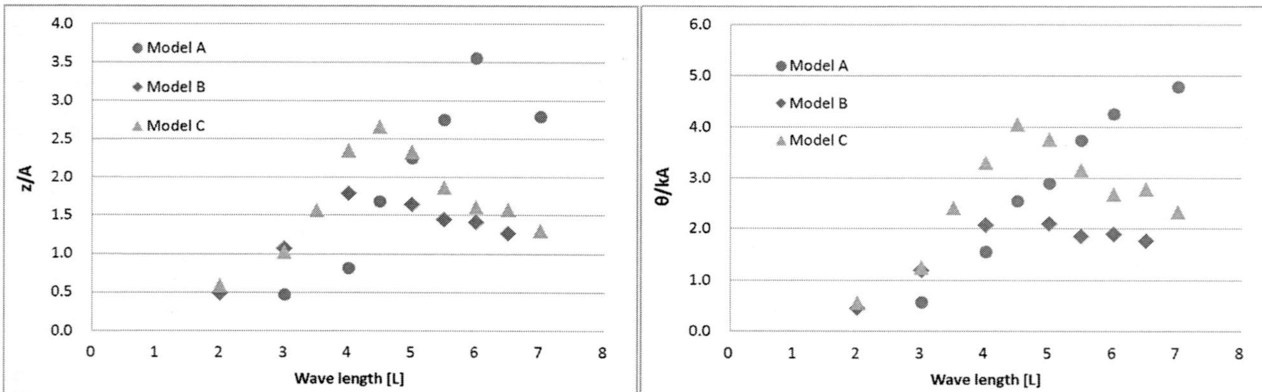

Fig. 1.3.8. Heave (left) and pitch (right) transfer functions of models A, B, and C at F_n 1.44 (towing rod 4.4kgf)

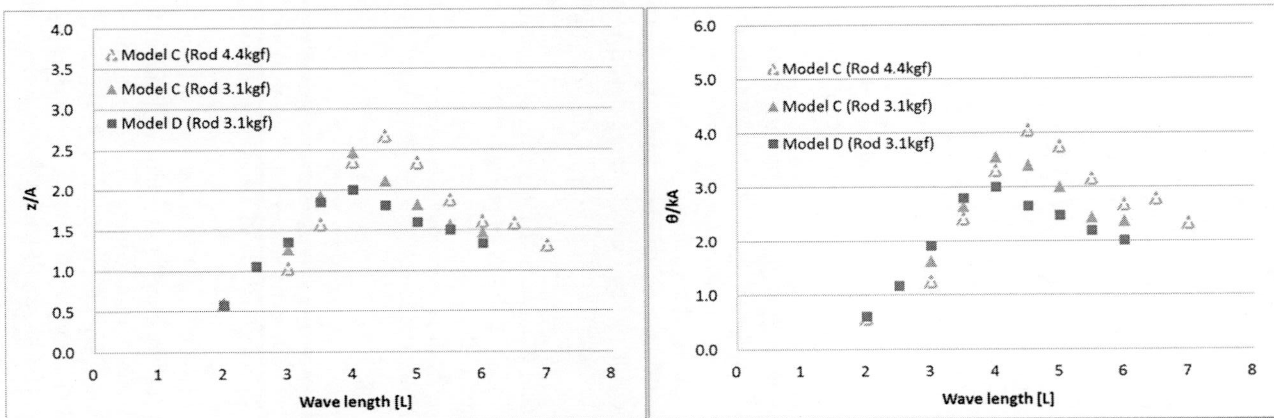

Fig. 1.3.9. Heave (left) and pitch (right) transfer functions of models C and D at F_n 1.44 (towing rod 3.1kgf)

model A. The sea-keeping performance of model C is in the middle between that of model A and that of model B.

2. C/D model (weight of towing rod: 3.1kgf)
The weight of the towing rod becomes lighter for the tests of models C and D. As the towing rod is lighter, the resonant frequency gets higher due to the decrease of inertia of the towing rod and the counterweight.

Model D has a larger length-to-beam ratio than model C. Thus, the sea-keeping performance of model D is better than that of model C, as shown in Fig. 1.3.9.

1.3.5. Conclusions

Four planing hulls that have almost the same displacement are designed, and high-speed model tests were performed to confirm resistance and sea-keeping performance of the designed hulls.

Small deadrise angle and length-to-beam ratio lead to good resistance performance, and large deadrise angle and length-to-beam ratio make the vessel create small motions in waves.

In particular, forebody shapes have a strong influence on the sea-keeping performance of the model. When small wave exciting forces are acted on the forebody of the hull due to large deadrise angles, the bow hardly comes out of the water, and the model ship shows good sea-keeping performance.

In this study, model C has a hull form that is a compromise between models A and B. Model D shows improved resistance and sea-keeping performance compared with model C. More hull form optimization will be carried out to improve both resistance and sea-keeping performance of the planing hull.

References

[1-3-1] G. Fridsma, "A Systematic Study of the Rough-Water Performance of Planing Boats," Stevens Institute of Technology, Davidson Laboratory Report 1275, 1969.

[1-3-2] D. J. Kim, Y. G. Kim, S. H. Kim, & S. Y. Kim, "An Experimental Study on the Hydrodynamic Performance of Planing Vessels in Calm Water and in Waves," Proceedings of Annual Autumn Meeting of Korea Towing Tank Conference, 2012.

[1-3-3] D. J. Kim, S. Y. Kim, Y. J. You, K. P. Rhee, S. H. Kim, & Y. G. Kim, "Design of High-Speed Planing Hulls for the Improvement of

Resistance and Sea-Keeping Performance," *International Journal of Naval Architecture and Ocean Engineering* 5 (1), 2013.

[1-3-4] D. Savitsky, "Hydrodynamic Design of Planing Hulls," *Marine Technology* 1 (1), pp. 71–95, 1964.

[1-3-5] D. Savitsky, "Sea-Keeping Considerations in Design and Operation of Hard Chine Planing Hulls," Royal Institution of Naval Architects, pp. 55–59, 1979.

1.4

Simulation of Transverse Motion and Dynamic Stability for High-Speed Planing Craft (2001)

Xu Lixin and Yung S. Shin

American Bureau of Shipping, USA

B	Beam
C_1, C_2	Zero-pressure points on the right and left sides, with transverse coordinates of Y_{C1} and Y_{C2}, respectively
C_2	Ratio of the wetted width on the left-side bottom to that on the right side
C_E	Damping coefficient in the equation of roll motion
$C.G.$	Center of gravity
C_I	$I_G/\rho B^4$, inertia coefficient
C_M	$W/\rho g B^2$, mass coefficient
C_R	Total restoring moment per degree in roll, normalized to $\rho V^2 B^2$
C_{R0}	Hydrostatic restoring moment per degree in roll, normalized to $\rho V^2 B^2$
C_{R1}	Hydrodynamic restoring moment per degree in roll, normalized to $\rho V^2 B^2$
C_{RG}	Roll moment per degree in roll due to the boat weight, normalized to $\rho V^2 B^2$
C_v	$U_A/\sqrt{gB}$, speed coefficient
G	gB/V^2, the non-dimensional acceleration of gravity
h_G	H_G /B, non-dimensional height of the C.G. above the keel
I_G	Moment of inertia with respect to a longitudinal axis passing through the C.G.
t	Time

U_A	Advance speed of the planing hull
V	Local effective vertical velocity valid for a sectional impact solution
W	Hull weight per unit length
X	Longitudinal coordinate in the direction of advance
Y	Transverse coordinate
Z	Vertical coordinate
α	Trim angle, in degrees
β	Deadrise angle, in degrees
β_1, β_2	Effective deadrise angle in the right and left sides, respectively
θ	heel angle, in degrees
$\dot{\theta}$	Angular velocity in roll
$\ddot{\theta}$	Angular acceleration in roll
ρ	Water density
τ	Vt/B, non-dimensional time
τ'	Equivalent time corresponding to two-dimensional impact solution for hull section at x

Fig. 1.4.1. “Chine Walking” (photographed by J. Palfy)

1.4.1. Introduction

The planing hull is a form of the oldest, most effective, and most popular high-speed craft, and has always been the point of interest in practices of hydrodynamic design. With applications from small pleasure boats to large military patrols, planing boats possess increasingly higher power and speed. When planing boats operate in a seaway, however, bottom impact shock and instability cases may lead to extreme motion and cause the boat to go out of control or capsize (Codega and Lewis 1987). Motion control and stability of high-speed planing craft therefore becomes more and more important. As reported in Blount and Codega (1992), "The problem is not well understood and accepted guidelines do not exist which will ensure adequate dynamic stability." Extensive studies have been carried out in the past decades: most focused on the coupled motions of pitch and heave, or on "porpoising" instability problems—for instance, Perring and Glauert (1933), Payne (1974, 1984), Martin (1978), Thornhill et al. (2000), and others. The success and significance of the above mentioned works and others are apparent; meanwhile, transverse stability problems such as chine walking are also of great importance for the design and operation of high-speed planing craft. Currently, "No accepted guidelines are available for predicting the conditions which results in chine walking" (Blount and Codega 1992), and few works have previously reported on the transverse stability of high-speed planing boats. Recently, Lewandowski (1995, 1996, and 1997) proposed a method to evaluate the dynamic roll stability for hard-chine planing craft; still, the difficulties of understanding and determining the coefficients in his equation of motion might limit its application in practice. Also uncertain is the accuracy in prediction of roll restoring moments due to applying an assumption of two sides acting independently of each other even if the boat is under asymmetric (heeled) conditions—that is, if interaction between the two asymmetric sides was neglected in his works. On the other hand, the significance of incorporating asymmetric effects in the general hydrodynamic impact/planing model has been demonstrated by Xu et al. (1998a) and Xu and Troesch (1999). For asymmetric (or heeled) planing problems, therefore, it is understood that it may not be appropriate to employ on each side the formula from symmetric flows. The present study hence intends to propose a direct method for an effective engineering solution that includes the interaction between two asymmetric sides when considering high-speed steady planing associated with transverse roll motion. As solving the roll motion from an equation of motion in the transverse plane, the criteria of transverse stability are also developed for potential applications in the practical design and operation of high-speed planing boats.

With similar features between the two-dimensional impact flow and the cross sectional behaviors of high-speed planing, the hydrodynamic physics of steady planing surfaces are often simulated using a two-dimensional water impact theory via the slender body method. Zhao et al. (1997) provided a "2.5" dimensional nonlinear theory for predicting hydrodynamic pressure and lifting force on high-speed planing boats, including effects of flow separation from chines. Xu and Troesch (1999) presented a two-dimensional method for simulating the steady planing surface of a prismatic hull with hard chines and for predicting hydrodynamic pressure/force and also transversely restoring moments in asymmetric (heeled) conditions. Employing a two-dimensional impact theory previously developed (Xu et al. 1998a), the dynamic profiles of the wetted area on the planing hull, or the non-zero pressure surface, was accurately predicted for transversely symmetric and asymmetric planing, and also verified by experimental results (Xu and Troesch 1999).

Furthermore, for transversely asymmetric problems, Xu et al. (1998b) successfully solved this for the transverse restoring moments and roll motion coupled with vertically downward water entry of a prismatic hull under heeled conditions. Good agreements were obtained between the theoretical predictions and drop test results; however, the angular motion and acceleration were slightly under-predicted. It has been thought that the under-prediction was caused because the angular velocity of roll motion is assumed to be small and is neglected compared with body boundary conditions for calculating the restoring moment.

Overall, the previous works have provided fundamental insights into and are a practical tool for simulation of roll motion and dynamic stability in the transverse plane. Alongside basic investigations into prismatic hulls, the present study is taken as a first step to solving the chine-walking problems of high-speed planing craft. An equation of roll motion in the transverse plane is established, incorporating effects of both rolling angle and angular velocity. The dynamic coefficients in the equation of motion are herein reformulated based on similar methods in Xu et al. (1998b). The transverse roll motion can therefore be solved at each heeled position. The dynamic stability of transverse roll motion is then discussed in the paper. Although some results shown in the present study are for cases of prismatic hulls, the equation of motion, the general method, and, moreover, the transverse stability criteria are actually developed to extend directly to non-prismatic hulls. They should certainly be suitable for practical applications of planing craft design and operation.

1.4.2. Value of the mome equation of roll motion

Consider high-speed planing in calm water, where the hydrodynamic pressure is dominant on the planing surface. Total weight of the planing boat is mainly supported by the hydrodynamic forces instead of the buoyancy from displacement, and the draft is actually very shallow in the dynamic dominant region (before the chine is wetted). Transverse dynamic stability of the planing boat is related to the speed, and basically depends on the heeled position, wetted area, and variation of the center of gravity (C.G.) in the transverse direction. Viscosity in the form of skin friction herein is considered to be of little influence in the purely transverse roll motion. Under high-speed planing conditions, the tangential velocity on body surface is dominant, and the friction force actually does not contribute to the rolling moment. Figure 1.4.2 shows a hull section of the planing boat under heeled position, where it represents the heel angle. C_1 and C_2 are zero-pressure points on the two sides, respectively; the wetted area is defined within these two points. The hydrodynamic pressure distribution on the wetted area, *p(y)*, shows the maximum peaks and high derivatives closely near the zero-pressure points. In the figure, *V* is the local effective vertical velocity valid for the sectional impact solutions—the conversion between the time-dependent impact solutions and steady planing flows will be described in a later section. During steady planing, the effective vertical velocity *V* depends on the local trim angle.

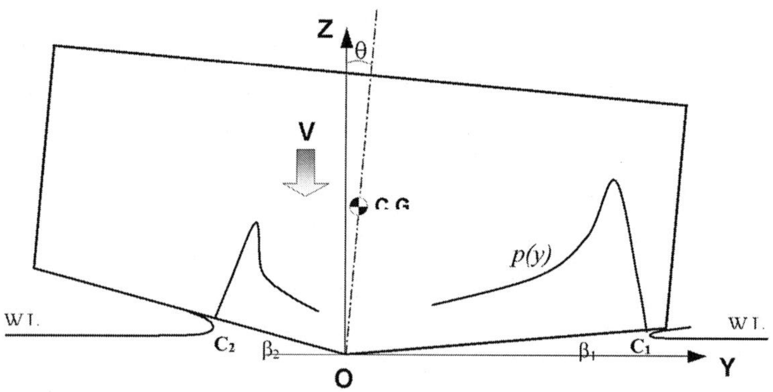

Fig. 1.4.2. Hull section under heeled planing (chine-dry)

The roll motion was solved in a constant meant draft, following Newton's Law, that is:

$$I_G\ddot{\theta} = W H_G \sin\theta - M_R(\theta) \qquad (1.4.1)$$

where the heel angle of the boat at planing and the time t is implicit in the equation. I_G is the moment of inertia with respect to a longitudinal axis passing through the center of gravity. The first term in the right-hand side (RHS) of equation 1.4.11 is the rolling moment due to the boat weight itself with transverse variation of the center of gravity from the keel. W here is the weight of the hull, and H_G is the height of the center of gravity above the keel at zero-heel position. The second term in RHS, M_R (θ), represents the total restoring moment including contributions of the hydrostatic and hydrodynamic forces. More specifically, the hydrostatic moment is due to the asymmetric distribution of hydrostatic pressure on the body wetted-surface. The hydro dynamic moment is due to planing induced dynamic pressure on the wetted surface, which can be determined using the similar method of Xu et al. (1998b). Assumptions of inviscid and incompressible flow are applied for the solutions. The influence of angular velocity is also incorporated in the hydrodynamic moment. The restoring moments here are calculated with respect to the keel.

The above equation of roll motion in the transverse plane is rewritten to be non-dimensional, that is:

$$C_I\ddot{\theta} + C_E\dot{\theta} + C_R\theta = 0 \qquad (1.4.2)$$

where the non-dimensional time $\tau = Vt/B$ is also used implicitly in the above equation. B here denotes the beam of the hull section. C_I in (2) is the inertia coefficient, that is:

$$C_I = \frac{I_G}{\rho B^4} \qquad (1.4.3)$$

where ρ is the density of the water.

The term of angular velocity here is separated from the dynamic restoring moment, with C_E denoting the coefficient of extra restoring moment due to $\dot{\theta}$ (the term of $C_E\dot{\theta}$ is actually a reduction on the total restoring moment due to $\dot{\theta}$ with the opposite direction of θ). This term is obtained by including $-y\dot{\theta}$ in the downward direction, with contribution of the angular velocity, in satisfaction to the kinematic body boundary condition and also in the Bernoulli's equation for calculation of dynamic pressures on the body contour. Then

$$C_E = \frac{1}{3}(\cos\beta_1 + c_2{}^3 \cos\beta_2) \tag{1.4.4}$$

where c_2 is the ratio of the wetted width on the left-side bottom to that on the right side—that is, $-Y_{C2} / Y_{C1}$, as shown in Fig. 1.4.2. β_1 and β_2 represent the effective deadrise angles on the left and right sides, respectively. Fig. 1.4.3 shows the typical values of C_E versus the heel angles for various deadrise hulls. As can be seen, C_E is determined only by the geometric form of the hull section and the difference of C_E is actually very small between hulls with different deadrises—that is, C_E is relatively insensitive to the deadrise angle. For a specific hull, moreover, C_E decreases very slowly as the heel angle increases. It is suggested that in practical design this parameter may be taken as a constant as for consideration of transverse stability in roll.

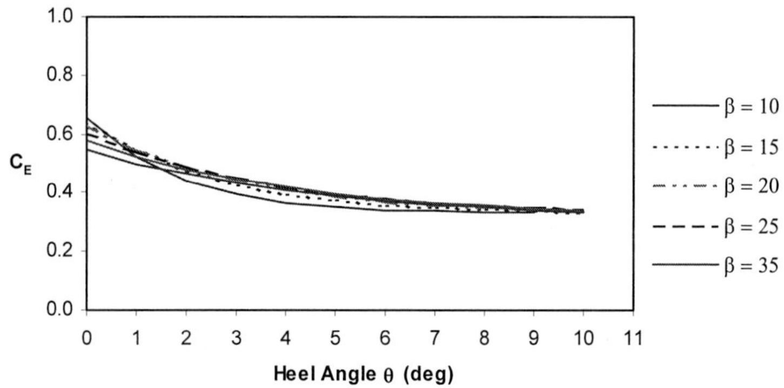

Fig. 1.4.3. The coefficient of the extra restoring moment due to $\dot{\theta}$ is the deadrise angle of a prismatic hull, in deg.

C_R in equation 1.4.2 is the non-dimensional restoring moment per degree in roll, excluding the influence of $\dot{\theta}$. The C_R consists of three components: the hydrostatic moment C_{R0}, the hydrodynamic moment C_{R1}, and the rolling moment due to the boat weight itself, i.e., C_{RG}. The moments are normalized to $\rho V^2 B^2$.

The non-dimensional hydrostatic restoring moment C_{R0} is obtained in the following:

$$C_{R0} = \frac{G}{3\theta}(\tan\beta_1 - c_2^{\ 3}\tan\beta_2) \qquad (1.4.5)$$

where $G = gB/V^2$ is the non-dimensional acceleration of gravity.

The non-dimensional dynamic restoring moment is calculated by integrating the dynamic pressure on the wetted surface, that is:

$$M_{R1} = \frac{1}{2}\int_{Y_{C2}}^{Y_{C1}} C_p(y)y\,dy \qquad (1.4.6)$$

where $C_p = (p - p_\infty)\big/\frac{1}{2}\rho V^2$ is the dynamic pressure coefficient as in Xu et al. (1998a), excluding the influence of angular velocity in roll. The C_{R1} is then obtained as $C_{R1} = M_{R1}/\theta$.

Furthermore, rewriting the roll moment into a non-dimensional form due to the boat weight, C_{RG} is given as follows:

$$C_{RG} = -C_M G \cdot h_G \frac{\sin\theta}{\theta} \qquad (1.4.7)$$

where C_M is mass coefficient which is defined as

$$C_M = \frac{W}{\rho g B^2} \qquad (1.4.8)$$

h_G is the height of the center of gravity above the keel, normalized to B.

For commonly used prismatic hull sections, the above three components for the coefficient of restoring moment in the equation of motion (1.4.2) are calculated and presented here in Fig. 1.4.4–7. Fig. 1.4.4 shows the non-dimensional dynamic restoring moments for various hull sections with deadrise angles ranging from 10 to 35 degrees. The restoring moment is presented in a function of heel angle; the results here might suggest that the restoring moment per degree in roll is rather insensitive to the heel angle changing. An average n_t can be assumed for each hull. Fig. 1.4.5 gives the restoring moment due to hydrostatic forces on hull section; the values of C_{R0} are close to the different deadrise hull forms shown here. In particular, compared with the corresponding hydrodynamic component, C_{R0} is extremely small for flatter bottom forms. Also, for each hull section, with different heel angles the change of the hydrostatic moment per degree is slow. The combined restoring moments are shown in Fig. 1.4.6, in summation of the hydrostatic and hydrodynamic components. Fig. 1.4.7 is for the roll moment due to the hull weight, in a form of $-\frac{C_{RG}}{C_M G \cdot h_G}$, which is obviously close to a small constant of $\pi/180$.

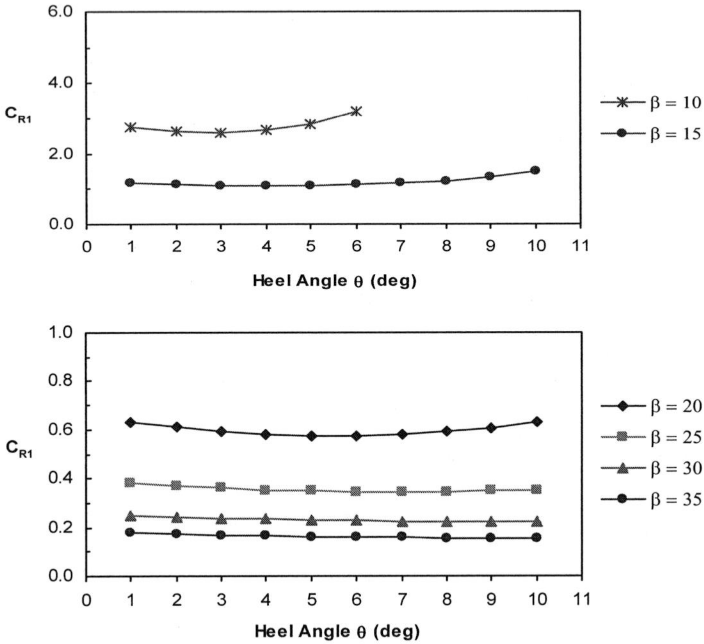

Fig. 1.4.4. The non-dimensional hydrodynamic restoring moments per degree in roll. $\dot{\theta}$ is the deadrise angle of a prismatic hull, in deg.

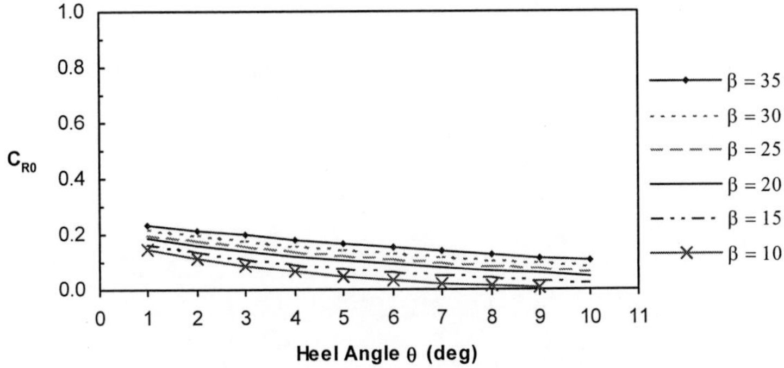

Fig. 1.4.5. The non-dimensional hydrostatic restoring moments per degree in roll. $\dot{\theta}$ is the deadrise angle of a prismatic hull, in deg.

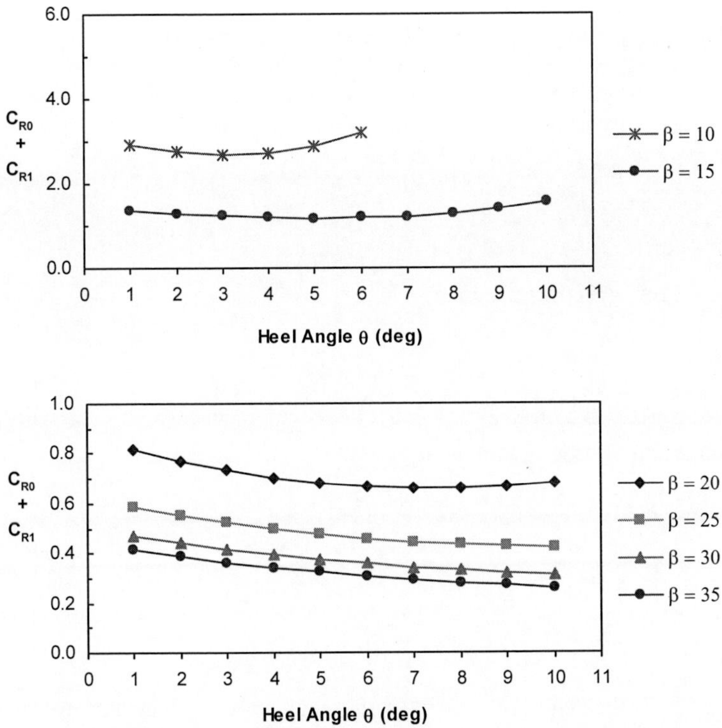

Fig. 1.4.6. The combined restoring moments per degree in roll. $\dot{\theta}$ is the deadrise angle of a prismatic hull, in deg.

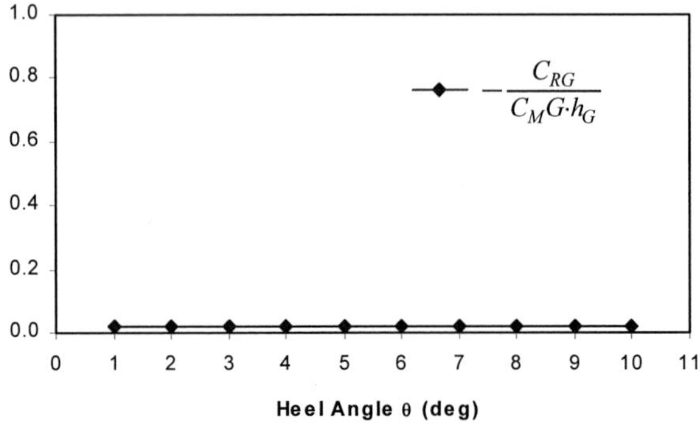

Fig. 1.4.7. The roll moment per degree, due to transverse variation of the gravity centre.

1.4.3. Criteria of dynamic stability in a transverse plane

As the coefficients in the equation of motion can be regarded as constants approximately from 1.4.2, dynamic stability requires:

$$C_E^2 - 4C_I C_R < 0 \quad (1.4.9)$$

or

$$C_E^2 - 4C_I(C_{R0} + C_{R1} + C_{RG}) < 0 \quad (1.4.10)$$

The following is thus obtained:

$$-C_{RG} < C_{R0} + C_{R1} - \frac{C_E^2}{4C_I} \quad (1.4.11)$$

Since $\sin\theta/\theta \leq \pi/180$ in 1.4.7, it closely approaches to $\pi/180$ with an assumption of little perturbation,

$$C_M G \cdot h_G < \frac{180}{\pi}\left(C_{R0} + C_{R1} - \frac{C_E^2}{4C_I}\right) \quad (1.4.12)$$

or

$$h_G < \frac{180}{\pi C_M G}\left(C_{R0} + C_{R1} - \frac{C_E^2}{4C_I}\right) \quad (1.4.13)$$

where h_G is the height of the center of gravity above the keel at zero heeled position. C_M is the mass coefficient and G denotes the non-dimensional acceleration of gravity. The terms of restoring moments in the above actually includes effects of speed, hull form, and the heeled position.

In the case of a flat deadrise hull section, C_{R0} is much smaller compared to C_{R1}, and may be assumed to be negligible in 1.4.14. Furthermore, as can be seen from Fig. 1.4.2, C_E varies very slightly as the heel angle increases and always reaches its maximum position, which equals to $\frac{2}{3}\cos\beta$. Thus, 1.4.14 can be simplified as follows:

$$h_G < \frac{180}{\pi C_M G}\left(C_{R0} + C_{R1} - \frac{\cos^2\beta}{9C_I}\right) \qquad (1.4.14)$$

where β is the deadrise angle of the planing hull section.

The solutions to the roll motion equation (1.4.2) use $\theta = e^{rt}(a_1 \cos\omega t + a_2 \sin\omega t)$, where $r = -C_E / 2C_I$. The constants, a_1 and a_2, can easily be obtained with initial conditions as $\theta(0) = \theta_0$, $\dot{\theta}(0) = 0$. The roll motion therefore becomes a decaying oscillation with an initial amplitude of θ_0. The rate of decay of the rolling amplitude is directly related to C_E, which is determined by the geometric form of the hull section. Furthermore, in the vicinity of the resonant encounter frequency in waves, a well-damped mode can be defined considering the equivalent damping ratio of oscillation.

1.4.4. The wetted area and dynamic restoring moment along a prismatic hull

As previously mentioned in relation to the similar features of flow physics, steady planing can be simulated using the two-dimensional impact flow, in which the longitudinal coordinate (in the direction of advance speed) is converted into a time variable, as in the two-dependent solution of cylinder impact. Assumptions of inviscid and incompressible flow, and also slenderness of hull body, are applied. Similar to Xu and Troesch (1999), the cross sectional behavior of steady planing is then approximated by two-dimensional cylinder impact solutions, under a coordinate conversion between time and space, as follows:

$$x = U_A \tau' \qquad (1.4.16)$$

In equation 1.4.16, x represents the longitudinal coordinate in the direction of advance, U_A is the advance speed of the planing hull, and τ' is an equivalent time corresponding to a two-dimensional impact solution for the hull section at x. The local vertical impact velocity, $V(\tau')$, which is valid for the sectional solution, therefore depends on the local trim angle *(x)*, i.e.,

$$V(\tau) = U_A \tan\alpha(x) \quad (1.4.17)$$

The above conversion and also the coordinate system applied here is described in Figure 7, which shows a prismatic hull with hard-chines at steady planing under heeled conditions. The heel angle can be either zero or non-zero. The hull is assumed to be of constant deadrise on the bottom, of constant beam, and with a constant trim angle of keel. Thus, the valid $V(\tau')$ here is a constant for the cylinder impact solutions to simulate the prismatic planing flows. In general, two different flow phases, chine-dry (CUW) and chine-wetted (CW), exist on each side of the planing surface. A–A, B–B, and C–C are three typical sections representing various impact stages—CUW, CW_1 (the chine-wetting occurs on one side only (e.g., the side with smaller deadrise angle as in Figure 7), and CW_2 (the chines are wetted on both sides), respectively. The impact model utilized here is obviously of Type A (Xu, Troesch, and Vorus 1998); that is, the flow attaches the bottom contour until separating off at the two chines. The wetted area is defined as a region of non-zero dynamic pressure within the zero-pressure points. It is also called the pressure surface—the dynamic pressure is dominant in this area. Profiles of the wetted areas of cases with various trim angles are solved according to Xu et al. (1998a) and Xu and Troesch (1999), as shown in Figures 8 to 10; also calculated are the dynamic restoring moments under three different speed coefficients—i.e., $C_v = 3.0$, 4.0, and 5.0. C_v here represents the speed coefficient, as defined in the following:

$$C_v = \frac{U_A}{\sqrt{gB}} \quad (1.4.18)$$

It is understood that the hull is in fully planing condition when $C_v > 1.5$; the dynamic lifting force is dominant in this speed range.

Calculation results are obtained for a prismatic hull with 20° deadrise under steady planing conditions. Figures 1.4.9–11 show cases of 3°, 6°, and 8° trim angles, respectively, showing the wetted area profiles and the dynamic restoring moments per degree in roll under various speed coefficients. In the case of a small trim angle, the results clearly show a longer wetted length on keel and also a lower restoring moment. This is due to the factor of $\tan\alpha$ in equation 1.4.17, which results in a smaller equivalent vertical impact velocity for the small trim angle case. Since the hydrodynamic lifting force depends on the equivalent vertical velocity, the planing efficiency is obviously lower in boats with a too small trim angle.

On the other hand, the planing resistance could be higher due to the larger wetted surface.

Furthermore, as can be noted in Figs 1.4.9–11, between two transitions of chine-dry to chine-wetted the negative moments apparently decrease the transverse stability. The negative moment however is small as shown in the results. Table 1.4.1 gives the total dynamic restoring moments by integrating the sectional moments along the hull length. It clearly indicates that the total moments are positive. The results of different speed coefficients also suggest that the transverse stability be better with faster planing speed. The ratio of restoring moments under different speeds is a square of the speed ratio.

Table 1.4.1. Total dynamic restoring moment per degree in roll (20° prismatic hull at steady planing)

Trim angle°	**Speed coefficient,** $C_v = U_A / \sqrt{gB}$		
	3.0	**4.0**	**5.0**
3°	0.2736	0.4864	0.7600
6°	0.5788	1.0289	1.6078
8°	0.7744	1.3768	2.1513

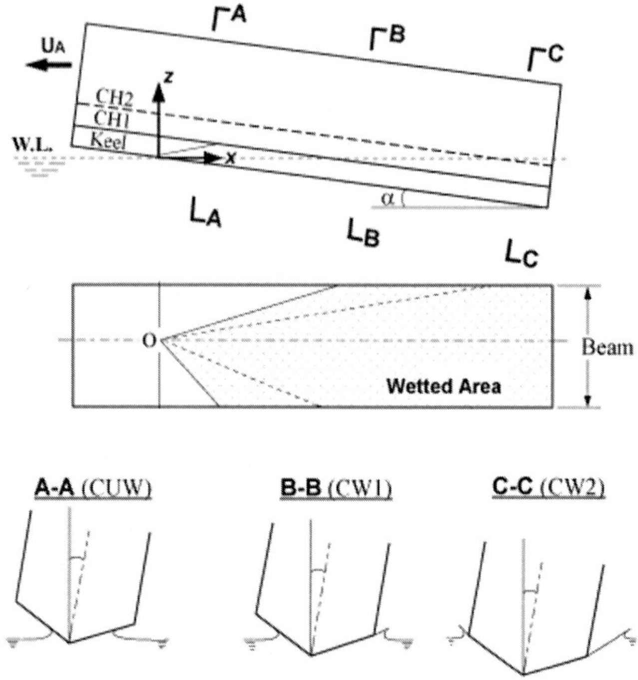

Fig. 1.4.8. Prismatic hull at steady planing under heeled condition

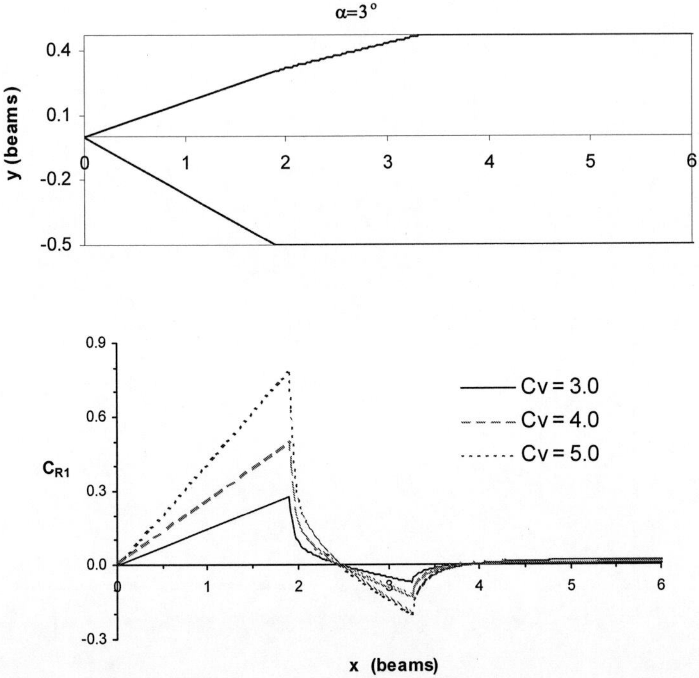

Fig. 1.4.9. Wetted area and dynamic restoring moment per degree in roll with trim angle $\alpha = 3^{\circ}$

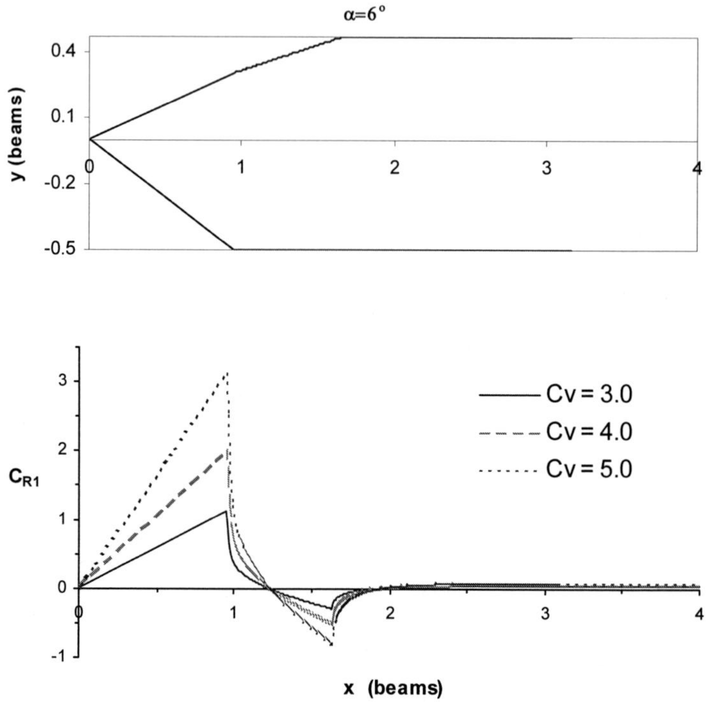

Fig. 1.4.10. Wetted area and dynamic restoring moment per degree in roll with trim angle $\alpha = 6^{\circ}$

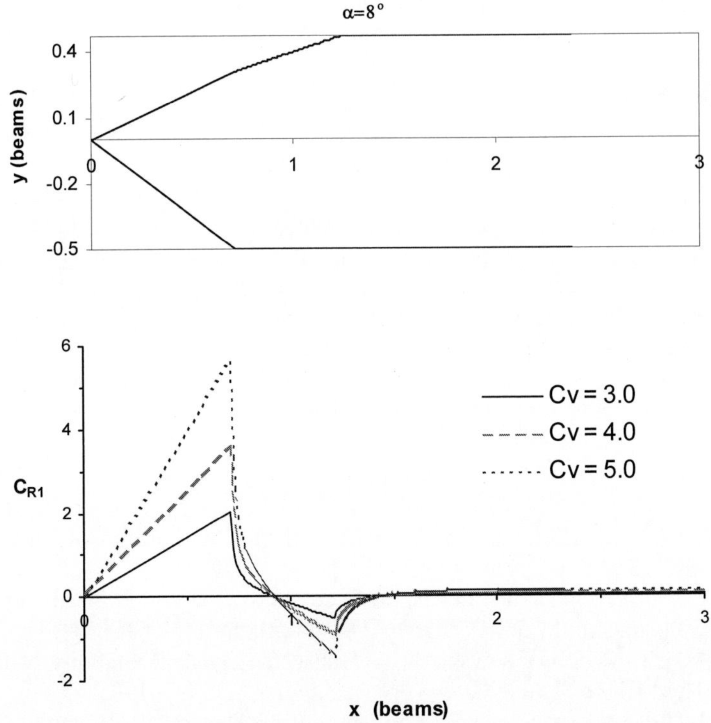

Fig. 1.4.11. Wetted area and dynamic restoring moment per degree in roll with trim angle $\alpha = 8^{\circ}$

1.4.5. Conclusions

The study has been carried out to provide a practical method for prediction of the transverse roll motion and restoring moments for high-speed planing craft. The equation of roll motion is developed employing a two-dimensional asymmetric planing model, incorporating the effects of hull form and flow separation at hard-chines. The dynamic restoring moment is influenced by planing speed and trim angle, and the damping coefficient of roll motion is directly related to the geometric form of the hull. While solving roll motion in the transverse plane, stability criteria are developed to provide a height requirement on the boat's center of gravity. Although some example results shown here are for prismatic hulls, the work appears to be suitable as a general method for simulating roll motion in the transverse plane, and also transverse dynamic stability in practice.

Furthermore, the method itself is developed to extend directly to non-prismatic hulls. Further verification and extension for more general applications in planing boats are to be presented in subsequent publications.

Acknowledgements

The authors would like to thank the management of the American Bureau of Shipping for their support of this paper; we especially thank Dr D. Liu and Dr J. Spencer of ABS for their encouragement. The authors are grateful to Prof. A. W. Troesch of the University of Michigan for his constructive suggestions and general instructions in the present study. The opinions expressed here are those of the authors and do not necessarily reflect those of the organization that they are representing.

References

[1-4-1] D. L. Blount & L. T. Codega, "Dynamic Stability of Planing Hulls," *Marine Technology* 29 (1), pp. 4–12, 1992.

[1-4-2] L. Codega and J. Lewis, "A Case Study of Dynamic Instability in a Planing Hull," *Marine Technology* 24 (2), pp. 143– 63, 1987.

[1-4-3] E. M. Lewandowski, "The Transverse Dynamic Stability of the Hard-Chine Planing Craft," Proceeding, PRADS '95, 1995.

[1-4-4] E. M. Lewandowski, "Prediction of the Dynamic Roll Stability of Hard-Chine Planing Craft," *Journal of Ship Research* 40 (2), pp. 144–48, 1996.

[1-4-5] E. M. Lewandowski, "Transverse Dynamic Stability of Planing Craft," *Marine Technology* 34 (2), 1997.

[1-4-6] M. Martin, "Theoretical Determination of Porpoising Instability of High-Speed Planing Boats," *Journal of Ship Research* 22 (1), pp. 32–53, 1978.

[1-4-7] W. G. A. Perring & H. Glauert, "Stability on the Water of a Seaplane in the Planing Condition," *Aeronautical Research Council*, TR, 42, 1933.

[1-4-8] P. R. Payne, "Coupled Pitch and Heave Porpoising Instability in Hydrodynamic Planing," *Journal of Ship Research* 8 (2), 1974.

[1-4-9] P. R. Payne, "On the High-Speed Porpoising Instability of a Prismatic Hull," *Journal of Ship Research* 28 (2), pp. 77– 89, 1984.

[1-4-10] E. Thornhill, B. Veitch, & N. Bose, "Dynamic Instability of a High-Speed Planing Boat Model," *Marine Technology* 37, (3), pp. 146–152, 2000.

[1-4-11] L. Xu, A. W. Troesch, and W. S. Vorus, "Asymmetric Vessel Impact and Planing Hydrodynamics," *Journal of Ship Research* 42 (3), pp. 187–98, 1998.

[1-4-12] L. Xu, A. W. Troesch, & R. Peterson, "Asymmetric Hydrodynamic Impact and Dynamic Response of Vessel," Proceedings, OMAE '98, 1998.

[1-4-13] L. Xu & A. W. Troesch, "A Study on Hydrodynamics of Asymmetric Planing Surface," Proceedings, FAST '99, 1999.

[1-4-14] R. Zhao, O. M. Faltinsen, & H. A. Haslum, "A Simplified Nonlinear Analysis of a High-Speed Planing Craft in Calm Water," Proceedings, FAST '97, 1997.

1.5

SERIES-TEST RESEARCH ON THE LINEAR FORM OF DOUBLE-STEP PLANING BOATS (2014)

SHEN XIAOHONG

China Ship Development and Design Center

1.5.1. Overview

The double-step planing boat is excellent in both rapidity and stability of longitudinal motion; in recent years it has been widely used in military and civil markets at home and abroad and it possesses broad market prospects. The research reveals that the main reason for the boat's super high-speed performance is its unique hull configuration—two steps are longitudinally set at the bottom of the boat.[1-5-1] Therefore, an understanding of the impacts of step parameters (longitudinal position and altitude of steps) of the double-step planing boat on its resistance performance helps to improve its form and performance. However, due to its unique characteristics of the double-step planing boat, there is as yet no effective theoretical method at home or abroad to calculate its resistance performance. On this basis, the paper adopts a model-test method to develop research on the impacts of step parameters of the double-step planing boat on its resistance performance.

First, the paper selects several step parameters through research and various demonstrations, thereby determining the quantity of series models of the double-step planing boat, and then it carries out a towing test on a series of models according to the test contents. Through the analysis of test results, the fundamental law of impacts of step parameters on high-speed section resistance as well as their interactions is discovered.

1.5.2. Test model and test contents

1. Test model

The step parameters of the double-step planing boat mainly refer to the fore step length L_2, after step length L_1 and the height h of two steps (normally their heights are equal); therefore, the selection of step parameters takes these three varying factors into account. If each factor selects three different values, then 27 ship models will be required,[1-5-2] which is inconvenient to achieve due to cost considerations. If each factor selects two different values, then eight ship models will be required, but we would still be unable to find out the changed rule due to the small number of values. Therefore, this paper adopts one new method to determine the selection of step parameters—that is, by applying the geometrical relation via overall step length L_0 and step angle Q_o (the connecting lines between the lower-edge points of two steps and those of the stern enclosed plate on any buttock line are on the same straight line, and the included angle between the straight line and the base line is the step angle), the values of fore step length L_2, after step length L_1, and step height h could be obtained. In this way, for step parameters of the double-step boat, the change of three factors could be converted into the change of two factors. Consequently, the step parameters of test models selected in this paper are determined by two factors: overall step length L_0 and step angle Q_0.

If the two factors overall step length L_0 and step angle Q_0 are known (see Fig. 1.5.1), other characteristic parameters of the linear form can be obtained as follows:

$L_1 = L_0 \times 1/(2 - Q_1/Q_0)$

$L_2 = L_0 - L_1$

$h = L_2 \times \tan(Q_0)$

For the series models of the double-step planing boat, the length L, chine breadth B, molded depth D, and deadrise angle of hull bottom β (β unchanged from midship to stern) are identical; the step plane shown as a V-shape and the lower edge of the fore and after steps of each buttock line as well as that of stern enclosed plate are shown as a straight line.[1-5-3] Step parameters of the series of linear forms of the double-step planing boat are shown in Table 1.5.1.

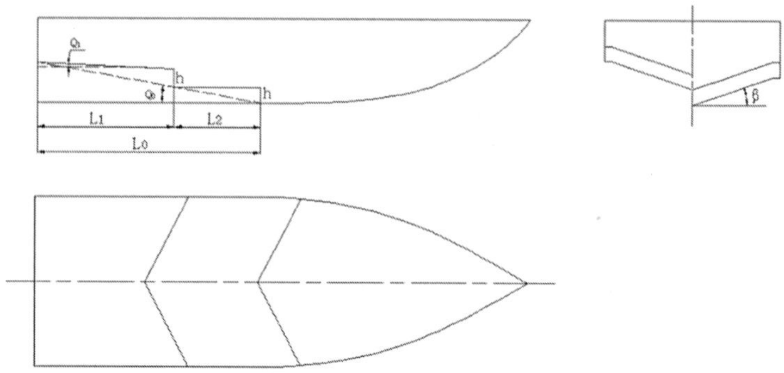

Fig. 1.5.1. Schematic diagram of the double-step planing boat

Table 1.5.1. Step parameters of the series of linear forms of the double-step planing boat

No.	L_0/L	Q_0/deg.	Q_1/deg.	L_1/L	L_2/L	h/L
A_1	0.458	0.81	0.3	0.281	0.177	0.0027
A_2	0.458	1.11	0.3	0.265	0.193	0.0036
A_3	0.458	1.41	0.3	0.256	0.202	0.0050
B_1	0.485	0.81	0.3	0.298	0.187	0.0027
B_2	0.485	1.11	0.3	0.281	0.204	0.0041
B_3	0.485	1.41	0.3	0.272	0.213	0.0055
C_1	0.431	0.81	0.3	0.264	0.167	0.0023
C_2	0.431	1.11	0.3	0.249	0.182	0.0036
C_3	0.431	1.41	0.3	0.241	0.190	0.0045

2. Test contents

The speed points of the towing test on series models are identical, and each model test selects the same five states—that is, by changing the three longitudinal position states of C.G. (marked as X_1, X_2, X_3, and $X_1<X_2<X_3$, C.G. [center of gravity] longitudinal position refers to the distance to the stern enclosed plate) at the same displacement (marked as Δ_2), as well as changing the three displacement states (marked as Δ_1, Δ_2, Δ_3, and $\Delta_1<\Delta_2<\Delta_3$) at the same C.G. longitudinal position (marked as X_2).

1.5.3. Processing and analysis of a series of model test results

1. The impacts of overall step length
When step angle Q_0 is relatively large (Q_0=1.11, Q_0=1.41), the overall step length L_0 and step angle Q_0 will interact with each other; when step angle Q_0 is the smallest, the weight and C.G. will also interact with each other; at different step angles Q_0, the change rules of overall step length L_0 will be inconsistent, and quite the contrary sometimes.

When step angle Q_0 is large, the high-speed resistance of model A_3 featuring medium overall step length L_0 is the least; however, that of model B_3 featuring the largest overall step length L_0 is the greatest, and the changes of weight and C.G. exert little influence on the high-speed section trends of curves.

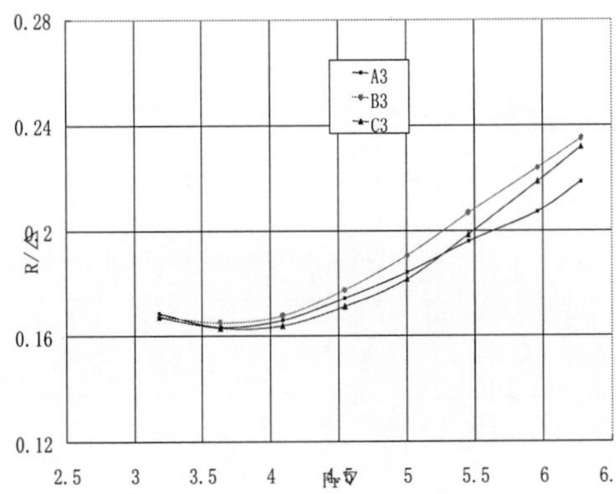

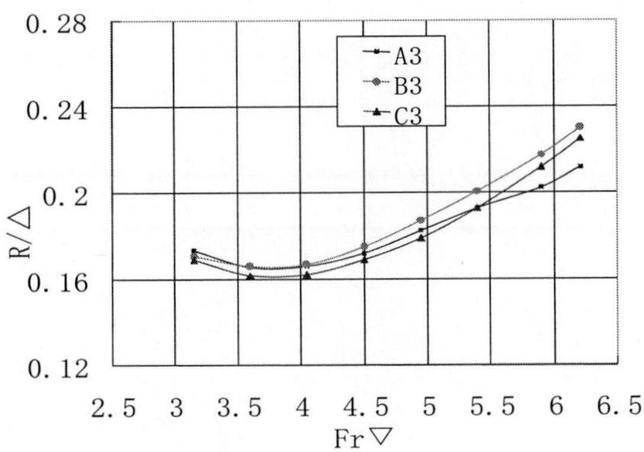

$\Delta=\Delta_1$, $X_g= X_2$ (A_3, B_3, C_3)
$\Delta=\Delta_2$, $X_g= X_2$ (A_3, B_3, C_3)

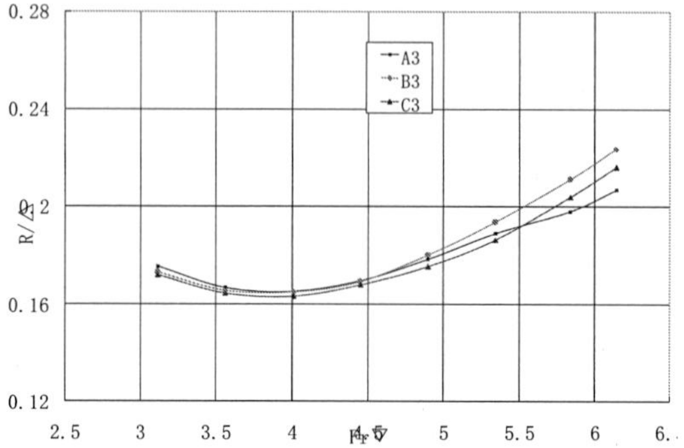

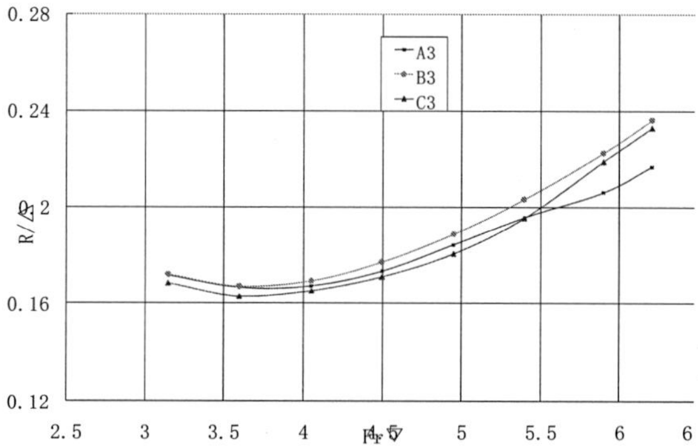

$\Delta = \Delta_3$, $X_g = X_2$ (A_3, B_3, C_3)

$\Delta = \Delta_2$, $X_g = X_3$ (A_3, B_3, C_3)

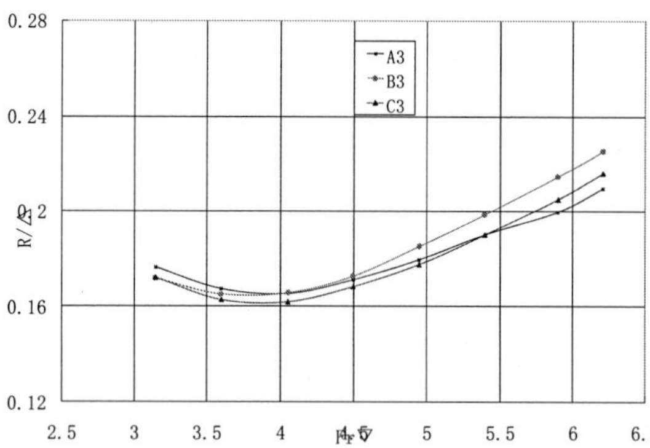

$\Delta = \Delta_2$, $X_g = X_1$(A_3, B_3, C_3)

When step angle Q_0 is medium, and overall step length L_0 is decreasing, the high-speed resistance will increase; however, the high-speed resistances at L_0/L=0.458mm and L_0/L=0.485mm are very approximate (except at the highest speed of 13.8m/s). While the weight changes, the influence exerted on the high-speed section trends of curves is relatively small; when the C.G. is set back, it will exert some influence on the high-speed section trends of curves.

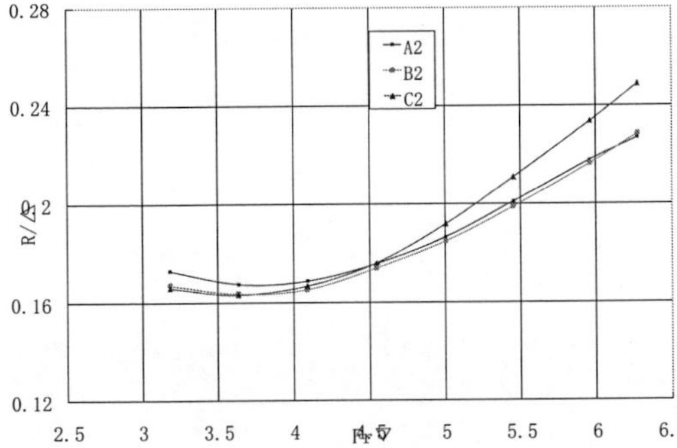

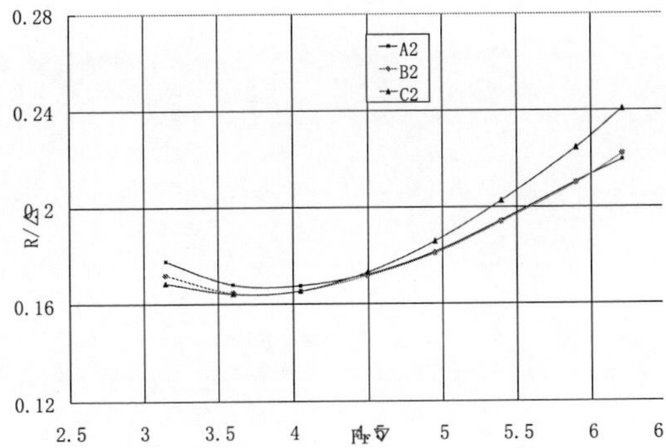

$\Delta=\Delta_1$, $X_g= X_2$ (A_2, B_2, C_2)
$\Delta=\Delta_2$, $X_g= X_2$ (A_2, B_2, C_2)

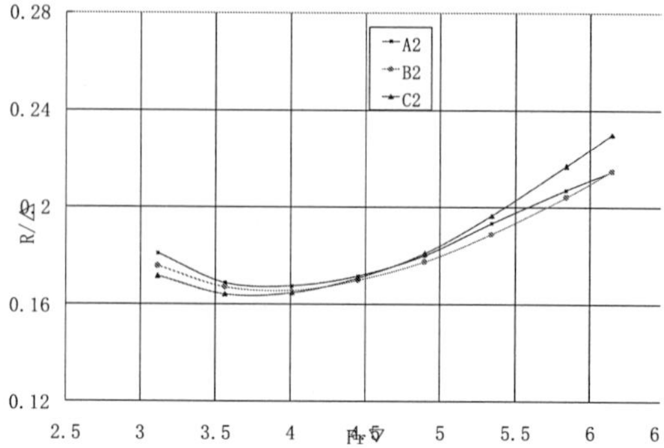

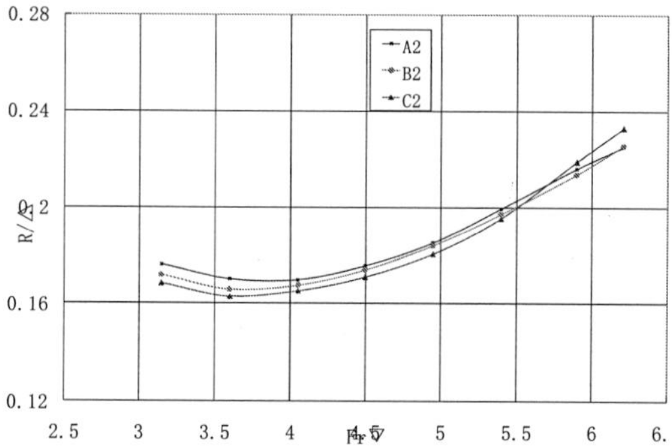

$\Delta = \Delta_3$, $X_g = X_2$ (A_2, B_2, C_2)

$\Delta = \Delta_2$, $X_g = X_3$ (A_2, B_2, C_2)

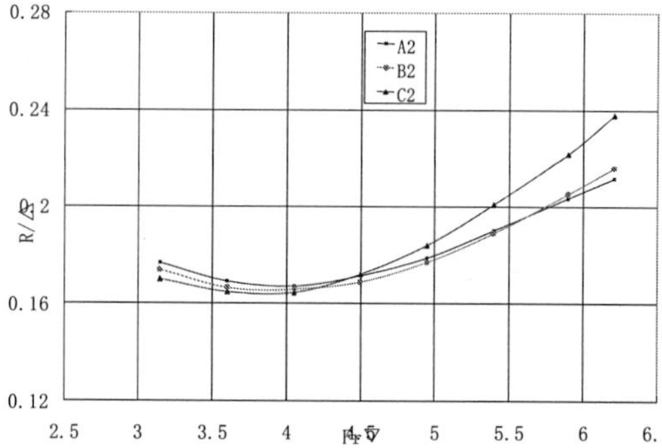

$\Delta = \Delta_2$, $X_g = X_1$ (A_2, B_2, C_2)

When step angle Q_0 is small, the weight will be small, and the high speed resistance of model C_1 featuring the smallest overall step length L_0 is less than that of model A_1, which features medium overall step length L_0. However, with the gradual increase of the weight, from the curve trends, it can be seen that the resistance of C_1 at a high-speed section exceeds that of A_1 much faster, and when the weight increases to a certain value, the resistance performance of C_1 will be clearly weaker than that of A_1. Although the resistance performance of B_1 featuring the largest overall step length L_0 is good, its longitudinal stability is the weakest.

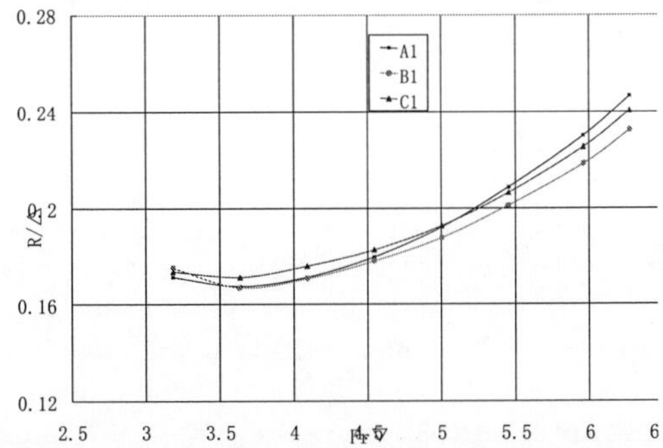

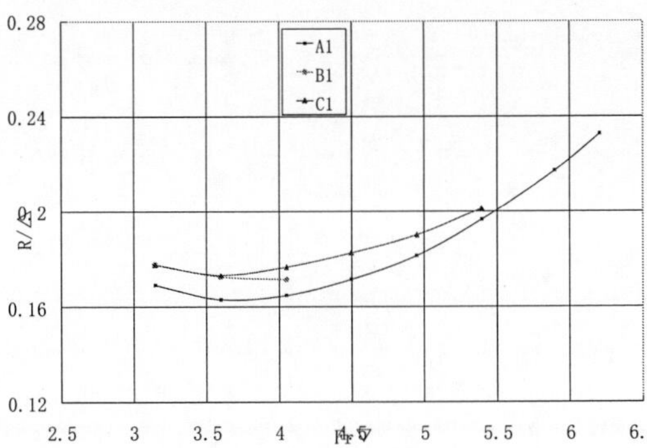

$\Delta = \Delta_1$, $X_g = X_2$ (A_1, B_1, C_1)

$\Delta = \Delta_2$, $X_g = X_2$ (A_1, B_1, C_1)

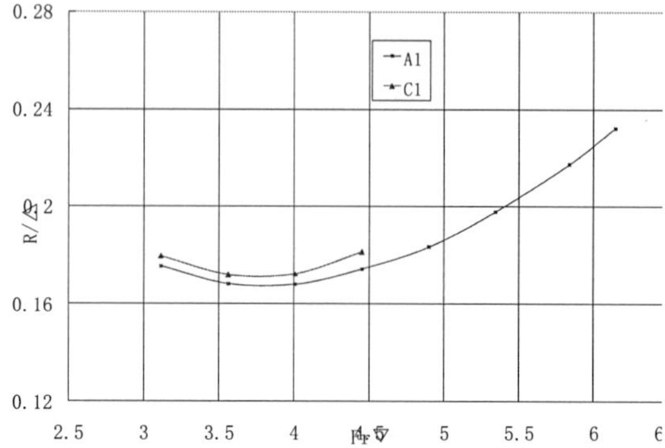

$\Delta=\Delta_3$, $X_g= X_2$ (A_1, C_1)

When step angle Q_0 is small, and the C.G. is set forward, with the decrease of overall step length L_0, the high-speed resistance will increase. Therefore, in the former condition, the overall step length L_0 selected must be relatively large. From the comprehensive consideration of longitudinal stability, overall step length L_0/L=0.458mm matches more closely with a smaller high-speed resistance.

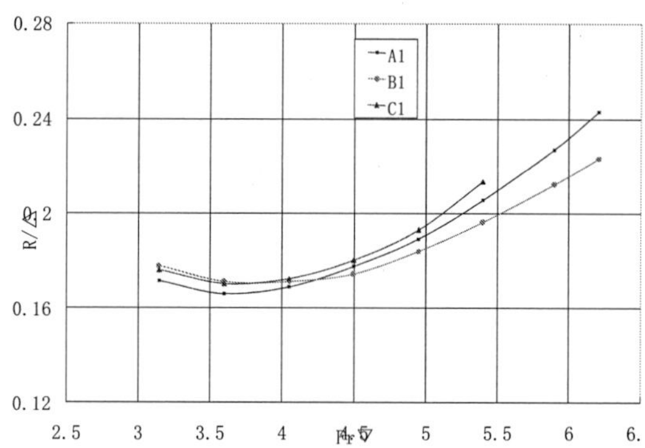

$\Delta=\Delta_2$, $X_g= X_3$ (A_1, B_1, C_1)

2. The influence of step angle

When the overall step length L_0 is relatively large (L_0/L=0.458mm, L_0/L=0.485mm), the overall step length L_0 and step angle Q_0 will interact with each other; when step angle Q_0 is the smallest, the C.G. will also interact; at different overall step lengths L_0, the changed rules of step angle Q_0 will not be completely consistent, quite the contrary sometimes.

When the overall step length is L_0/L=0.458mm, and step angle Q_0 is large, the high-speed resistance will decrease, with no influence from the C.G. setting forward or the change of weight on the graphic trends of the high-speed section. The influence from the C.G. setting back is relatively obvious, and the two curves with larger step angles are very approximate at the high-speed section.

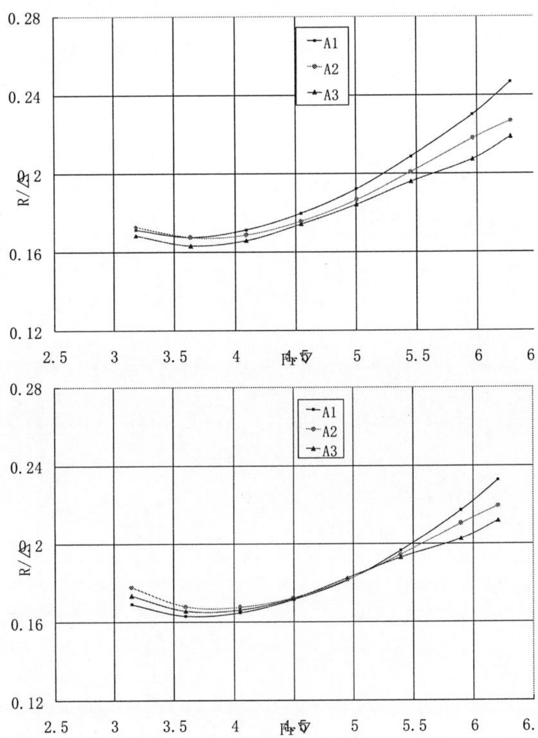

$\Delta=\Delta_1$, $X_g= X_2$ (A_1, A_2, A_3)
$\Delta=\Delta_2$, $X_g= X_2$ (A_1, A_2, A_3)

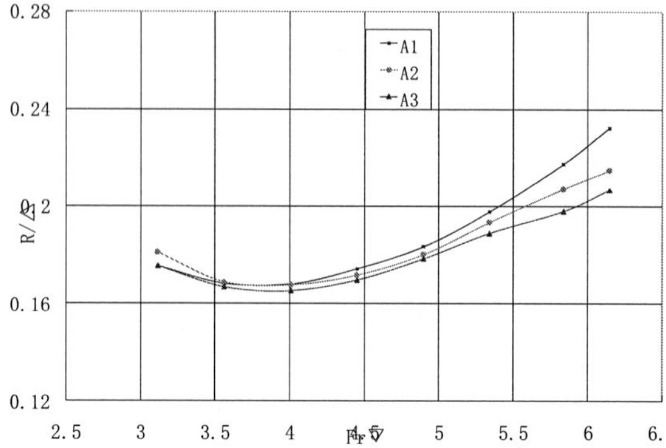

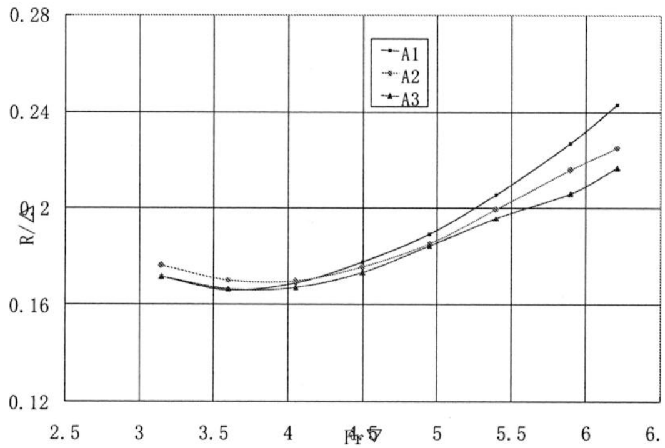

$\Delta = \Delta_3$, $X_g = X_2$ (A_1, A_2, A_3)

$\Delta = \Delta_2$, $X_g = X_3$ (A_1, A_2, A_3)

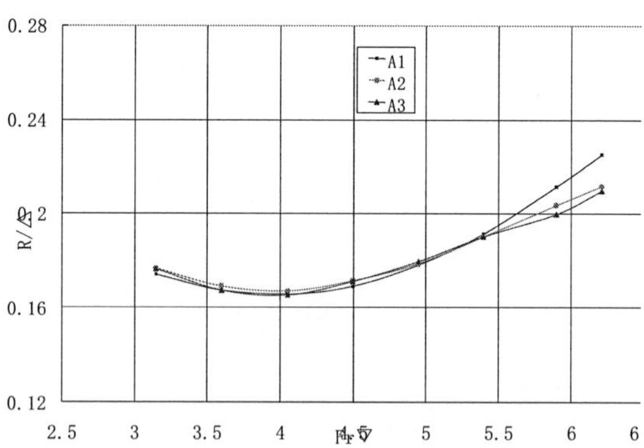

$\Delta = \Delta_2$, $X_g = X_1$ (A_1, A_2, A_3)

When the overall step length is L_0/L=0.485mm, under light-weight conditions with satisfying longitudinal stability, the high-speed resistance of model B_1 featuring the smallest step angle Q_0 is greater than that of model B_2 featuring the medium step angle Q_0, and that of model B_3 featuring the largest step angle Q_0, is the greatest, whose rule is inconsistent with the situation of the overall step length of L_0/L=0.458mm.

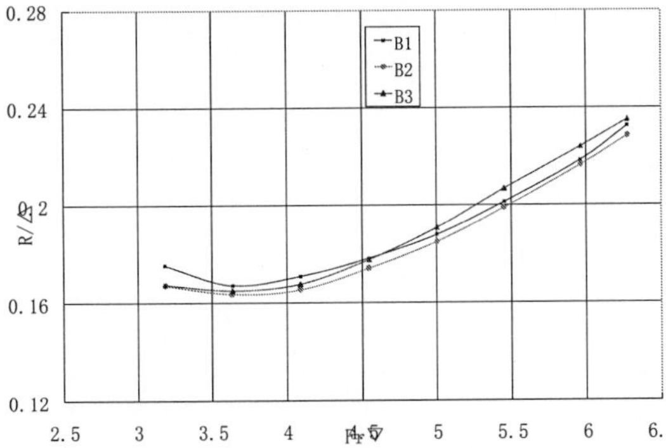

$\Delta=\Delta_1$, $X_g=X_2$ (B_1, B_2, B_3)

When the overall step length is L_0/L=0.485mm, under heavy-weight conditions with satisfying longitudinal stability, then step angle Q_0 is large, then the high speed resistance will increase, completely opposite to the rule at the overall step length L_0/L=0.458mm, and the change of C.G. position will exert no influence on the curve trends. Nevertheless, the change of weight (compared with relatively light weight) exerts relatively great influence on the high-speed resistance curve rules of models at different step angles Q_0 (overall step length L_0/L=0.485mm).

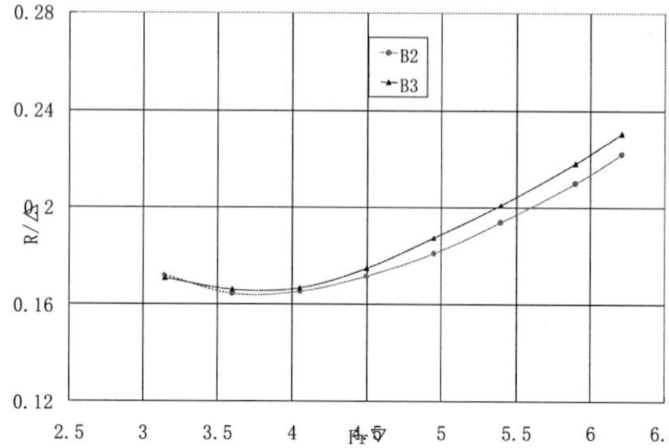

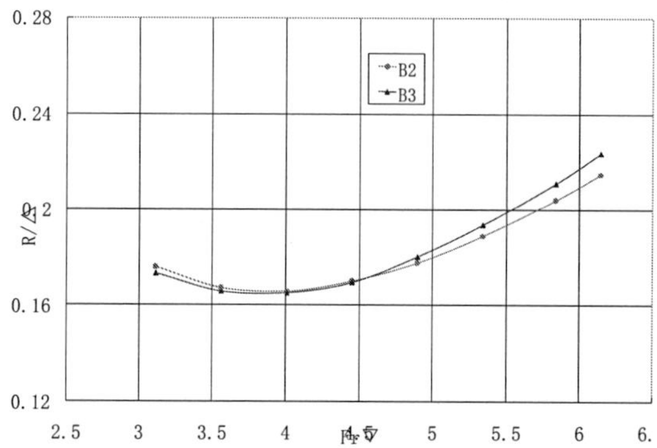

$\Delta = \Delta_2$, $X_g = X_2$ (B_2, B_3)
$\Delta = \Delta_3$, $X_g = X_2$ (B_2, B_3)

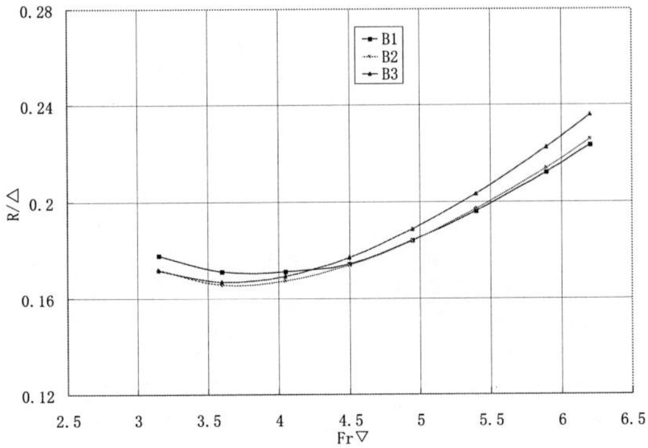

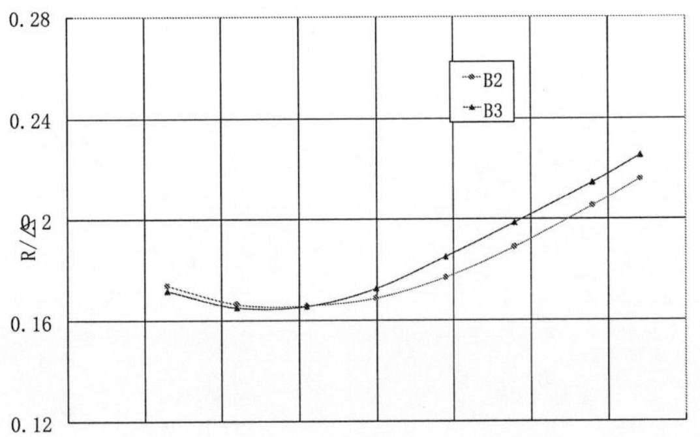

$\Delta=\Delta_2$, $X_g=X_3$ (B_1, B_2, B_3)
$\Delta=\Delta_2$, $X_g=X_1$ (B_2, B_3)

When the overall step length is $L_0/L=0.431$mm, under light-weight conditions with the C.G. setting forward and satisfying longitudinal stability, then step angle Q_0 is large and the high-speed resistance will decrease. This is consistent with the rule at $L_0/L=0.458$mm. However, under light-weight conditions, with the C.G. setting back and satisfying longitudinal stability, the high-speed resistance of model C_1 featuring the smallest step angle Q_0 is less than that of model C_2 featuring the medium step angle Q_0, and that of model C_3 featuring the largest step angle Q_0 is the least. This rule is inconsistent with the situation of overall step lengths of $L_0/L=0.458$mm and $L_0/L=0.485$mm. Therefore, at an overall step length of $L_0/L=0.431$mm with light weight, the change of C.G. position will exert obvious influence on the curve trends.

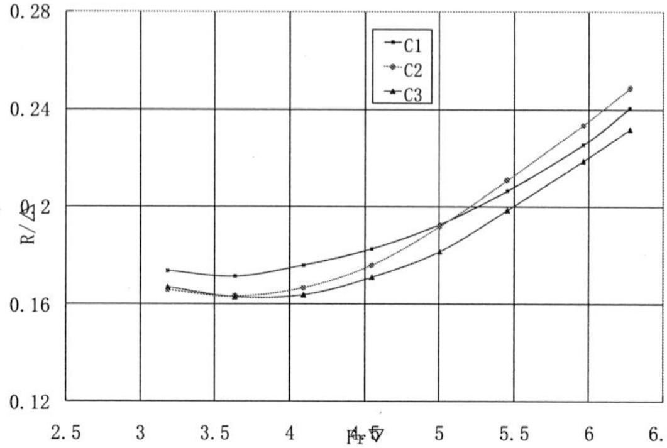

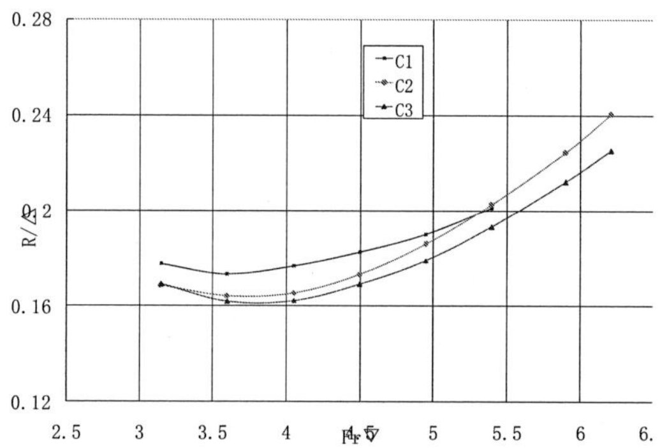

$\Delta = \Delta_1$, $X_g = X_2$ (C_1, C_2, C_3)

$\Delta = \Delta_2$, $X_g = X_2$ (C_1, C_2, C_3)

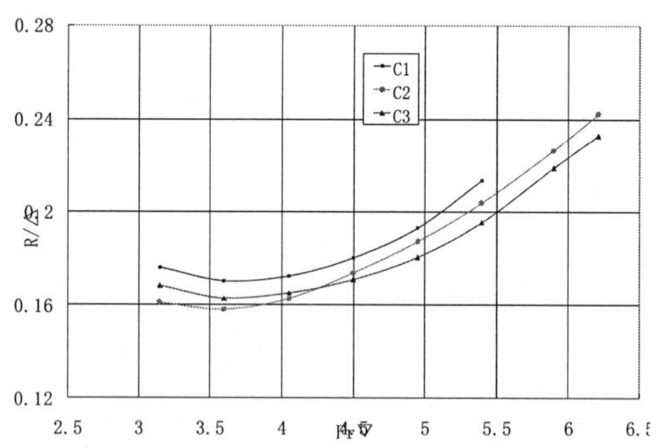

$\Delta = \Delta_2$, $X_g = X_3$ (C_1, C_2, C_3)

When the overall step length is $L_0/L=0.431$mm, under heavy-weight conditions with satisfying longitudinal stability, then step angle Q_0 is large and the high-speed resistance will decrease. This is consistent with the rule at $L_0/L=0.458$mm.

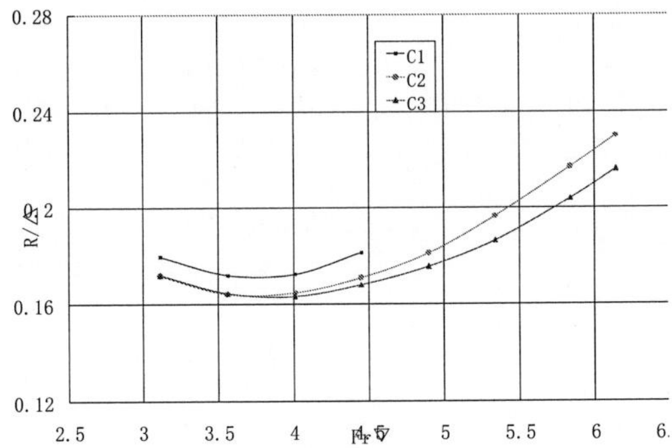

$\Delta=\Delta_3$, $X_g= X_2$ (C_1, C_2, C_3)

1.5.4. Conclusion

Through the analysis on the influences of four factors such as weight, C.G. longitudinal position, overall step length L_0, and step angle Q_0 exerted on resistance performance, it could be seen that the four factors exert great influences, and they interact with one another. The change of any model with the change of weight and C.G. longitudinal position shows the same rule, which is identical to the conventional planing boat.[1-5-4] But the rule of the influence of the changes of overall step length L_0 and step angle Q_0 exerted on resistance performance is relatively complex, showing that the interaction of influences among step parameters is great. At overall step lengths of $L_0/L=0.485$mm and $L_0/L=0.431$mm, the longitudinal stability of the two models B_1 and C_1 featuring small step angles are relatively weak; at overall step lengths of $L_0/L=0.458$mm, though the longitudinal stability of model A_1 featuring a small step angle is good, it is nevertheless weaker compared with the resistance performances of the two models A_2 and A_3, which feature relatively large step angles. Thus, in the design of double-step planing boat, it would be appropriate to select step angle Q_0 larger.

References

[1-5-1] Dong Zushun, *Speedboat Dynamics*, Wuhan: Huazhong University of Science and Technology Press, 1994.

[1-5-2] Wang Yuan, "Homogeneous Design—One Experimental Design Method," *Selected Papers on Homogeneous Design*, 1997 (2): pp. 1–4.

[1-5-3] Xia Xiang, "Research on Operating Condition Matching Characteristic of Propulsion System of Super High Speed Double-Step Planing Boat," *Chinese Journal of Ship Research* 3 (2): pp. 39–42, 2008.

[1-5-4] Yang Haiyan & Hua Yuqun, "Model Test on Hull Performances of Planing Boat," *International Ship Engineering*, 304 (6): pp. 1–6, 2004.

1.6

STUDY ON LINE OPTIMIZATION OF M-FORM HULLS (2014)

HUANG WUGANG & SHEN XIAOHONG

China Ship Development and Design Center

Introduction

Single-hull planing craft, air cushion vehicles, and catamarans all belong to dynamically supported craft, among which catamarans are of the static floating type. Under certain navigational conditions, they will exploit their respective advantage, but they cannot simultaneously possess the following merits: high speed, low power, wave-damping, good sea-keeping ability, high stability, and economic benefits. The M-form hull combines features of regular single-hull planing craft, high-speed catamarans, and air cushion vehicles, and integrates the advantages of these three types. Through the combination of fluid mechanics and aerodynamics, it shows excellent performance in wave-damping, effective load, maneuverability, and high-speed navigation capability during against-weather conditions, all of which enable it to be one of the most competitive new types of all.[1-6-1] [1-6-2]

The study of the hydrodynamic capability of high-speed vessels has always been a difficult point of research, among which research on resistance is the most important.[1-6-3] If an M-form hull of optimum resistance could be achieved, it would have a great influence on the development of new-type, high-speed, multi-purpose, and large-scale combat ships and transportation vessels.

1.6.1. Main dimensions of the original hull form and coefficient analysis of the original hull form

1. Main dimensions of the original hull form
For main dimensions of original hull form M-1, see Table 1.6.1; for the body plan, see Fig. 1.6.1.

Table 1.6.1 Main dimensions of original hull form M-1

Item	M-1
Overall length/m	20.0
Breadth/m	6.2
Molded depth/m	2.6
Displacement/t	30
Main hull half-breadth/m	1.45
Air flue width/m	1.45

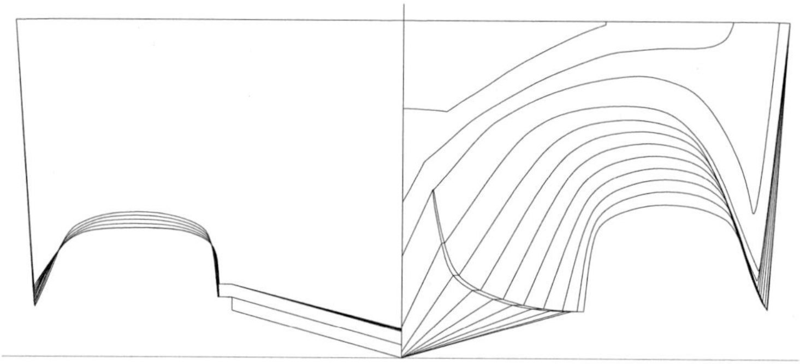

Fig. 1.6.1. Body plan of original hull form M-1

2. Coefficient analysis of original hull form
For the M-form hull, the main hull is like that of a regular planing craft, for which breadth is one of the most important parameters for resistance performance. Breadth includes main hull breadth and air flue width, among which the former has a great influence on the hydrodynamic performance of the M-form hull and the latter has a great influence on the hydrodynamic performance of the bidirectional airflow. From wing theory it is known that the lift coefficient of a wing with a finite aspect ratio is related to relative span, that is, aspect ratio, which can be expressed as follows:[1-6-4]

$$C_Y = 2\pi\alpha \frac{A}{A+2} \qquad (1.6.1)$$

It should be noted that aspect ratio A=span/chord length—this is, the ratio between the width of the wing and chord length. This just means the reciprocal of wetted length ratio $\lambda = l / B$; thus, substitute $\lambda = 1 / A$ and the drag lift of planing plate at finite width will be derived.

$$C_L = C_Y / 2 = \pi\alpha \frac{A}{A+2} = \pi\alpha \frac{1}{1+2\lambda} \qquad (1.6.2)$$

From the above expression, it can found that when the width of the planing surface decreases, the wetted length ratio λ will increase and the lift coefficient will decrease. Generally speaking, increasing the main hull breadth will be helpful in improving planing performance. Analyzed from the perspective of wing theory, the aspect ratio will be increased in correspondence with the increase of breadth; thus, the lift coefficient will be increased. It should be pointed out that although the increase of breadth will be beneficial to improvement of hydrodynamic performance, it will decrease the navigation trim angle and increase the friction resistance. Especially when the center of gravity is kept at the same position, increase of breadth will lead to increase of resistance. Actually, the breadth of a regular planing craft is within a certain range. Under normal conditions, the L/B of planing craft is within a range of 3 to 7. The L/B of the original hull form of M-form hulls will be taken to be 6.9:1, and 1:1 will be taken to be the ratio between the main hull half-breadth and the air flue width.

1.6.2. Optimized hull form plan

M-form hulls will be optimized in molded lines by employing theories on regular planing craft[1-6-3] [1-6-4] [1-6-5] and meshless numerical simulation.[1-6-1] [1-6-6] [1-6-7] In consideration of the influence of the ratio of the air flue and main hull on high speed and sea-keeping, the most appropriate breadth and ratio between air flue and main hull will be taken. The determination of an optimized plan depends on the requirements for high speed and sea-keeping: when the overall length is 20m and the displacement is 30t, theoretical analysis and optimized calculation will be done for the following three typical plans:

1. M-1: original hull form
2. M-2: on the basis of molded lines of M-1, increase the main hull breadth by 0.6m, decrease 0.3m for each of the two air flues; the breadth and other appendage will remain the same.

3. M-3: on the basis of molded lines of M-2, increase the breadth by 1.0m, increase 0.5m for each of the two air flues; the main hull breadth and other appendage will remain the same.

Table 1.6.2. Three types of M-form hull

Item	M-1	M-2	M-3
Overall length/m	20.0	20.0	20.0
Beam/m	6.2	6.2	7.2
Molded depth/m	2.6	2.6	2.6
Displacement/t	30	30	30
Main hull half-breadth/m	1.45	1.75	1.75
Air flue width/m	1.45	1.15	1.65
L/B	6.9	5.7	5.7
Ratio between main hull half-breadth and air duct width	1	1.5	1.1

For the main dimensions of three types of M-form hull, see Table 1.6.2. Through theoretical analysis and optimum calculation, a resistance test will be done for the above-mentioned three plans in the high-speed hydrodynamic laboratory tank of the China Special Vehicle Research Institute. According to the test result, an optimum plan will be chosen. For the model of one type of M-form hull, see Fig. 1.6.2. As theoretical calculations and numerical methods are still in their infancy for research on the hydrodynamic performance of high-performance ships, model tests are still an indispensable method.[1-6-8]

Fig. 1.6.2. Test Model of an M-form hull

1.6.3. Result analysis of the model test

1. Model resistance

For the drag–lift ratio of models of three types of M-form hull under the same conditions, see Table 1.6.3. For the contrast curve of Rm/△m–$F_{N\nabla}$, see Fig. 1.6.3., in which Rm is the model resistance, △m is model displacement, and $F_{N\nabla}$ is volume Froude number.

Table 1.6.3. Drag–lift ratio of models of three types of M-form hull

$F_{N\nabla}$	Rm/△m	Rm/△m	Rm/△m	Relative Change /%	Relative Change /%
	M-1	M-2	M-3		
	(1)	(2)	(3)	[(2)-(1)]/(1)	[(3)-(1)]/(1)
0.88777	0.06134	0.06544	0.07063	6.69	15.14
1.33165	0.14413	0.15061	0.15580	4.50	8.09
1.77554	0.16220	0.16184	0.15400	-0.22	-5.06
2.21942	0.16206	0.15731	0.13808	-2.93	-14.79
2.66331	0.15616	0.15047	0.13161	-3.64	-15.72
2.79647	0.15853	0.15083	0.13081	-4.86	-17.48
3.10719	0.16350	0.14888	0.13218	-8.94	-19.15
3.55108	0.17113	0.15335	0.14903	-10.39	-12.92
3.99496	0.17531	0.16739	0.16040	-4.52	-8.50%
4.43884	0.18603	0.17603	0.16465	-5.38	-11.49
4.88273	0.19978	0.17783	0.16645	-10.99	-16.68

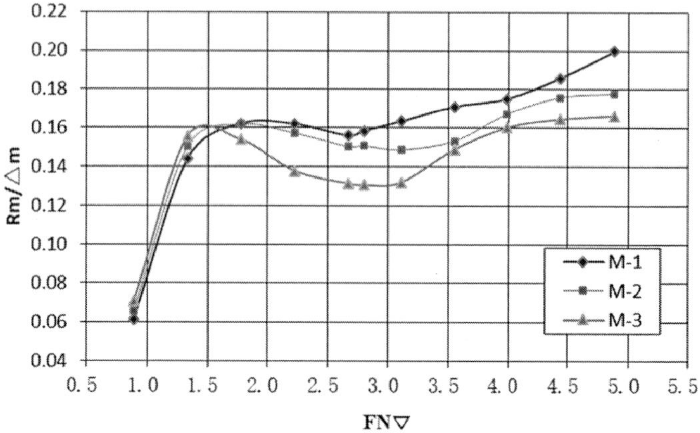

Fig. 1.6.3. Dimensionless contrast of model resistance of three M-form hull plans

2. Test result analysis

1. According to the test results, at high speed, the resistance of M-2 and M-3 decreases compared with the original ship M-1. When the $F_{N\nabla}$ is 2.79647 (at the speed of 30kn), the resistance of M-2 decreases by 4.86% and the resistance of M-3 decreases by 17.48%. As the main hull breadth of M-1 is smaller than those of M-2 and M-3, it will not have enough planing hydrodynamic lift at high speed, which will lead to poor resistance performance. Comparing M-2 and M-3, although the main hull breadths are the same and the planing hydrodynamic lifts are the same, the air flue width of M-3 is larger with a greater aerodynamic lift; thus, M-3 has the better resistance.
2. From the test it was discovered that all three plans have good transverse stability and good sea-keeping performance. In addition, M-1 and M-2 have good high-speed longitudinal stability; however, M-3 is poor in this aspect and slight porpoising occurs at 11m/s and 12m/s. M-form hulls mainly take advantage of hull form to catch air and reduce water resistance and speed up due to the lift from static buoyancy, hydrodynamics, and aerodynamics. For M-3, as its air flue is wider than those of M-1 and M-2, greater aerodynamic lift will be provided during high-speed navigation, but the instability of large-range air flow during high-speed navigation will lead to poor longitudinal stability.

1.6.4. Conclusion

The change of breadth and the ratio of the air flue and main hull will generate an influence of more than 10% on the resistance of the M-hull form. Taking appropriate breadth and ratio between the air flue and main hull will have a great influence on the resistance and longitudinal stability of M-form hulls, to which special attention will be paid.

References

[1-6-1] Wang Qingxu & Yue Lei, "The Hull Characteristics and Numerical Simulation without Grids of M-Form Hull," 2013 China International Boat Show & HPMV Conference, E12-1–5.

[1-6-2] Liu Qian, Pang Liguo, & Lei Yunhong, "Study on Operation Principle and Technical Characteristics of Trimaran Wave-Damping Planing Craft," *Jiangsu Ship* 17 (1): pp. 6–7, 2000.

[1-6-3] Dong Zushun, *Speedboat Dynamics*, Wuhan: Huazhong University of Science and Technology Press, 1994.

[1-6-4] Shao Shiming & Wang Yuncai, *Speed Boat Dynamics*, Shanghai: Shanghai Jiao Tong University Press, 1990.

[1-6-5] Zhao Lianen, *Hydrodynamic Principles and Design of High-performance Ship*, Harbin: Harbin Engineering University Press, 2009: pp. 212–16.

[1-6-6] Liu Xin, "Study of Meshless Numerical Method," Nanjing: Nanjing University of Aeronautics and Astronautics, 1998: pp. 10–15.

[1-6-7] Chen Hui & Chen Keqiang, "Numerical Simulation of M-Form Hull Resistance," *Marine Engineering* 41 (5): pp. 51–53, 2012.

[1-6-8] O. M. Faltinsen, *Hydrodynamics of High-Speed Marine Vehicles*, Cambridge: Cambridge University Press, 2005: pp. 123–46.

1.7

THE HULL CHARACTERISTICS AND NUMERICAL SIMULATION WITHOUT GRIDS OF M-FORM HULLS (2013)

WANG QINGXU AND YUE LEI

China Ship Development and Design Center

Preface

The M-form hull, a newly developed high-performance hull form, is a combination of conventional planing hulls, planing hulls with channels, and high-speed ships with multiple bodies. The hull aggregates all the advantages of these three hull forms to a certain extent and combines hydrodynamic force with aerodynamic force. The mixtures of gas and water in the two-sided channels play a role in reducing frictional resistance and enhancing stability.

The M-form hull consists of three components: one main body in the middle of the hull, and two outriggers on both sides. There are two longitudinal circular-section channels between the main body and the outriggers. The section of the hull is similar to the M-type.[1-7-1]

1.7.1. Hull characteristics and applicable uses of M-form hulls

1. A unique shape available for general arrangement

The M-form hull consists of one main hull and two outriggers, one on either side. The unique shape, which makes the hull very different from other yachts, could be considered a highlight in recent yacht designs. Compared with a high-speed vessel of the same displacement, the breadth of the M-form hull is bigger and could provide a bigger deck area available for general arrangement. Thus, the hull form could be applicable for yachts and has the advantage of greater comfort. The British Ice Marine company has developed a series of racing boats and yachts based on the M-form hull, which has garnered great attention in many international racing boat competitions and yacht exhibitions.[1-7-2]

2. Small stern wave and sailing track
The M-form hull originated from the small sight-seeing boats used in the Venice River in Italy. The hull form was developed as a wave-avoiding hull in order to weaken waves and avoid eroding and breaking the riverbed of the Venice River, damage to which was being caused by the waves of high-speed boats.[1-7-3] Because of the existence of the main body and outriggers, the stern waves generated by the main body are essentially absorbed into the channels instead of being diffused to the two sides, as in the Kelvin wave system. The stern G-track of the M-form hull is more linear. This characteristic shows that the M-form hull could be widely used in freshwater environments, which would lessen erosion of riverbeds and coastal areas.

3. High stability
The outriggers on either side and the mixture of air and water in the channel play a role in supporting the main body and give high stability to the M-form hull. The model test proves that the M-form hull could retain excellent stability and its maximum volume Froude number could reach 8.0–9.0, which greatly exceeds that of conventional high-speed vessels.

4. Protection for the main body of two-sided outriggers
The two-sided outriggers are narrow in breadth and lie at the right and left sides of the main body hull. Important parts such as crew arrangements and the main engine are in the main body hull. The two-sided outrigger could shelter the main body hull partly or completely, which provides good protection for the main body hull. This advantage makes it useful for military use and gives the hull better bulletproofing capability. Most M-form hulls are made of fiber-reinforced plastic and high-intensity carbon fiber to make them stronger.

5. Little noise and good concealment
The geometrical shape of the M-form hull not only can capture the bow wave of the central main body hull but also can avoid the obvious "fragmentized wave light" generated in the position of the stem of a conventional hull. The fragmentation of the bow wave could easily be the main visible light source for vessels while sailing at night. The channels of the M-form hull could absorb the bow wave, which aids concealment and makes it difficult for the boat to emerge as a target. This makes the hull

type reasonably fit to be the basic hull form for military hulls. In 2006, the company that built the M-hull built and tested the new-concept double M-form hull "Stiletto" in the United States. The Stiletto performed very well in concealment. The appearance of the Stiletto attracted wide attention from many countries and was acclaimed as the one of the best global inventions of 2006.[1-7-4]

1.7.2. Resistance test on the M-form hull model

In reference to excellent foreign M-form hulls, the hull lines were designed. The hull model was made and a model test was carried out in the tank of the China Special Vehicle Research Institute. The tank is 550m long, 6m wide, and 5m deep and the maximum test velocity is 30m/s, which makes it the longest and fastest model-dragging tank in China.[1-7-5] The model test was carried out on July 23, 2012. The sailing model is shown in Fig. 1.7.1. The model is 4m long, 180kg in displacement, and 2.3 deg. in initial heel.

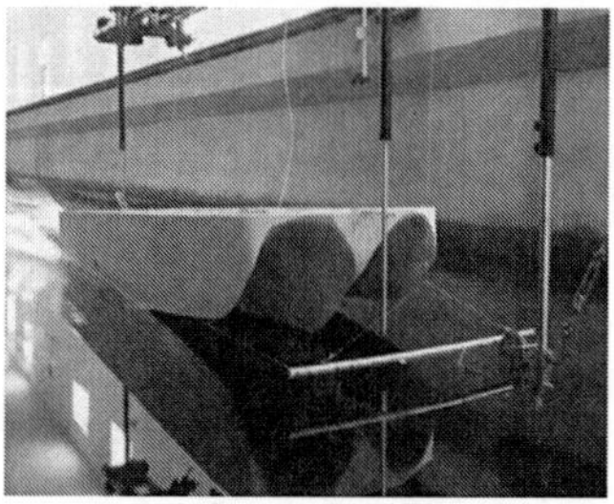

Fig. 1.7.1. The sailing attitude of the hull model at a velocity of 8m/s

In the process of dragging the hull model, the stern waves are all absorbed into the channels and have only a slight impact on the two sides of the tank. The heave detected has an increasing current while the speed is rising. When the Froude number is less than 3.0, the trimming angle decreases as the speed increases. When the Froude number is more than 3.0, the trimming angle increases as the speed increases. The maximum value of the trimming angle is about 5 degrees. The resistance values at all speeds were drawn as a resistance curve and were compared with the resistance values of a catamaran planing hull with a displacement of 143.5kg and of the same size. The transverse axis stood for the volume Froude number and the vertical axis stood for the resistance of one unit displacement. The comparison figure is shown in Fig. 1.7.2.

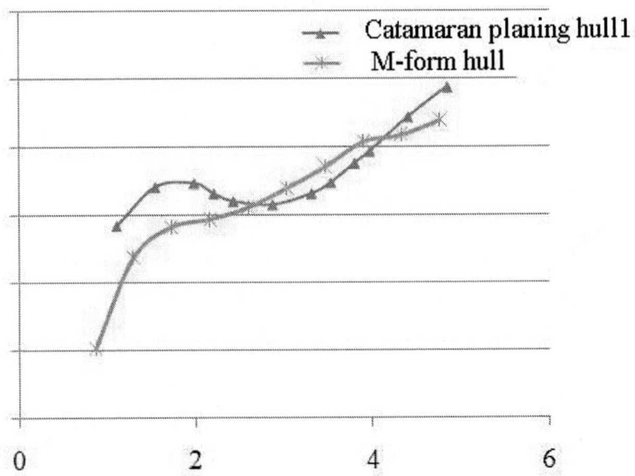

Fig. 1.7.2. A comparison between an M-form hull and a catamaran planing hull

According to the comparison of the resistance curve, it can be seen that in general the resistance curve of the M-form hull has a smaller wave than the catamaran planing hull. The resistance curve becomes more like a line after passing the resistance peak. The resistance curve of a catamaran planing hull decreases at first and increases with the shape of the parabola after passing the resistance peak. The M-form hull has a resistance peak with the volume Froude number of approximately 2; however, it has a less obvious resistance peak and has a lower resistance of one unit displacement. When the Froude number is between 3.0 to 4.5, the resistance of the M-form hull exceeds that of the catamaran planing hull. When the Froude number exceeds 4.5, the M-form hull has a lower resistance of one unit displacement.

1.7.3. Numerical simulation without grids

1. Brief introduction to no-grid software

The no-grid method is a numerical method that does not require one to define grids in advance to build a discrete system equation on the question field. One of the earliest no-grid methods was Smooth Particle Hydrodynamics (SPH), proposed by Lucy and Moraghan in 1977, which continues to be widely used in research on celestial body physics. In 1994,

Professor Belytschko corrected the Diffuse Element Method and came up with the Element Free Galerkin Method. After that, the no-grid method achieved an obvious and rapid development. In the no-grid method the coupling units are not used to disperse the question field, and the problem caused by using the unit in finite element analysis can be solved naturally.[1-7-6]

FPM (Finite Point set Method) is a CFD numerical method without grids and points that represents the newest development direction of fluent numerical simulation. The greatest difference between the finite point set method and other conventional numerical methods such as the finite element method, finite volume method, and finite difference method is that there is no need to carry out partition grids to disperse the differential equation. Compared with the Fluent and CFX software another advantage of FPM software is that it involves less computer time. The aim of using FPM software to simulate the M-form hull numerically is to explore methods for numerical simulation calculations in high-performance, high-speed multihull vessels.

2. The result of numerical simulation

The surface model of the M-form hull was built in Rhino software and was imported into FPM software. According to the parameters, such as displacement, draft, and speed, and after choosing the particle density, boundary conditions, and calculation model, the calculation was carried out. The resistance curve and pressure distribution graph is shown in Figs 1.7.3 and 1.7.4.

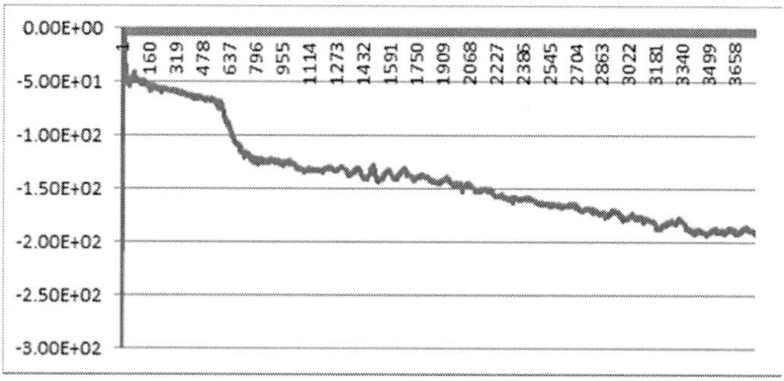

Fig. 1.7.3. Resistance curve of M-form hull at 5m/s

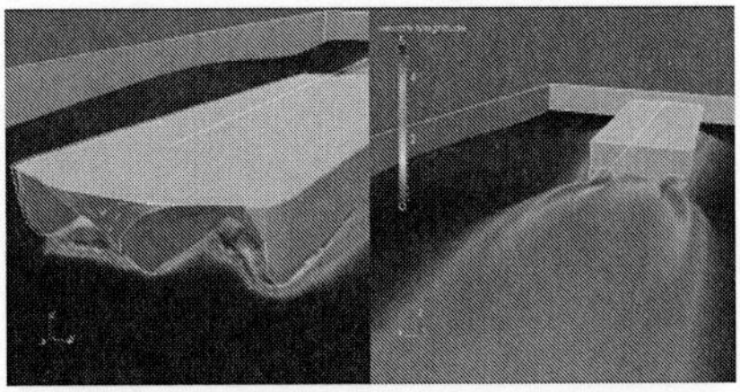

Fig. 1.7.4. Pressure distribution graph of M-form hull at5 m/s

Three speed points—low, medium, and high—were chosen to conduct the resistance calculations of the M-form hull. The three speeds are 2m/s, 5m/s, and 8m/s. In the calculation, the resistance curve was convergent after iterations of 1,500 times and 3,500 times, respectively at 2m/s and 5m/s. But, at 8m/s, the resistance value still had an obvious surge and could not get convergent after iterations of 5,000 times. The resistance calculated by FPM software was compared with the resistance in the model test; the error table is shown in Table 1.7.1.

Table 1.7.1. The comparison table between the numerical simulation and model test

Velocity/m/s	2	5	8
Test value/kgf	8.29	23.45	29.6
Calculation value/kgf	7.85	21.42	Not convergent
Error	5.3%	8.7%	--

It can be seen from Table. 1.7.1 that at 2m/s and 5m/s, the error between calculation value and test value is less than 10%, which is acceptable for high-speed multihull planing hulls. But at the speed of 8m/s, the resistance curve could not get convergent after many iterations and the calculation value could not be achieved. Thus, it is suggested that FPM software be used to predict model resistance at low and medium speeds; or, in addition, that the particle amount and the setting of parameters are optimized to peruse the method to predict the resistance at high speed of the M-form hull.

1.7.4. Conclusion

The unique shape and dynamic characteristics of the channels make the M-form hull suitable for extensive use. Its advantages make the hull very suitable for the hull platform to be used for yachts, racing boats, sightseeing boats on inner rivers, and military vessels. First, the M-form hull lines were designed. The model was made and a resistance test was carried out. From the comparison with a catamaran planing hull, it could be seen that the M-form hull has a superior performance on resistance peak and value, with a volume Froude number passing 4.5. With a volume Froude number between 3.0–4.0, the resistance of the M-form hull and catamaran planing hull are little different. No-grid software has had a considerable evolution in the field of computer dynamics today. Three speed points at low, medium, and high were chosen to conduct the resistance calculation of an M-form hull with FPM software. The result shows that the resistance results at low and medium speeds using FPM software is acceptable. The method for calculating the resistance at high speed should be explored.

References

[1-7-1] Zou Jin, Wang Qingxu, Shi Shengzhe, & Zhou Jie, "The Brief Introduction of High-Speed Trimaran Planing Hull and the Preliminary Study of the Longitudinal Stability," *Applied Mechanics and Materials* 117–19: pp. 559–64, 2011.

[1-7-2] Lorne Campbell, "Boat Designs" (2007-10-20)/[2009-06-03].

[1-7-3] Liu Qian, Pang Liguo, and Lei Yunhong, "The Sailing Principles and Technical Features of Wave-Reducing Trimaran Planing Hull," *Jiangsu Ship* 17 (1): pp. 6–7, 2000 (in Chinese).

[1-7-4] Zhao Lian'en, *Dynamic Principle and Design of Hulls with High Performance*, Harbin: Harbin Engineering University Press, 2009: pp. 212–16 (in Chinese).

[1-7-5] The Introduction of China Special Vehicle Research Institute (2008-10-20)/[2010-04-03].

[1-7-6] Liu Xin, *Research on Numerical Method without Grids*, Nanjing: Nanjing University of Aeronautics and Astronautics Press, 1998: pp. 10–15 (in Chinese).

1.8

A Probe into the Turning Heel of Deep-V Vessels (2000)

Zhang Chengxia

Hudong Shipbuilding Group, Shanghai, China

1.8.1. Preliminary analysis

Fluid dynamic analysis of the steady turning of the displacement type hard-chine deep-V hull form reveals that its behavior is similar neither to that of hard-chine planing boats nor to that of round-bilge displacement type vessels.

When turning at high speed, a planing boat presents a large inward heel because the water flow velocity outside is bigger than that inside and lateral drifting causes additional hydrodynamic forces, thus leading to the outside dynamic lifting force being bigger than the inside one. As regards round-bilge vessels, they present larger outward heeling due to the centrifugal forces during high-speed turning together with smaller hydrodynamic forces that are due to the hull from and flow velocity and at a nearly horizontal direction, as shown in Fig. 1.8.1.

The deep-V hull form of our design belongs to the hard-chine type, but its Froude number is comparatively small, near that of high-speed displacement type. There exists a fair portion of upward component forces in the hydrodynamic forces exerted vertically on the inclined surfaces of V-type shell plates. These upward component forces produce a righting moment to reduce the big outward heeling of the round-bilge vessel to some extent, but they will not cause a big dynamic lifting force for the sake of other factors, such as speed. Therefore, its motion state (including heeling) should differ from those of planing boats and round-bilge vessels, and might be somewhere between those of the above-mentioned craft.

To further understand the background mechanisms and to improve our knowledge of the underlying law, lateral dynamic forces when turning should be analyzed, and the relations between heeling and forces moments should be determined and verified through tank testing.

1.8.2. Theoretical Probing

1. Hydrodynamic forces and moments exerted on the hull
Hydrodynamic forces and moments exerted on the hull have something to do with the shape of the hull, motion characteristics, medium property, and boundary conditions. With regard to a particular type of vessel moving in a particular medium, it is considered that only the following elements are involved:[1-8-1]

$$X=X(u, v, r, \dot{u}, \dot{v}, \dot{r}, n, \dot{n}, \delta, \dot{\delta})$$
$$Y=Y(u, v, r, \dot{u}, \dot{v}, \dot{r}, n, \dot{n}, \delta, \dot{\delta})$$
$$N=N(u, v, r, \dot{u}, \dot{v}, \dot{r}, n, \dot{n}, \delta, \dot{\delta}) \qquad (1.8.1)$$

In the above X, Y is the component of resultant external forces in the moving co-ordinates; N is the moment of resultant external forces about the vertical axis passing the center of gravity of the vessel; u,v is the velocity component respectively in the direction of axes x and y; r is angular velocity of turning; n is turning speed; δ is the rudder angle.

If turning speed n is neglected, since the rudder maneuvering process is of short duration, $\dot{\delta}$ can also be neglected.

If we use Taylor expansion on multivariate functions, then we have

$$X= X(u_1, v_1, r_1, \dot{u}_1, \dot{v}_1, \dot{r}_1, \delta_1)+\frac{\partial X}{\partial u}\Delta u+\frac{\partial X}{\partial v}\Delta v+\frac{\partial X}{\partial r}\Delta r+\frac{\partial X}{\partial \dot{u}}\Delta \dot{u}+$$
$$\frac{\partial X}{\partial \dot{v}}\Delta \dot{v}+\frac{\partial X}{\partial \dot{r}}\Delta \dot{r}+$$
$$\frac{\partial X}{\partial \delta}\Delta\delta+\ldots+\frac{1}{n!}\left(\frac{\partial}{\partial u}\Delta u+\frac{\partial}{\partial v}\Delta v+\frac{\partial}{\partial r}\Delta r+\frac{\partial}{\partial \dot{u}}\Delta \dot{u}+\frac{\partial}{\partial \dot{v}}\Delta \dot{v}+\right.$$
$$\left.\frac{\partial}{\partial \dot{r}}\Delta \dot{r}+\frac{\partial}{\partial \delta}\Delta \delta\right)^n\times X+\ldots$$

$$Y=Y(u_1, v_1, r_1, \dot{u}_1, \dot{v}_1, \dot{r}_1, \delta_1)+\frac{\partial Y}{\partial u}\Delta u+\frac{\partial Y}{\partial v}\Delta v+\frac{\partial Y}{\partial r}\Delta r+\frac{\partial Y}{\partial \dot{u}}\Delta \dot{u}$$
$$+\frac{\partial Y}{\partial \dot{v}}\Delta \dot{v}+\frac{\partial Y}{\partial \dot{r}}\Delta \dot{r}+$$

$$\frac{\partial Y}{\partial \delta}\Delta\delta+\ldots+\frac{1}{n!}\left(\frac{\partial}{\partial u}\Delta u+\frac{\partial}{\partial v}\Delta v+\frac{\partial}{\partial r}\Delta r+\frac{\partial}{\partial \dot{u}}\Delta \dot{u}+\frac{\partial}{\partial \dot{v}}\Delta \dot{v}+\frac{\partial}{\partial \dot{r}}\Delta \dot{r}+\frac{\partial}{\partial \delta}\Delta\delta\right)^{n}\times Y+\ldots$$

$$N=N(u_1, v_1, r_1, \dot{u}_1, \dot{v}_1, \dot{r}_1, \delta_1)+\frac{\partial N}{\partial u}\Delta u+\frac{\partial N}{\partial v}\Delta v+\frac{\partial N}{\partial r}\Delta r+\frac{\partial N}{\partial \dot{u}}\Delta \dot{u}+\frac{\partial N}{\partial \dot{v}}\Delta \dot{v}+\frac{\partial N}{\partial \dot{r}}\Delta \dot{r}+$$

$$\frac{\partial N}{\partial \delta}\Delta\delta+\ldots+\frac{1}{n!}\left(\frac{\partial}{\partial u}\Delta u+\frac{\partial}{\partial v}\Delta v+\frac{\partial}{\partial r}\Delta r+\frac{\partial}{\partial \dot{u}}\Delta \dot{u}+\frac{\partial}{\partial \dot{v}}\Delta \dot{v}+\frac{\partial}{\partial \dot{r}}\Delta \dot{r}+\frac{\partial}{\partial \delta}\Delta\delta\right)^{n}\times N+\ldots \qquad (1.8.2)$$

In the above equation, $X(u_1,v_1,r_1,\dot{u}_1,\dot{v}_1,\dot{r}_1,\delta_1)$, $Y(u_1,v_1,r_1,\dot{u}_1,\dot{v}_1,\dot{r}_1,\delta_1)$, and $N(u_1,v_1,r_1,\dot{u}_1,\dot{v}_1,\dot{r}_1,\delta_1)$ are the values of the functions respectively at the expansion point of $(u_1,v_1,r_1,\dot{u}_1,\dot{v}_1,\dot{r}_1,\delta_1)$.

$$\Delta u = u - u_1 \quad \Delta v = v - v_1 \quad \Delta r = r - r_1$$
$$\Delta \dot{u} = \dot{u} - \dot{u}_1 \quad \Delta \dot{v} = \dot{v} - \dot{v}_1 \quad \Delta \dot{r} = \dot{r} - \dot{r}_1$$
$$\Delta\delta=\delta-\delta_1 \qquad (1.8.3)$$

The derivatives in equation 1.8.2 all take the values at the expansion point.

The equations can be simplified by means of the linear method (depending on the calculation state and the extent of approach to the expansion point), or by taking zero for some items (depending on the specific moving state).

When turning steadily, v and r are constants, $\dot{v}=\dot{r}=0$. Under the condition of linearization assumption, and only taking into consideration the lateral action, the hull motion equations can be simplified as:

$$-(\partial Y/\partial v)\, v+(mu_1-\partial Y/\partial r)\, r=(\partial Y/\partial \delta)\delta_0 \qquad (1.8.4)$$
$$-(\partial N/\partial v)\, v+(mx_G u_1-\partial N/\partial r)\, r=(\partial N/\partial \delta)\delta_0 \qquad (1.8.5)$$

in which m is the mass of the vessel and x_G is the coordinate of the vessel's center of gravity.

In regard to the above equations, the analysis is as follows: $(\partial Y/\partial v)\, v$ and $(\partial Y/\partial r)\, r$ are hydrodynamic forces pointing to (turning) inner side and acting on the shell of the vessel; mu_1r is the hull's inertial force pointing to the outer side and acting on the vessel's center of gravity; $(\partial Y/\partial \delta)\delta$ is the rudder force pointing to the outer side and acting on the rudder.

Here $\delta=\delta_0$.

For calculation and prediction, their relations in conventional expression ought to be worked out.

2. Conventional equations of relations[1-8-3]

(1) Fundamental formula

If the swash on the rudder surface due to the inclination of the vessel is neglected, the entrained water and viscous forces are also neglected.

When the rudder is steered starboard and the vessel is turning steadily to starboard, the various forces acted on it are shown in Figs 1.8.2 and 1.8.3.

From equation 1.8.4, the following equation can be derived (positive and negative signs as shown in the figures, taking starboard as positive):

$R_Y—N_Y—P_Y=0$ (1.8.6)

From equation 1.8.5, the following can be derived (for the point between H and E, taking the clockwise as positive, for the righting moment M_θ, during the heeling, taking outward heeling as positive):

$M_{RYZ}--M_{NY}+M_{PY}+M=0$ (1.8.7)

R_{YZ} passes point H, taking the moment about point H; next, equation 1.8.7 is reduced as:

$M= M_{NY}--M_{PY}$ (1.8.8)

In the above three equations, R_Y is the component of hydrodynamic force at direction Y; N_Y is the component of the vessel's centrifugal force at direction Y; P_Y is the component of rudder force at direction Y; M_{RYZ} is the lateral moment of hydrodynamic force; M_{NY} is the centrifugal moment of the vessel; M_{PY} is the rudder moment; M_θ is the righting moment of the vessel.

$M_\theta=D\overline{MG}\sin\theta$ (1.8.9)

$M_{NY}=N'_Y\overline{GH}\cos\theta$ (1.8.10)

$M_{PY}=2P_Y\overline{HE}$ (1.8.11)

Since $N'_Y=N\cos\Phi$, in the case of a single rudder configuration $P_Y=P_N\cos\delta$, while $P_N\approx P$, so,

$M_{NY}=N\cos\Phi\overline{GH}\cos\theta$ (1.8.12)

$M_{PY}=2P\cos\delta\overline{HE}$ (1.8.13)

In the above formulae, D is the displacement of the vessel, $\overline{MG}$ is the initial metacentric height of the vessel, θ is the heeling angle of the vessel, N is the centrifugal force of the vessel when turning, Φ is the drifting angle, $\overline{GH}$ is the height of the gravity center above point H, P is rudder force, δ is rudder angle, $\overline{HE}$ is the height of the rudder force's center above point H.

(a) Calculation of vessel's centrifugal force when turning

$N=DV_t^2/gr$ (1.8.14)

in which V_t is the vessel's speed along the tangent line of the track of the vessel's gravity center. According to Фирсов's formula of velocity reduction for a vessel in turning, the following can be obtained:

$V_t=V_0 th(\overline{r}/2.45)$ (1.8.15)

In this equation, V_0 is the initial speed of the vessel when turning, g is gravitational acceleration; $\overline{r}$ is the relative radius of curvature of the track of vessel's gravity center, $\overline{r}=r/L$; r is the radius of curvature of the track of the vessel's gravity center, and L is the length of the vessel.

(b) Calculation of drifting angle

No formula of drifting angle is found in the reference literature. As shown is Fig. 1.8. 4, drifting angle Φ can be calculated with $l(\overline{GC})$ and r—the radius of curving of the track of the vessel's gravity center—as follows:

$\Phi=arc\ sin\ (l/r)$ (1.8.16)

Making reference to some literature, l/L=0.342 can be taken.

Note: the tests show that the drifting angle of deep-V type vessels when turning is less than that of conventional round-bilge vessels (refer to the "Testing" section of this paper).

(c) Calculation of rudder force P

$P=(1/2)\rho SV_t^2C_L$ (1.8.17)

in which ρ is the density of water, S is the area of rudder, and C_L is the lift coefficient. When C_L is calculated, the acting rudder angle δ' should be taken in consideration of the effect of the drifting angle on the flow direction and of water flow reduced by the streamlining of the vessel's longitudinal centerline plane; refer to the following table and its note:

Table 1.8.1.

Relative radius of curvature	$\overline{r}=r/L$	1	1.5	2	2.5	3
Reduction of drifting angle	Φ'(°)	36.5	27	20.5	17	14

Note: the aggregately affected reduction angle $\Phi=\Phi/2$(°)

Substitute equation 1.8.14 in equation 1.8.12, equation 1.8.17 in equation 1.8.13, and, then, equations 1.8.12, 1.8.13, and 1.8.9, in equation 1.8.8, thus

$$D\,\overline{MG}\,\sin\theta=(D\,V_t^2/gr)\cos\Phi\,\overline{GH}\,\cos\theta-\rho S\,V_t^2 C_L\cos\delta\,\overline{HE} \qquad (1.8.18)$$

and the following relation can also be established:

$$\overline{GH}+\overline{HE}=Zg—Zp \qquad (1.8.19)$$

Therefore, critical point H can be determined.

(2) Example of calculation (omitted).

3. Determination of the position of action point K for the lateral hydrodynamic force exerted on the shell plate

(3) Longitudinal position of point K

Point C in Figure 5 is a pivot center. When turning, the vessel drifts laterally while self-rotating about point C. Forward of point C, the hydrodynamic force directs outward; behind point C, the hydrodynamic force directs inward. The pressure change is assumed to be linear along the length of the vessel. The force-acting center should be the product of pressure intensity and area center. The lead in pressure factor is W_F, the area factor is WS, and the center of W_FW_S is the longitudinal position X_K of the hydrodynamic action center:

$$X_K=\frac{\int_F^A W_F W_S X dX}{\int_F^A W_F W_S dX} \qquad (1.8.20)$$

(a). Pressure factor W_F

Given a nominal value of any point along the length of a vessel, the nominal values of other points can be calculated. For the sake of simplification, given $W_F=-1$ at point F, calculate the nominal values at each station (X_1).

(b) Area-factor W_S

Take the force acting length below the waterline at each station (X_1) as the approximate nominal value of W_S (expansion at the direction of height, any unit of length will do):

$$X_K=\frac{\int_F^A W_F W_S K_1 Kx\Delta X}{\int_F^A W_F W_S K_1 \Delta X} \qquad (1.8.21)$$

in which K_1 is length factor and Kx is distance factor.

In addition, if Xi is given as the position of each station and the span between stations is $\Delta X=1$, then

$$X_K=\frac{\int_F^A W_F W_S K_1 Kx}{\int_F^A W_F W_S K_1} \qquad (1.8.22)$$

(4) Height of point K

According to the Пєршиц formula, when $2.5\leq B/d\leq 0.80$,

$Z_K/d=4-B/d+0.02(B/d-5.35)^3$ (1.8.23)

in which, Z_K is the vertical distance from the action point of lateral hydrodynamic force to the vessel's bottom, B is the vessel's beam, and d is the vessel's draft.

(5) Example of calculation (omitted).

4. Lateral hydrodynamic force R_{YZ}

(1) Substitute various composition elements into equation (1.8.6):

$R_Y=N_Y+P_Y=(D\,V_t^2/gr)\cos\Phi\cos\theta+2(1/2)S\,V_t^2C_L\cos\delta$ (1.8.24)

According to the results of the above calculation, determine point K. Link point K and point H, obtain the action line of lateral hydrodynamic force. The angle formed by KH and the horizontal line is the elevation angle ζ of the hydrodynamic force.

According to the equations $\cos\zeta=R_Y/R_{YZ}$ and $\mathrm{tg}\zeta=R_Z/R_Y$, R_{YZ} then R_Z can be calculated.

(2) Example of calculation (omitted).

1.8.3. Prediction and Design Control

By means of testing and sorting out the calculation data, certain parameters can be obtained to make some preliminary predictions.

For example, in order to calculate static heeling angle θ when turning, equation 1.8.18 can be changed into:

$A \sin \theta = B \cos \theta - C$ (1.8.25)

in which $A = D\,\overline{MG}$; $B = (D\,V_t^2/gr)\cos\Phi\,\overline{GH}$; $C = \rho S\,V_t^2 C_L \cos\delta\,\overline{HE}$.

When Zh is known—that is, $\overline{GH}$ and $\overline{HE}$ are known—we can solve equation 1.8.25 to obtain the following:

$$\theta_{1,2} = arc\ tg\left(\frac{2AB \pm \sqrt{4A^2B^2 - 4(A^2 - C^2)(B^2 - C^2)}}{2(A^2 - C^2)}\right) \qquad (1.8.26)$$

Through using these, during the concept design the heeling state of vessels of different alternatives under different conditions, β, δ, V_0 and Z_g, can be predicted tentatively, and can also be selected as required to bring about design control. Further information will be presented separately in another paper. In this thesis, only an example comparison between the calculation results and test results of static heeling angle θ when turning is given in the table below (deep-V type model β=25°, speed V_0=2.4m/s, the negative value of angle means inward heeling):

Rudder angle δ /deg.		20	25	30	35
θ	calculation value/deg.	0.165	-0.168	-0.444	-0.515
	testing value/deg.	-0.520	-0.765	-0.990	-1.165
	Error/deg	0.685	0.597	0.546	0.650

As can be seen in the above table, the error of θ between calculation value and testing value is only 0.6° approximately.

1.8.4. Testing

Among the tank tests on the maneuverability of deep-V-type and round-bilge-type models in the comparison of conventional and non-conventional items and hydrodynamic measurement, tests relating to heeling are those with variable parameters of deadrise angle of hull bottom (β), speed (V), rudder angle (δ), height of gravity center (Zg) of vessel, ratio of length to beam (L/B), drifting angle (Φ), turning angular velocity (r), and heeling angle (θ). The results of these have verified some conclusions derived from analysis and theoretical probing, and also have provided a number of laws for deep-V-type hull forms.[1-8-4] [1-8-5]

1. When turning, the steady heeling of the deep-V hull form falls within the range of planing boats with big inward heeling (the inward angle is more than 15° at high speed with a big rudder angle); round-bilge vessels also have big outward heeling (the outward heeling angle is more than 10° at high speed with a big rudder angle). Furthermore, the heeling angle is as little as ±2°. Even for extreme hull forms tested under extreme conditions (β=35°, V=30kn, δ=35°), the heeling angle still falls between 3.6° (inward heeling) and 1.5° (outward heeling). The heeling angle v_s time curve is shown in Fig. 1.8.6.

2. The relation between θ and β: the whole tendency is that while β increases, θ may tend to be inward, and be more obvious at big δ.

3. The relation between θ and δ: in general, when δ is small (δ<10°) θ tends to outward when δ increases, after δ is over 12°, it tends to be inward, which can be seen clearly in θ-δ diagram (see Fig. 1.8.7).

Plotting also the θ-δ diagram of conventional round bilge vessels in Fig. 1.8.7, it can be seen that the θ of deep-V type vessels is very small in comparison with that of conventional round bilge ones.

4. The relation between θ and Z_g: enlarging the varying scope of Z_g (from lightest weight of a real ship to three times maximum displacement), making the turning test and the zigzag maneuver test at a speed of 22kn, and taking measurement of heeling, it is shown that Z_g has an influence on the initial heeling (which is much less than that for a round bilge and disappeared immediately), but has little influence on steady heeling. The heeling angles are all very small.

5. The relation between θ and V: generally, the curve of θ-V inclines more when V increases; that is to say, when δ<15°, heeling is mostly outward, and the outward heeling at a larger V (30kn) is bigger than that at a smaller V (18kn). At δ>20°, mostly, heeling becomes inward and the inward heeling is bigger at a bigger V. The intersection point of the curves falls at the side of a smaller δ (about 15°) when β is big (35°), and falls at the side of a bigger δ (about 20°) when β is small.

6. The relation between θ and Φ: generally, Φ is about 10° when a conventional round-bilge vessel is turning at full rudder; however, since the action point of outward force of a deep-V type vessel is at a position a little aft than that of a conventional round-bilge one, the Φ is thus impelled to reduce. As is shown in the hydrodynamic measurement, Φ is only 6° or so for three models of deep-V hull forms (β=10°, 25°, 35°).

7. Several hydrodynamic forces have derivatives in hydrodynamic measurement test of maneuverability:

(a) The derivative of zero dimensional sideward force with respect to the turning angular velocity *Y'r* is bigger than that of a conventional round-bilge vessel. This is because the above mentioned V-shape shell of deep-V type vessels causes a R_{YZ}.
(b) Since the action of R_Z is bigger than that of conventional round-bilge vessel, the K'_V—the derivative of the zero dimensional heeling moment with respect to sideward velocity—is weakened, thus θ is reduced.
(c) A bigger R_Z and the influence of the action point enlarged the *N'r* and *N'v*, the derivatives of a zero-dimensional yawing moment with respect to turning speed and sideward velocity, respectively; as a result, it led to the reduction of Φ and θ.
(d) The deeper the V shape of the deep-V hull form—that is, the bigger the angle of β—the bigger the derivatives *N'r*, *N'v*, *K'v*, and *Y'v*, except that *Y'r* is reduced a little.

1.8.5. Conclusion

The theoretical probing of the excellent performance shows that there exists almost no heeling during the displacement type deep-V hull form turning. From this a number of mathematical equations of force and moments in relation to the deep-V specific heeling have been derived. Preliminary predictions using these equations have reached results extremely near to those in real cases.

A wide range of model tests of deep-V hull forms have verified the theoretical analysis and derived relational equations of the parameters concerned and their tendency of variations. The results of the hydrodynamic derivatives obtained during the hydrodynamic force measuring of maneuvering tests have been analyzed and a conclusion has been reached that is consistent with the material mentioned above.

Acknowledgements

The author wishes to express his appreciation to Mr Gong Wenhuang (Hudong Shipbuilding Group) and Mr Li Dingzun and Shen Ding'an (China Ship Scientific Research Center), for their advice during the writing of this paper.

References

[1-8-1] Wu Xiuheng, Zhang Lewen, & Wang Renkang, *Maneuverability and Seakeeping Ability of Ships*, Renmin Jiaotong Publishing House, 1988: pp. 7–31

[1-8-2] Б В Мирохин., "Влияние угла крена на циркуляцию судна ," *Судостроение*, 1964 (6): pp. 9–16.

[1-8-3] Zhang Chengxia, "Lateral Hydrodynamic Force in Turning of Deep-Vee Hull Forms," *Shipbuilding of China*, 1997 (3): pp. 9–15

[1-8-4] Shen Ding'an & Gan Pinzhang, "Maneuverability Tests of Self-propulsion Model for a Frigate of Variable β," Report of China Ship Scientific Research Center, 1994.

[1-8-5] Shen Ding'an & He Chunrong, "Hydrodynamic Measurement of Maneuverability Tests for Series Models of Deep-Vee Hullform," Report of China Ship Scientific Research Center, 1997.

[1-8-6] Edward M. Lewandowski. "Trajectory Predictions for High Speed Planing Craft," *International Shipbuilding Progress*, 41 (426): pp. 137–48, 1994.

[1-8-7] Feng Tiecheng, Zhu Wenwei, & Gu Shuhua, *Ship Maneuver and Swinging*, National Defence Publishing House, 1989.

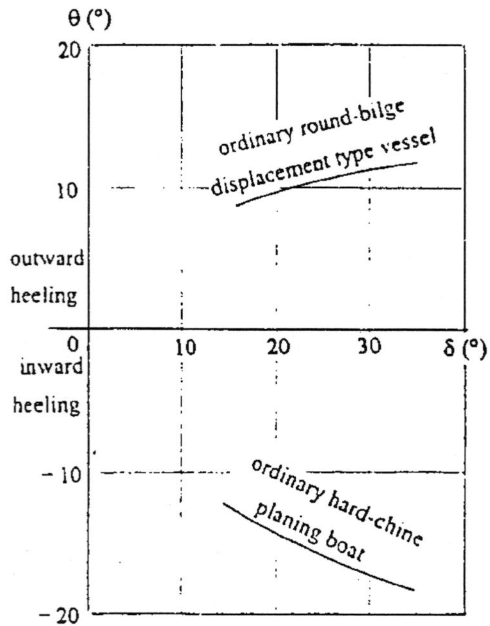

Fig. 1 Heeling Angles When Turning at High Speed

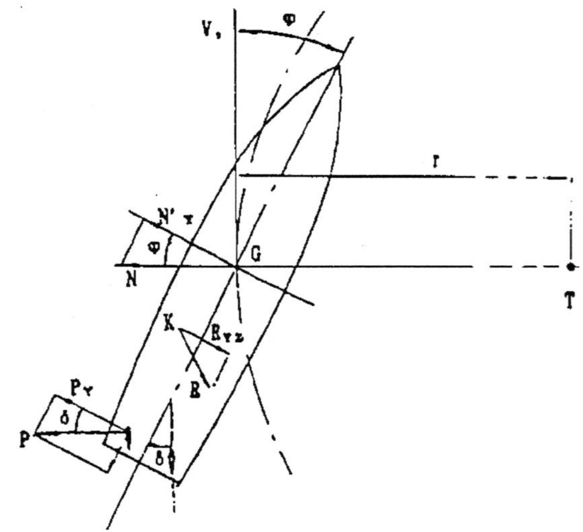

Fig. 2 Forces Acted on the Vessel When Turning

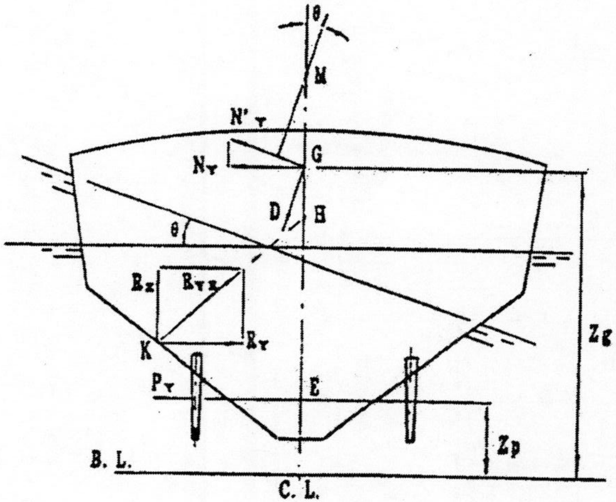

Fig. 3 Lateral Forces When Turning

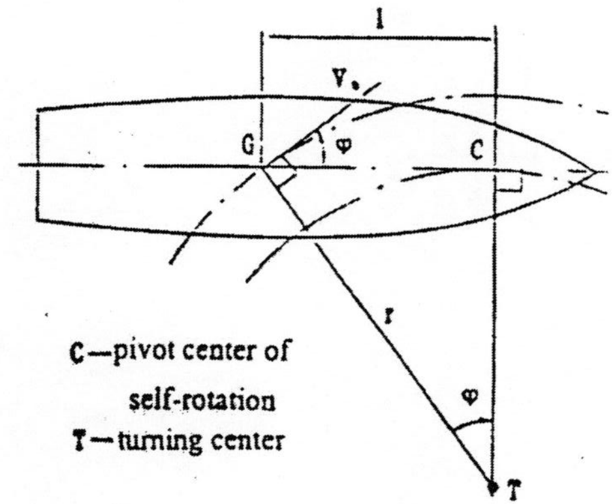

Fig. 4 Relation between *l* and *r*

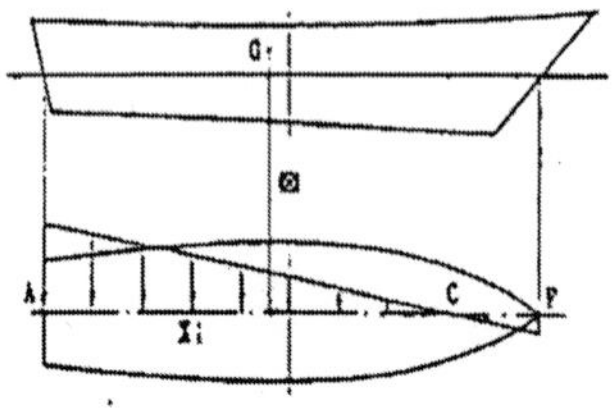

Fig. 5 Lateral Hydroresistance When Turning

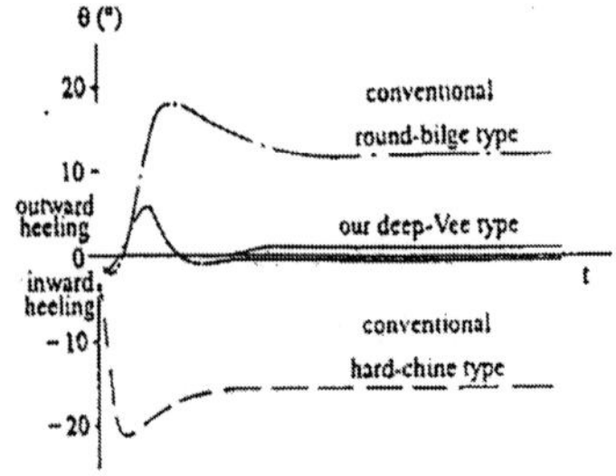

Fig. 6 Relation between θ and Time

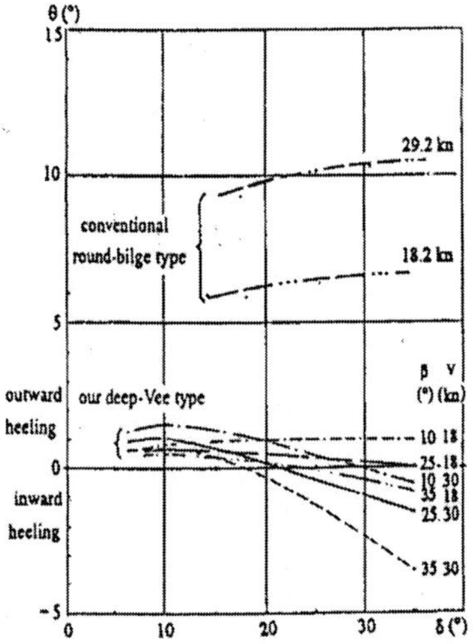

Fig. 7 Relation between θ and δ under Different β and V

1.9

Novel Design and Hydrodynamic Optimization of a High-Speed Hull Form (2005)

L. Koh,[1] C.-E. Janson,[2] M. Atlar,[1] L. Larsson,[2] E. Mesbahi,[1] & C. Abt[3]

[1] University of Newcastle upon Tyne, UK
[2] Chalmers University of Technology, Sweden
[3] FRIENDSHIP-Systems GmbH, Germany

Nomenclature

B/T	Breadth-to-draft ratio
C_P	Prismatic coefficient
Δ	Displacement
η_H	Hull efficiency
Fn	Froude number, $V/\sqrt{L_{WL}xg}$
F_P	Forward perpendicular
GM	Metacentric height
KG	Centre of gravity
L/B	Length-to-beam ratio
L_{BP}	Length between perpendiculars
LCG	Longitudinal centre of gravity
L_{OA}	Length overall
R_F	Frictional resistance
R_{WP}	Wave pattern resistance
R_T	Total resistance
(1-t)	Thrust deduction factor
w	Wake fraction

1.9.1. Introduction

Deep-V hull forms have demonstrated competitive powering in rough seas and favourable sea-keeping and manoeuvring when compared with round-bilge hull forms at equal or similar length/beam ratios and displacements.[1-9-1] However, all the deep-V hull forms in the market are constructed for operation at higher Froude numbers (F_n>0.55) and at smaller displacements (Δ<3,900 t).

A deep-V hull form has sections with a minimum of 17° to 20° of dihedral with the horizontal plane at the transom. The transom dihedral is the principal element influencing the deep-V hull shape but there are at least 30 additional parameters, which influence the accurate formulation of a deep-V hull form design.[1-9-1]

The deep-V hull form design is illustrated[1-9-2] in its most simple form by:

- A triangular midship section with chines/knuckles in the waterline
- A constant midship section over more than 60% of the length between L_{BP}
- A bow form made up of simple surfaces and with chines/knuckles in the upper works
- 80% of the hull above the waterline formed by flat surfaces

The elements influencing the formulation of a deep-Vee hull form design are:[1-9-2]

- Deadrise angle
- Length-to-beam ratio (L/B)
- Bottom curvature
- Keel rise
- Longitudinal centre of gravity (LCG)
- Spray rails

Both references 1-9-1 and 1-9-2 concluded through their experimental investigations that deep-V hull forms display more favorable sea-keeping performance than an optimum round-bilge hull form. A deep-V hull form also has less added resistance than a round-bilge hull form due to lower motions, less yaw, and better course keeping ability in rough seas.[1-9-3] From model tests,[1-9-3] it has been discovered that a deep-V hull form can have a much lower probability of slamming than an excellent round-bilge hull form with comparable displacement if it is combined with a so-called anti-slamming bow feature. However, a deep-V hull form has a higher calm water resistance than a round-bilge hull form due to greater wetted surface area and a fully immersed transom stern.[1-9-3]

Deep-V hull forms, with their high dihedral angles and constant midship sections over more than 60% of the length, demonstrate favorable directional stability and excellent maneuverability even at high speed, with a reduction of around one third of the turning diameter compared to a round-bilge hull form. The deep-V hull form, with its flat hull surfaces and constant transverse sections for the majority of the hull length, is simpler to construct and less costly compared with a round-bilge hull form.[1-9-1]

The deep-V hull forms, designed or constructed in the industry, operate at Fn > 0.55 as shown in Figs 1.9.1–3. The statistics were compiled from information gathered for the deep-V ships designed or built by major shipyards. The statistical trends from the figures reveal the upper and lower boundary lines and the optimum L/B and B/T ratio of the deep-V hull forms. The upper boundary line in Fig. 1.9.1 reveals the feasible ship length against the maximum designed Fn, and the lower boundary lines in Figs 1.9.2 and 1.9.3 reveal the minimum designed F_n.

The major shipbuilders of large commercial deep-V vessels (L>95m) are Chantiers de l'Atlantique "Corsair" (24.14%), Fincantieri "MDV" (31.03%), Izar "Alhambra & Mestral" (13.79%), Mitsubishi Heavy Industries (3.45%), and Rodriguez "TMV" (27.59%). The major shipbuilders of small commercial and naval vessel (L<95m) are Abeking and Rasmussen, Austal, Chantiers de l'Atlantique, Kockums, Lurssen, Rodriguez, and Wavemaster. Abeking and Rasmussen built the pioneer commercial deep-V ship, the SAR 33-Patrol Craft, in 1976. Ever since, the deep-V hull form has gained huge popularity for high-speed application in naval vessels and commercial ferries. The Visby class corvette from Kockums and the Sea Blade (LCS) from Lockheed Martin are the latest examples of deep-V naval vessels and Gotland II is the latest high-speed deep-V ferry currently being built at Fincantieri. Taurus (high-speed car ferry) is the largest deep-V vessel ever constructed, with a displacement of 3900 tonnes and L_{OA} of 145.6m. Fincantieri delivered a family of four vessels in 1998.

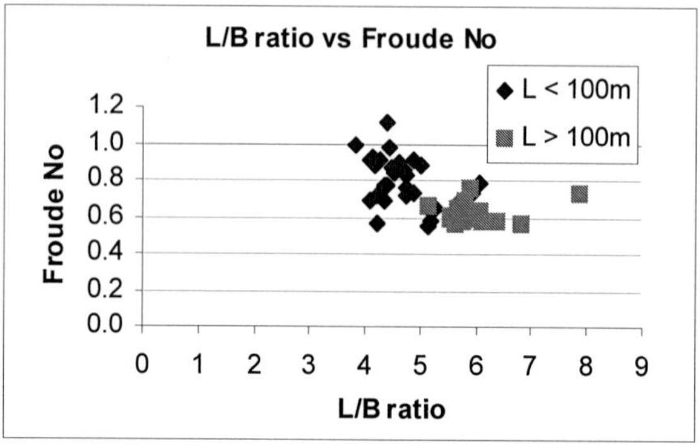

Fig. 1.9.1. L_{OA} versus F_n for deep-V hull forms

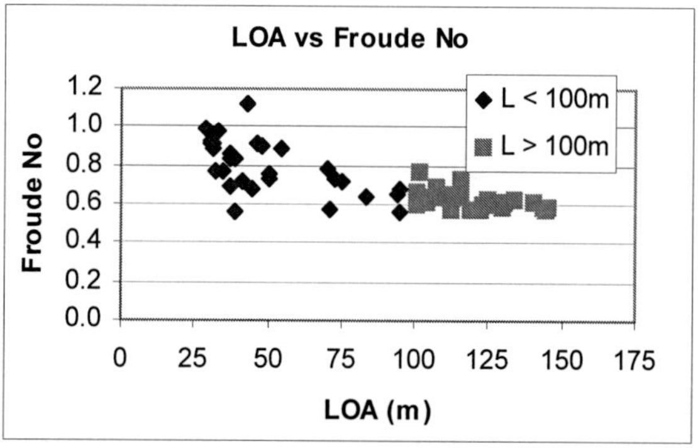

Fig. 1.9 2. L/B ratio versus Fn for deep-V hull forms

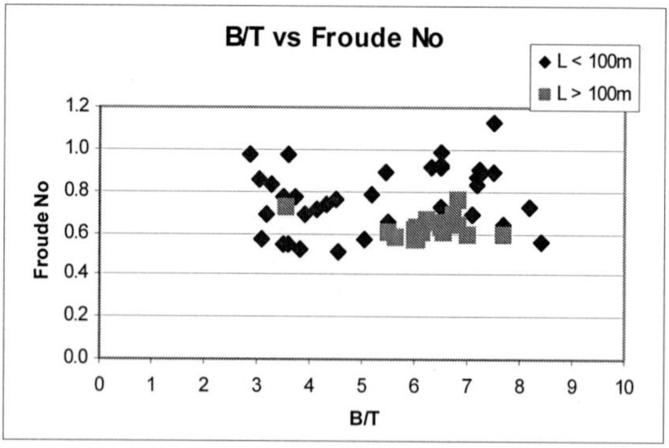

Fig. 1.9 3. B/T ratio versus F_n for deep-V hull forms

In the present project, a novel deep-V hull form is designed on the basis of a deep-V hull form; for the first time, this form features a bulbous bow and a modified aft end and transom stern, and is designed for operation at low Froude numbers (F_n=0.2 & 0.388) with a displacement of 16,000t. The deep-V hull form on which the design was based was obtained from Hydro Research Systems S.A. and provided the basis for the ongoing research program on deep-V hull-form developments at the University of Newcastle upon Tyne.

The hydrodynamic optimization scheme consists of three main components. First, a CAD modeler for modeling design geometry, an efficient CFD code, and last an optimization tool to solve the non-linear optimization problem created by the objective function and constraints.

The novel deep-V (NDV) hull form is modeled and designed using the Friendship CAD modeler. The CAD modeler is also utilized for parametric changes in the automated optimization.

A direct use of the Artificial Neural Network (ANN) technique is introduced for hull form hydrodynamic optimization. The use of ANN has been evinced to carry some advantages such as a direct search for the global optimal design and a relatively more efficient optimization method. The ANN technique will be compared with a gradient-based Method of Moving Asymptotes, MMA, which uses an iterative convex approximation method, for efficiency and performance. A Ro-Ro hull form, with a length of 123m, is used as an example for the comparison.

The wave pattern resistance is computed using the non-linear potential flow in the SHIPFLOW program developed at Flowtech International AB. Two different techniques for wave pattern resistance prediction, pressure integration, and transverse wave-cut analysis are compared for robustness to be included in the automatic hull form optimizations.

The optimization procedure involving parametric modeling, the non-linear panel method, MMA and ANN is described in more detail in Section 1.9.2. A pre-study for the comparison of different optimization techniques, MMA and ANN, and for using a ro-ro hull form are described in Section 1.9.3. The new design of the deep-V hull form and optimization of the bulbous bow and aft stern for two different Froude numbers (F_n= 0.2 & 0.388) is described in Section 1.9.4. Finally, some conclusions are presented in Section 1.9.5.

1.9.2. Optimization tools

A hydrodynamic optimization scheme consists of three main components. First, a CAD modeler for modeling design geometry, an efficient CFD code, and last an optimization tool to solve the non-linear optimization problem created by the objective function and constraints.

1. Parametric modeling

A parametric approach has a natural advantage in the context of optimization. The approach contains a few variables, but carefully selected form parameters generate desirable shapes, leaving enough freedom for shape modification.

(1) Friendship CAD modeler

Initially developed at TU Berlin, the FRIENDSHIP-Modeller (Form parameter oriented ship modeler)[1-9-4] is commercialized by Friendship Systems, Potsdam. The program provides methods for the parametric design of ship hull form and is pertinent for both manual and fully automated hydrodynamic optimization of hull form as shown in 1-9-5.

The current version of the friendship modeler allows a number of different types of ships to be represented by different sets of parameters. Different generation approaches take into account the need to best address specific hull properties in the context of shape variation. Most commercial vessels can be distinguished by their characteristic sectional shape, either being

- Dominated by positional information, or
- Dominated by integral information

Conventional round bilge designs represent the latter type of hull shapes. Without consideration of the dry part of the section, a great number of different shapes are available featuring the same beam and draft while changing their sectional area. For this type of hull a parameterized sectional area curve is usually applied, which utilizes a predefined sectional area curve and generates sections being constrained to the local sectional area, see Fig. 1.9.4. From a small set of positional information—for example, the design waterline and the centerplane curve—most of the hull shape can be developed. The ro-ro design in Fig. 1.9.5 represents a design from that hull category.

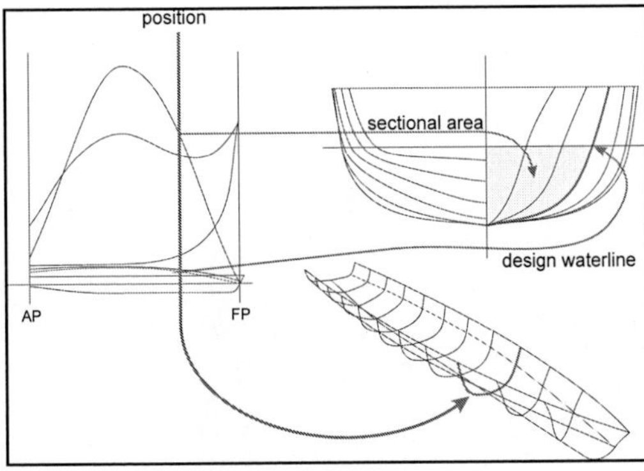

Fig. 1.9.4. Frame construction utilizing the sectional area

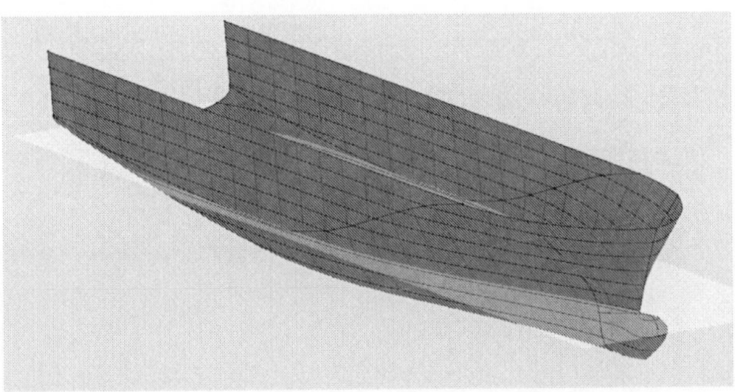

Fig. 1.9.5. Initial ro-ro design from the optimization

An example for hull shapes being dominated by positional information is the hard chine design presented in this paper. The shape of a section is almost fully defined by the position of keel, deck, and knuckle. Modifiers for the rail/bevel shape then allow minor modifications providing only very limited freedom in the sectional area. Putting a constraint on that area would quickly lead to deteriorated shapes. To comply with the need of addressing shape variations without deteriorating the shape, an additional parameter for the local camber of the lower part of the section has been made available. Fig. 1.9.6 illustrates a variety of sectional shapes of the deep-V design for different values of the camber (0%, 2%, and 4%). The spray rail—which turns into a bevel in the aft part of the hull—has been parameterized by its local absolute angle and its depth, hinged at its upper position defined by the knuckle parameters (see Fig. 1.9.7). The last two parameters complete the set of eight parameters for the bare hull section definition of the deep-V.

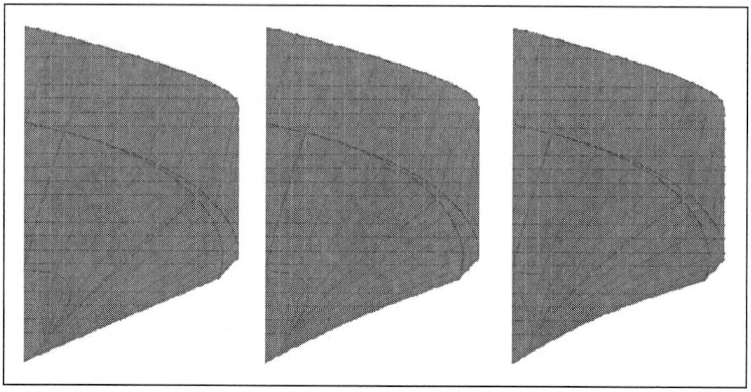

Fig. 1.9.6. Midship section with different camber values

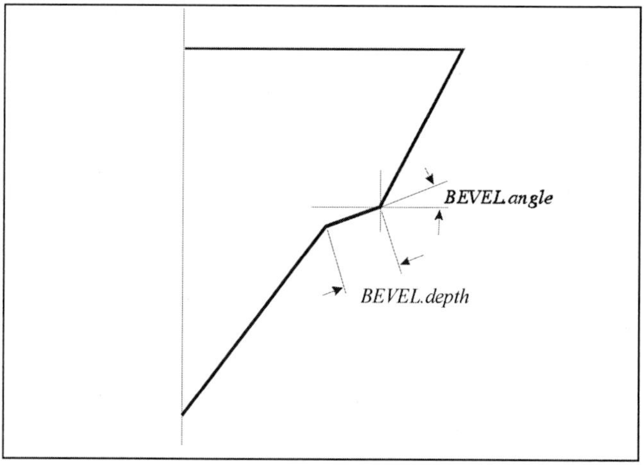

Fig. 1.9.7. Rail/bevel parameters

As a special feature of this research work, a new parametric model for the bulbous bow was developed. The bulbous bow shape, from the forward perpendicular to its tip, is guided by four curves, namely the lateral curve at top and bottom, and the curve of maximum beam—that is, its beam and vertical position. Each of the four curves provides an extensible set of parameters for its definition. A minimum set of only four parameters is sufficient to specify a suitable bulb design with a good range of variability. Fig. 1.9.8 depicts the bulb and a variation of only one parameter. In total a set of 13 bulb parameters has been made available, including the lateral area of the bulb offering enough freedom for local shape variation.

Fig. 1.9.8. Single parameter bulb variation

2. Non-linear panel method

The non-linear panel method (potential flow), in the SHIPFLOW program, is used to compute the wave pattern resistance (objective function) by a transverse wave cut technique. The program is jointly developed by Flowtech International AB and Chalmers University of Technology.

There have been a number of suggested methods of theoretical wave spectrum analysis based on longitudinal wave cuts or transverse wave cuts of the wave spectrum (through measurements or computation) to calculate the wave resistance coefficient.[1-9-6] In addition, the use of pressure integration over the hull length is also a common technique to predict wave resistance. The pressure integration technique is sensitive to local irregularities of the panels describing the hull surface. However, the wave pattern is less sensitive to these irregularities. Wave cut analysis should therefore be a more robust technique than pressure integration and particularly important in the context of automatic shape optimizations.[1-9-7]

Transverse wave cut

The wave cut analysis predicts the wave pattern resistance by measuring the wave profile along a straight track through a wave spectrum, either parallel to the ship's direction of motion (longitudinal wave cut) or perpendicular to it (transverse wave cut). The wave profile tracks must be located at some distance from the hull where non-linear effects are assumed to be small, so as to be consistent with the linear theory used for the wave cut analysis methods. The wave profile data can either be taken from measurements in a towing bank or from CFD. The wave profile data from LWC (longitudinal wave cut) are preferred in experiments[1-9-7] since the profiles are relatively easy to measure. An analytical extension to infinity must however be added.

There is no requirement for any analytical extension in the transverse wave cut method. The wave cut can be truncated at any length outside the Kelvin wedge, since beyond such a point there are no more waves and the free-surface is unaffected. The transverse wave profile can easily be simulated in CFD but it is more complicated to measure it in a towing tank.[1-9-7]

In SHIPFLOW, the multiple TWC technique (MTWC) is used to predict the wave pattern resistance (R_{WP}). The MTWC technique utilizes the elevation of several, non-equidistant transverse wave cuts behind the ship. TWC's can be truncated to any length outside the Kelvin wedge, whereas LWC's have to be extended far downstream to obtain the required information. Thus the TWC's require lower dimensions of the free-surface grid and hence reduce computational time in CFD programs than LWC's. The TWC technique is a suitable tool for optimization using wave pattern resistance since they are robust and shown to be able to predict the same ranking for ship variants as the pressure integration and the towing tank measurements.[1-9-7]

3. Optimization methods

The Ro-Ro bulbous bow is optimized for resistance using two different techniques, the gradient-based Method of Moving Asymptotes (MMA) and an Artificial Neural Network (ANN) approach, for comparison of efficiency and performance. The deep-V aft end and transom stern are optimized using MMA and the bulbous bow using ANN approach.

(1) Method of moving asymptotes

The gradient-based method of moving asymptotes (MMA)[1-9-8] is an iterative convex approximation method for solving non-linear constrained optimization problems. It uses an inverse approximation of the values and first partial derivatives with respect to the design variables of the objective function and the constraint functions to formulate the convex sub-problem.

A convex sub-problem is solved using a dual formulation and no second derivative is required to formulate the sub-problem.

MMA was initially developed for structural optimization but can be applied to CFD problems provided that the objective function, constraints, and their gradients with respect to the design variable changes are computed. The schematic diagram of a MMA automated optimization of the bulbous bow involving shipflow and friendship is shown in Fig. 1.9.9.

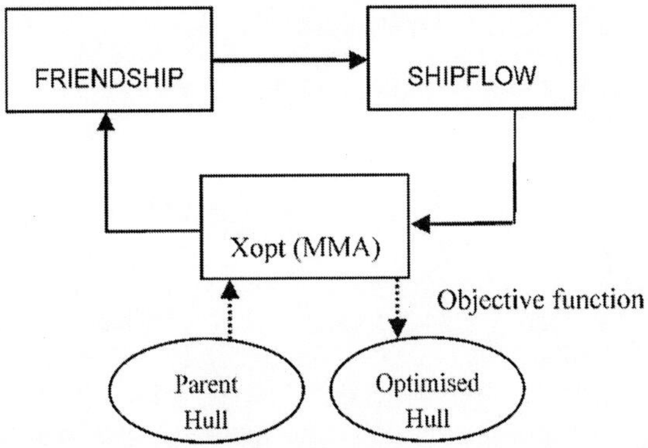

Fig. 1.9.9. Flowchart of the MMA optimization system

(2) Artificial neural network

Artificial neural networks (ANNs) are simplified models of the biological nervous system. They are massively parallel-interconnected networks of neural computing elements, called artificial neurons. A neural network is not programmed like traditional mathematical models or algorithms and it has the ability to respond to input stimuli and learn to adapt to the environment (learn by experience). For a more detailed explanation of the method, see 1-9-9.

ANNs have the capability to identify and learn the functional relationships between multiple inputs and outputs of any non-linear multi-dimensional system, as shown in Fig. 1.9.10. Thus, they have become very prevalent for solving problems that are difficult to solve by traditional means.

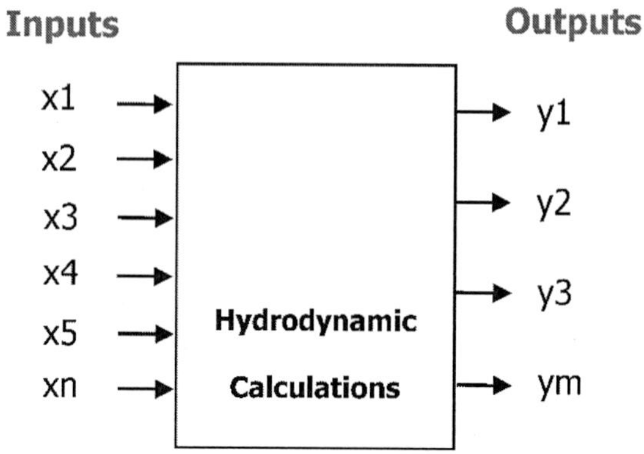

Fig. 1.9.10. Non-linear multi-dimensional system (hydrodynamic calculations)

With a sufficient number of sets of design variables and corresponding responses, ANN can learn the relationship between the sets, which subsequently outlines an explicit formula for the objective function. A direct use of artificial neural network (ANN) technique for multi-objective hydrodynamic optimization was first attempted in 1-9-10. The use of ANN has been evinced to carry some advantages, such as the direct search for global optimal design and relatively more efficient optimization methods with a multitude of repeated usage or further changes to physical design constraints.

In this study, the design variables for the ro-ro vessel are listed in Table 1.9.1, which represents the bulbous bow parameters in the optimization space. The corresponding response is the wave pattern resistance. The design variables for the deep-V bulbous bow in the optimization space are listed in Table 1.9.3. The corresponding response is the sum of the frictional resistance (R_F), using ITTC-57, and the wave pattern resistance (R_{WP}) computed at two Froude numbers (0.2 and 0.388).

Static artificial neural networks are capable of storing data during the learning process and subsequently reproducing this data during the recall process. A schematic representation of the feedforward associative network is shown in Fig. 1.9.6. The ability as an optimization tool in minimizing the wave resistance of a ship hull was introduced by 1-9-11. ANNs' generalization capability is able to extend to valid extrapolation in some cases, which is the focus of their utilization for mapping real-valued patterns from n-dimensional space to real-valued patterns in m-dimensional space (i.e., multi-dimensional input/output data sets, $\underline{X} = (x_1, x_2, ..., x_n)^T$ and $\underline{Y} = (y_1, y_2, ..., y_m)^T$ respectively) as:

$$f:\Re^n \rightarrow \Re^m \quad (1.9.1)$$

where f is a continuous function representing the relationship between multi-dimensional input/output data sets.

With multi-layer feedforward network (MLFF) generalization and mapping capability, a cluster of 70% arbitrarily selected input values ($x_1, x_2, \ldots, x_6$) were used to generate a set of output values (y_1 and y_2). Of the data sets, 15% were used to train a fully connected MLFF with the sigmoid function as its activation function. The remaining 15% of the data sets were reserved for cross validation of the trained network and to measure its generalization ability for unseen data sets. After successful training and cross validation, the MLFF model can, effortlessly and competently, replace the hydrodynamic calculations as represented in Fig. 1.9.12. This allows the MLFF model to compute the output values in a significantly shorter time, in comparison to the actual software.

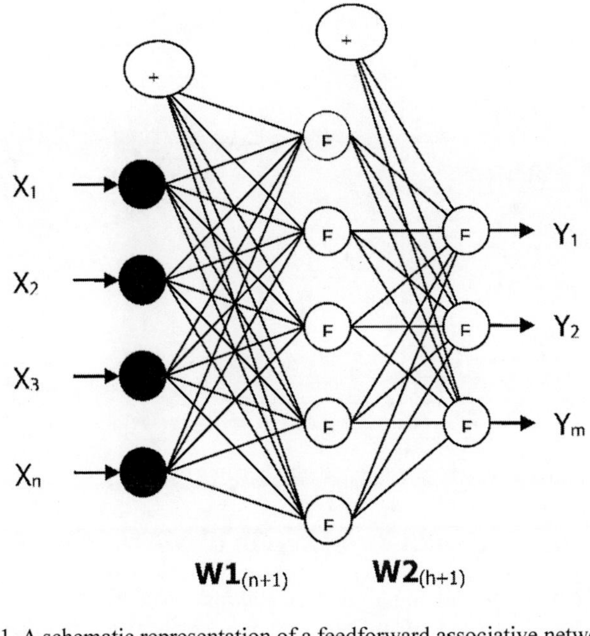

Fig. 1.9.11. A schematic representation of a feedforward associative network

Using MLFF, ANN also provided us with a single functional relationship in equation 1.9.2 for further analysis and simple hard/software practical implementations.

$(y_1, y_2, \ldots, y_m) = f(x_1, x_2, \ldots, x_n)$ (1.9.2)
where x_n is measured data and f is the non-linear relationship between sensor readings and themselves. In this case study n=6. Function f in equation 1.9.1 can be represented as:

$$y_m = \mathrm{sig}\left[\sum_{i=1}^{h} \mathrm{sig}\left(\sum_{j=1}^{n} x_j \mathbf{W}_{i,j}^{1} + c_i\right)\mathbf{W}_{m,i}^{2} + b_m\right] \quad (1.9.3)$$

where h, m, and n are hidden, output, and input nodes, respectively; $\mathbf{W}^1, \mathbf{W}^2, c$ and b are network weights, which are frozen after training; and *sig*(.) is the sigmoidal function presented in equation 1.9.4:

$$sig(\mathrm{x}) = \frac{1}{1 + \mathrm{e}^{-\mathrm{x}}} \quad (1.9.4)$$

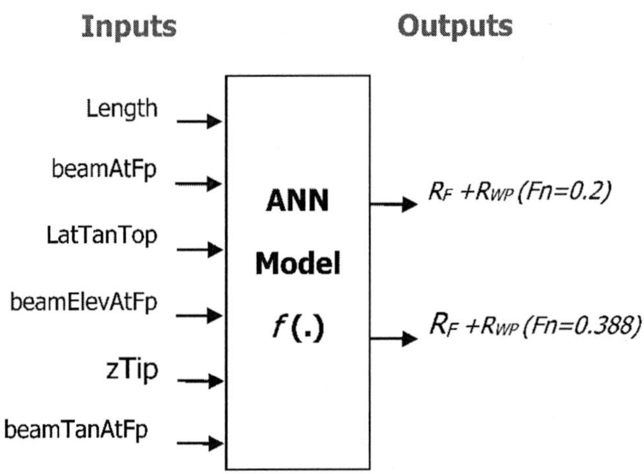

Fig. 1.9.12. ANN model substituting hydrodynamic calculations

1.9.3. Pre-study comparison of MMA and ANN

In this section, the artificial neural network (ANN) approach and the gradient-based method of moving asymptotes (MMA) optimizations are compared for performance and efficiency. Design charts are utilized to plot the resistance results (response surface) from the pressure integration and transverse wave cut and compared for robustness for optimization. The ro-ro hull form is used as an example for the comparison.

1. Ro-ro hull form

The ro-ro hull form has a length of 123m, a maximum beam of 19.5m, a draft of 5m, and a displacement of 6,681t, as shown in Fig. 1.9.5. The hull form has a pronounced bulbous bow whose top is slightly below the design waterline. The ro-ro bulbous bow parameters are used for the optimization for comparing the two different optimizers.

2. Design charts for ro-ro hull form

Design charts, such as a response surface, are capable of demonstrating the change in the preferred performance indicators from the parent hull form over a matrix of any two parameters changes. Thus the influence of hull form parameter changes can be promptly represented and interpreted by a hull designer (see 1-9-10 and 1-9-12).

The design charts (response surfaces) for the investigated ro-ro bulbous bow parameters are computed by two different computation techniques for the resistance: pressure integration and transverse wave cut, as shown in Figs 1.9.13a–15b. From the figures, the response surfaces from the transverse wave-cut technique are smoother and contain less irregularity. It can be summarized as the most robust technique for hydrodynamic optimization and ranking of designs as compared with the pressure integration.

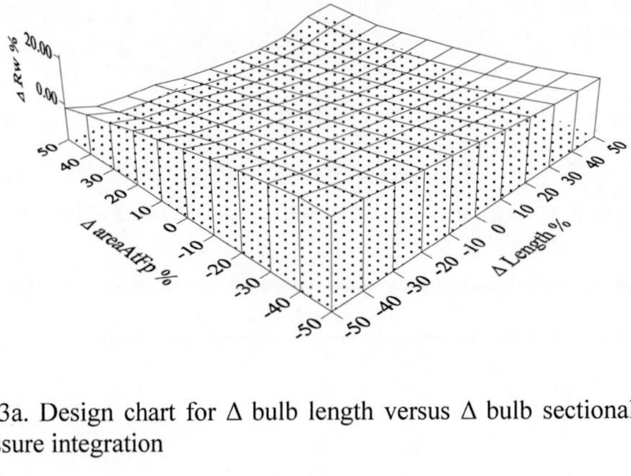

Fig. 1.9.13a. Design chart for Δ bulb length versus Δ bulb sectional area at F.P. using pressure integration

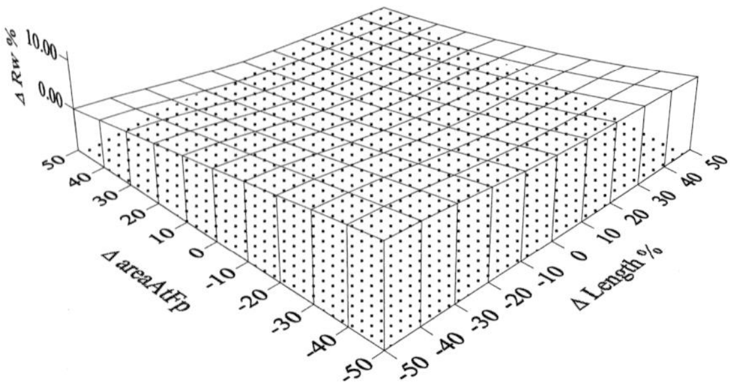

Fig. 1.9.13b. Design chart for Δ bulb length versus Δ bulb sectional area at F.P. using transverse wave cut

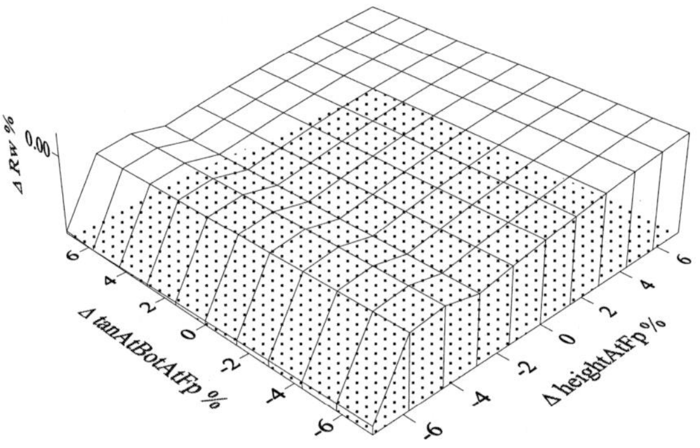

Fig. 1.9.14a. Design chart for Δ bulb max height at FP versus Δ bulb angle at base at F.P. using pressure integration

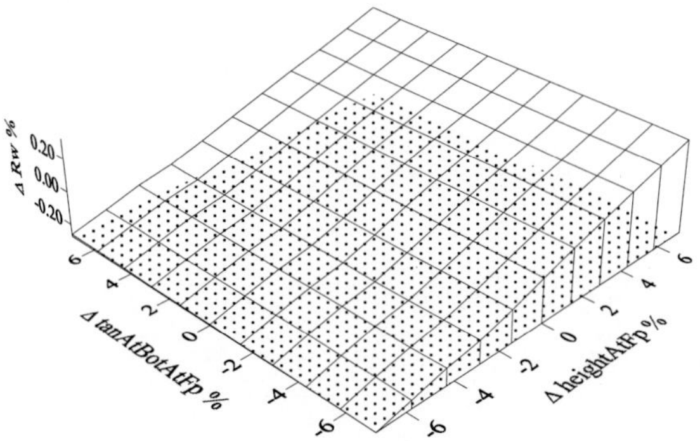

Fig. 1.9.14b. Design chart for Δ bulb max height at FP versus Δ bulb angle at base at F.P. using transverse wave cut

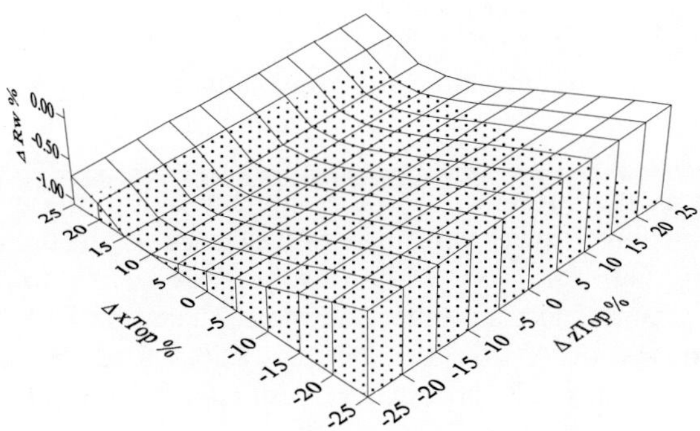

Fig. 1.9.15a. Design chart for Δ bulb vertical position of the max horizontal top versus Δ bulb position ahead of F.P. of max horizontal top using pressure integration

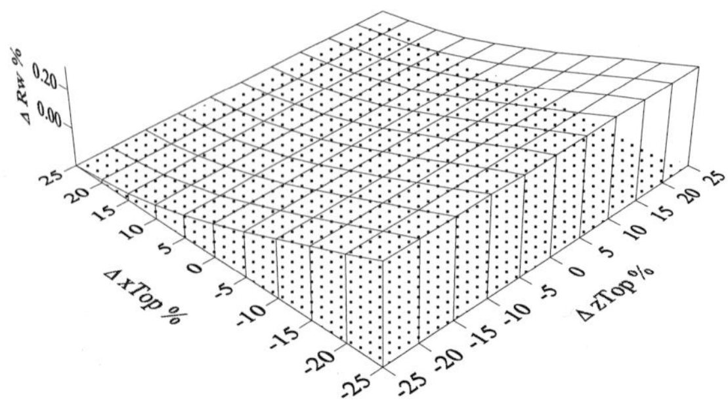

Fig. 1.9.15b. Design chart for Δ bulb vertical position of the max horizontal top versus Δ bulb position ahead of F.P. of max horizontal top using transverse wave cut

3. Optimization of ro-ro bulbous bow

The results for the optimization of a ro-ro hull form with wave pattern resistance as the objective function are summarized in Table 1.9.1. The wave patterns of the optimized hull forms are shown in Fig. 1.9.16.

From the results in Table 1.9.1, the optimum hull form, produced by MMA, has a slightly lower wave pattern resistance, R_{WP}, compared with the optimum hull form by ANN (8.90% and 8.77%, respectively); however, it also produces a smaller hull form. The errors of the objective function results produced from the ANN regression-based optimization are relatively small compared with the actual results from SHIPFLOW.

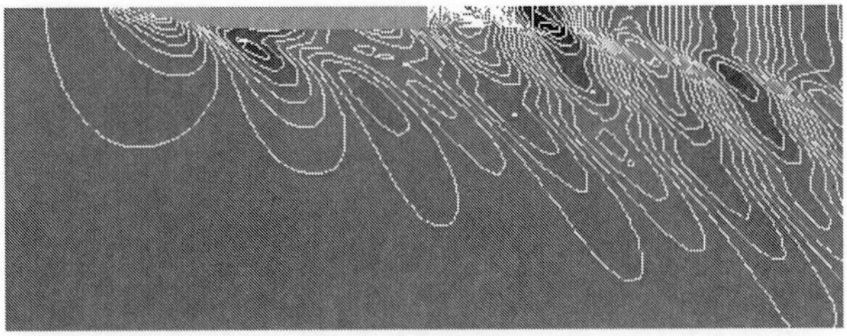

Ro-ro original

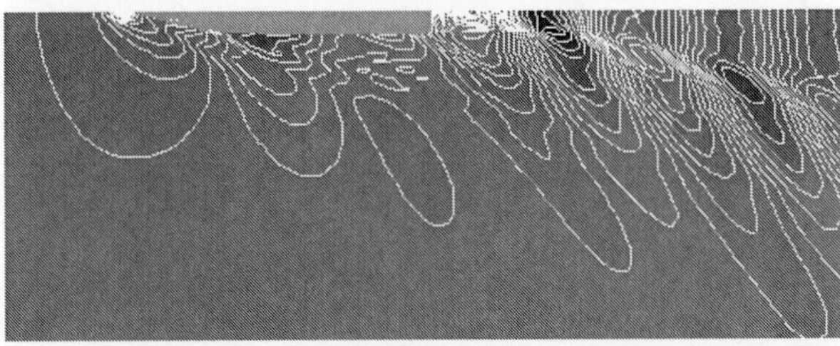

Ro-ro MMA

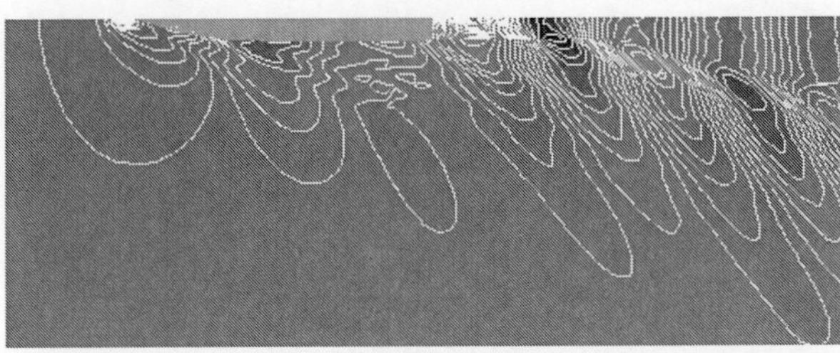

Ro-ro ANN

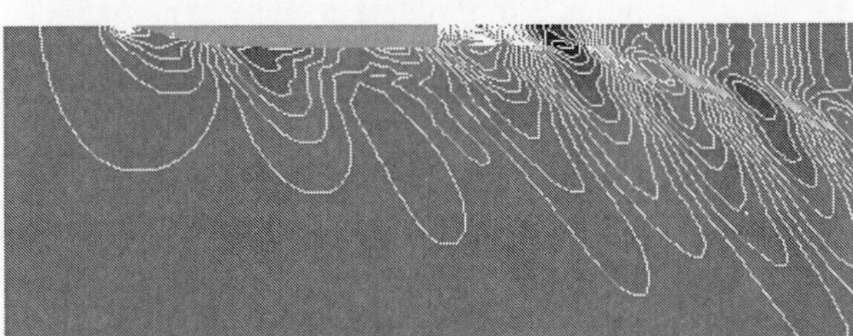

Ro-ro Practical

Fig. 1.9.16. Comparison of wave patterns using SHIPFLOW for a ro-ro hull form

Table 1.9.1 Summary for optimization of a ro-ro bulb

	Ro-Ro Parent	Ro-Ro MMA	Ro-Ro ANN	Ro-Ro Practical
Δ /t	6,681	6,730 (+0.73%)	6,755 (+1.1%)	6,751 (+1.05%)
Length/m	4.250	6.375	6.375	6.375
Area at Fp/m^2	8	12	12	10
Height at Fp/m	4.500	4.1860	4.18605	4.5
Tan at Bot. at Fp /deg.	45.000	47.8673	41.8605	47.000
zTop/m	-0.080	-0.07	-0.10	-0.10
xTop/m	2.890	3.200	2.1675	2.1675
R_{WP}/kN [gain]	120.684	109.948 [8.90%]	110.097 [8.77%]	113.243 [6.17%]
R_T/kN [gain]	345.058	334.553 [3.04%]	334.799 [2.97%]	337.844 [2.09%]
ANN R_{WP}/kN [error]	120.692 [0%]	110.985 [0.94%]	110.055 [0.04%]	113.561 [0.28%]

4. Conclusions of the pre-study

MMA is considered to be a reliable method, which in most cases generates a sequence of steadily improved solutions.[1-9-8] The solution generated after 10–15 iterations is considered acceptable since the marginal improvement has become small between iterations. ANN can be used for multi-objective hydrodynamic optimization.[1-9-10] The advantages are a direct search for the global optimal design and a relatively more efficient optimization method with a multitude of repeated usage or further changes to physical design constraints.

The computations for the Ro-Ro bulbous bow optimization using MMA took three days. Gradient-based methods such as MMA can only be expected to find local minima in the design space. The optimization was therefore started from three initial designs in order to search the design space for the global optimum. A total of nine days were used for the global search.

The systematic variation to generate the ANN database and training took nine days. The optimization using the direct search always obtains the global optimal hull form parameters in the search space. Therefore, the two methods are comparable in terms of efficiency.

MMA performed slightly better than the ANN in terms of the objective function results (8.90% and 8.77% respectively). The displacement was smaller for the MMA result as shown in Table 1.9.1. The errors of the objective function results produced from regression-based ANN optimization are relatively small as shown in Table 1.9.1.

The advantages and disadvantages of the ANN method compared with MMA are:

- NN always obtains global optimal hull-form parameters in the design space by a direct search
- It is useful for multi-objective optimization
- It is a relatively more efficient optimization method with a multitude of repeated uses or further changes to physical design constraints
- ANN is only a semi-automated optimization instead of being fully automated like MMA
- The errors are relatively small in a regression-based optimization

1.9.4. Novel deep-V hull-form design

The basis deep-V hull form provided by Hydro Research Systems is redesigned with a bulbous bow, a modified aft end, and a displacement of 16,000t for operation at a lower Froude number (F_n=0.2 and 0.388). The deep-V aft end and transom stern, due to design constraints and computation efforts, are optimized and finalized prior to the bulbous bow optimization.

1. Modified aft end and transom stern

A deep-Vee hull form has unfavourable frictional resistance and calm water resistance compared to a comparable round-bilge hull form mainly due to a larger wetted surface, relatively higher prismatic coefficient (C_P), and fully immersed transom stern.

A transom stern reduces resistance when compared with a cruiser stern at high Fn. At lower Fn, the full transom stern will become wet[1-9-13] and the following stern waves will change into forward-oriented breaking waves and increase the resistance drastically.[1-9-14] Slamming rarely occurs for this type of stern and the flowlines largely follow the hull form. However, for $F_n < 0.5$, a fully immersed transom stern should be modified in order to achieve favorable resistance performance.[1-9-14]

The deadrise angle and keel rise on the stern is considered for the modification of the deep-V aft end and transom stern. The selected deadrise angle will have an impact on the beam and draft required to obtain the desired hull volume and the keel rise will affect the sea-keeping abilities, directional stability, and resistance. The parent deep-V hull form for the optimization is designed, from a basis deep-V hull form, with a bulbous bow and a modified aft end and transom stern. The modification of the deep-V aft end and transom stern are determined by the design constraints on displacement and propulsion systems. All the aft end designs were optimized by MMA and finalized prior to the bulbous bow optimization.

The free variables and constraints for the deep-V aft-end design are listed in Table 1.9.2 and Fig. 1.9.17. The results for the optimized deep-V aft end are shown in Table 1.9.3 and the optimized aft end design as shown in Fig. 1.9.18.

Table 1.9.2. Summary of the stern parameters for deep-V

Stern Parameters	Lower Bound	Parent	Upper Bound
Transom-Draft	0.31	0.33% of design draft	0.35
CPC-xPos aft base	95	100m	105
Knuckle plan-xPos max. beam aft	30	35m	40
Knuckle plan-transom beam	0.932	0.975% of max. beam	1.0
CPC-tan at stern	-1.0	0.0 deg.	2.0
CPC-rad at aft base	350	500m	700
Camber at stern	0%	2 %	2.5%

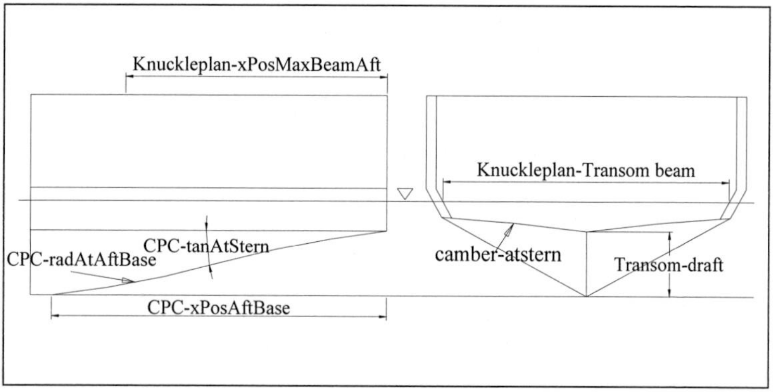

Fig. 1.9.17. Stern parameters for deep-V hull forms

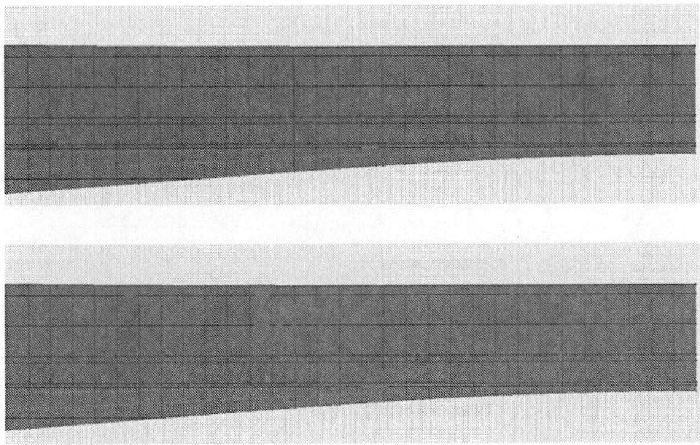

Fig. 1.9.18a. Side-view of the parent (top) and novel deep-V with optimized aft end for F_n 0.2 and 0.388 (bottom)

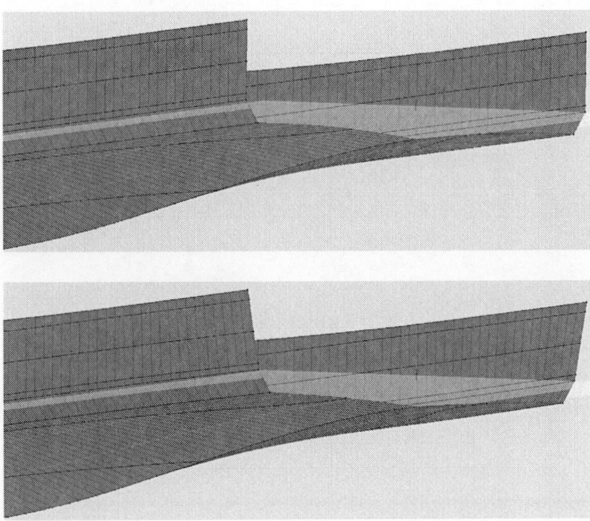

Fig. 1.9.18b. Perspective view of the parent (top) and novel deep-V with optimized aft end for F_n 0.2 and 0.388 (bottom)

Table 1.9.3. Summary for optimization of deep-V aft end

	NDV Parent	NDV Optimized Aft End
Δ /t	16,016	15,818 (-1.23%)
Transom-draft/%	0.33	0.31
CPC- x Pos aft base/m	100	100
Knuckle plan-xPos max. beam aft /m	35	40
Knuckle plan—transom beam /%	0.975	0.932
CPC-tan at stern/deg.	0.0	-0.5
CPC-rad at aft base/m	500	350
Camber at stern/%	2	1
R_{WP} (kN) at F_n 0.2 [gain]	293.38	273.39 [6.81%]
R_T (kN) at F_n 0.2 [gain]	679.58	656.95 [3.33%]
R_{WP} /kN at F_n 0.388 [gain]	509.78	482.60 [5.33%]
R_T /kNat F_n 0.388 [gain]	1863.98	1827.20 [1.97%]

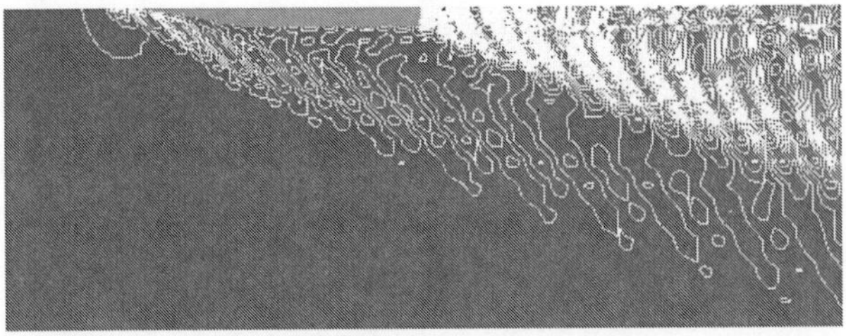

Novel parent deep-V at F_n 0.2

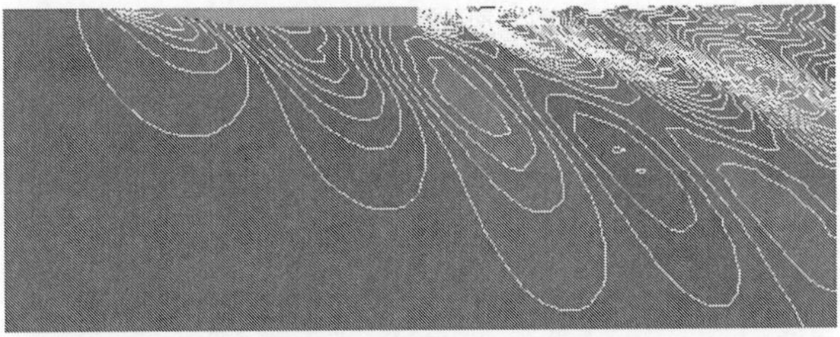

Novel parent deep-V at F_n 0.388

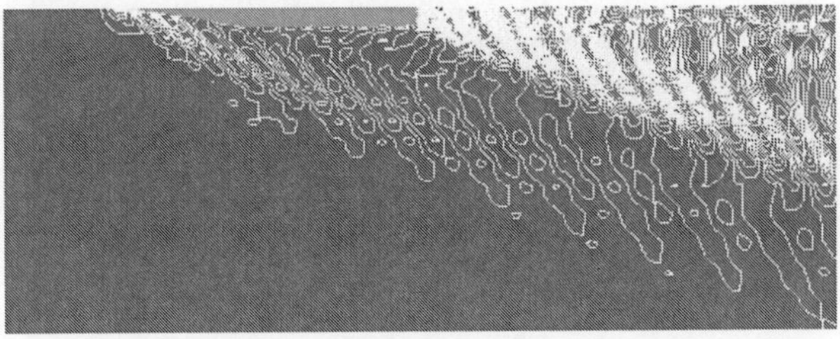

Novel deep-V with optimized aft end at F_n 0.2

Novel deep-V with optimized aft end at F_n 0.388

Fig. 1.9.19. Comparison of wave patterns using SHIPFLOW for deep-V aft end

2. Optimization of deep-V bulbous bow

The introduction of a bulbous bow by introducing moderate changes to the hull form geometry can significantly improve ship resistance.[1-9-15] The bulbous bow reduces the wave resistance by attenuation of the bow-wave system through hydrodynamically varying the velocity field in the vicinity of the rising waves. In addition, the bulbous bow also reduces the viscous resistance by smoothing the flow around the bow region. The bulbous bow also has an indirect impact on the hull efficiency (η_H), i.e., thrust deduction (1-t) and wake fraction (w). It also has a direct impact on the wake distribution in the propeller plane.[1-9-16] Consequently, the favorable effect of a bulbous bow depends on its size and the position and form of the bulbous bow.[1-9-16] For a ship at Fn > 0.26, a bulbous bow reduces power, more than lengthening the waterplane, by projecting the length of the bulbous bow. However for $F_n < 0.24$, the equivalent increase in length is always less than the length of the bulbous bow and for $F_n > 0.3$, the favorable bulbous bow effect may be not be achieved by lengthening the waterplane.[1-9-13]

The deep-V hull form is optimized for two Froude numbers (0.2 and 0.388) by using the multi-objective Artificial Neural Network (ANN) approach. The bulbous bow parameters are optimized to achieve the minimum powering in a voyage (fuel consumption) for the two speeds. The weighing factors of 0.3 and 0.7 are assigned to F_n 0.2 and F_n 0.388, respectively, which are used to determine the fraction of the operating time (voyage).

The objective function for the optimization is:

Minimize:

$$R_{T\,0.2} \times V_{0.2} \times [T \cdot W] + R_{T\,0.388} \times V_{0.388} \times [T\,(1-W)] \qquad (1.9.5)$$

Subject to:

$$(R_{T\,0.388} \times V_{0.388})_{\text{Optimised}} < (R_{T\,0.388} \times V_{0.388})_{\text{Parent}}$$

$$(R_{T\,0.2} \times V_{0.2})_{\text{Optimised}} < (R_{T\,0.2} \times V_{0.2})_{\text{Parent}}$$

where:

W = Fraction of operating time

R_T = Total resistance (kN)

V = Speed (m/s)

T = Operating time (day or hour)

The flowchart of semi-automated ANN hydrodynamic optimization of the bulbous bow involving SHIPFLOW and FRIENDSHIP is shown in Fig. 1.9.20. First, the optimization of the modified aft and transom stern is investigated using the initial bulbous bow. As the parameters for the modified aft and transom stern are constrained by the propulsion units and displacement, the parameters and changes for the optimization are limited.

Using this process—separating the bulbous bow and stern—also reduces the optimization and computation time.

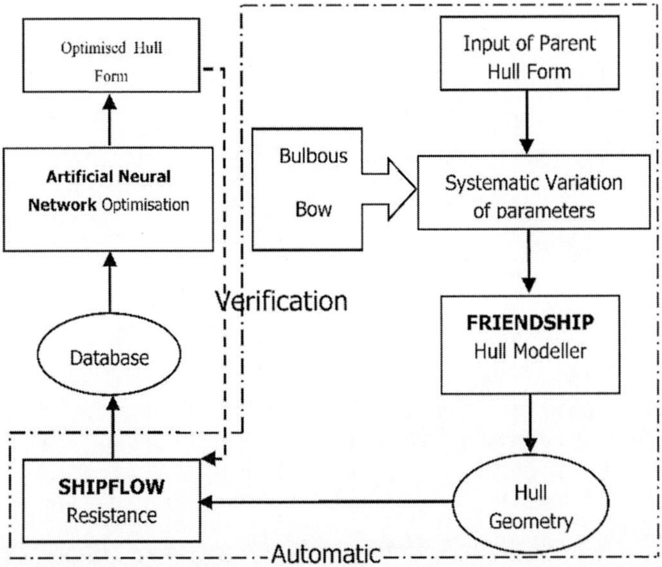

Fig. 1.9.20. Flowchart of ANN hydrodynamic optimization

The free variables and constraints for the deep-V bulbous bow design are listed in Table 1.9.4 and Fig. 1.9.21.

Table 1.9.4. Summary of the bulbous bow parameters for deep-V

Bulb Parameters	Lower Bound	Parent	Upper Bound
Length/m	8.16	10.2	12.24
Beam at Fp/m^2	3.2	4.0	4.8
Beam Elev at Fp/m	2.4	4.0	5.6
LatTan Top/deg.	3	5	7
zTip/m	-4.0	-2.0	0.0
Beam Tan at Fp/deg.	0	5	10

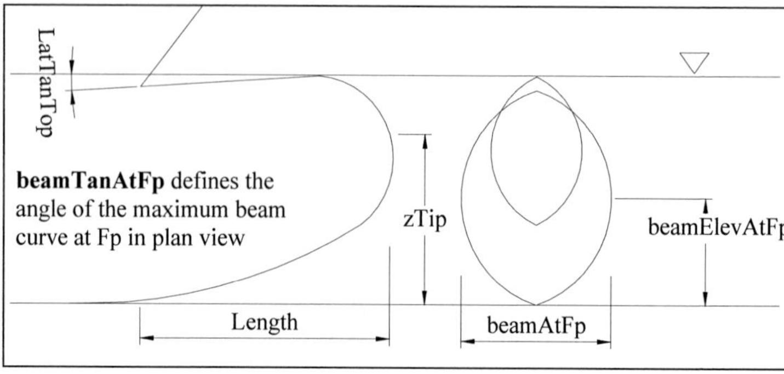

Fig. 1.9.21. Bulb parameters for deep-V hull forms

3. Design charts for deep-v bulbous bow

The design charts (response surfaces) for the investigated bulbous bow parameters as shown in Figs 1.9.22–24. The investigation of the design charts for the deep-V bulbous bow are summarized in Tables 1.9.5–7. The total resistance is computed using the sum of the frictional resistance (R_F), using ITTC-57, and the wave pattern resistance (R_{WP}) as computed by SHIPFLOW.

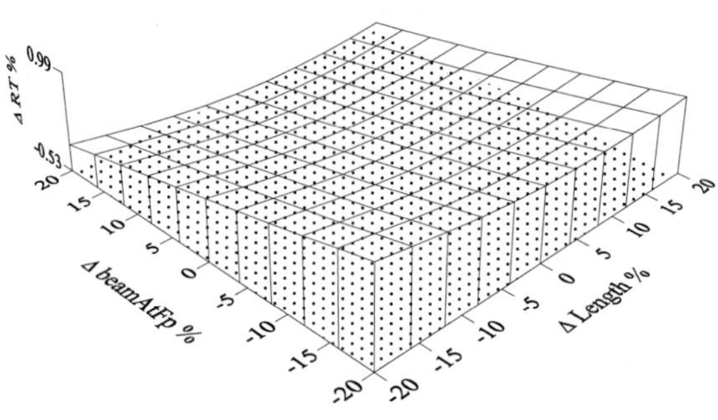

Fig. 1.9.22. Design chart for Δ length versus Δ beam at Fp (F_n 0.388)

Table 1.9.5. Summary of design chart 1 for F_n 0.388

Parameter Investigated		R_T
Δ length	Optimum	+/-
	Trend	+/-
	Sensitivity	Medium
Δ beam at Fp	Optimum	+20 %
	Trend	+
	Sensitivity	High
Δ ultimate response %		-0.534

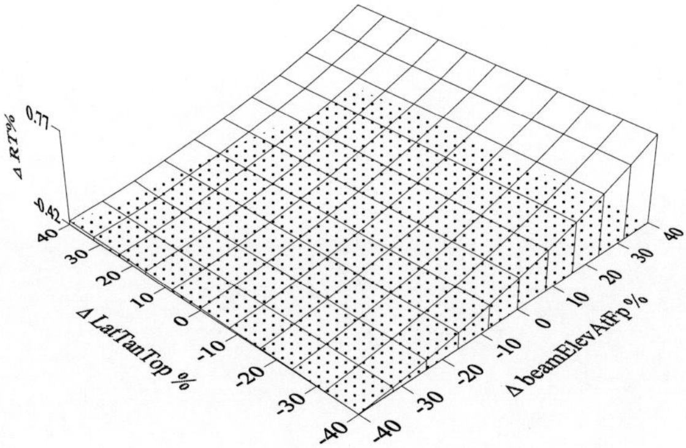

Fig. 1.9.23. Design chart for Δ beam Elev at Fp versus ΔL at Tan Top (F_n 0.388)

Table 1.9.6. Summary of design chart 2 for F_n 0.388

Parameter investigated		R_T
Δ beam Elev at Fp	Optimum	-40 %
	Trend	-
	Sensitivity	High
Δl at Tan Top	Optimum	-40 %
	Trend	-
	Sensitivity	Low
Δ Ultimate response %		-0.420

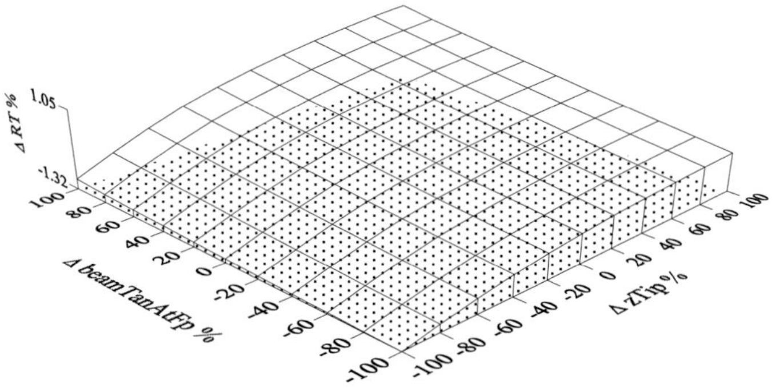

Fig. 1.9.24. Design chart for Δz Tip versus Δ beam Tan at Fp (F_n 0.388)

Table 1.9.7. Summary of design chart 3 for F_n 0.388

Parameter investigated		R_T
Δz Tip	Optimum	-100 %
	Trend	-
	Sensitivity	High
Δbeam Tan at Fp	Optimum	-100 %
	Trend	-
	Sensitivity	Low
Δ Ultimate response %		-1.317

1.9.5. Results

The results for the optimization of the deep-V bulbous bow for two Froude numbers are summarized in Table 1.9.8. The parent design, with the original bulbous bow, and the optimized deep-V bulbous bow are shown in Fig. 1.9.26, and the wave patterns for the optimized bulbous bow are shown in Fig. 1.9.25.

The optimized bulbous bow, produced for two Froude numbers, has achieved lower resistance (0.33% and 1.54%) than the parent design in both F_n. The table also shows that the optimized bulbous bows for a single Fn will have poor performances in the other Fn. The errors produced from the ANN regression-based optimization are again relatively small compared with the actual results from SHIPFLOW as shown in Tables 1.9.1 and 1.9.8.

Table 1.9.8. Summary for optimization of deep-V bulb

	NDV_1	NDV_2 F_n 0.2	NDV_3 F_n 0.388	NDV_4 Opt for 2 F_n
Δ/t	15,822	15,841 (0.12%)	16,074(1.59%)	15,963 (0.89%)
Length/m	10.2	8.16	10.52	10.52
Beam at Fp/m^2	4.0	3.2	4.8	4.126
Beam Elev at Fp/m	4.0	2.4	2.4	2.4
Lat Tan Top/deg.	5	3	7	7
zTip/m	-2.0	-2.526	0	0
Beam Tan at Fp/m	5	3.158	0.526	2.632
R_{WP} (kN) at F_n 0.2 [Gain]	274.20	263.34 (3.96%)	278.08 (-1.4%)	271.787 (0.88%)
R_T (kN) at F_n 0.2 [Gain]	658.50	645.64 (1.95%)	663.97 (-0.8%)	656.33 (0.33%)
R_{WP} (kN) at F_n 0.388 [Gain]	482.112	492.214 (-2.1%)	442.157 ([8.29%)	453.039 (6.03%)
R_T (kN) at F_n 0.388 [Gain]	1,826.65	1,829.79 (-0.2%)	1,792.27 (1.88%)	1,798.46 (1.54%)
ANN R_T /kN ANN P_E /kW at respective F_n (Error)	1827.18 32,899.3 (0.03%)	644.83 5971.05 (0.13%)	1,793.85 32,299.0 (0.09%)	-

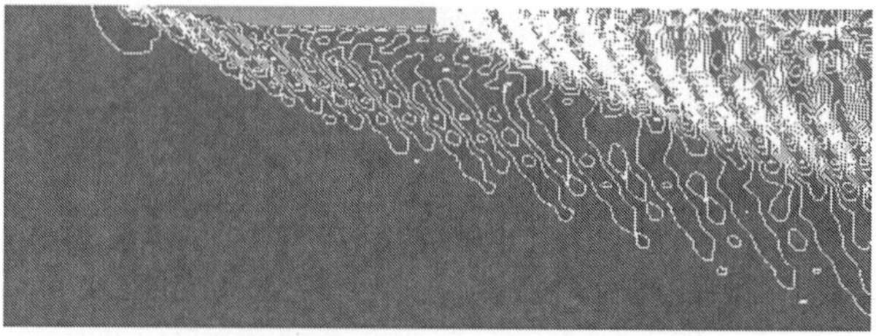

Novel deep-V_1 with original bulbous bow at F_n 0.2

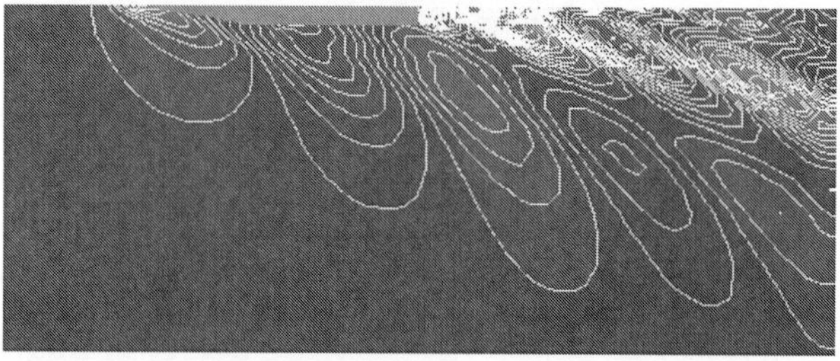

Novel deep-V_1 with original bulbous bow at F_n 0.388

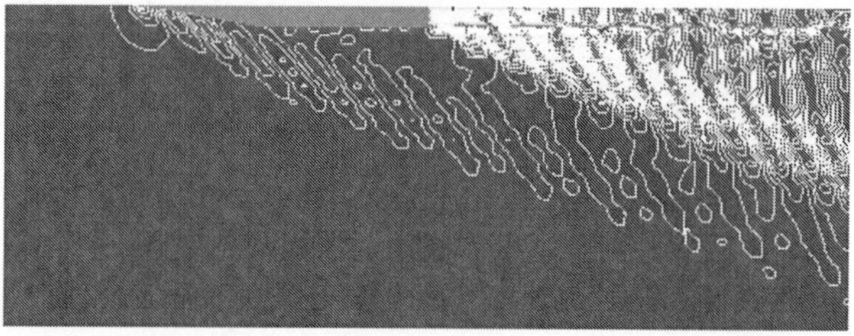

NDV_2 optimized bulb for F_n 0.2 at F_n 0.2

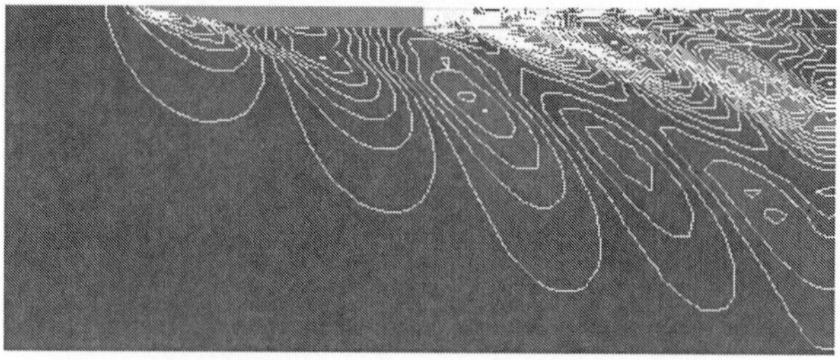

NDV_2 optimzed bulb for F_n 0.2 at F_n 0.388

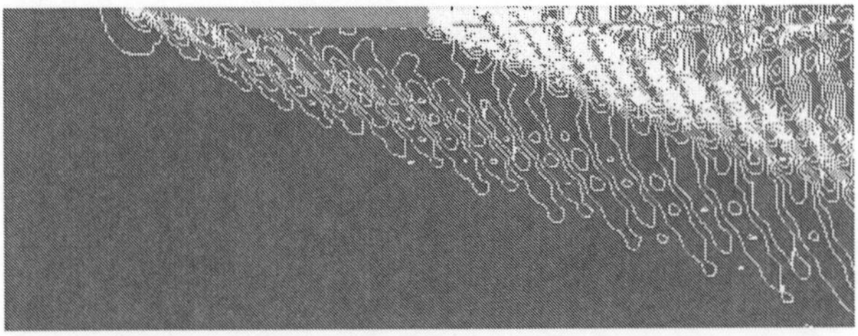

NDV_3 optimized bulb for F_n 0.388 at F_n 0.2

NDV_3 optimized bulb for F_n 0.388 at F_n 0.388

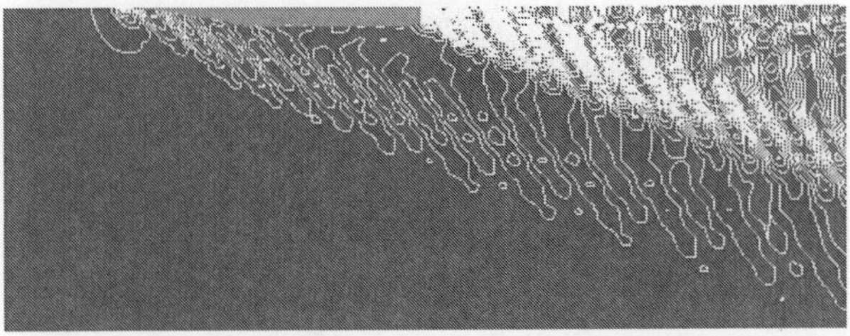

NDV_4 optimized bulb for both F_n 0.2 and 0.388 at F_n 0.2

NDV_4 optimized bulb for both Fn at Fn 0.388

Fig. 1.9.25. Comparison of wave patterns using SHIPFLOW for deep-V hull forms

Fig. 1.9.26. The parent (top) and novel deep-V with optimized bulbous bow for F_n 0.2 and F_n 0.388 (bottom)

1.9.6. Conclusions

A novel deep-V hull form has been designed for the first time with a bulbous bow and a modified aft and transom stern for operation at lower Froude numbers (F_n=0.2 and F_n=0.388) and with a displacement of 16,000t.

A multi-objective optimization of a bulbous bow for two Froude numbers can be achieved with relatively better resistance than the parent design over the two speeds.

The transverse wave cut (TWC) technique seems to be more robust than pressure integration. This is particularly important in the context of automatic shape optimizations.

The performance and efficiency of ANN regression-based optimization is also comparable to a gradient-based method like MMA. The errors of the objective function results produced from regression-based ANN optimization are relatively small. ANN optimization is useful for multi-objective optimization and is a relatively more efficient optimization method with a multitude of repeated usage or further changes to physical design constraints.

Acknowledgements

The collaboration for this research between University of Newcastle upon Tyne, Chalmers University of Technology, and FRIENDSHIP-Systems is gratefully acknowledged. The principal author would like to thank Dr Keunjae Kim of FLOWTECH International AB and Mr Hans Liljenberg of SSPA for their technical advice on hull form design. Additionally, thanks to Mr Erbil Serter of Hydro Research Systems S.A. for the collaboration and ALFGAM Optimering, Sweden, for providing the Xopt, MMA optimization program. Lastly, much gratitude to Singapore Technologies Marine and Mr Ching Eng Tan for their faith and support.

References

[1-9-1] E. H. Serter, "Hydrodynamics and Naval Architecture of Deep-Vee Hull Forms," Research-Development-Designs Istanbul, 1993.

[1-9-2] A. Coscia, I. Elice, L. Mangili, L. Rossi, & M. Parodi, "Application of Deep-V Hull Forms to Fast Naval Crafts," *La Marina Italiana* 2: pp. 23–30, Italy, 1998.

[1-9-3] J. W. Kehoe, K. S. Brower, & E. H. Serter, "The Deep-V Hull Form—Improves Seakeeping and Combat System Performance," *Naval Engineers Journal* 99 (3), May 1987.

[1-9-4] S. Harries & C. Abt, "Formal Hydrodynamic Optimization of a Fast Monohull on the Basis of Parametric Hull Design" Fast'99, Seattle, Washington, August 1999.

[1-9-5] S. Harries, "Parametric Design and Hydrodynamic Optimization of Ship Hull Forms," Phd thesis, Technical University of Berlin, 1998.

[1-9-6] K. W. H. Eggers, S. D. Sharma, L. W. Ward, "An Assessment of Some Experimental Methods for Determining the Wavemaking Characteristics of a Ship Form," *Sname Transactions* 75, 1967.

[1-9-7] C. E. Janson & D. Spinney, "A Comparison of Four Wave Cut Analysis Methods for Wave Resistance Prediction," *Ship Technology Research* 51, 2004.

[1-9-8] K. Svanberg, "Convex Approximation Methods for Structural Optimization," Comett Seminar on Computer Aided Optimal Structural Design, Germany, May 1992.

[1-9-9] E. Mesbahi, "Artificial Neural Networks—Fundamentals," Optimistic Optimization in Marine Design, 39th Wegemt Summer School, Berlin, Germany, March 2003.

[1-9-10] L. Koh, M. Atlar, P. N. H. Wright, E. Mesbahi, & D. Cooper, "An Inverse Design Procedure for Hydrodynamic Optimisation of

High Speed Hull Form Using Commercial Software," RINA Conference On High Speed Craft, Nov 2004.

[1-9-11] D. B. Danisman, E. Mesbahi, M. Atlar, & Ö. Gören, "A New Hull Form Optimisation Technique for Minimum Wave Resistance," 10th Intl. Congress of the International Maritime Association of the Mediterranean (Imam 2002), Crete, May 2002.

[1-9-12] P. N. H. Wright, "The Preliminary Design of Catamarans for Seakeeping and Resistance," Phd thesis, University of Newcastle upon Tyne, 2004.

[1-9-13] H. Schneekluth & V. Bertram, *Ship Design for Efficiency and Economy*, 2nd ed., Oxford: Butterworth-Heinemann, 1998.

[1-9-14] T. Yamano, T. Ikebuchi, & I. Funeno, "Stern Waves Consisting of Forward-Oriented Breaking Waves and the Remaining Following Waves," *Journal of Marine Science and Technology* 6 (1): pp. 13–22, 2001.

[1-9-15] J. W. Hoyle, B. H. Cheng, B. Hays, B. Johnson, & B. Nehrling, "A Bulbous Bow Design Methodology for High-Speed Ships," *Sname Transactions* 94: pp. 31–56, 1986.

[1-9-16] A. M. Kracht, "Design of Bulbous Bows," *Sname Transactions*, 86: pp. 197–217, 1978.

1.10

MODEL EXPERIMENTAL INVESTIGATION ON RESISTANCE OF AXE BOW SHIPS (2011)

WEI SONG, ZENG QING, GAO XIAOPENG AND DONG ZUSHUN

Naval University of Engineering, China

1.10.1. Abstract

The axe bow boat is a new kind of ship that has emerged in recent years. It is mainly used to improve navigability in severe sea conditions when navigating at high speed. Its characteristics are as follows:

1. Extremely sharp bow; outside, the section has almost no drift angle; the bow waterline width is very small
2. The bow vessel wall is almost vertical; the waterline up and down are the same; it has a high sheer bow; the perpendicular of the stem has been extended to a considerable depth below the baseline
3. The bottom of the center line from the ship has been extended to the lowest point of the stem to allow maximum draft. TU Delft Ship Hydrodynamic Laboratory studied and found that the hydrodynamic lift and vertical motion acceleration of the ship at high speed have a negative influence on sea-keeping and rapidity performance

The ESC (Enlarged ship concept) has greatly improved the previous two negative effects; to achieve better results, they proposed the concept of an axe bow. To study the hydrodynamic performance of the new type of ship, we have produced a model of 2.94 meters. Its hydrodynamic performance is tested in related experiments. According to the analysis of the ship model experimental data, the sea-keeping performance of the axe bow ship is rather better than that of a general ship; furthermore, its high-speed resistance performance is rather good compared with a typical U-hull ship.

1.10.2. The characteristics of the axe bow ship

The axe bow ship is a new type of ship that has emerged in recent years which is mainly used to improve the navigability of the ship at high speed in severe sea conditions. The British navy's *Intrepid*, which dominated the ocean in World War One, had its shadow.[1-10-1] The Ship Hydrodynamic Laboratory of TU Delft made a study and found that hydrodynamic lift and vertical motion acceleration have a tremendous negative influence on the rapidity performance and sea-keeping behavior of a ship at high speed. Their ESC (enlarged ship concept) has greatly improved the negative impact of the previous two problems, and has achieved good results in practical applications. According to their published data, an example is the extended 54% of the ESC 4,000 ship, with cost increases of 6%, operating costs decreased by 7%, 67% increase in transport efficiency, and operability increase of 68%.[1-10-2] To attain better results and radically decrease the negative influence of hydrodynamic lift and vertical motion acceleration at the front of a ship, they also proposed the concept of the axe bow. The typical lines plan is as follows:

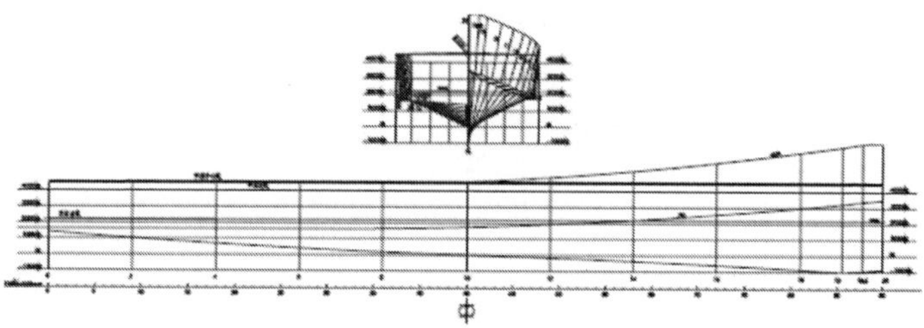

Fig. 1.10.1. Hull lines plan[1-10-2]

The ship has the following characteristics:

1. Extremely sharp bow, outside the section has almost no drift angle; bow waterline width is very small
2. Bow hull wall is almost vertical, the waterline up and down are the same; a high sheer bow; the perpendicular stem has been extended to considerable depth below the baseline
3. The bottom of the center line from the ship has been extended to the lowest point of the stem, to allow maximum draft.[1-10-2]

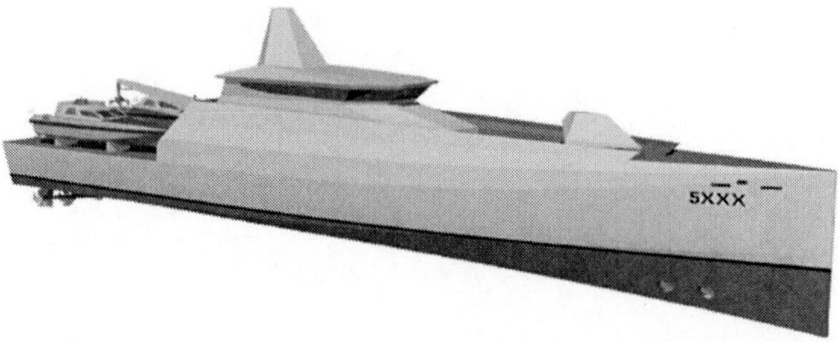

Fig. 21.10.2. New patrol ship hull according to axe bow concept[1-10-3]

1.10.3. Axe bow ship model

To study the hydrodynamic performance of the new type of axe bow ship, we have produced a model of 2.94 meters. Its hydrodynamic performance and related experiments were performed at the Aviation Industry Corporation of China Special Vehicle Research Institute. Trailer speed ranged between 0.1–22 m/sec. Speed stability and accuracy was better than 0.2%. The experimental speed was 1–6 m sec. The model's main parameters are as follows:

Main parameters:	
Length	2.94m
Width	0.58m
Length of waterline	2.875m
Width of waterline	0.459m
Displacement	47kg
C_p	0.789
C_b	0.206
C_m	0.521
C_{wp}	0.760

1.10.4. Sea-keeping performance of axe bow ship

According to the model test results and practical effects obtained by the Fast Crew Supplier series of ships of the Dutch Damen company, the ship-bow slamming vertical acceleration greatly abased, the pitch substantially reduced, and the lost speed in waves also significantly improved when using the axe bow ship in severe sea conditions.[1-10-4] When the wave height is 150mm in regular wave conditions and the ship model speed is 4.7 m/s, figs 1.10.3–5 show the ship's bow, the bow, and the amidships acceleration and pitch motion response.

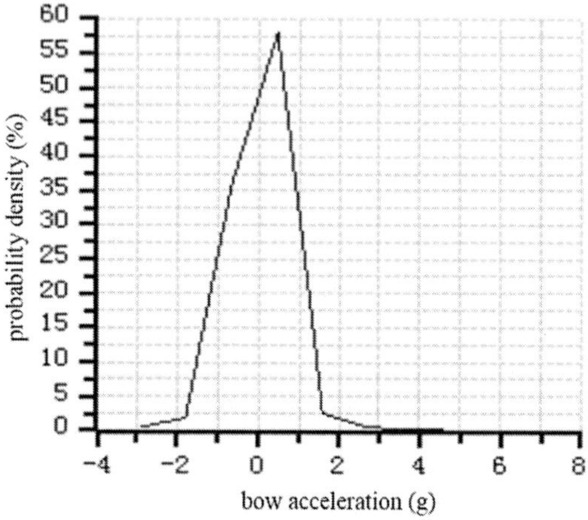

Fig. 1.10.3. Bow acceleration probability density curve

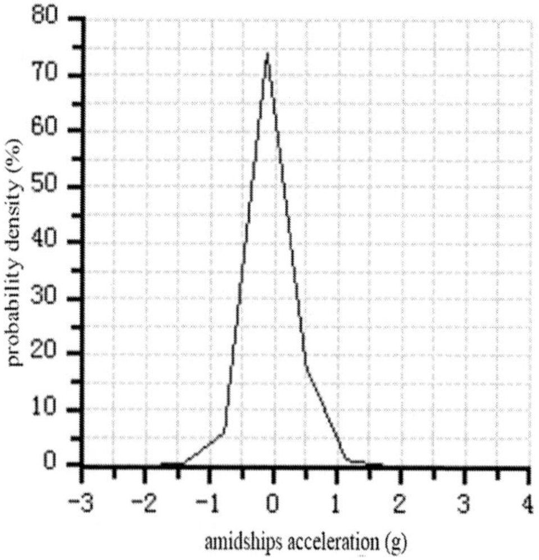

Fig. 1.10.4. Amidships acceleration probability density curve

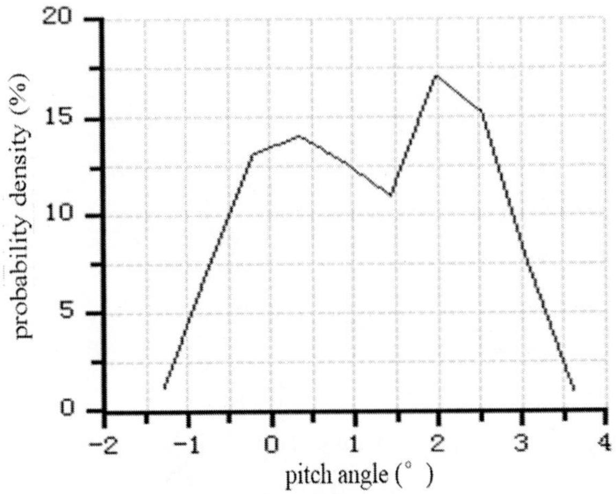

Fig. 1.10.5. Probability density curve of pitch angle

We can see from the above that the ship has very small motion response in waves; therefore, it can greatly improve the ship's seaworthiness in waves.

1.10.5. Experimental ship model resistance

For better rapidity performance and sea-keeping, the hull in the middle and rear part has been designed with a deep-V hull. In order not to undermine its hydrodynamic characteristics, axe bow ships are usually more slender and light, which means that compared to the same length conventional ship the available space is smaller—a 50m-long axe bow ship and a 43–44m-long conventional boat have almost the same carrying capacity. But this can increase the wet surface area of the hull, thereby increasing the friction; it can be seen from the heave curve in the following diagram. Due to the bow buoyancy being smaller than that of a conventional ship, the ship has a tremendous trend under the bar. It also increased the frictional resistance at low speeds, but this phenomenon is ameliorated at high speed by the course of the dynamic lift effect.

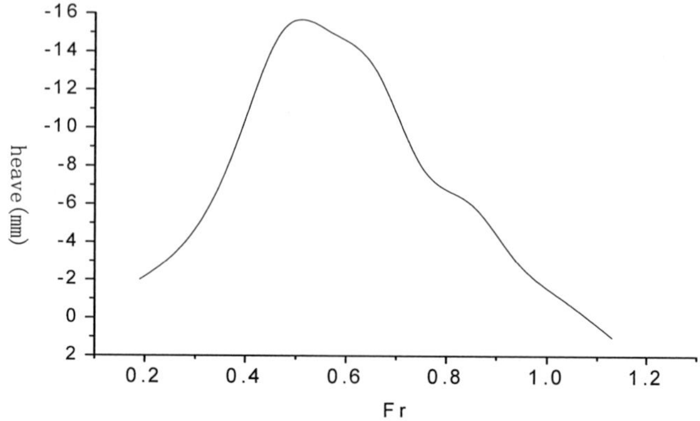

Fig. 1.10.6. Heave curve of axe bow ship model in still water

The residual resistance curve of an axe bow ship translated from model experimental data is shown below:[1-10-5] [1-10-6] [1-10-7]

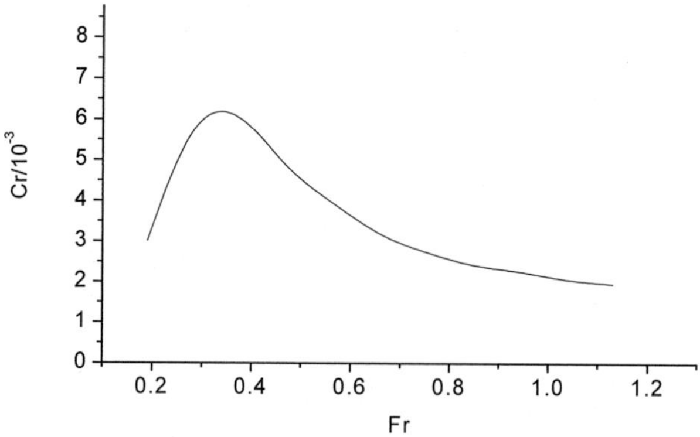

Fig. 1.10.7. Curve of residual resistance coefficient of the axe bow ship

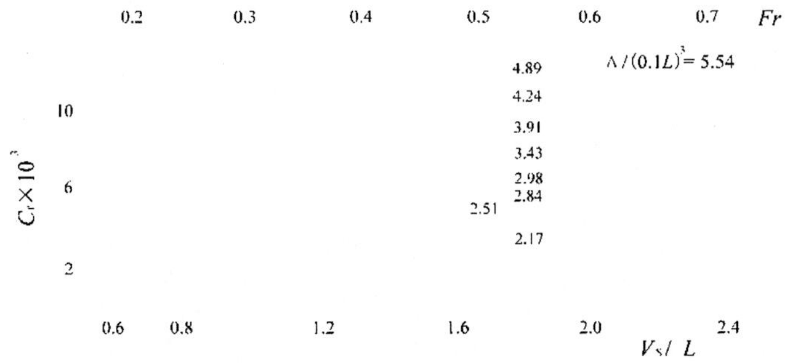

Fig. 1.10.8. Residual resistance coefficient curve of Nordstrom U-type ships

Because the displacement length coefficient is an important parameter for the residual resistance coefficient (the coefficient is composed of Δ and L, in the case of constant displacement), the increase of length can reduce the residual resistance. High-speed displacement vessels are generally U-shaped (round bilge type ship) and V-shaped (knuckle-type ship). Considerable research data shows that the round bilge type of resistance has better performance than the knuckle type in the same displacement length coefficient from the view of resistance. To assess the resistance performance of the axe bow ship, we take the ship coefficient of a typical Sweden Nordstrom U-boat as a reference of residual resistance.[1-10-5] The C_∇ of the experimental ship model is 1.97; with the interpolation of the Nordstrom curve, the figure shows that compared with the Nordstrom U-type ships, the axe bow ship reaches its resistance peak at a lower Fr, but the residual resistance coefficient drops rapidly with the number of Fr larger than 0.7, which is the most suitable time for ship sailing. With a small Fr, the length of the ship has less influence on residual resistance, which was not clear at this time as it was dominated by friction, so there is no advantage to the ship at low speed.

In 2006, the Dutch started the offshore patrol vessel project and the Netherlands Defence Materiel Bureau made a comparative analysis of bulbous bows and axe bows. It found that the former had better performance in calm sea conditions, but there was no advantage in severe sea conditions. As the speed of patrol ships is not high, the Dutch-class patrol boats finally adopted the traditional bow shape.

1.10.6. Conclusion

1. The Axe bow boat is a new type of high-performance ship that has unparalleled advantage in sea-keeping performance in severe sea conditions.
2. Compared with the Nordstrom U-type boat's residual resistance coefficient, which has excellent resistance performance, the resistance of axe bow ships not only does not deteriorate but also has certain advantages when the Fr is greater than 0.7.
3. This is because the hull has a large wet surface area and for some other reasons. In the case of a small Fr, the resistance performance is not ideal; we can conclude this because the Netherlands' ultimately chose a traditional bow for its patrol ships.

References

[1-10-1] Lu Qiang, "Dutch High-Performance Cutting Edge Bent Bow," *Weapon of Knowledge*, 2009.

[1-10-2] Wu Nanhuo, "Hull Lines Design of a 45m High-Speed Platform Vessel based on Axe Bow Concept," Shipbuilding of Guangdong K 2009, "Grinding the Bow" or "How to Improve the Operability of Fast Monohulls," *International Shipbuilding Progress* 53 (2006): pp. 281–310.

[1-10-4] J. L. Gelling & Yue Weiyang, translation, "Sea Axe Bow Yacht"—Hull New Trend.

[1-10-5] Shao Shiming, Zhao Lian'en, & Zhu Nianchang. "Ship Resistance," Beijing: National Defence Industry Press, 1995.

[1-10-6] Jia Jingbei, Zong Zhi, Ni Shaoling, Shi Huaqiang, Tang Xiaoyu, Zheng Jinwei, & Gao Yun. "Resistance Study on Trimaran Model Test," *Ship Mechanics*, 2009.

[1-10-7] Li Peiyong, Qiu Yongming, Gu Mintong, & Wang Wenfu, "Trimaran Model Test of Resistance," *Shipbuilding of China*, 2002.

[1-10-8] Zhao Lian'en & Han Duanfeng, *Hydrodynamic Theory and Design of High Performance Ship*, Heilongjiang: Harbin Engineering University Press, 2007.

[1-10-9] "Sea Axe Bow Yacht—Axe Type Bow: A New Trend Yacht," *Yacht*, 2009.

1.11

An Axe Bow Ship of Excellent Sea-Keeping (2012)

Wu Qingwei, Zeng Qing, Gao Xiaopeng and Dong Zushun

Naval University of Engineering, China

1.11.1 Introduction

Ship transport is one of the oldest forms of transportation. For the economic prosperity of humankind, it made a great contribution to the progress of civilization. A long time ago, people began to study the performance of ships in still water, especially fast sailing performance. However, a large number of facts indicate that a ship that performs excellently in still water is not necessarily good in waves.[1-11-1] As ships often sail in rough seas, sea-keeping became an important criterion to measure the sailing performance of modern ships. Its pros and cons were directly related to the livability and navigation of ship security. So more and more attention had been paid to studying the sailing performance of ships in waves, and designing a ship with good sea-keeping has being a goal pursued by many people. It is useful to find a ship design that can quickly improve the performance of the hull, without sacrificing speed and increasing operating costs. The Damen Shipbuilding Group in cooperation with Delft University of Technology Department of Ship Hydrodynamics in the Netherlands first proposed an "Enlarged Ship Concept," which referred to the ESC concept boat. On the basis of the ESC concept boat, they created two new hull forms: one is a high-speed wave-piercing concept boat (WPC), which can pierce waves so as to limit vertical acceleration; the other is an axe bow concept ship (ABC).[1-11-2] This ship's bow looks like a sharp blade, and is named after the silhouette of an axe.

The axe bow ship emerged in recent years. As a new type of ship its sea-keeping performance has greatly improved. In 2003, the Damen Project Study Group tested three models—ESC concept ship, WPC, and

ABC—in waves, ABC's highest level of vertical acceleration is 50% lower than the ESC concept ship, while the ship's vertical acceleration of the ESC concept is 50% lower than a conventional ship.[1-11-3] These comparison results have a close relationship with the characteristics of the axe bow ship. The axe bow ship's main characteristics[1-11-4] can be listed as follows:

1. The bow is extremely sharp, and each section's floating angle is very small or almost zero
2. The sidewall of the bow is almost vertical; the waterline is basically the same from top to bottom
3. The waterline width of the bow is very small, the entire bow section forms a very narrow V-shape, and the largest draft is at the lowest point of the first column
4. The first column has been extended to a considerable depth below the baseline; the bow can easily cut into the water and it is rarely carried out of the water
5. The bow is generally high and very steep; it has increased buoyancy when the bow pitching is less and does not gradually increase

1.11.2. Model experiment

To study the hydrodynamic performance of the axe bow ship, we made a number of improvements based on the original model. First, to reduce the hull's longitudinal roll, we installed a trim tab in the stern at the bottom of the boat. Second, to prevent the splash of water from being too high and improve the resistance properties of the hull, two parts of splash bars were installed on the bow above the waterline. Third, to improve the lateral stability of the hull, we abated the depth of the vertical stem appropriately, which extended to a depth below the baseline. The wave test in regular waves was carried out in the Aviation Industry Corporation of China Special Vehicle Research Institute. The towing tank of the laboratory is 510m × 6.5m × 5.0m, the trailer speed range is 0.1–22m/s, and it has speed stability and accuracy of 0.2%. The wave is set to a height of 50mm; during the series of wavelengths tests, the speed was 4 m/s and 6 m/s. The axe bow ship model parameters are shown in Table 1.11.1.

Table 1.11.1. Axe bow ship model parameters

Displacement /kg	Captain of ship/m	Width of ship/m	Waterline length/m	Waterline width/m
88	3.22	0.71	3.12	0.59

Fig. 1.11.1. Axe bow ship model

1.11.3. Sea-keeping data analysis

1. The sea-keeping prediction method of real ships

In this paper, we use the spectral analysis method to predict the performance of the axe bow ship in irregular waves. The ITTC two-parameter spectral formula was applied in its general form as:

$$S_{\xi}(\omega) = \frac{A}{\omega^5} \exp\left\{-\frac{B}{\omega^4}\right\}$$

Where: $A = \frac{173 \bar{\varsigma}^{2}_{w/3}}{T_1^4}$, $B = \frac{691}{T_1^4}$

$$T_1 = 2\pi \frac{m_0}{m_1} = 3.86\sqrt{\bar{\varsigma}_{w/3}}$$

Spectral centroid cycle:

m_0、m_1 : Spectral density function of 0 and first-order spectral moments.[1-11-6]

According to the calculation procedure of ship swinging in the literature,[1-11-7] the axe bow ship's double amplitude statistics in irregular waves were forecast, such as longitudinal roll, heave, and vertical acceleration.

2. Comparison of axe bow ships and conventional ships

To analyze the sea-keeping performance of the axe bow ship in irregular waves, we use the spectral analysis method to transform the experimental data on regular waves and the significant wave height of 2m of irregular waves. We then compare the forecast data with the data of a scale conventional ship; the comparative results are as follows:

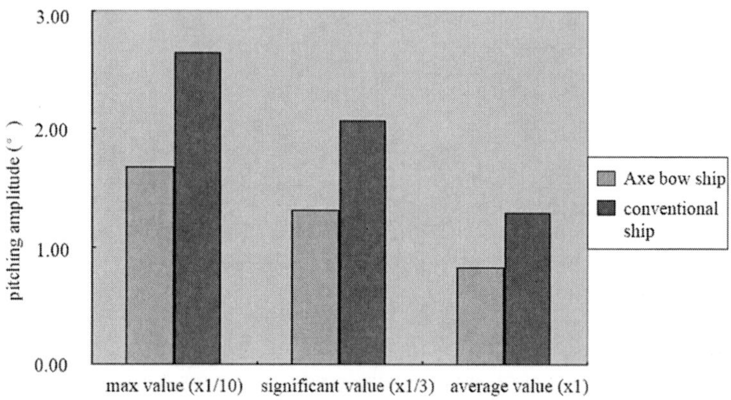

Fig. 1.11.2. Pitching amplitude contrast

It can be seen from Fig. 1.11.2 that the axe bow ship's pitching amplitude is reduced by about 37% compared with a conventional ship.

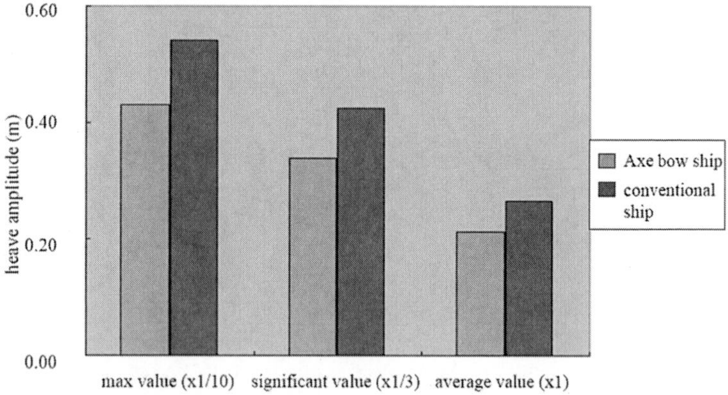

Fig. 1.11.3. Heave amplitude contrast

It can be seen from Fig. 1.11.3 that the axe bow ship's heave amplitude decreases by about 20% compared with a conventional ship.

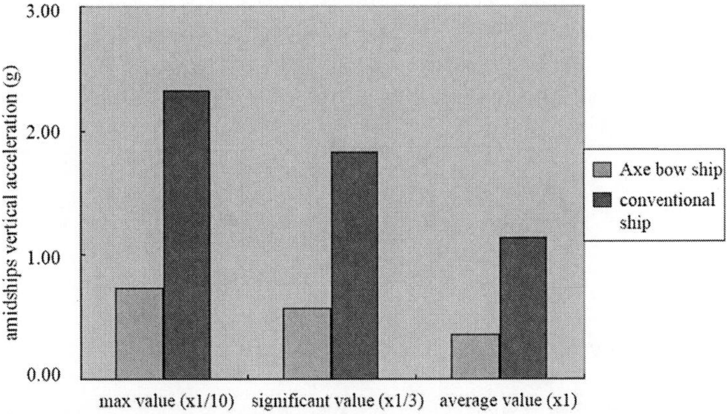

Fig. 1.11.4. Amidships vertical acceleration contrast

It can be seen from Fig. 1.11.4 that compared with conventional ships, the axe bow ship's amidships' vertical acceleration is reduced by about 68%.

In the comparison between these three aspects, the axe bow ship showed clear superiority. Referring to the resistance performance, the axe bow ship hull is relatively slender. It not only increased the length of the waterline of the hull, but also increased the wet surface area. And the hull's frictional resistance was increased. When it sails at a low speed, friction is the dominant factor in the total resistance, thus the resistance properties of axe bow ships are, relatively, little worse. However, when it sails at a higher speed, especially with a Fr greater than 0.7, the axe bow ship's resistance performance would be greatly improved, making it even more excellent than the resistance performance of Nordstrom U-shaped boats.[1-11-4]

1.11.4 Conclusion

1. As a new type of high-performance ship, the axe bow ship has excellent sea-keeping performance in head sea conditions. Compared with conventional ships, the axe bow ship's pitching

amplitude is reduced by 37%, heave amplitude is reduced by 20%, and amidships' vertical acceleration is reduced by 68%.

2. When the axe bow ship sails at a low speed, its resistance properties are little worse, but when it sails at a higher speed, especially with a Fr greater than 0.7, the axe bow ship's resistance performance would be greatly improved, making it even more excellent than the resistance performance of Nordstrom U-shaped boats.

References

[1-11-1] Wu Xiuheng, *Maneuverability and Seakeeping of the Ship*, Communications Press, 1999.

[1-11-2] J. L. Gelling, Yue Weiyang, trans., "Sea Axe How—Hull New Trend," 2009.

[1-11-3] Lu Qiang, "Dutch High-Performance Cutting Edge Bent Bow," *Weapon of Knowledge*, 2009.

[1-11-4] Wei Song, Zeng Qing, Gao Xiaopeng, & Dong Zushun, "Model Experiment Investigation on Resistance of Axe Bow Ship," The 16th China International Boat Show & High Performance Marine Vehicles Conference 2011.

[1-11-5] J. A. Keuning, "The Nonlinear Behavior of Fast Monohulls in Head Waves." PhD thesis, TU Delft, 1994.

[1-11-6] Sheng Zhenbang & Liu Yingzhong. *Principle of the Ship*, Shanghai Jiaotong University Press, 2010.

[1-11-7] Feng Tiecheng, Zhu Wenwei, & Gu Shuhua, *Ship Maneuvering and Sway*, National Defence Industry Press, 1989.

[1-11-8] Wu Suhuo. "Hull Lines Design of a 45m High-speed Platform Vessel Based on Axe Boat Concept," *Design & Development*, 2009.

1.12

Hydrodynamic Analysis of the Performance of Stern Flaps in a Semi-displacement Hull (2003)

Marcos Salas and José Rosas

Institute of Naval and Maritime Sciences and Faculty of Engineering Sciences, University Austral of Chile

Nomenclature

L_f	Flap chord length
α	Flap angle
L_{PP}	Length between perpendiculars
B_T	Transom beam
Δ	Ship displacement
τ	Trim, in degrees
Vs	Speed, in knots
F_N	Froude number
$F_N\nabla$	Displacement Froude number
C_R	Residual resistance coefficient
λ	Scale factor
η_h	Hydrodynamic efficiency

1.12.1. Introduction

A stern flap is an appendage built in the form of a plate that extends aft of the transom in an angle relative to the buttock plane of the ship. Its interaction with the hull modifies the ship running trim, reduces propulsion resistance, and increases maximum attainable speed. The critical parameters for the stern flap geometry design are: chord length (L_f), flap angle (α) referenced to an extension of the local buttock plane angle, and flap span across the transom.[1-12-1] Fig. 1.12.1 shows these critical parameters in a stern flap of a military vessel. Secondary design aspects are: determination of platform shape, transverse thickness variations, and the detailed fairing into the hull with special attention to the outboard edges. Generally, a simple radius corner is used to simplify design and construction.

The principal benefits that stern flaps produce are: reduction in powering resistance (experience has shown that this reduction is between 5% to 12), an increase in maximum attainable speed, achieved beneficial propulsion interaction, and the modification of transom wave systems.

The reasons for the beneficial interaction due to the presence of a stern flap are different, depending whether the flaps are fitted on a small craft or on a larger vessel. All stern flaps, independently of the size of the vessel, cause a force aft of the transom, modify the ship trim, and modify the pressure distribution beneath the aft portion of the hull.[1-12-1] On small planing crafts, a stern flap affects the running trim angle by 4° to 5°.[1-12-2] This variation is the principal reason for the reduction in resistance on this type of hull.

In contrast to the planing hull case, a stern flap affects the trim angle by 0.1° to 0.3° on displacements vessels. This amount of trim change does not produce significant reductions in resistance. The principal powering benefits on these vessels are attributable to the induced change in the flow field around the propeller. This flow field change reduces the drag on the stern zone and modifies the wave resistance of the ship.[1-12-1]

1.12.2. Hydrodynamic principles

The hydrodynamic phenomena that are present on a vessel due to the presence of a stern flap are a subject under discussion. Thorough the combination of modern techniques of numerical flow computations and model experiments, these phenomena have been identified, and they should be viewed only as a general representation. The principal hydrodynamic mechanisms are discussed in references 1-12-1, 1-12-3, and 1-12-4:

1. Afterbody flow modifications

This causes the principal enhancement performance on a displacement hull. The stern flap reduces the flow velocity and increases the dynamic pressure under the hull. Model experiments have shown that this flow variation can affect an extension of 5% to 15% of L_{PP} in the stern zone. Furthermore, the increment in the pressure field causes a greater lift force that produces a positive component in the movement direction of the ship.

The stern flap increases the exit flow velocity just under its trailing edge in comparison with the transom without a flap. This increased velocity reduces the flow separation velocity and promotes a cleaner flow separation. All this produces a reduction of the viscous pressure resistance.

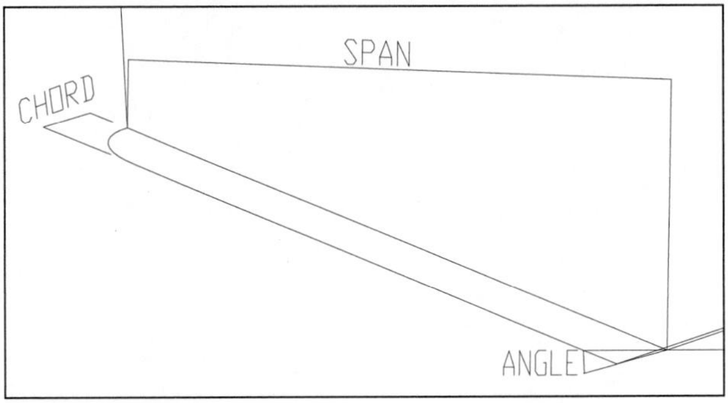

Fig. 1.12.1. Characteristics of stern flaps

2. Wave system modifications

Localized flow around the stern zone without a stern flap is affected by lost energy due to the presence of various mechanisms, such as eddy making, turbulence, and generation of "white water," which may not propagate to the far-field energy zone. At equivalent speeds, a stern flap affects the stern zone flow by reducing height, slope, wave breaking, and the amount of "white water" in the trailing waves. Model experiments have shown a reduction of wave heights in the near-field stern wave and far-field wave energy systems. Fig. 1.12.2 shows differences on the transom wave system with and without the presence of a stern flap on a patrol craft.

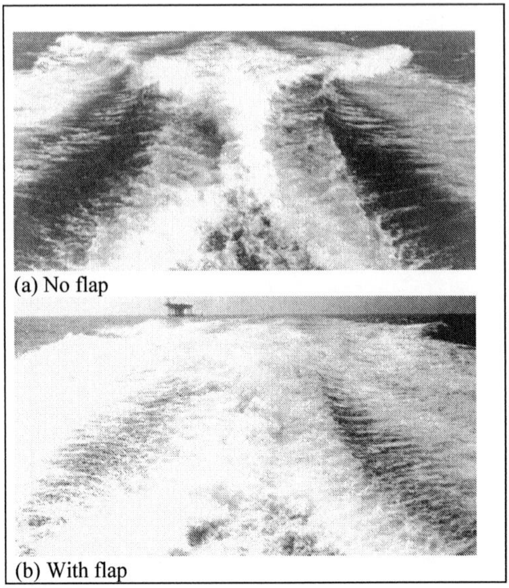

Fig. 1.12.2. Transom wave system with and without a stern flap. Courtesy: US Navy

3. Lift and drag forces
A stern flap is a producer of lift and drag forces at all speeds and conditions. However, the beneficial interaction with the hull and propeller reduce the ship powering resistance. Drag and lift forces increase with increasing chord length, span, angle (trailing edge down), and depth of immersion. Lift forces are greater than drag in all speed conditions.

4. Secondary hydrodynamic effects
Other important mechanisms are ship length increase, beneficial propulsion interaction, reduced ship sinkage, and running trim modification. These last two mechanisms are the key to maximizing performance on planing hulls.

1.12.3. Design considerations

Stern flaps are a viable mechanism to enhance powering performance of ships that operate at speeds corresponding to F_N>0.2.[1-12-3] At low speeds, there is a penalty in powering resistance and the stern flap is totally submerged and wet. When the speed increases, the flow is transitional, periodically breaking free of the transom and flap and then coming back to wet it again. At high speed, the flow detaches cleanly and breaks away from the trailing edge of the flap producing a significant reduction in resistance.[1-12-5]

Generally, a longer chord or an increase in flap angle produces greater efficiency at high speeds, but the penalty at low speeds increases and the crossover speed point is delayed to a greater ship speed.[1-12-6] The design of a stern flap is a compromise of the operational and economical requirements of a vessel, and the selection of the critical parameters must be studied in all speed ranges. It is recommended to design stern flaps that reduce resistance over the speed/time/mission profile.

Several stern flap configurations have been tested for US Navy military vessels at DTMB. They recommend designing flaps within the following ranges:[1-12-] [1-12-7]

- Chords between 0.5 to 2.5% L_{PP}
- Flap angles between 10° trailing edge upward (negative angles) to 20° trailing edge downward. To the authors' knowledge, there are no records of stern flaps being installed with negative angles in full scale ships.
- Span must extend the maximum reasonable width across the transom, without affecting the wake of the corners and without requiring significant curvature around the bilge radius. In some cases, it has been found that a span approaching half of the transom width is optimal.

1.12.4. General description of the vessel

To assess the hydrodynamic performance of the existent flap design recommendations, six stern flap configurations were designed within the recommended ranges and model tested on a semi-displacement hull. The chosen vessel was the pilot craft "Aguja" of the Panama Canal Authority, a V-hull with twin shafts and 29.5″ (75 cm) diameter fixed-pitch propellers. Hull sections of the craft are shown in Fig. 1.12.3; main dimensions are given in Table 1.12.1.

Table 1.12.1. Main dimensions of the tested vessel

Length between perpendiculars (L_{PP})	15.14m
Length of waterline (L_{WL})	14.89m
Beam of waterline (B_{WL})	4.00m
Draught (T_m)	0.83m
Displacement (Δ)	20.22t
Deadrise angle	18°
Top speed (V_s)	20kn

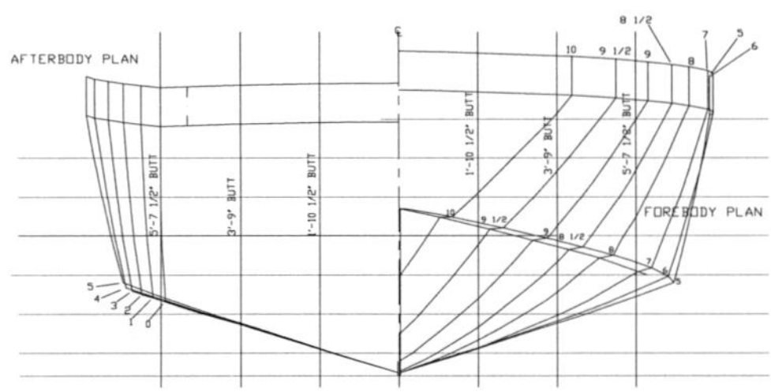

Fig. 1.12.3. Sections of tested craft

The craft has a bottom extension on the transom. This extension works as a stern flap, and its principal geometric characteristics are given in Table 2.12.2.

Table 1.12.2. Characteristics of existent stern flap

Chord length (L_f)	0.15m (1.0% L_{PP})
Span	3.66m (100% B_T)
Angle (α)	0°

1.12.5. Description of model and experimental procedure

A wood model, of scale factor $\lambda = 16$, representing the pilot craft *Aguja* was constructed for this project. No appendages were installed on the model, and all the experiments were done to the full load displacement of 20.22 tons. Experiments were conducted for the conditions of the hull with no flap, with six different stern-flap designs, and with the existent bottom extension to measure the actual hydrodynamic performance of the vessel, in the speed range of 14kn to 22kn.

Tests were carried out in the towing tank of the Institute of Naval and Maritime Sciences at the University Austral of Chile.

Results from the model experiments were extrapolated to the full scale craft applying the Froude Method and the ITTC-57 ship model correlation line's correlation allowance of ATTC of 0.0004 over the friction coefficient of the prototype and 0.0 over the friction coefficient of the model. All data presented in this investigation corresponds to the full scale craft, floating in smooth, deep salt-water conditions, with a uniform standard temperature of 15°C.

To induce turbulent flow along the length of the model hull, 12 turbulence stimulator studs (1/8″ diameter) were placed aft of the stem at approximately 10% of the waterline length, spaced 1.4cm apart, according to Hughes's recommendations.

1.12.6. Stern flap series selection

The tested stern flap configurations were designed following the existing recommendation about the ranges of effectiveness of the critical design parameters. Each flap design was tested in three angles: at 0, 5, and 10 degrees. The principal dimensions of the tested configurations are given in Table 1.12.3.

Table 1.12.3. Principal dimensions of tested configurations.

Flap #	Chord /m	Chord /% L_{PP}	Span /m	Span /% B_T	Area /m^2
1	0.38	2.5	3.66	100	1.33
2	0.38	2.5	1.83	50	0.64
3	0.23	1.5	3.66	100	0.82
4	0.23	1.5	1.83	50	0.40
5	0.078	0.5	3.66	100	0.28
6	0.078	0.5	1.83	50	0.14

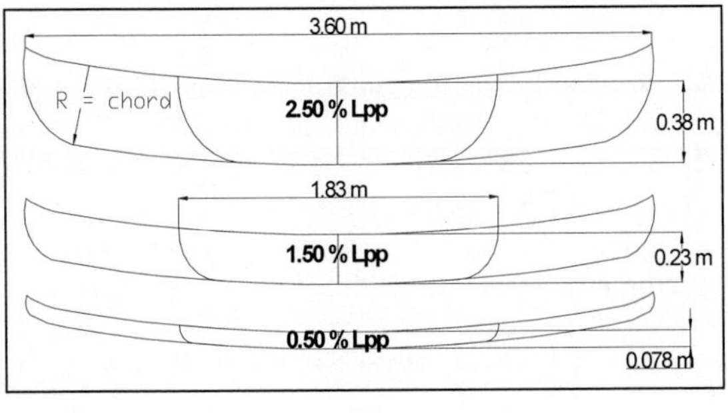

Fig. 1.12.4. The geometry of the tested stern flaps

The selection criteria for the optimum flap of the analyzed vessel are: (1) to maximize reduction in ship resistance in the speed range of 20–22kn; (2) to avoid any increase in ship resistance in the speed range of 14–20kn.

1.12.7. Stern flap performance

1. Resistance

The curves of hydrodynamic resistance of the vessel obtained from the model tests are shown in Figs 1.12.5 and 1.12.6, with the results expressed in the non-dimensional form of resistance divided by displacement (R/Δ). In each figure, the resistance curve of the hull without flap was drawn to compare it with hull-plus-flap conditions. It can be seen that the curves are basically similar and the presence of stern flaps reduced considerably the resistance at a wide range of speeds. All tested flaps had a good performance at low speeds.

Generally, for both spans, flap angles of 10° were the most efficient for all tested chords at low speeds. At the same time as the speed increases, the performance of lower flap angles increases too; at high speeds, flap angles of 0° were the most efficient, except for the flap of a chord length of 1.5% L_{PP}, which had its best performance at an angle of 5° (Fig. 1.12.5.).

Flaps of spans of 100% B_T were the most efficient at all analyzed speeds. Generally, the tested flaps began to lose efficiency at equivalent speeds of $1.82<F_{N\Delta}<2.15$ (18–22kn).

2. Residual resistance coefficient

Residual resistance includes viscous and wave phenomena representing the highest loss of energy in this type of vessel. It is an important aspect in the study of the hydrodynamic performance of stern flaps. Figs 1.12.7 and 1.12.8 show the residual resistance coefficients curves against speed for all tested flaps.

Generally, all curves have the same trend: higher flap angles reduce a vessel's efficiency at high speeds. Flaps of span 100% B_T produced higher reductions than spans 50% B_T at all analyzed speeds. The greatest reduction was achieved by the flap of a chord length of 2.5% L_{PP}, a span of 100% B_T, and an angle of 0°, at an equivalent speed of F_N= 1.95 (19.5kn), reducing it approximately 9.75%.

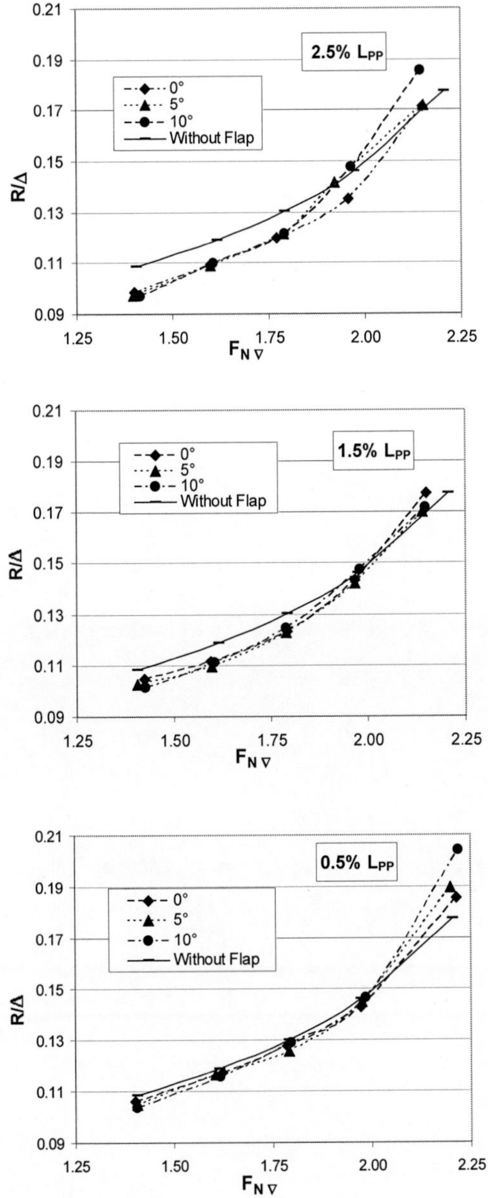

a)

b)

c)

Fig. 1.12.5. Resistance/displacement—flaps of span 100% B_T

At high speeds, the higher reductions are achieved by flaps with an angle of 0°. The only exception was the flap of a chord length of 1.5% L_{PP}, a span of 100% B_T, and an angle of 10° (Fig. 1.12.7.b), which changes its trend at F_N~2.0, increasing its performance but producing a higher C_R than the hull with no flap.

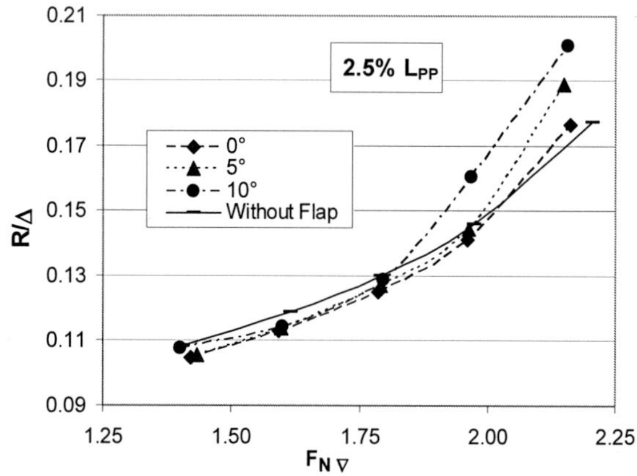

a)

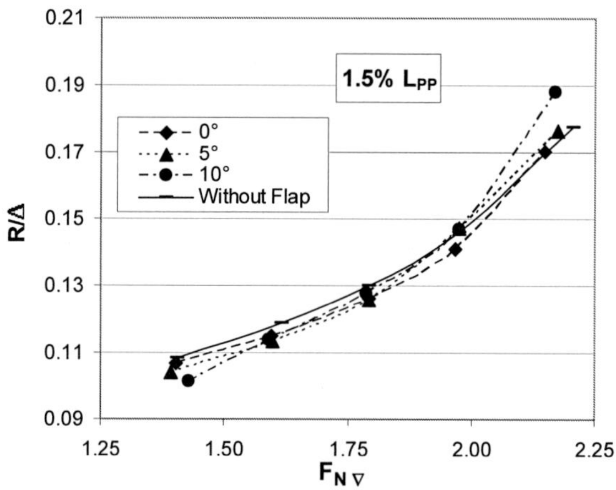

b)

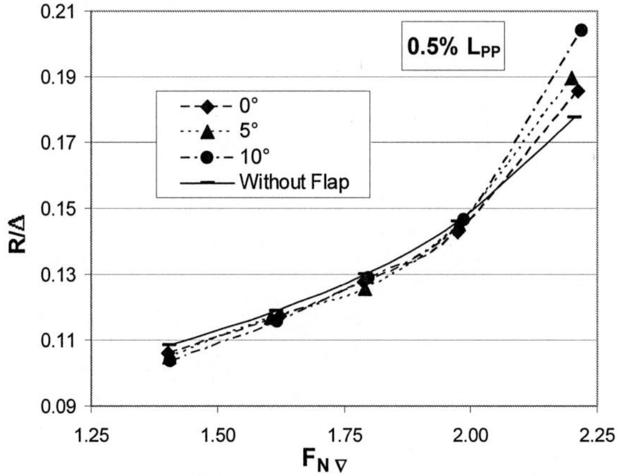

c)

Fig. 1.12.6. Resistance/displacement—flaps of span 50% B_T.

For spans of 50% B_T, the lowest values of C_R were produced by flaps at angles of 0° at equivalent speeds of $2.05<F_N<2.15$ (20–21.5kn), and the highest reduction was in the order of 6.75%, at an equivalent speed of F_N–1.4.

Flap efficiency values are closely similar to C_R reductions. Table 1.12.4 gives the optimal flaps for both spans at all speed ranges.

Table 1.12.4. Optimal flaps in all speed range

	Span 100% B_T			Span 50% B_T		
V/ Kn	L_f /%L_{PP}	angle /deg.	η_h /%	L_f /%L_{PP}	angle /deg.	η_h /%
14.5	2.5	5	9.5	1.5	10	6
16	2.5	5	7.7	1.5	5	3.8
17.5	0.5	5	7.8	0.5	5	5.1
19	2.5	0	6.9	1.5	0	3.1
20.5	2.5	0	4.5	1.5	0	2.8

3. Increments in maximum attainable speed

The resistance reduction implies an increment in the maximum attainable speed of the vessel. The greatest increment was achieved by the flap with a chord length of 2.5% L_{PP}, a span of 100% B_T, and an angle of 5°. Tables

1.12.5 and 1.12.6 give the increment of speed in all analyzed speeds for both studied flap spans.

Table 1.12.5. Increments of speed for flaps 100% B_T

V no flap/kn	Flap type	V with flap/kn	Increment/%
14.5	2.5% L_{PP} 5°	16	12.1
16.0	2.5% L_{PP} 5°	17.32	8.3
17.5	0.5% L_{PP} 5°	18.35	4.9
19.0	2.5% L_{PP} 0°	19.9	4.7
20.5	2.5% L_{PP} 0°	20.85	1.7

Table 1.12.6. Increments of speed for flaps 50% B_T

V no flap/kn	Flap type	V with flap/kn	Increment/%
14.5	1.5% L_{PP} 10°	15	6.8
16	1.5% L_{PP} 10°	16.78	4.9
17.5	2.5% L_{PP} 5°	17.89	2.2
19	1.5% L_{PP} 0°	19.47	2.5
20.5	1.5% L_{PP} 0°	20.79	1.4

The selection criteria for the optimum flap have been established previously. Considering these requirements, the flap of a chord length of 2.5% L_{PP}, a span of 100% B_T, and an angle of 0° was selected as the optimum flap of the tested configurations.

4. Ship running trim

Figs 1.12.9 and 1.12.10 shows the trim angle curves of the tested flaps. All stern flaps reduced the trim angles (bow down) considerably, except for the flap with a chord length of 0.5% L_{PP}, a span of 50% B_T, and an angle of 0°.

Trim angle curves have the same trend for flaps with a span of 100% B_T for the analyzed values of chord lengths and angles (Fig. 1.12.9). These flaps produced higher differences in trim change than the flaps with a span of 50% B_T. The greatest change was produced by the flap of a chord length of 2.5% L_{PP} at an angle of 10°, affecting the original trim by 1.9° and giving a bow down trim of 1.25°.

Different trends can be observed for flaps of a span of 50% B_T (see Fig. 1.12.10). All flaps of chord lengths of 2.5% L_{PP} increased the trim angle (bow up) at equivalent speeds of $F_N<1.65$, giving a bow down trim at higher speeds. The remainder of the tested flaps reduced the trim angle (bow down) at all speeds. The greatest change in trim was achieved by the flaps of chord lengths of 2.5% and 1.5% L_{PP} at 10°, giving a bow down trim to 0.5° and affecting the original trim by 1.15°.

Comparing these results with resistance analysis, it can be seen that no similarity exists between the trend of the trim changes and resistance reductions. This indicates that both parameters are independent in the hydrodynamic performance of this vessel. It is evident that variations in resistance are influenced by flow changes in the stern zone.

1.12.8. Actual hydrodynamic performance of the vessel

Additional experiments were done for the condition of the hull with the existent bottom extension, in order to measure the actual hydrodynamic performance of the vessel. The results were compared with the efficiency of the conditions of a hull with no flap and with the optimum selected flap (chord of 2.5% L_{PP}, span of 100% B_T, and angle of 0°), as is shown in Fig. 1.12.11. It can be seen that the bottom extension has a good performance at all analyzed speed range, but its efficiency is lower than the optimum flap. However, the performance for this two-condition trend is similar at high speeds and both lose efficiency at approximately F_n–2.1 (21kn). Fig. 1.12.12 shows the trim changes for the three conditions.

1.12.9. Conclusions

Model tests were done to evaluate the hydrodynamic performance of six stern flap configurations on a semi-displacement hull, in the speed range of 14kn to 22kn. The tested vessel was the pilot craft *Aguja*, which is the property of the Panama Canal Authority. The principal geometric characteristics were designed following existent recommendations, specially from DTMB, to compare its effectiveness at the project and design stages.

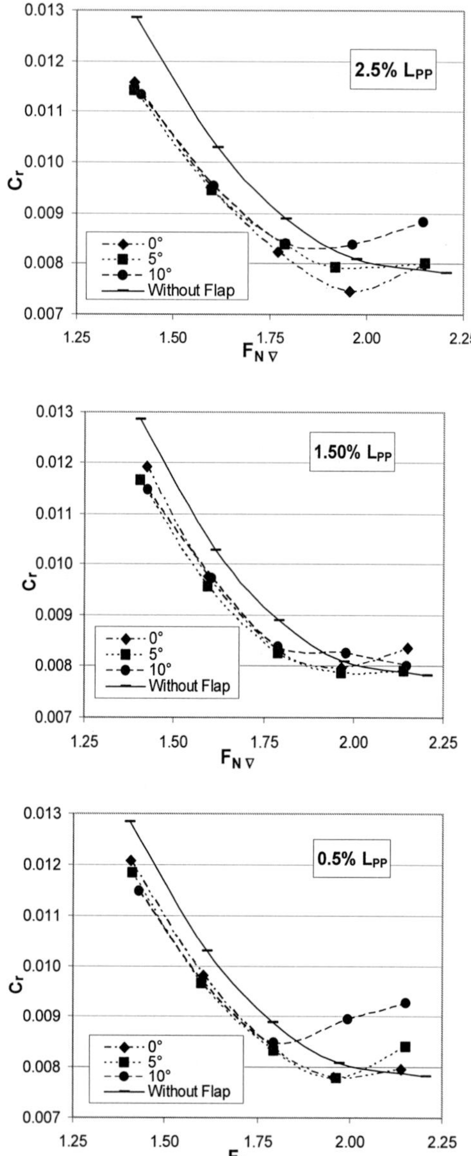

a)

b)

c)

Fig. 1.12.7. Residual resistance coefficients—flaps of span 100% B_T

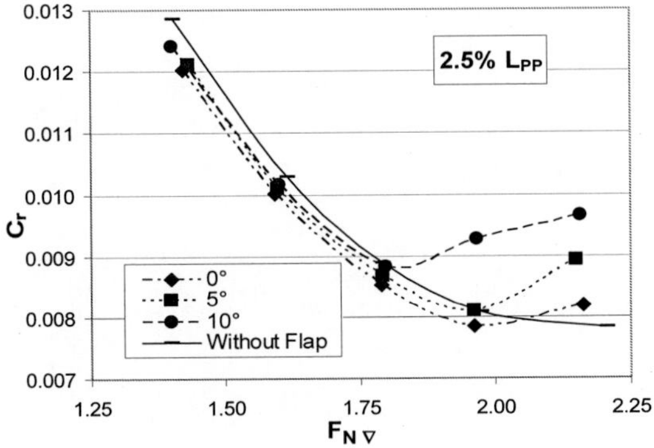

a)

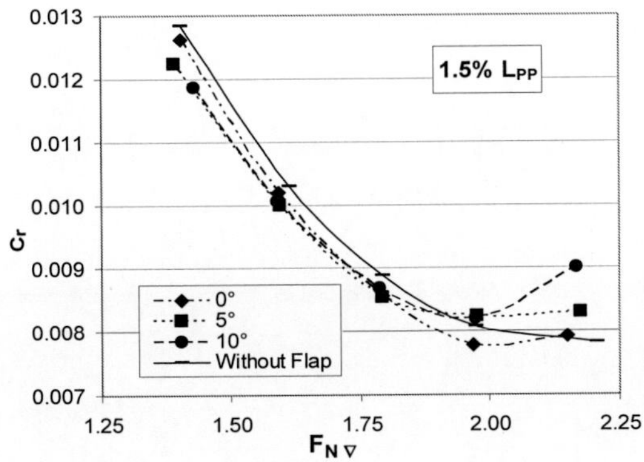

b)

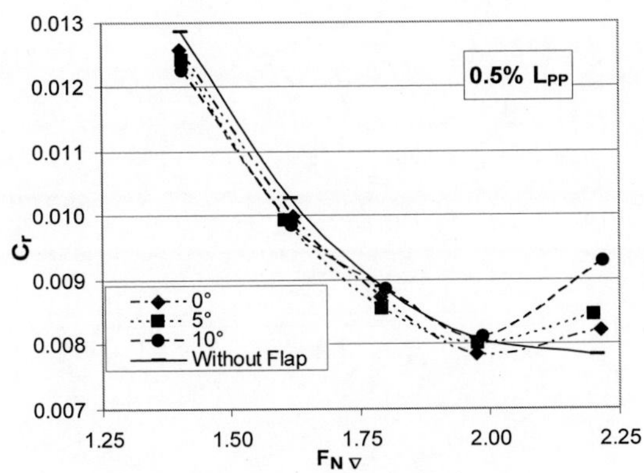

c)

Fig. 1.12.8. Residual resistance coefficients—flaps of span 50% B_T

The results obtained from the model tests show that:

1. All tested flaps reduced resistance at a wide range of speeds.
2. Generally, the loss of efficiency began at equivalent speeds of $1.82<F_N<2.15$ (18–22kn).

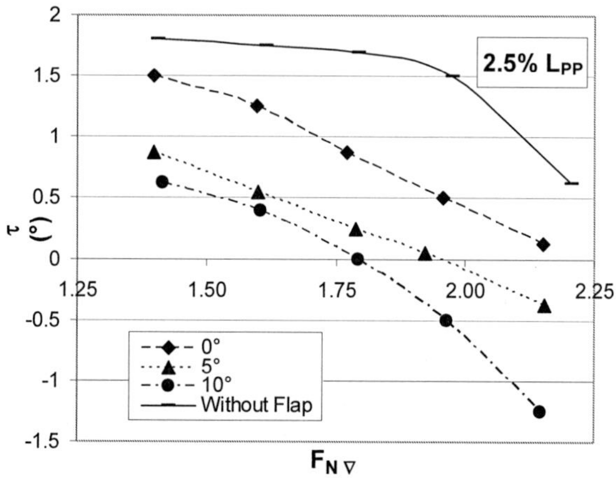

a)

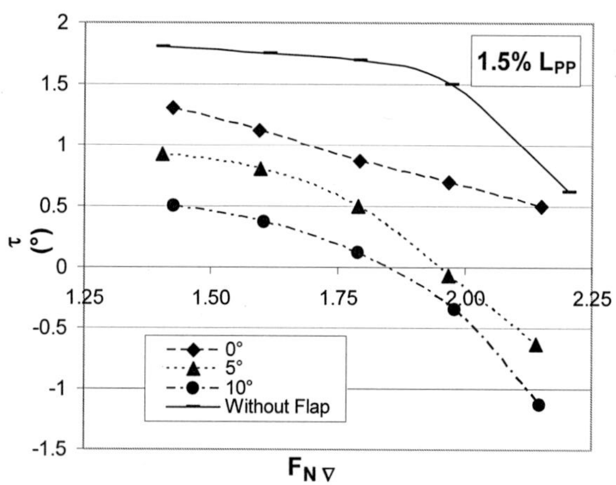

b)

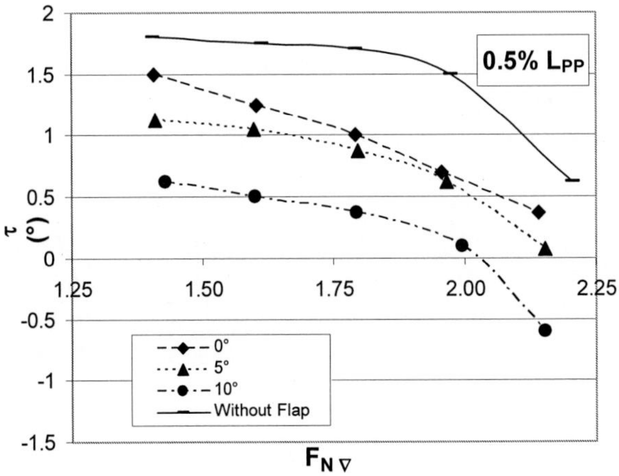

c)

Fig. 1.12.9. Trim changes—flaps of span 100% B_T

3. Generally, higher values of flap angles are the most efficient at low speeds. At the same time as speed increases, lower flap angles increase their performance.
4. Flap spans of 100% B_T were the most efficient at all analyzed speeds.

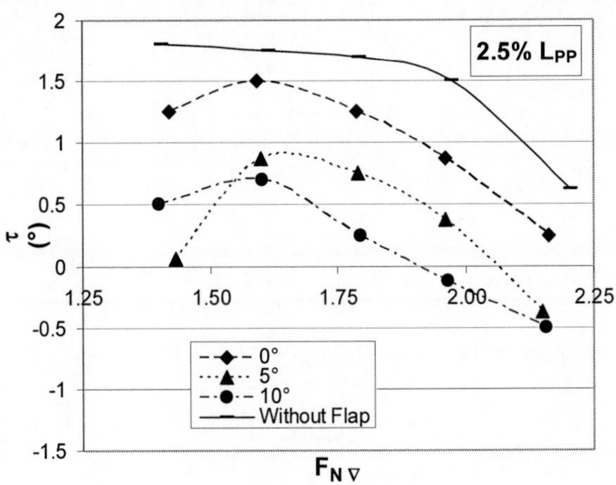

a)

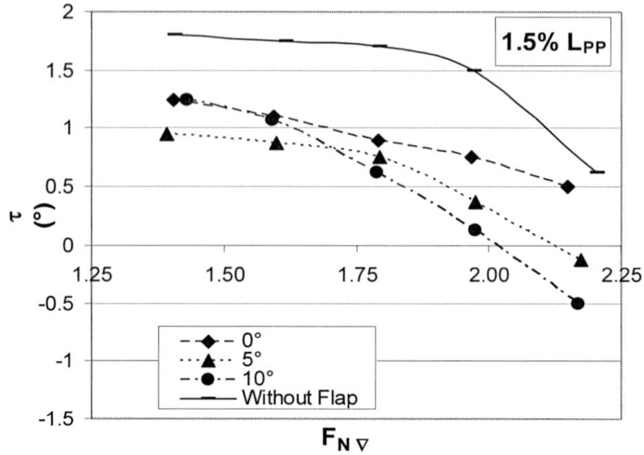

b)

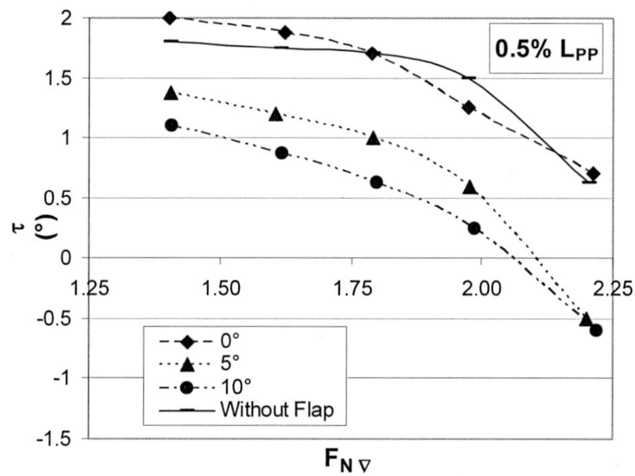

c)

Fig. 1.12.10. Trim changes—flaps of span 50% B_T

5. No relation was established between trim changes and resistance reductions. It is suggested that benefits to hydrodynamic performance are caused by flow changes in the stern zone. It is recommended to study these changes by CFD analysis.

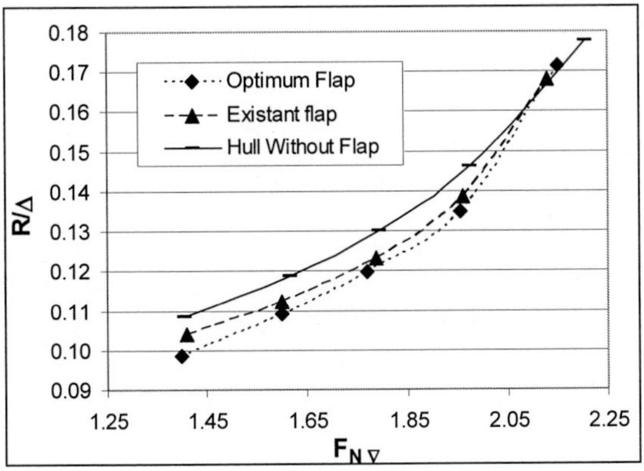

Fig. 1.12.11. Resistance/displacement for optimum flap and existent bottom extension

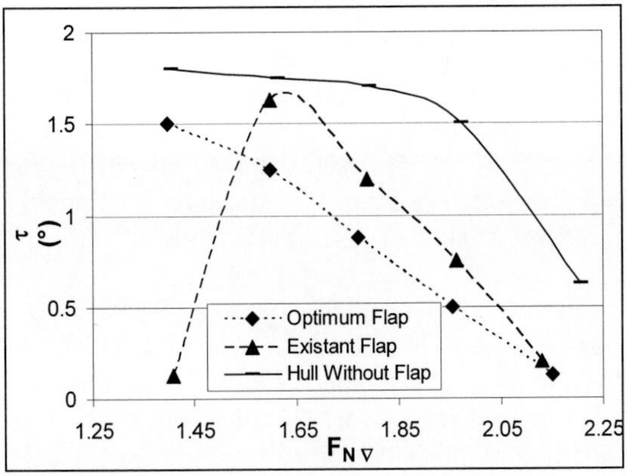

Fig. 1.12.12. Trim changes for optimum flap and existent bottom extension

6. The efficiency of the tested configurations shows a good level of agreement with the existent recommendations. However, according to the trend of resistance curves from model experiments, it is possible that higher values of chord lengths and negative flap angles could produce better hydrodynamic performances. To measure the actual hydrodynamic performance of the vessel, the hull was tested with the existent bottom extension. The results were compared with the conditions of a hull with no flap and with the selected optimum flap. The results show that the extension has a

good performance, but its efficiency is lower than the optimum flap. However, its performance tends to be similar at equivalent speeds of F_n–2.1 (21kn).

Acknowledgments

The authors wish to express their gratitude to Mr Dominic Cusanelli (DTMB) and Dr Adrian Millward (University of Liverpool). Special gratitude is due to Mr Oscar Amaris from the Panama Canal Authority for granting access to information about pilot craft *Aguja*.

References

[1-12-1] D. Cusanelli, G. Karafiath, & Cheng Wen Lin, "Stern Wedges and Stern Flaps for Improved Powering Performance—U.S. Navy Experience," SNAME Annual Meeting, September 29–October 2, 1999, Baltimore, MD.

[1-12-2] A. Millward, "Effect of Wedges on the Performance Characteristics of Two Planing Hulls," *Journal of Ship Research* 20 (4), December 1976.

[1-12-3] D. Cusanelli & G. Karafiath, "Integrated Wedge-Flap for Enhanced Powering Performance," Fourth International Conference on Fast Sea Transportation FAST 1997, July 21–23, 1997, Sydney, Australia.

[1-12-4] D. Cusanelli, "Stern Flap Powering Performance on the PC1 Class Patrol Coastal, Full Scale Trials and Model Experiments?" Patrol '96 Conference, December 1996, New Orleans, Louisiana.

[1-12-5] D. Cusanelli, "U.S. Coast Guard Island Class 110 WPB: Stern Flap Evaluation and Selection (Model 5526)," NSWCCD-50-TR-1999/061, 1999.

[1-12-6] D. Cusanelli & K. Forgach, "Stern Flaps for Enhanced Powering Performance," Twenty-Fourth American Towing Tank Conference, November 2–3, 1995, College Station, Texas.

[1-12-7] D. Cusanelli & G. Karafiath, "Stern Flaps," Professional Boatbuilder, April–May 2001.

[1-12-8] Rosas Rodríguez & José Alberto, "Theoretical and Experimental Analysis of the Hydrodynamic Performance of Stern Flaps on High Speed Vessels" (in Spanish), degree thesis project, University Austral of Chile, July 2002.

[1-12-9] A. Millward & M. Sparkes, "An Investigation of the Performance of a Tri-hedral Planing Hull," IMarE Transactions 109 (3), pp. 301–11, July 1997.

1.13

INTRUDER: A DEVICE TO IMPROVE HULL PERFORMANCES (2006)

A . MANCINI & A . MORICONI

INSEAN—Italian Ship Model Basin, Italy

Nomenclature

AP	Planing bottom area/m2
AR	Aspect ratio
BPT	Transom breadth
*B*PX	Maximum breadth over chines/m
CF	Frictional resistance coefficient
CR	Residuary resistance coefficient
CT	Total resistance coefficient
Fn_{∇}	Volumetric Froude Number
g	Acceleration of gravity/9.803 m/s2
LPR	Projected chine Length/m
LWL	Length of Waterline/m
PE	Effective power/kW
Rn	Reynolds Number
RT	Total Resistance/N
S	Area of wetted surface/m2
Si	Protrusion of intruder/m
V	Speed of the model m/s
β	Deadrise angle of planing bottom/deg.
βT	Deadrise angle at transom/deg.
λ	Scale ratio
∇	Displacement Volume/m3
ν	Kinematic viscosity/N s m-2
Δ	Displacement/N
ϑ	Static trim angle/deg.
τ	Dynamic trim angle/deg.

1.13.1. Introduction

A small vertical thin plate positioned under the transom jutting out the bottom a few millimeters is what in today's naval world is very often called an intruder. The hydrodynamic effect it causes enables it to be used instead of the more ordinary flaps. Like these, it can stabilize the hull and correct the trim, but with the advantage that the performances are often better.

Until now in the literature there has been scarce information about a quantitative evaluation of its advantages. To add something more to what today is known about the hydrodynamic effects caused by this device, various comparative tests with and without intruders have been performed at the Italian Ship Model Basin using three planing and semi-planing models.

1.13.2. Origin and history of intruders

This "intruder" does not yet have a unique definition: sometimes it is also named an "interceptor" or something else. It comes from a device used in aeronautics called a "Gurney flap," after racing car driver Dan Gurney who at the end of the 1960s installed it on his cars to increase ailerons lift. From the 1970s it was studied and developed in aeronautics research,[1-13-1] and the first applications were on the wings of a DC10. It was a vertical plate positioned on the trailing edge of the wing, jutting out a small way, which generally included from 1% and 2% of the wing chord.

Experimental studies[1-13-2–4] have proved that the Gurney flap positioned on traditional profiles can determine, for a specific angle of attach, an increase of the pressure on the face and, above all, an important decrease on the top surface—that is, a high increase of the lift—that for some profiles can exceed 50%.[1-13-5] Besides, if the protrusion of the plate is less than 2% of the chord, the increase of resistance is negligible, because the device is completely dipped in the boundary layer. For the same lift coefficient produced, the increase of Gurney flap protrusion delays separation, causing a further increase of the profile efficiency.

These effects are caused by an alteration of the Kutta-Joukowsky condition (flow uniformity downstream the profile): the Gurney flap produces a little upstream separation region, and two downstream controrotating vortices, and then the chord profile is longer near the trailing edge.

This vorticity, added to the peculiar profile, brings about a higher velocity gradient between the face and the top surface of the profile, and produces an increase of pressure gradient and then of lift.[1-13-6]

A Gurney flap can even be used as a stabilizer: several of them, appropriately positioned, have been put on many helicopter types.[1-13-7] The device has been used in the naval field for many years on deep-V high-speed monohulls, such as the *Acquastrada* ferryboat, as a trim corrector, stabilizer, and backing device for steering the ship; in this use, it has been named an "interceptor." Today, with some changes, it is often used on planing and semi-planing high-speed small vessels to improve their performances; in this use, it is often named an "intruder."

1.13.3. Tested models

At the Italian Ship Model Basin (INSEAN), an intruder has been tested on three models—two standard planing ones and one catamaran—that for simplicity we will call model A, B, and C, with scales of 5.5, 10, and 8, respectively.

The L_{PR}/B_{PX} ratios (projected chine length/maximum breadth over chines) of the first two tested models are typical of a 10–15m-long yacht and a 20m-long large yacht, respectively. We have to underline that these ratios are relative to the planing surface determined by the chine, as is usual for planing hulls. Model C, which is very different from the others, is a particular case, but it has shown similar results.

Table 1.13.1. A, B, and C models' dimensionless parameters

	A	B	C*
L_{PR}/B_{PX}	2.88	4.72	12.09
$A_P/\nabla^{2/3}$	6.15	3.54	-----
β	16°	14.8°	14.5°
β_{TT}	12.7°	7.9°	14.5°
$L_{WL}/\nabla^{1/3}$	4.28	5.65	5.87

* Parameters related to only one hull (L_{WL} related to tests without intruder)

In Table 1.13.1 typical dimensionless parameters are reported for the three models.

During the tests, resistance and heaves have been measured for different speeds in accordance with INSEAN standard procedures.[1-13-8] To make the comparison easier, the obtained results have appropriately been made dimensionless. In terms of speed, the volumetric Froude number has been calculated:

$$F_{n\nabla} = \frac{V}{\sqrt{g\nabla^{1/3}}}$$

where V is speed, g is gravity acceleration, and ∇ is the displacement volume. The resistance has been simply divided by displacement, obtaining the resistance coefficient $R_T/$.

1.13.4. Fixed intruder

1. Experimental data

Model A has been tested from 2.15 to 3.45 volumetric Froude number (without an intruder at ϑ=0°; with an intruder both at ϑ=0° and at ϑ=1.15° [trim by the stern], using the ratio $S_i/L_{PR}=1.03\times10^{-3}$, where S_i is the intruder protrusion).

In Fig. 1.13.1, the R_T/Δ and dynamic trim angleτ curves versus the volumetric Froude number are represented.

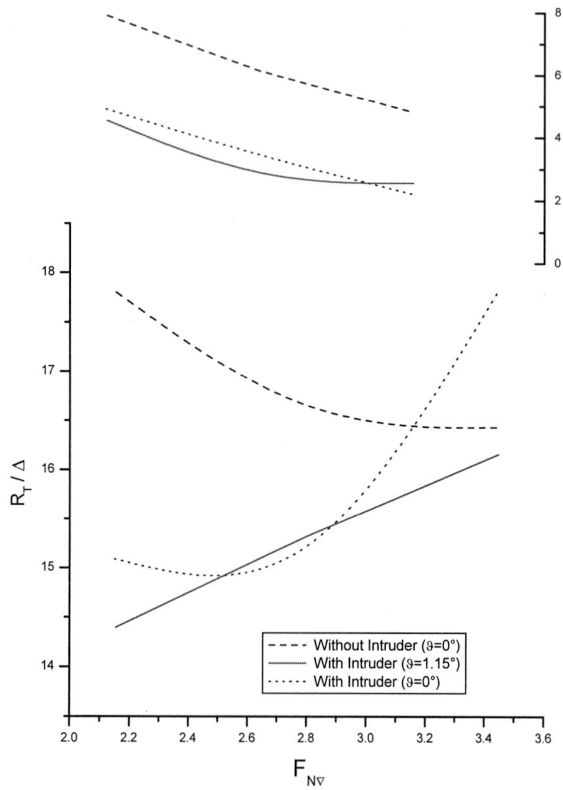

Fig. 1.13.1. Model A—experimental data, with and without intruder

Comparing the two broken lines (ϑ=*0°*), the advantage of the intruder is very clear at lower volumetric Froude numbers, but it decreases when the speed increases, as far as $F_{n\nabla}$=3.1, where there is a reversal of the trend. In the same figure the continuous line represents the coefficient resistance curve obtained at ϑ=*1.15°*: the starting positive trim (by the stern) causes a higher improvement at very low speeds, about 19% at $F_{n\nabla}$=2.2, an advantage that decreases if the speed increases, remaining obvious for all the tested field anyway.

Tests on model B have been carried out from $F_{n\nabla}$=1.73 to $F_{n\nabla}$=2.67. Without the device it has been tested at ϑ=-0.5° (trim corresponding to the smallest resistance at $F_{n\nabla}$=1.9), while with an intruder it has been tested at ($S_i/L_{PR}=0.76\times10^{-3}$) at ϑ=*0°*.

The curves trend in Fig. 1.13.2 is slightly different from the preceding case: the improvement starts from 9.5% at $F_{n\nabla}$=1.7 and increases and grows steadily, about 17%, for the remaining tested $F_{n\nabla}$ (from 2.0 to 2.7).

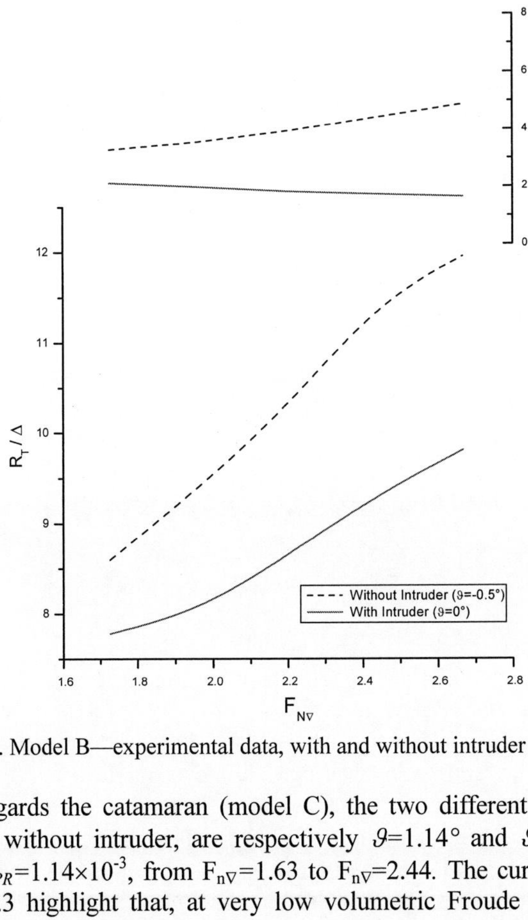

Fig. 1.13.2. Model B—experimental data, with and without intruder

As regards the catamaran (model C), the two different tested trims, with and without intruder, are respectively ϑ=1.14° and ϑ=2.12°, and, with S_i/L_{PR}=1.14×10^{-3}, from $F_{n\nabla}$=1.63 to $F_{n\nabla}$=2.44. The curves shown in Fig. 1.13.3 highlight that, at very low volumetric Froude numbers, the intruder causes a slight worsening, a trend that soon afterward changes when the speed increases, getting on for 10% at $F_{n\nabla}$=2.4.

In the described figures, the dynamic trim angle curves are also shown, always versus volumetric Froude number: in all tests with an intruder, τ is less.

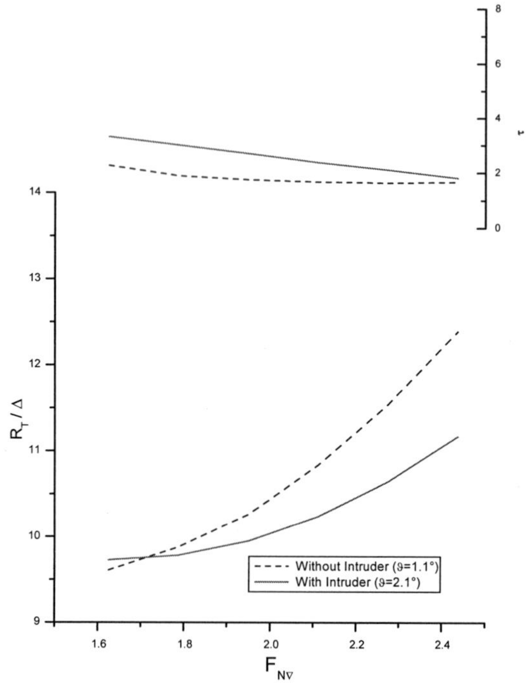

Fig. 1.13.3. Model C—experimental data, with and without intruder

2. Data analysis

Analyzing the obtained experimental results, it is necessary to notice that the volumetric Froude number fields are very different, because they correspond to the real functioning speeds of full-scale ships. For example model A—short and wide—has been tested up to $F_{n\nabla}$=3.4 (the starting planing $F_{n\nabla}$ is about 2.5), experiments with model B—longer and narrower—have been done up to $F_{n\nabla}$=2.6, while the catamaran has been tested up to $F_{n\nabla}$=2.44. These differences do not prevent us from drawing comparisons and common conclusions, such as the clear advantage of using an intruder, even if in percent it is very different in the three cases, where the volumetric Froude number varies.

There are no in-depth studies that justify the positive effect of an intruder from a quantitative point of view. But we may suppose the qualitative principle of operation: on the analogy of the Gurney flap[1-13-9] in the aerodynamic field, we may think that a small separation region forms upstream of the intruder. It is a kind of dynamic wedge that modifies significantly the pressure field just below the bottom of the hull, producing greater hydrodynamic support.

This hypothesis is supported for example by Fig. 1.13.4, which shows model C at $F_{n\nabla}$=2.44 without and with an intruder: a small plate jutting about one thousandth of the hull length raises the stern about 5% of its own length.

Fig. 1.13.4. Model C profile of stern wave at $F_{n\nabla}$=2.44 without (upper photograph) and with intruder (lower photograph)

The overall effect is comparable to that produced by big flaps, but the added resistance is much lower. In fact, an intruder—even if it is perpendicular to the flow—works very deep in the boundary layer. Tests have shown that both the wet length and the surface are usually higher (even more than 20%); but the variation of length changes the friction resistance coefficient C_F very little in terms of percent, whereas the increase of surface conditions greatly affects the total resistance coefficient C_T, which decreases considerably, where the C_F calculated in accordance with ITTC '57[1-13-10] is:

$$C_F = \frac{0.075}{(Log_{10}R_n - 2)^2}\,.$$

Mentioning

$$C_T = C_F + C_R = \frac{R_T}{\frac{1}{2}\rho S V^2}$$

it is evident that the advantage caused by an intruder is due to a decrease of the residuary resistance coefficient (C_R), which, besides, is very important for a high-speed hull, producing upward of 50% of the total resistance.

This kind of analysis is confirmed by experimental observations; in Fig. 1.13.5, for instance, it is clear that when using an intruder the waves are lower, which generally means lower C_R.

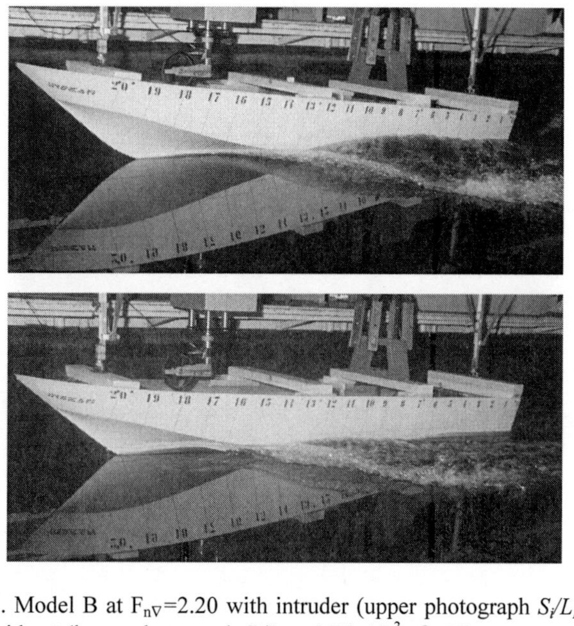

Fig. 1.13.5. Model B at $F_{n\nabla}$=2.20 with intruder (upper photograph S_i/L_{PR}=0, ϑ=-*0.5°*) and without (lower photograph $S_i/L_{PR}=0.76\times10^{-3}$, ϑ=0°)

From the obtained graphs, it is clear that the positive effect of an intruder is more evident in the preplaning range ($1<F_{n\nabla}<2.5$). Looking at curves, three models have differences in behavior, mainly because they have quite dissimilar geometries and so they plane at different volumetric Froude numbers. Model A, for example, at $F_{n\nabla}$=2.2, is in the preplaning range; therefore, an intruder produces nearly a 20% gain soon after full planing starts, and the advantage decreases quickly. Model B, which is longer and narrower, has a larger preplaning range; indeed, the advantage caused by the intruder extends over $F_{n\nabla}$=2.2. If model B tests had been performed also at higher volumetric Froude numbers, the two curves would have got nearer to each other and afterward intersected; however, test speeds had been chosen in accordance with the real speed of corresponding full scale ships.

1.13.5. Mobile intruder: comparison with flaps

1. Experimental data

As has already been said, intruder use can greatly improve hull performance, but it is very often difficult to foresee its efficacy; above all, it can considerably depend on its protrusion, especially for hulls that work over a large speed range. It is sometimes suitable to use a variable protrusion of the device, as a function of speed, analogously to flaps, whose pressure plane is movable.

To evaluate the efficacy of intruder protrusion variation and to compare it with a couple of flaps, some tests have been performed on model A. It has been chosen because of its very large speed range. Tests have been done at $F_{n\nabla}$=2.15, 2.80, and 3.45, and at $\vartheta=0°$, varying both intruder protrusion and flap angle. As regards tests with intruders, they have been performed also at $\vartheta=1.15°$. The flaps used were commercial ones with AR=2 (width=27% of B_{PT}, chord=2.3% of L_{PR}).

Experimental data are gathered in the graphs in Fig. 1.13.6 where, for every tested $F_{n\nabla}$, R_T/Δ variations in percent are shown, in regard to starting conditions without intruder nor flaps at $\vartheta=0°$. In the same figure, dynamic trim angles in degrees are represented versus flap angle or intruder protrusion S_i/L_{PR} (upper curves).

It is evident that an intruder is more efficacious than flaps. It is useful to underline that the protrusions of an intruder are always lower than 20mm for full-scale ships; over that value a high decrease in performance is generally found.

At $F_{n\nabla}$=2.15 flap use improves hull performances, up to about 5% at their maximum testable angle of incidence, that is, 8°. At the same volumetric Froude number, improvement caused by an intruder is about 17%. With model trim by stern ($\vartheta= 1.15°$), intruder efficacy produces the maximum resistance reduction—about 20%—but not at the maximum intruder protrusion. In fact, if S_i/L_{PR} increases a little more, R_T/Δ increases a lot. It's interesting to notice that if the resistance decreases the dynamic trim angle generally decreases too: from 8° of the starting common condition without flaps or intruder, to 7° with flaps, and to about 4° with an intruder at minimum R_T/Δ.

$F_{n\nabla}$	Tests with flaps (θ=0°)	Tests with intruder (θ=0°)	Tests with intruder (θ=1.15°)
2.15	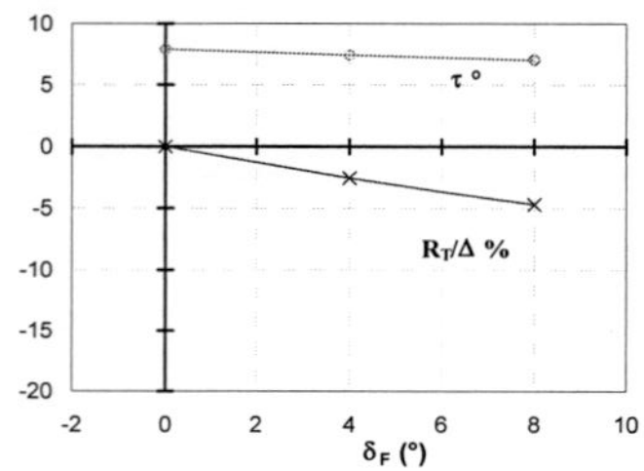	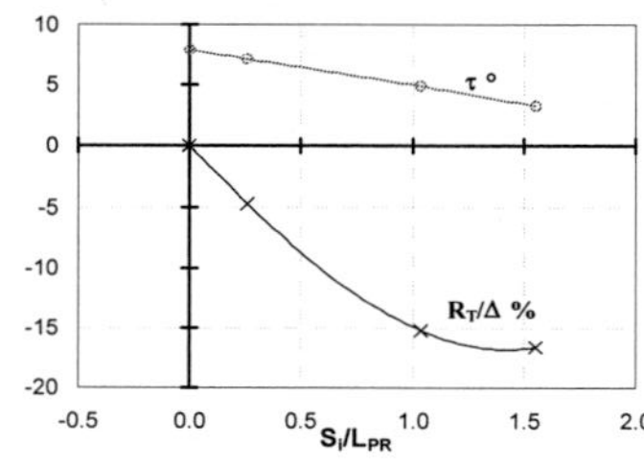	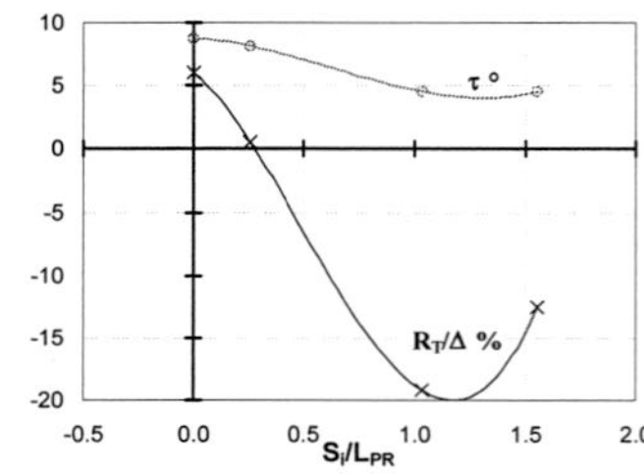
2.80	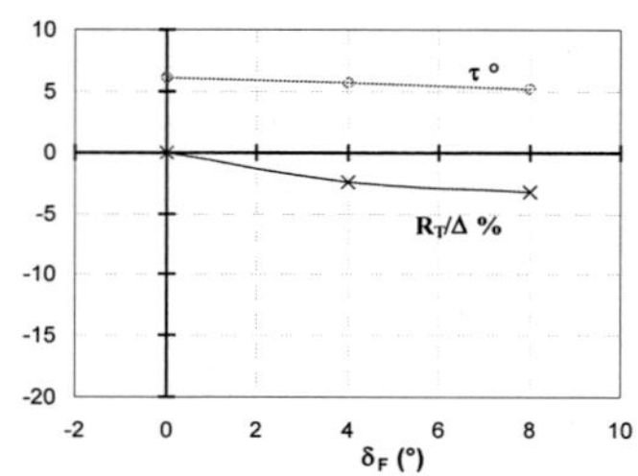	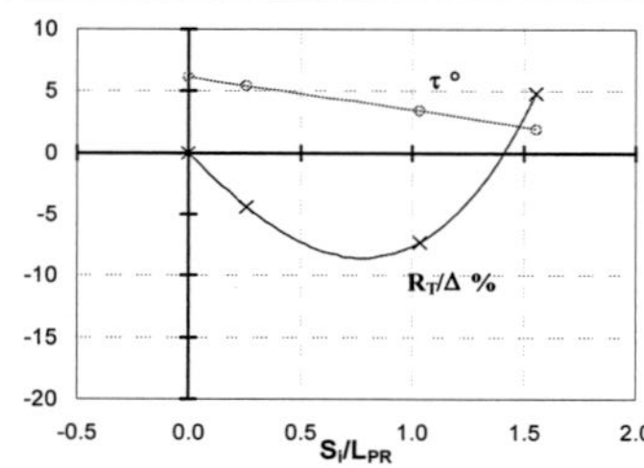	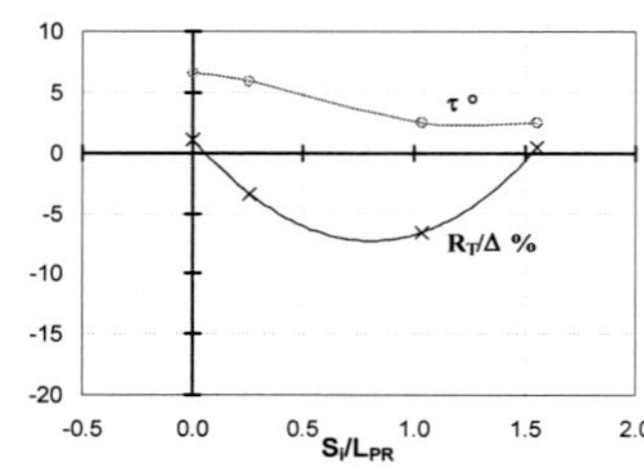

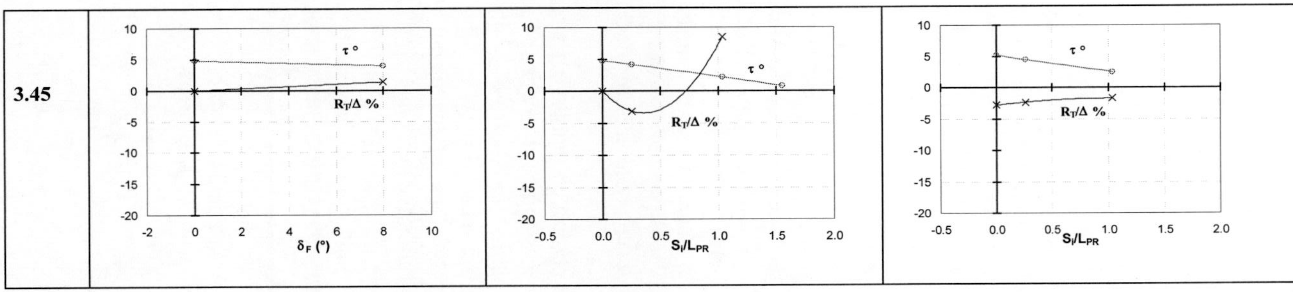

Fig. 1.13.6. Model A: comparative tests, flaps/intruder

At $F_{n\nabla}$=2.80, performances improve up to about 3% with flaps, whereas resistance decreases about 8.5% at $\vartheta=0°$ and about 7% at $\vartheta=1.15°$ with an intruder; the corresponding τ is 4°. Also, in this condition, it is evident that if the intruder protrusion keeps on increasing, resistance increases too.

At $F_{n\nabla}$=3.45, resistance worsening is obtained by flaps, whereas by intruder use, at ϑ=0°, resistance decreases about 3%, and the dynamic trim angle τ is 4° again. At ϑ=1.15°, R_T/Δ reduction—about 2%—is present anyway because of trim by stern condition, and it persists even if there is little intruder protrusion. Also, in this case, the minimum R_T/Δ dynamic trim angle is 4°. If protrusion increases, τ still decreases but R_T/Δ increases.

2. Data analysis

In the graphs of Fig. 1.13.6 it is interesting to note that the τ of about 4°—corresponding to the minimum R_T/Δ—is the angle of minimum resistance obtained by Savitsky for prismatic planing hulls[1-13-11] and in DTMB Systematic Series 62[1-13-12] with similar geometrical characteristics. Besides, it is appropriate to remark that the dynamic trim angles obtained by the model during the basin tests, for reasons that are not well understood but are caused by scale effect, are generally higher than the corresponding full-scale ship values.[1-13-13]

Some graphs show that the minimum resistance condition is verified at different S_i/L_{PR} values if the $F_{n\nabla}$ changes.

On ships with a large speed range, remote controlled mobile intruders are advantageous; they can be used like flaps to stabilize the boat and to improve performances. Plate protrusion—higher in the pre-planing phase—must be reduced when the hump speed gets near.

The photographs in Fig. 1.13.7 show the behavior of the model at $F_{n\nabla}$=3.45 with ϑ=0°. In the first, without an intruder, the keel comes out of the water at section 7 whereas in the second one, with $S_i/L_{PR}=0.26\times10^{-3}$, it comes out further on (at section 8), and the hull is more horizontal in comparison with the precedent condition. In this last case we have about the minimum resistance and the dynamic trim angle is 4°. The following photographs show tests with $S_i/L_{PR}=1.03\times10^{-3}$ and $S_i/L_{PR}=1.55\times10^{-3}$: because of the excessive protrusion, the point where the keel comes out of the water moves toward the bow, the dynamic trim angle decreases, and the waves are higher and more visible. The result is a quick increase in resistance and a decrease in the dynamic trim angle in comparison with its optimum value. These phenomenon have already been pointed out by studies on planing prismatic hulls[1-13-11] and by systematic series[1-13-12].

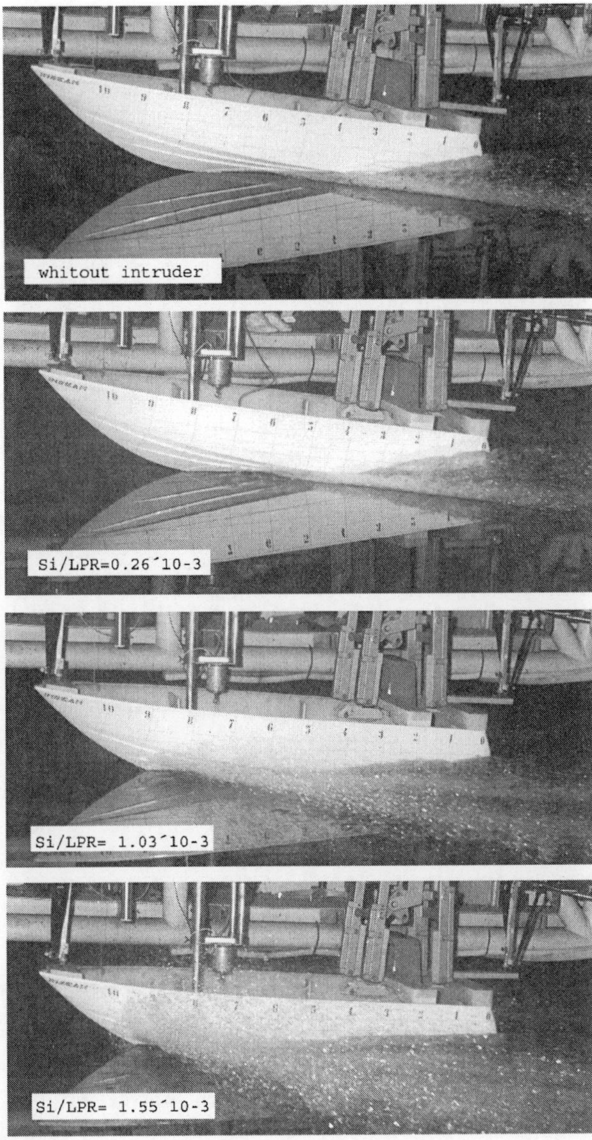

Fig. 1.13.7. Differences of behaviour of model A at $F_{n\nabla}$=3.45, ϑ=$0°$ with and without intruder

1.13.6. Conclusions

The tests performed at INSEAN on three different models make clear that an intruder can induce significant changes in dynamic trim angle and can lift the stern. These changes can reduce waves and then hull resistance. On the other hand, an excessive protrusion, or its use if unnecessary, can strongly increase resistance. Furthermore:

1. A fixed intruder is useful above all on large planing and semi-planing hulls, generally ones with a high L_{PR}/B_{PX} (≥4) or the ones that navigate at volumetric Froude numbers lower than 2.5 (hulls that never navigate completely planing). In these cases, an intruder is useful to correct an over-high trim by stern, which sometimes is characteristic of boats that tend "to sit on the waves" they produce. Its installation on these hulls is very cheap because it consists of a plain plate positioned under the transom, jutting out by just a few millimeters. Experimental results have shown a resistance reduction even greater than 20%. The protrusion must be carefully chosen through practical testing.
2. A mobile intruder is recommended in particular for ships with a larger speed range ($F_{n\nabla}$>2–2.5). In these cases, the plate protrusion has to be shortened, because an intruder reduces the dynamic trim angle too much. On such boats, which typically have low *L/B* ratios, it is often enough to install commercial devices, even if they have a partial length in comparison with the transom one.
3. An intruder is useful to correct the behavior of a boat that navigates trimmed by stern too much and to improve its performance. Moreover, experimental tests have shown that—in these cases—the efficacy of this device increases and it is valid at a greater range of speeds. Therefore, if an intruder is used, it can sometimes be useful statically to trim the boat by stern.
4. By experimental data analysis intruder is generally more efficacious than flaps, both as regards improvement of performances and as regards the larger functioning speed range. In fact to obtain the same effect by flaps as one can get from an intruder—with respect to dynamic stern support and improvement to the trim—big flaps with high angles of incidence are indispensable, but they cause a high resistance increase and reduce (until they undo them) the advantages they produce. Using an intruder, instead, a few millimeters protrusion is enough to obtain

the same dynamic effects, against a negligible added resistance because it completely works in the boundary layer.

5. Matched intruder and flap devices have already been tested and they are spreading, with an intruder placed downstream of flaps on their trailing edge or upstream. In this last case, to be practical, flaps are not coplanar with the bottom but are a few millimeters lower.[1-13-14] If the two devices are joined, it is necessary to balance their characteristics very well. The rapid spread of the intruder, caused by its cheapness and its simple installation and use, is giving rise to some changes to its morphology, and then of its working, to adapt it to particular hull forms or appendages and/or to existing flaps. Therefore further studies and tests are useful and desirable to advise planners and market about the working and the potential of this device, as well as about its correct installation.

References

[1-13-1] R. H. Liebeck, "Design of Subsonic Airfoils for High Lift," *Journal of Aircraft* 15 (9), September 1978.

[1-13-2] B. L. Storms & C. S. Jang, "Lift Enhancement of an Airfoil Using a Gurney Flap and Vortex Generators," *Journal of Aircraft* 31 (3), June 1994.

[1-13-3] D. Jeffrey, X. Zhang, & D. W. Hurst, "Aerodynamics of Gurney Flaps on a Single-Element High-Lift Wing," *Journal of Aircraft* 37 (2), 2000.

[1-13-4] J. Zerihan & X. Zhang "Aerodynamics of Gurney Flaps on a Wing in Ground Effect," *AIAA Journal* 39 (5), May 2001.

[1-13-5] J. Katz & R. Largman, "Effect of 90 Degree Flap on the Aerodynamics of a Two Element Airfoil," *Journal of Fluids Engineering*, 111: pp. 93–94, 1989.

[1-13-6] R. W. Prouty, "Aerodynamics," *Aviation Today—Rotor & Wing*, October 2003.

[1-13-7] P. Roesch & A. Vuillet, "New Designs for Improved Aerodynamic Stability on Recent Aerospatiale Helicopters," Proceeding of the 37th Annual Forum, American Helicopter Soc., 1981.

[1-13-8] IX ITTC, "Ninth International Towing Tank Conference—Proceedings", Paris 1960.

[1-13-9] W. G. Sewall, R. J. Mcphee, & J. C. Ferris, "Wind Tunnel Test Results of Airfoil Modifications for the EA-6B," *AIAA Paper* 87–2359, August 1987.

[1-13-10] VIII ITTC, "Eighth International Towing Tank Conference," Madrid, 1957.

[1-13-11] D. Savitsky, "Hydrodynamic Design of Planing Hulls," *Marine Tecnology*, 1964.

[1-13-12] E. P. Clement & D. L. Blount, "Resistance Tests of a Systematic Series of Planing Hull Forms," *Transaction SNAME*, 71, pp. 491–579, 1963.

[1-13-13] C. Pensa, "La Sperimentazione in Vasca Navale delle Imbarcazioni da Diporto," SEA-MED 2004, Messina, July 2004.

[1-13-14] J. F. Tsal, J. L. Hwang, & S. K. Chou, "Study on the Compound Effect of Interceptor with Stern Flap for two Fast Monohulls with Transom Stern," FAST 2003, Ischia (Italy), October 2003.

1.14

ESTIMATION OF EFFECT AND SIZE ON SHIP INTERCEPTORS (2014)

YANG SHUAI, XIE WEI AND LI HUIMIN

China Ship Development and Design Centre

1.14.1. Outline

To improve the ship's navigational state and reduce resistance and motion in waves, from the 1950s on, a plate with an angle has been located in the boat tail that extends backward, installed along with the boat's bottom. It is fixed on the tail sealing plate by a plurality of transoms, which is called the horizontal tail plate (or trim tab), also known as pressure wave plate. It is called a wedge when it is moved to the front of the panel to the transom of the boat bottom.

In the 1980s, interceptors came out when people extended the flat downward along the transom, also called the barrier or spoiler. Above all, therefore, there are three lift plates at the transom that are used to improve ship navigational conditions and drag reduction: the trim tab, tail wedge, and interceptor, as shown in the diagram in Fig. 1.14.1.

At present, they can be integrated into the control unit so that the angle and height are adjustable, via hand control or automatic control depending on choice.

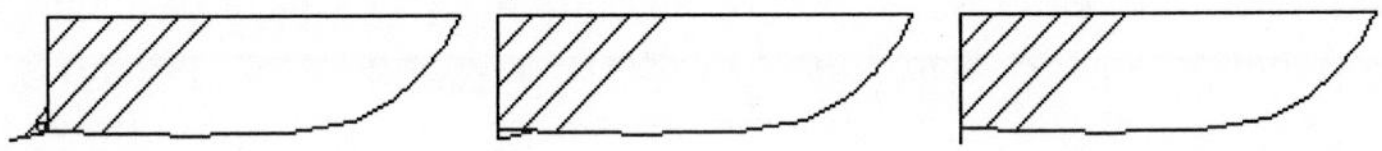

Fig. 1.14.1. Sketch of three kinds of lift plate at the transom: *left*, trim tab; *center*, wedge; *right*, interceptor

1.14.2. The role of the aft lift plate

The stern plate regardless of type has the following effect:

1. For high-speed ships (especially for lift support ships, with a volume Froude number $Fr_\nabla \geq 2.5$ at design speed), it can produce tail lift and moment and solve the effects of pitch that is too large because of ship slipping, leading to effects such as influence on the driver's sight, the launch angle of weapon being too large, and the main-engine overloading due to resistance increase.
2. For those ships having a Froude number between 1.5 and 2.5, its role is mainly to reduce resistance.
3. It has a greater role in inhibiting ship's longitudinal instability (such as dolphin movement).
4. It can improve ship sea-keeping alone or in combination with a hydrofoil as a control panel of the ride control system on a ship.

The following is a comparison between the two of them:

The trim tab has the following characteristic compared to the wedge: it can reduce wave resistance by increasing the virtual length of the ship. It can reduce the ship's motion response in storms by increasing the ship's damping. It increases the torque of the fore pitch because the hydrodynamic action center is at the back of the ship. And it will not affect the spacing between the propeller and ship bottom (the wedge will reduce this distance if it is located on the above board of the propellers). Nevertheless, its structure is more complex and its weight is increased by installing it. The wedge will increase the ship's resistance because it increases the friction area. Reversing resistance will also be increased; and with water jet propulsion, the negative interference will be larger due to the wedge.

Interceptors have the following characteristics compared with trim tabs: Their resistances are almost the same because the interceptor's height is small. The interceptor's structure is simple, it is light weight, it is convenient to install, and it is low cost. Interceptors do not affect the reverse flow of water jet propulsion. An interceptor's damping in waves is small. Generally the interceptor's control scope is restricted to prevent the lift being too big and instability on roll.

If we only want to solve the ship's navigational state and resistance in still water, the better choice will often tend to be to use interceptors rather than trim tabs because it is very difficult to make their installation accurate (due to welding deformation). It is also very troublesome when a ship changes the trim tab angle, and this problem will be more prominent especially when the transverse ramp angle of the transverse profile is greater than 0°. The use of the interceptor will be the ship's propulsion by water jet, in order to prevent reverse flow interference.

1.14.3. Action principle of interceptors

Interceptors have been used in many ships. They can reduce resistance, which has been proven through model tests and real-ship tests. It is particularly effective for those medium speed ships where the principal dimensions (L, B), weight, and center of gravity do not match. In a comparison between a bare hull and a hull with an interceptor (see Fig. 1.14.2), we can see that the interceptor can reduce the resistance to about 85%[1-14-1] at various speeds. However, this does include its mechanism of action.

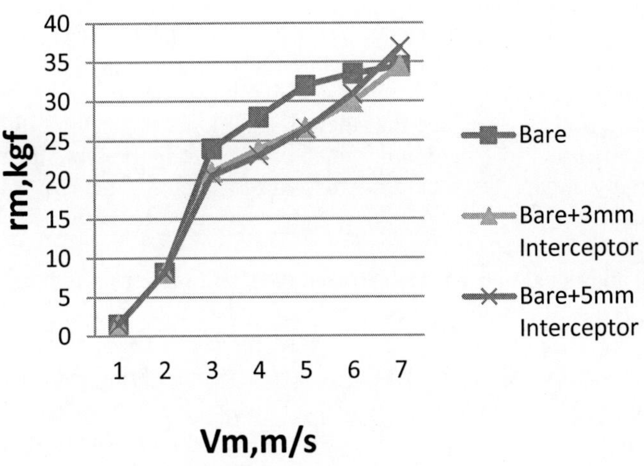

Fig. 1.14.2. The effect of an interceptor on resistance

1. On the basis of foreign research, Figure 1.14.3 shows the pressure distribution and flow caused by an interceptor, a trim tab, and a wedge at the bottom of a ship.

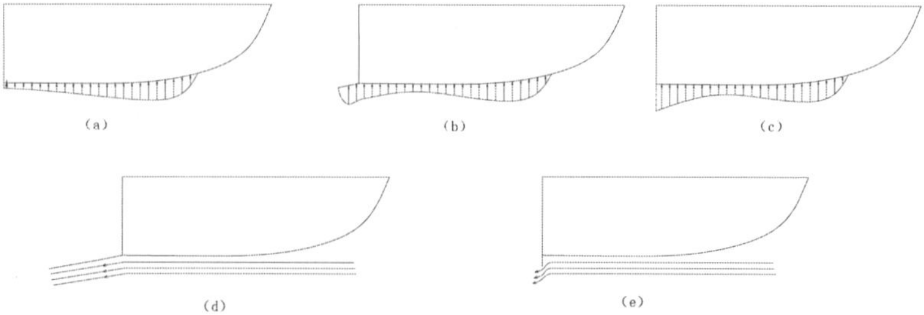

Fig. 1.14.3. Schematic diagram of pressure distribution and flow

From the pressure distribution it can be seen that the bottom pressure distribution will be redistributed for ships with a tail lift panel. Ships equipped with a trim tab will form a hump (see Fig. 1.14.3[b]). It will form a mutation peak shape before the interceptor (see Fig. 1.13.4[c]), which is similar to the wing tip resistance of a baffle on airplane wings. Downward sloping on the flow of the interceptor causes bigger change than does a trim tab (see Fig. 1.14.3[e]). Fig. 1.14.3(a) shows the pressure distribution of a bare hull. We can see that there is additional upper force for a ship with a lift panel to a bare hull, which can be got by pressure distribution shown by model tests or modern hydrodynamics. The two ways are difficult.

2. The additional upper force will lead the ship to fore roll, reduce the trim angle, reduce the bottom face angle, and then reduce the resistance of a ship (ΔR1). It produces lift, slightly reduces drainage volume, and decreases ship resistance (ΔR2). Of course, it also brings an increase in resistance (ΔX), when (ΔR1+ ΔR2) >ΔX, the whole ship resistance will be reduced; when (ΔR1+ ΔR2) <ΔX , the resistance of the ship increases.

1.14.4. Calculation of interceptor height

A large number of model tests show that the effect of the interceptor generated (within a certain range of Froude number) is the same as the effect of moving the ship's center of gravity forward. Its equivalent in the reduction of resistance is caused by the forward movement of a gravity center or generated by interceptor height h. Thus, we can obtain the approximate force and moment equations required by additional lift Y to further determine the desired height of the end plates, (see Equation 1.14.1).

$$h = Y \Big/ \frac{1}{2}\rho v^2 c_y b \qquad (1.14.1)$$

Where : Y —Lift generated by interceptor /kgf

v—Velocity of flow/m/s

c_y —lift coefficient of interceptor

b —The length of the interceptor (along Breadth)

ρ —water density/kgf.S^2/m^4

We can get the height of the interceptor when we know Y and c_y . The following is how to get Y and c_y

1. Getting Y by force and torque equation

After the boat has an interceptor, when sailing at a speed of V_0, generate lift tailgate Y; Δ_1 is the weight of the boat, Δ_2 is buoyancy. After ignoring the thrust and impact resistance, the force and moment equations can be written as follows:

$$\begin{cases} \Delta_1 = \Delta_2 + Y \\ \Delta_1 \times Xg_1 = \Delta_2 \times Xg_2 \end{cases}$$

Then $Y = (1 - \frac{Xg_1}{Xg_2})\Delta_1$ (1.14.2)

The role of additional lift is like the boat's displacement being reduced from Δ_1 to Δ_2, or the position of the center of buoyancy being moved from X_{g1} to X_{g2} (buoyancy centre moved ΔXg).

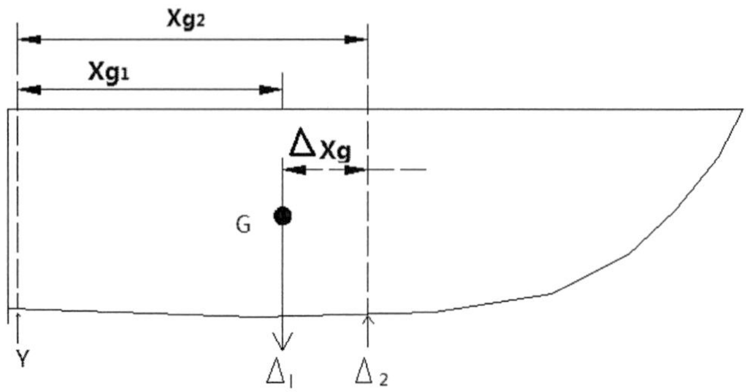

Fig. 1.14.4. Schematic diagram of stress

2. Getting the lift coefficient of an interceptor

Because of a medium-speed boats' longitudinal ramp angle γ (the mean of the longitudinal ramp angle of its keel and chine lines), the size and pitch angle is also very close; thus, the value is approximately fixed ($\sim\varphi$ can also be obtained if necessary) and the flow velocity interceptor can be determined approximately by the boundary layer velocity in formula 1.14.1. One can therefore use some of the ship model testing to determine the approximate value of C_y :

Such as: Ship model of a boat: Δ_m=203.7kg, Xg_m=1,133mm, h=3mm, or 5mm, B_T=735mm,β_T=22°

When Vm=3.0 , $\begin{cases} h_3 = 3mm \ \mathrm{r_m}\text{=21.4kgf} \\ h_5 = 5mm \ \mathrm{r_m}\text{=20.5kgf} \end{cases}$

The following value can be got from Fig. 1.14.4, which shows the resistance curves of the center of gravity position and resistance when Vm=3.0m/s:

$$\begin{cases} Xg_2\text{=1203mm(when } h_3 = 3mm) \\ Xg_2\text{=1265mm (when } h_5 = 5mm) \end{cases}$$

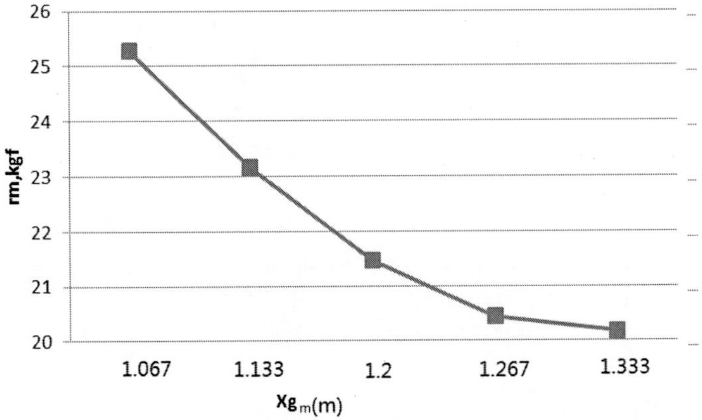

Fig. 1.14.4. The resistance curves at different center of gravity positions when Vm=3.0m

Then we can get $\begin{cases} Y_3 = 11.85\text{kgf} \\ Y_5 = 21.26\text{kgf} \end{cases}$ by (2) $Y = (1 - \frac{Xg_1}{Xg_2})\Delta_m$

There is obtained by Formula(1): $C_y = Y \Big/ \frac{1}{2}\rho v_c^2 h \frac{B_T}{\cos\beta_T}$

Among $h_3 = 3mm, h_5 = 5mm$, we require to calculate its v_c (see Appendix for calculation method of v_c)

So, $$C_{y3} = Y_3 \Big/ \frac{1}{2}\rho v_{c3}^2 h_3 \frac{B_T}{\cos\beta_T} = 21.94$$,

$$C_{y5} = Y_5 \Big/ \frac{1}{2}\rho v_{c5}^2 h_5 \frac{B_T}{\cos\beta_T} = 21.53$$

1.14.5. Description to determine interceptor height h

1. Application scope: mainly for medium-speed vessels ($Fr_\nabla = \frac{V}{\sqrt{g\nabla^{(1/3)}}} 0.9 \sim 2.9$) to reduce resistance.
2. Adjust the trim of ship ϕ_0 (especially for those ships that cannot make *Xg* forward by adjusting the arrangement).
3. When we have the resistance data at the different *Xg* of the model test of the ship, we can choose the point of the boat to be the resistance valley longitudinal center of gravity as Xg_2 to calculate Y. Without this curve, one can take the early center of buoyancy when the trim is zero degrees ($\phi_0 = 0$ °) as Xg_2 .
4. It is recommended to take around 21.73 of c_y when calculating.
5. Conduct an interceptor model test, if necessary, to further accurately determine h.
6. Use an adjustable interceptor boat; this method can determine the height adjustment range.
7. When used in a real ship, we should also consider the interceptor to flow as a velocity correction. The main reason for the ship sailing in the water is that the viscosity of the underwater hull forms a boundary layer, and the interceptor is within the boundary layer, which will reduce the flow. These calculations are in the appendix.

Appendix

The calculations of the average speed of the interceptor flow:

1. The vertical thickness δ of the boundary layer at the interceptor:

When the Reynolds number $\mathrm{Re} \geq 2\times10^{6}$ is used alongside the Songhai–Carmen formula to make the following calculation, we get:

$$\delta = [0.01954 - 0.00138\lg(v_0 \cdot L_{WL})]L_{WL}$$

Among: v_0—Velocity of flow/m/s

L_{WL} —the length of waterline/m

2. The direction of the boundary layer velocity distribution in perpendicular to the hull, can be the Von Karman formula:

$$v_y = v_0(\frac{y}{\delta})^{\frac{1}{11}}$$

y —the distance from the hull, and the vertical coordinate with velocity.

V_y —Velocity of flow at y/m

δ —thickness/m

Generally, $\delta > h$ (the height of interceptor)

$$v_c = \frac{1}{h}\int_0^h v_y dy$$

Integration:

$$v_c = \frac{v_0}{h}\int_0^h (\frac{y}{\delta})^{\frac{1}{11}} dy = \frac{h^{0.09091}}{1.09091}\delta^{\frac{1}{11}} v_0$$

1.15

Research on Model Tests for a 1,000t Hybrid Monohull with Stern Flap (2013)

Sun Shuzheng,[1 & 2] Miao Quanming,[2] Zhao Xiaodong[1] and Li Jide[1]

[1] Harbin Engineering University
[2] China Ship Scientific Research Center

1.15.1. Introduction

A hybrid monohull is one kind of high-performance vehicle. A ship's longitudinal motion performance can be greatly upgraded with a built-up appendage on the bottom of the bow, which can produce a large damping force (moment) that concludes obviously viscous damping.[1-15-1][1-15-2] Stern flaps started to be used on high-speed vessels in the 1970s, which turned out to be useful in saving energy. As the power output was equal, the design speed can be improved by more than one knot.[1-15-3] The research indicates that resistance of the main hull decreases by more than 4% under the cruising speed after the installment of the optimized stern flap; meanwhile, the resistance performance at high speed improves magnificently.[1-15-4] [1-15-5] However, the scale, angle, immerged depth, and other parameters of the stern flap all have an influence on ship resistance.[1-15-6] [1-15-7] The stern flap has been designed especially according to the characteristics of 1,000t class hybrid monohulls, aiming to improve their resistance performance. Model tests have been carried out to discover the impact of the scale, angle, and other parameters of the stern flap on resistance and sea-keeping performance.

A stern flap has been designed for 1,000t class hybrid monohulls V_1. In this paper, the maximum breadth of a stern flap on V_1 is 1.2m, the breadth is the same as that of a stern water plane, and the negative dihedral angle is 4°. The principal dimensions of a monohull are given in Table 1.15.1. A monohull was developed with a built-up appendage on the bottom of the bow, and the built-up appendage consists of a streamlined submerged body and a pair of bow fins fixed on the submerged body. Sketches of the built-up appendage and stern flap are given separately in figures 1.15.1 and 1.15.2.[1-15-8] [1-15-9]

Table 1.15.1. Principal hull form dimensions of two monohull composite ships

Hull form	Length of designed waterline/m	Breadth of designed waterline/m	Designed draught/m	Displacement /t
V_1	83.8	11	3.2	1,220

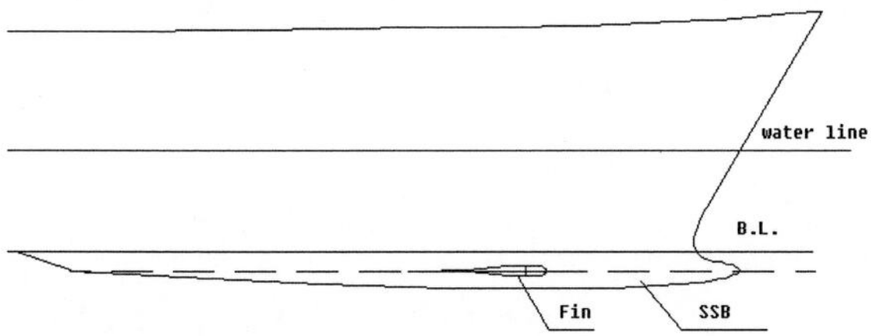

Fig. 1.15.1. Sketch of an appendage

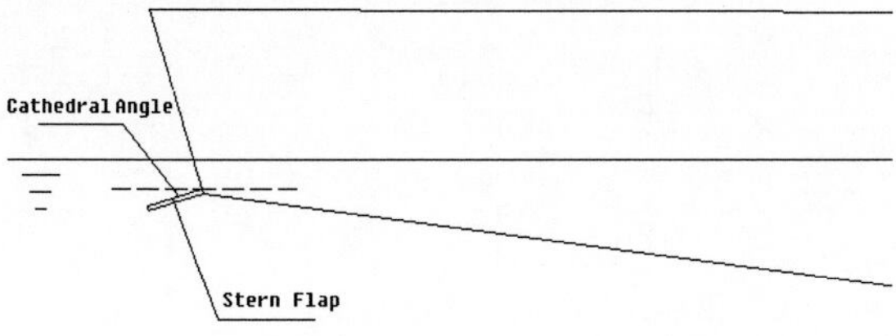

Fig. 1.15.2. Sketch of a stern flap

In this paper, a water tank experiment is carried out to study the retrofitting of a stern flap on a kiloton composite ship hull form. The code name of the scheme of the composite ship is V_1, the code name of the scheme of the ship with a stern flap is V_{1w}, the scale ratio of the model to the real ship is 1:25. The experiments include hydrostatic resistance experiments of the ship models before and after installing the stern flap and a regular wave experiment. The following details the water tank model experiments.

1.15.2. Hydrostatic resistance test

The first experiment is the hydrostatic resistance experiment on deep-V composite ships with and without a stern flap. First, statics and dynamics are adjusted according to the scale ratio of the model to the real ship. This is to ensure the similitude of the test models. Then the tests are carried out. The tests include model resistance, model pitch angle, and model heave curve of center of gravity. Resistance values with and without a stern flap are shown in Table 1.15.2. The hydrostatic resistance, model pitch angle, and model heave curve of the center of gravity of monohull composite ships with and without a stern flap are shown in Figs 1.15.3–5. The value of the model pitch angle is set as negative in the case of trim by head. The value of the heave of center of gravity is set as negative if the movement is downward.

Table 1.15.2. Hydrostatic resistance before and after installing a stern flap

Model speed/m/s	Real ship speed/kn	Resistance without stern flap /N	Resistance with stern flap/N	Ratio of the drag reduction/%
1.235	12	6.773	6.536	3.496
1.852	18	14.519	14.208	2.141
2.469	24	29.55	28.759	2.678
3.087	30	48.886	48.189	1.426
3.704	36	64.463	63.333	1.752
4.116	40	75.527	74.161	1.809
4.63	45	90.929	89.295	1.796

According to the still water-resistance model test results, after installing a stern flap, resistance of the V_1 decreases from 1.4 to 3.5%; thus, through the installation of a stern flap, a hybrid monohull has better resistance.

By comparing the trim curve and curve of center of gravity before and after the installation of a stern flap, it can be shown that the trim angle and center of gravity are reduced after the installation of a stern flap. Thus, the installation of the stern flap can effectively improve the hull sailing posture, thereby improving the stern wave by increasing the "virtual length" of the hull, so that the still water-resistance performance of a deep-V hybrid monohull would be enhanced.

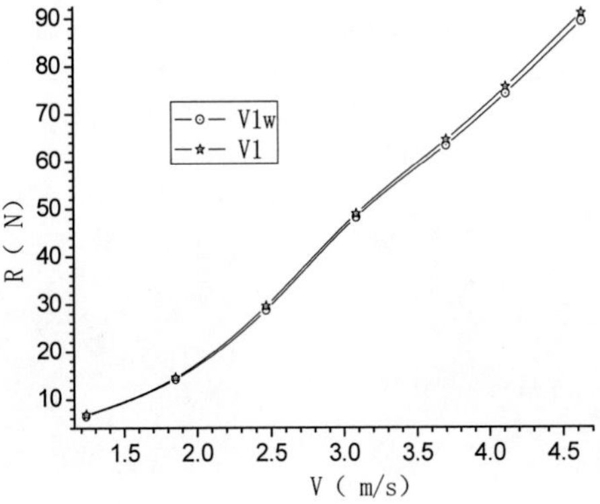

Fig. 1.15.3. Resistance of V_1 and V_{1w} with the same tonnage

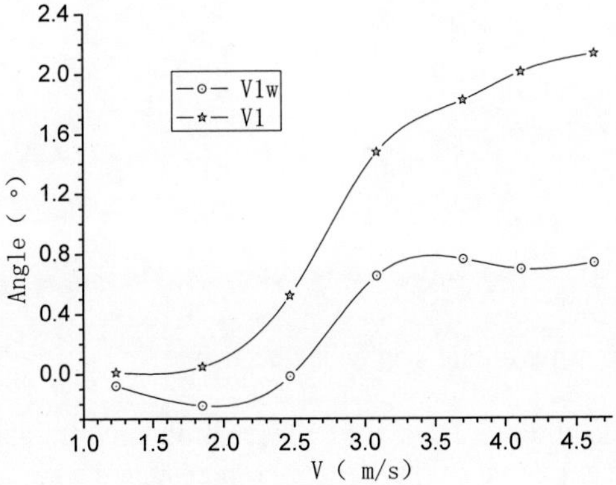

Fig 1.15.4. Comparison of trim angle of V_1 and V_{1w}

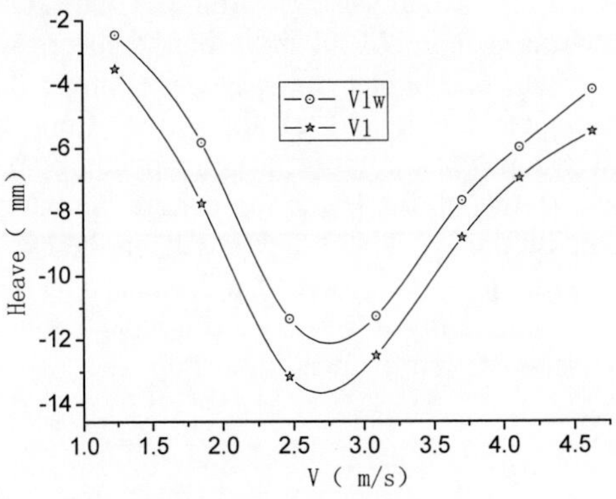

Fig. 1.15.5. V_1 comparison of heave of V_1 and V_{1w}

To analyze the impact of the stern flap on resistance, the RANS method is used to analyze the stern flow field and observe the stern wave before and after the installation of stern flap. Fig. 1.15.6 represents the stern wave of the central longitudinal section.

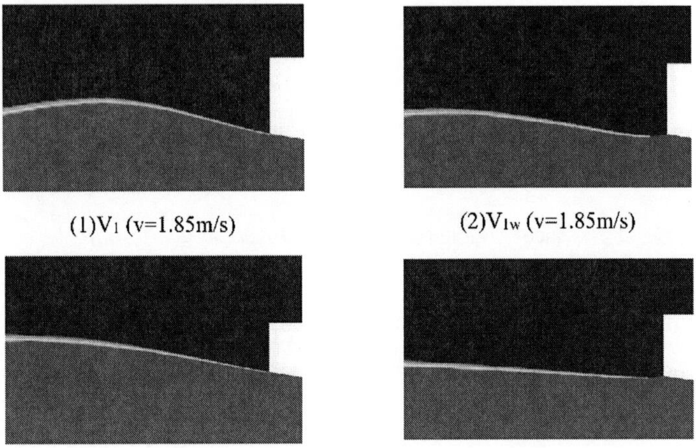

(1)V_1 (v=1.85m/s) (2)V_{1w} (v=1.85m/s)

(3)V_1 (v=3.08m/s) (4)V_{1w} (v=3.08m/s)

Fig. 1.15.6. Stern wave of mid longitudinal sections

Referring to the stern wave of the central longitudinal section (Fig. 1.15.6), because the two ship types are both designed with a transom stern, the stern stream line will extend along the transom stern section a certain distance behind the ship when sailing, forming a "chicken tail flow" and increasing the "virtual length" of the hull; then, the stern wake jumps and forms the stern wave. The higher the speed, the longer the "chicken tail flow" and the higher the height of the wave. Comparing the stern waveform of two ship types going at a corresponding speed, due to the installation of the stern flap the V_{1w} has a certain inhibitory effect on the wake; compared with the V_1, its "virtual length" in the rear of the hull has obviously increased, the wake waveform has become gentler, and the wave height has significantly lowered—these have a favorable effect on improving the resistance performance of the ship.

1.15.3. Regular wave test

The contents of the test measurement of the head sea in regular waves consist of: waves of resistance, pitch angle, heaving, vertical acceleration of bow 0.5 station, midship section 10 station, and stern 19.5 station. To meet the requirements of the linear theory, the wave height is set at 50mm. What follows are the response curves of each ship in head seas at different speeds in regular waves. Fig. 1.15.5 shows the response of a round bilge hull ship G_0 and a compound ship (V_{1w} for the compound ship with installation of semi-submersible body and tail-wave suppression plate, V_1 for the compound ship with installation of a appendage, V_1 for the compound ship with appendage and tail-wave suppression plate) in a head sea with a speed of 18kn (Fr=0.32) in regular waves. Fig. 1.15.7 shows the response of them at a speed of 24kn (Fr=0.43) and Fig. 1.15.8 shows their response at a speed of 35kn (Fr=0.539); the results have been processed with dimensionless treatment.

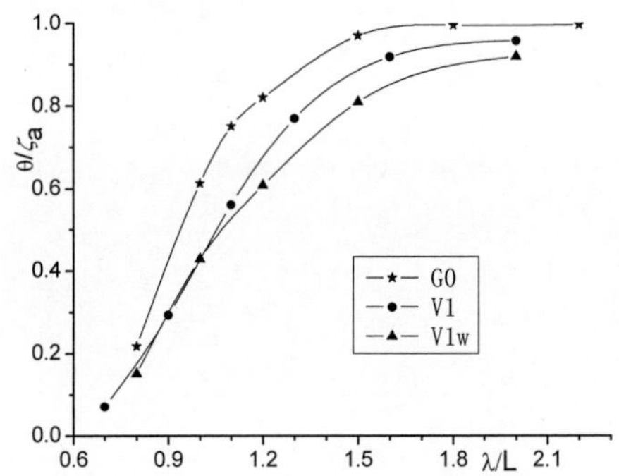

(a) Amplitude of pitch at 18kn

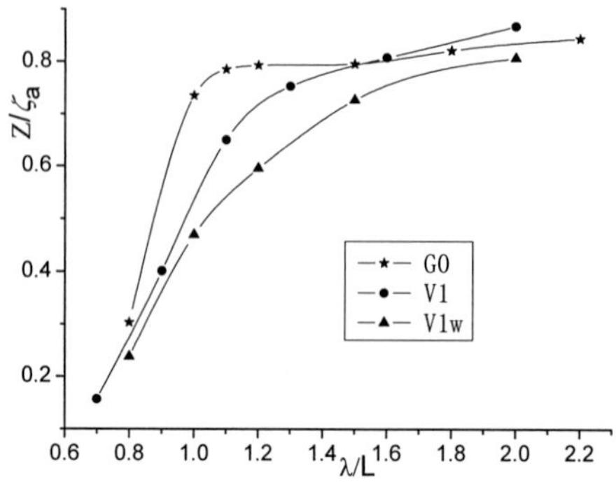

(b) Amplitude of heave at 18kn

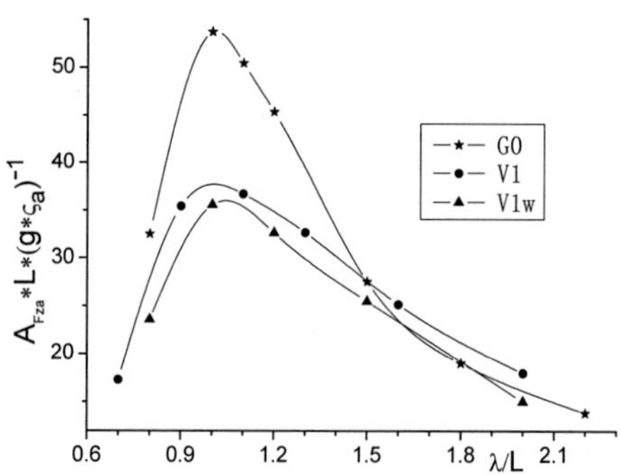

(c) Amplitude of bow vertical acceleration at 18kn

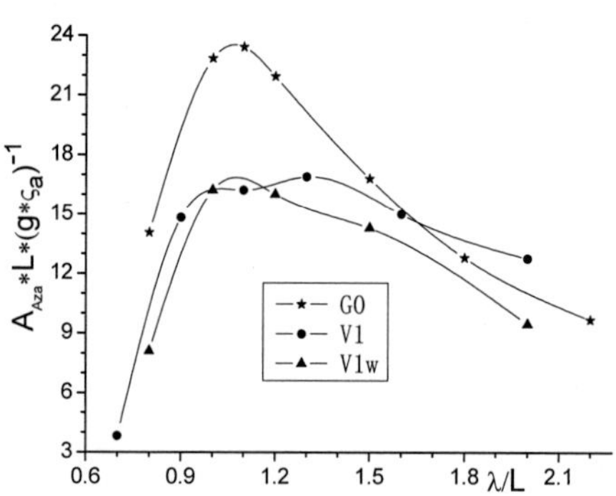

(d) Amplitude of aft vertical acceleration at 18kn

Fig. 1.15.7. Amplitude of motions at 18kn

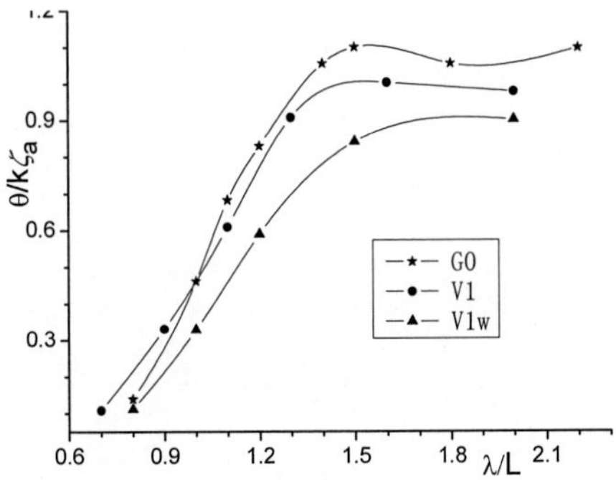

(a) Amplitude of pitch at 24kn

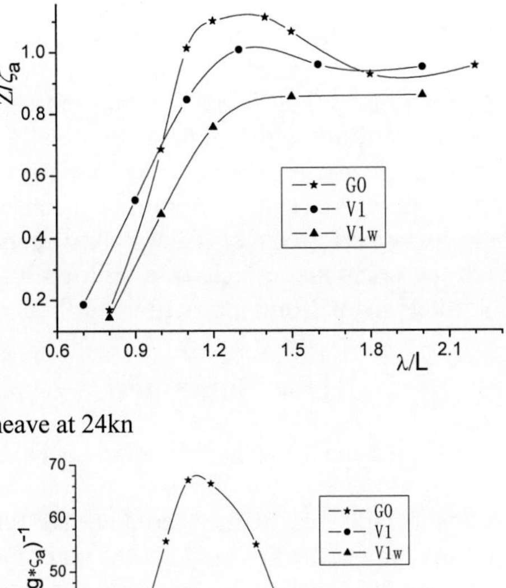

(b) Amplitude of heave at 24kn

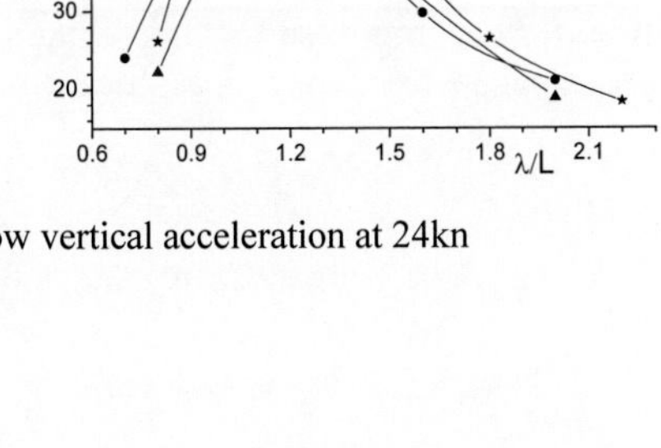

(c) Amplitude of bow vertical acceleration at 24kn

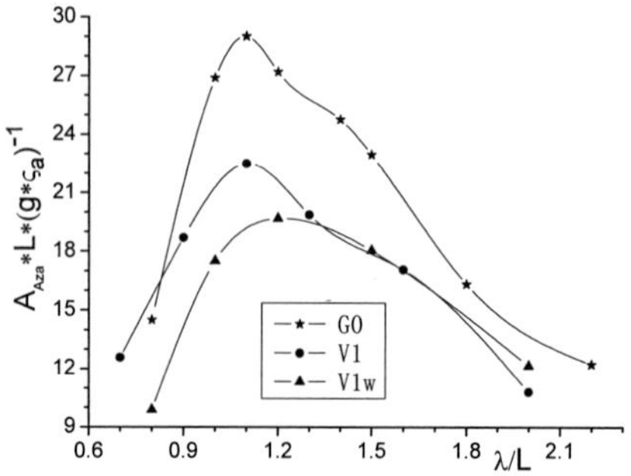

(d) Amplitude of aft vertical acceleration at 24kn

Fig. 1.15.8. Amplitude of motions at 24kn

Even for hybrid monohulls with no tailboard, regular wave test results show that its sea-keeping performance is greatly enhanced compared with the same tonnage round-bilge hull form. The hull motion response will be further reduced when retrofitting a tailboard. Especially near the peak of the resonance frequency, motion response will be significantly reduced. This shows that a tailboard can contribute to enhancing the hybrid monohull's sea-keeping performance. If the hybrid monohull has a stern flap, then through the combined effects of a built-up appendage and tern flap, the ship's sea-keeping performance will be further enhanced.

1.15.4. Summary

From the study above, we can draw the following conclusions:

1. By installing a tailboard in the stern, we can change the ship's sailing posture, adjust the flow field at the end of the ship, and then improve the ship's calm water-resistance performance.
2. The hybrid monohull's sea-keeping performance is better than that of the round bilge-hull form, and the hybrid monohull's sea-keeping will be further improved. By installing a tailboard near the peak of the resonance frequency, the motion response should be reduced.

References

[1-15-1] Cai Xingong, Li Jide, & Wang Jianfang, "Experimental Study on Built-up Stabilizing Appendage," *Shipbuilding of China* 2003 (9): pp. 50–57.

[1-15-2] Li Jide, Zhang Heng, & Zhao Xiaodong, "Experimental Study on 4000t deep-V Hybrid Monohull," *Journal of Ship Mechanics* 5: pp. 709–15, 2008.

[1-15-3] Cheng Mingdao, "An Application of Uniform Design on Stern Flap Design," *Shipbuilding of China* 1999 (5): pp. 9–14.

[1-15-4] Zhang Dayou & Zhang Yan, "On the Effect of Stern Plate upon Resistance Performance of the High-Speed Catamaran," *Ship & Ocean Engineering* 2006 (6): pp. 20–23.

[1-15-5] Lu Xiaoping, Li Yun, & Dong Zushun, "Comparison of Resistance for Several Displacement High Performance Vehicle," *Journal of Naval University of Engineering* 18 (1): pp. 34–41, 2006.

[1-15-6] Ji Hengteng, Chen Jiarong, & Li Wei, "Several Stern Modification Measures for Warships," *Chinese Journal of Ship Research* 6, 2006.

[1-15-7] Zhao Lianen, *The Principle and Design of High Performance Ships*, Harbin Engineering University Press, 2009.

[1-15-8] Wang Xujie, Sun Shuzheng, & Zhao Xiaodong, "Research on Hydrodynamic Performance of a Deep-V Hybrid Monohull with a Stern Flap," *Journal of Harbin Engineering University* 2012 (1): pp. 15–19.

[1-15-9] Wang Xujie, Sun Shuzheng, "Research on Model Test of Thousands-Tons Class High Seakeeping Performance Hybrid Monohull," *Journal of Ship Mechanics* 2011 (4): pp. 180–87.

1.16

Investigation of a Built-Up Appendage Applied to a High-Speed Round Bilge and a Deep-V Hull for Motion Performance Improvement (2014)

Sun Shuzheng, Li Jide and Zhao Xiaodong

Multihull Ship Technology, Key Laboratory of Fundamental Science for National Defence, Harbin Engineering University

1.16.1. Introduction

Longitudinal motion has a great impact on a ship's normal operation and a military vessel's work or operational effectiveness, as violent pitching will cause serious impacts due to green sea and slamming of ships, leading to significant threats to the use of equipment on deck, hull structural strength, and the safety of personnel. The vertical acceleration arising from pitching and heaving motions is also the main cause of high seasick rates among crew and passengers. Therefore, pitching stabilization techniques have been a problem needing to be addressed by shipbuilding engineers. However, pitching stability is very different from rolling: as the pitching damping force (torque) of the hull itself is huge, this means that a greater force (torque) will be required to further increase the damping and effectively resist the external force (torque), which may cause problems of strength and vibration. In addition, the peak value scope of pitching amplitude-frequency response is wider than the corresponding frequency band of rolling motion, while pitching stabilization acts more effectively only in the vicinity of the synchronous rolling range, and longer or shorter waves will significantly reduce the anti-rolling effects. These issues have caused great difficulties to pitching stabilization techniques. As a result, in order to stabilize the longitudinal motion of ships, research has been carried out for nearly a century, during which time ship researchers have proposed dozens of methods, carried out much theoretical research, and conducted numerous trials on models and actual ships. On the basis of hydrodynamic theory, passive fins or active fins have been fitted to the bow or stern, for the purpose of establishing a stable anti-pitching torque to resist the pitching torque generated by the waves; however, most of these have been proven to be a failure due to non-obvious effects, inadequate strength, hull vibration, and so on.

So far, anti-pitching devices that have been really applied to actual ships with obvious effects are not commonly seen, especially as applied to large-tonnage surface ships. Even those that have been reported in foreign literature are only those cases where boats below a tonnage of one hundred use a transom to stabilize pitching. In fact, research on controlling longitudinal motion was conducted almost simultaneously with those on rolling motions; and similar to controls on rolling motion, designers also equipped active or passive fins to the bow/stern. The stabilization effects through such installations were positive: the United States, Great Britain, and the former Soviet Union made comparative tests on actual ships, and the test results showed that the effect of the bow fin was more remarkable. But the key problem is that such installations have side-effects and cause problems, such as installation strength, hull vibration, and so on. For example, the former Soviet Union undertook many tests to install fixed bow fins to ships, and the results showed the longitudinal motion had been effectively stabilized; however, such a method has not been really applied to actual ships due to unavoidable problems like fin strength and hull vibration. Meanwhile, as the flow field of the stern has a thicker boundary layer, the effect of the stern fin is generally not as obvious as that of the bow fin. For bow-fin automation, a number of trials have also been conducted, and in some cases, its effect was better than that of fixed fins, but not very significantly, and the active fin would be subject to cavitation and fin angle saturation. Moreover, the limited space of the bow cabin has also brought many difficulties to fin automation, and the expenses for control devices are very high. In the 1980s, the Japanese installed a streamlined semi-submersible bow (SSB) on the hull bottom and bow of a boat below one hundred tonnage, which had effectively inhibited the longitudinal motion of the boat, especially the acceleration of the bow; yet, resistance had also increased by approximately 10% at higher speeds (SSB displacement accounted for 10% of the boat).[1-16-1]

1.16.2. Stabilization theory of a built-up stabilizing appendage

Over the past century, in order to improve sea-keeping capacity, people have made unremitting efforts and made gratifying achievements, especially in monohull displacement ships; but it is difficult to significantly improve sea-keeping capacities by improving hull forms, for which a number of special control techniques must be used—for example, fin stabilizers and anti-rolling tanks have been widely used on modern ships to effectively inhibit rolling motion. However, stabilization of longitudinal motion has not yet been achieved for any major breakthrough; in particular, no substantive progress has been made on medium- and large-sized surface ships, constraining the comprehensive upgrade of sea-keeping capacity. The biggest difficulty for pitching stabilization lies in the fact that a huge stable torque is required; thus, the appendage offering stable torques must withstand considerable hydrodynamic loads, making it difficult to meet the structural strength of the appendage and causing severe vibration of the hull. To solve the key techniques for pitching stabilization, we must first solve the problem that the anti-pitching appendage has to withstand excessive loads; furthermore, the issue of engineering feasibility should also be considered.

On the basis of hydrodynamic theory and composite technology, a built-up stabilizing appendage is a fluid-dynamic passive anti-pitching device, consisting of a streamlined submersible body and a pair of special bow fins mounted on both sides of the submersible body. The appendage is arranged near the bow bottom through the submersible body and the hull, as shown in Fig. 1.16.1. The semi-submersible body uses a NACA00 series airfoil, and numerical calculations show that the use of asymmetrical upper-lower portions for the cross section of the semi-submersible body is in favor of its speed-ability. Therefore, this paper employs an asymmetric semi-submersible body, of which the cross section is in the form of a asymmetric upper-lower semi-ellipse. The cross section of the bow fin uses a ЦАГИ airfoil in the form of triangular fins, and the leading edge of the bow fin is located near the largest cross section of the semi-submersible body.[1-16-2]

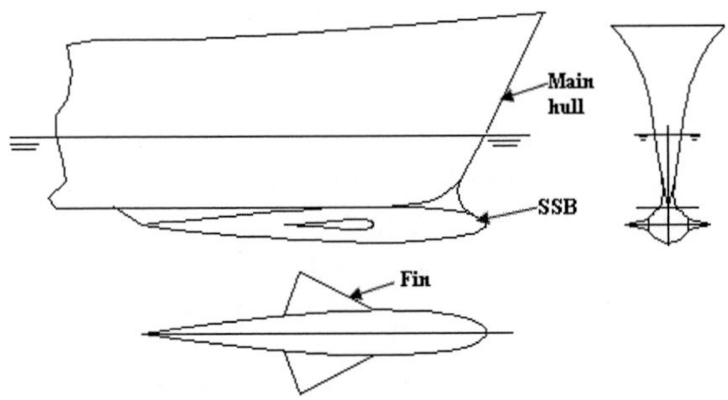

Fig. 1.16.1. Sketch of the structure and arrangement of the built-up appendage

When longitudinal motions are generated by ships navigating in waves, according to the physical characteristics of the damping force, the damping force of the pitching motion generated by the appendage should be always opposite to the direction of longitudinal motion, so as to stabilize the pitching motion of the ship. The appendage will generate enough torque for pitching stabilization, respectively borne by the submersible body and the bow fin. The submersible body mainly generates eddy current damping, while the bow fin mainly generates dynamic lift damping, thus ensuring the structural strength of the appendage meets the requirement of the engineering design. Meanwhile, the bow fin is arranged on the submersible body, and the submersible body is mounted at the bottom of the bow without water expelling from the bow fin, thus avoiding appendage and hull vibration caused by water slamming that has been reported in foreign countries. The new built-up stabilizing appendage has made a breakthrough in two key technologies that have been problematic for pitching stabilization technologies—namely, that the structural strength of the appendage fails to meet the requirements due to excessive fluid-dynamic loads on the appendage, and that severe hull vibration has been caused by appendage slamming.

1.16.3. Application of a built-up stabilizing appendage

1. Application to a 400t class Round Bilge Boat

Targeted at a 400t-class boat, the designed waterline (DWL) length is 60m, the DWL width is 8m, the designed draft is 1.5m, and the designed displacement is about 450t. The concept configuration is designed for a new built-up stabilizing appendage based on this boat, and the mounted semi-submersible body is 12m long, 1.4m wide, and 1.2m high, with a displacement of about 5.5t; the chord length of the bow fin is 3.1m, and the wingspan length is 1.2m, as shown in Fig. 1.16.2.

Theoretical and tank-model experimental methods were used to optimize the shape, principal dimensions, and hydrodynamic layout of the built-up appendage, to carry out the engineering design, construction design, and modification for installing the appendage on the boat, and finally to complete the validation on sea trials.

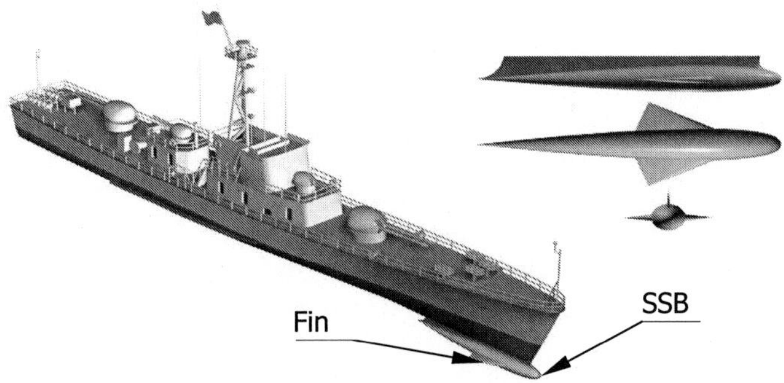

Fig. 1.16.2. 400t-class boat fixing built-up appendage

1. Tank model test

First, determine the allowed limited range of length, width, and height of the appendage according to the main factors, such as depth of the berthing harbor, alongside collision avoidance between the two boats, no anchoring operation influence, no change in the boat's floating state, and so on. Carry out tests on the calm water resistance and sea-keeping capacity on the three-meter model in a ship model towing tank in Harbin Engineering University. Fig. 1.16.3 shows a group of curves comparing calm water resistance between the boat with a built-up appendage and the original boat, in pitching, heaving and bow acceleration response among regular heading waves, of which the wave-to-course angle is 180° in head seas, 0° in stern seas, and 135° in bow seas.[1-16-3]

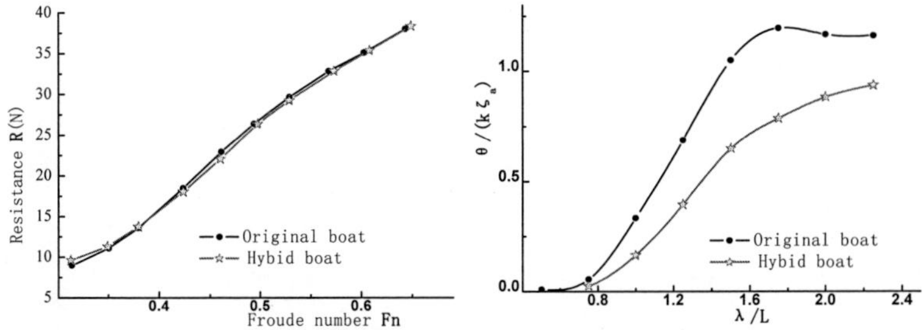

(a) Comparison of calm water resistance
(b) Comparison of pitch responses (head sea, 27kn) (head sea, 27kn)

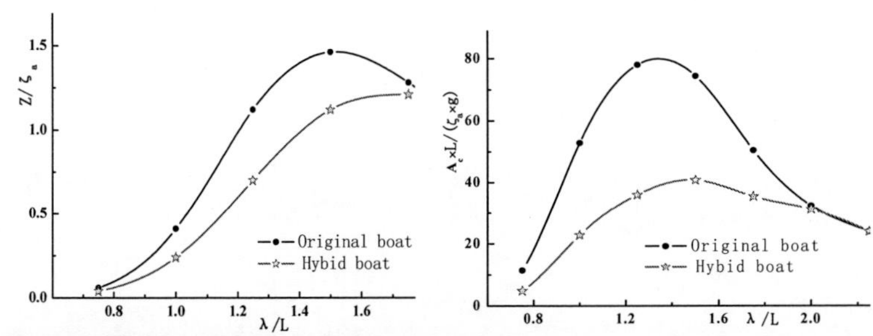

(c) Comparison of heave responses
(d) Comparison of bow acceleration responses (head sea, 27kn) (head sea, 27kn)

Fig. 1.16.3. Comparison of calm water resistance and motion response in regular waves

The test results indicate that the built-up appendage has almost no effect on the resistance of the original boat, and the pitching motion effect is remarkable.

2. Sea trial
To validate the test, after the transformation of the boat for about six months, a comparison test was conducted in respect of its speed-ability, maneuverability, sea-keeping capacity, and in respect of the effect of wave shock and vibration on the boats before and after mounting the built-up appendage in Qingdao waters between December 2001 and February 2002. The sea trials conclude that the built-up appendage has almost no impact on the speed-ability and maneuverability of the original boat, while the sea-keeping capacity has been significantly improved and the impact vibration has been significantly reduced.[1-16-4]

The sea-keeping test results of the boat fitted with the appendage S_1 and the original boat R_0 are as follows:

Sea state 3 for the first sea trial ($H_{1/3}$=1.06m): in head seas, S_1 pitching angle decreased by 29–40%, and the bow acceleration decreased by 52–68% (see Table 1.16.1 and Table 1.16.2 for details). The method for defining wave-to-course angle is the same as above.

Table 1.16.1. Motion decreasing effect of the trial in the 3rd sea state

Wave angles	Speed of original boat/kn	Speed of hybrid boat/kn	Motions	Decrease percentage/%
Head sea	18.9	18.9–19.0	Pitch angle	29.5
Head sea	18.9	18.9–19.0	Bow acceleration	61.5
Head sea	21.4	22.0	Bow acceleration	52.4
Bow oblique sea	18.0	18.5	Pitch angle	40.0
Bow oblique sea	18.0	18.5	Bow acceleration	68.7

Table 1.16.2. Motion decreasing effect of the trial in the 4th sea state

Wave angles	Speed of original boat/kn	Speed of hybrid boat/kn	Motions	Decrease percentage /%
Head sea	22	22	Pitch angle	24.7%
			Bow acceleration	51.6%
Bow oblique sea	14	14	Pitch angle	19.7%
			Bow acceleration	11.5%
Head sea	16 (unable to speed up)	22	Pitch angle	38.7%
			Bow acceleration	7.30%
Head sea	16 (unable to speed up)	24	Pitch angle	52.1%
			Bow acceleration	20.7%

Sea state 4 for the second sea trial ($H_{1/3}$=1.73m): R_0 only completed three sea trials, namely 22kn in stern seas, 14kn in bow seas, and 16kn in head seas. Prescribed by the regulations, the boat has reached the speed limit, and the accelerating test could not continue as the green sea and slamming were serious. For the safety of the boat, the sea-trial commander decided to stop R_0 testing and make a return voyage. S_1 successfully completed the sea-keeping tests of 22kn and 24kn in head seas according to the sea trial program.

The stern-sea and bow-sea tests showed that S_1 had obvious stabilizing effects on longitudinal motions. As R_0 could only reach a speed of 16kn, it could not keep pace with S_1 at a speed of 22kn or 24kn, but the sea trial results showed that S_1 had better stabilizing effects at high speeds. The boat was sailing very smoothly at a speed of 24kn in head seas, with only rare occurrences of waves or slamming.

S_1 could sail smoothly at high speeds in sea state 4. Field test results showed that the speed in waves was basically the same as the speed in calm waters as the engine ran at the same speed, from which it could be seen that the wave resistance was significantly reduced.

The sea-keeping and vibration tests of the two boats showed that S_1's vibration and impact performance were significantly improved. See the set of comparative data in Table 1.16.3.

Table 1.16.3. Comparison of impact of trial in the 4th sea state

Wave angles	Speed of original boat/kn	Speed of hybrid boat/kn	Impact probability of original boat	Impact probability of hybrid boat	Decrease percentage /%
Bow oblique sea	14	14	0.3551	0.1937	45.5
Head sea	16	22	0.1728	0.0384	77.8
Head sea	16	24	0.1728	0	100.0

Test results showed that the speed in waves was basically the same as the speed in calm waters as the engine ran at the same speed, and the wave resistance was significantly reduced; vibration tests showed that the vibration and impact performance of the boat with a built-up appendage was significantly improved.

This test was the world's first sea trial to successfully achieve longitudinal motion stabilization on a 450t-class boat.

2. Application to a 1,000t-class deep-V ship

In view of the longitudinal motion characteristics and boat shape features of 1,000t-class ships, a new 1,000t-class high-sea-keeping deep-V monohull ship was built by applying the technology of a built-up stabilizing appendage. The ship referred to the main dimension of a 1,000t-class round-bilge ship, of which the underwater part used a V-lined shape. Under the premise that the displacement and the parent ship remain basically unchanged, we appropriately increased the parallel middle body length, reduced the ship length, increased the ship width and the draft, reduced the immersion angle of the bow, increased the width of the stern, and selected the appropriate bottom angle. The stern used the square-stern shape to produce a 1,000t-class deep-V main hull. Then, the head of the main hull was fitted with the built-up stabilizing appendage to produce the 1,000t-class deep-V monohull composite ship. The DWL length of the main hull is 84m, the DWL width is 11m, the designed draft is 3.2m, and the displacement is about 1,200t; the semi-submersible body is 16m long, 2.4m wide, and 0.56m high, with a displacement of about 30t; the wingspan length of the bow fin is 1.08m, and the chord length is 5m.

The CFD method was used to optimize the principal dimension and the form of the semi-submersible body and the bow fin as well as the hydrodynamic layout of the main hull/submersible body/bow fin/transom plate, and preferably to work out a deep-V monohull ship solution, as shown in Fig. 1.16.4. The program was validated via the tank model test. The scale of the model was 1:25, and the test was conducted at the Towing Tank Laboratory of Harbin Engineering University. The test includes a calm-water resistance test and a sea-keeping test for a 1,000t-class deep-V monohull ship V_{g1} and a 1,000t-class round-bilge ship G_0. A comparison of displacement and calm-water resistance between V_{g1} and G_0 is shown in Fig. 1.16.4, and the test results of the regular waves in head seas at a speed of 24kn are shown in Fig. 1.16.5–7. The method for defining the wave-to-course angle is the same as above.[1-16-5] [1-16-6]

Fig. 1.16.4. 1000t-class hybrid deep-V monohull

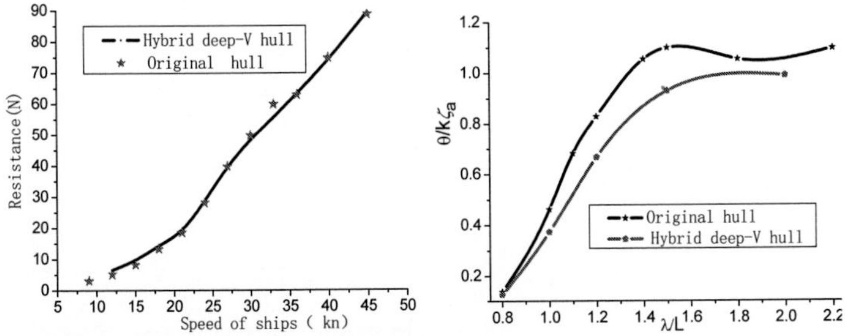

(a) Calm-water resistance of G_0 and V_{g1} models
(b) Amplitude of pitch response (24kn)

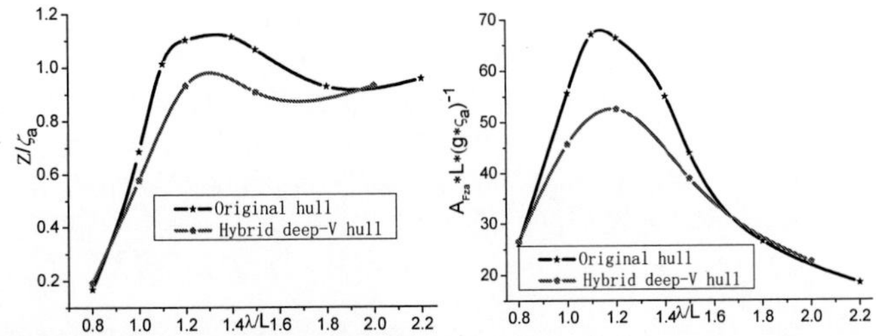

(c) Amplitude of heave response (24kn)
(d) Amplitude of bow acceleration response (24kn)

Fig. 1.16.5. Model test results

In sea state 6 ($H_{1/3}$=5m), the statistic results of the significant values for irregular wave tests on the hull motion of the two ships are shown in Table 1.16.4, and the statistical results of the wave frequency on deck are shown in Table 1.16.5.

Table 1.16.4. Significant value and decreasing percentage of motions in the 6th sea state

Ship model	Speed /kn	Pitch angle /deg.	Heave amplitude/m	Bow acceleration/m/s^2	Stern acceleration/m/s^2
G_0	18	3.1855	0.5435	4.552	2.422
	24	3.23095	0.6364	5.485	3.062
V_{g1}	18	2.82416	0.503	3.693	1.978
	24	2.67647	0.542	4.371	2.239

Table 1.16.5. Comparison of green water frequency

Sea state	Speed/kn	Frequency/Times/min	
		G_0	V_{g1}
6	18	4.76	0.51
	24	11.5	4.84

The tank model-test results show that the calm-water resistance performance of the 1,000t-class deep-V monohull ship with bow appendage is basically the same as that of the 1,000t-class round-bilge ship, and the sea-keeping capacity has increased substantially at a speed of 24kn in head seas. The stabilizing effects of all the motion response peaks were greater than 30%, and the average pitching stabilization effects were greater than 20%. The frequency of green water on deck was significantly decreased. It is clearly evident that the bow appendage has significantly improved the sea-keeping capacity of the 1,000t-class deep-V ship.

3. Application to a 4,000t-class deep-V ship

On the basis of the configuration technology of a 1,000t-class deep-V monohull ship and the characteristics of the decreasing longitudinal motion frequency of 4,000t-class ships, designs for a semi-submersible streamline, dimension, and main hull were significantly improved. Full play was given to the eddy current damping effect of the submersible body to generate adequate pitching stabilization torque and produce a 4,000t-class monohull deep-V ship, as shown in Fig. 1.16.6. The DWL length of the main hull is 125m, the DWL width is 12.5m, the designed draft is 4.5m, and the displacement is about 3,400t; the semi-submersible body is 25m long, 4m wide, and 1.5m high, with a displacement of about 62t; the wingspan length of the bow fin is 1.5m, and the chord length is 6m.

Fig. 1.16.6. Figure of a 4000t-class hybrid deep-V monohull

(1) Tank model test
As mentioned above, the optimization design was given to the primary hull and the appendage, and tank model tests were conducted on the optimized ship form. The reduced scale of the tank model test was 1:40, and the test consisted of a calm-water resistance test and a sea-keeping test of 4,000t-class deep-V monohull ship V_1 and 4,000t-class round-bilge ship S_0. The test results of model calm-water resistance and the test results of regular waves in head seas at a speed of 30kn are shown in Fig. 1.16.7. The method for defining the wave-to-course angle is the same as above.[1-16-7]

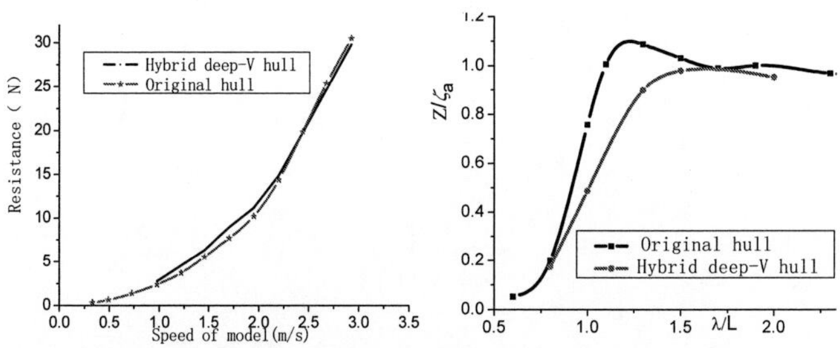

(a) Calm-water resistance of G_0 and V_{g1} models
(b) Amplitude of heave response (24kn)

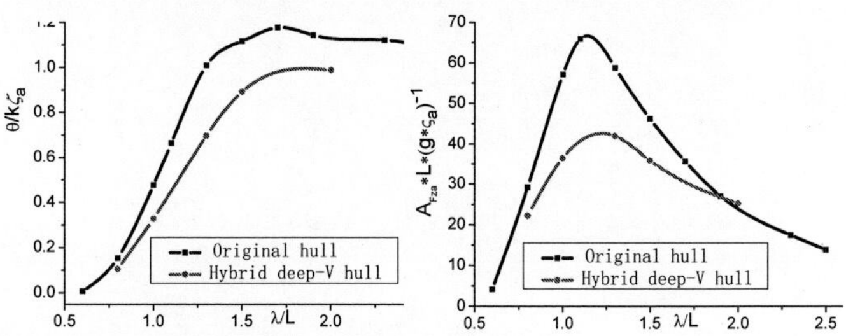

(c) Amplitude of pitch response (24kn)
(d) Amplitude of bow acceleration response (24kn)

Fig. 1.16.7. Model test results

It can be seen from the tank model-test results that, compared with S_0, V_1's calm-water resistance increased slightly at low and medium speeds (sailing speed <24kn), while the resistance decreased at high speeds (sailing speed >30kn), and the sea-keeping capacity increased substantially at a speed of 30kn in head seas. The stabilizing effects of all the motion response peaks were greater than 35%, and the average pitching stabilization effects were greater than 25%.

(2) Large-scale model test in actual sea environment conditions
To further verify the sea-keeping capacity of the ship with an appendage, large-scale models (reduced scale 1:19) of the round-bilge ship and the deep-V monohull ship underwent physical simulation tests under actual sea environment conditions in the waters of Xiaoping Island, Dalian City, in September 2009 to test the sea-keeping capacities at different speeds and in different wave directions in sea conditions equivalent to sea state 6. The test results are shown in Fig. 1.16.8, and the green sea of the models is shown in Fig. 1.16.9.[1-16-8]

(a) Original round bilge hull
(b) Hybrid deep-V hull

Fig 1.16.8. Large-scale model test in the sea area of Xiaoping Island, Dalian City, China

(a) Original round bilge hull
(b) Hybrid deep-V hull

Fig. 1.16.9. Green water pictures recorded by video camera

Fig. 1.16.10 shows the curves of pitching motion, rolling motion, and bow acceleration spectral density of the corresponding ships at speeds of 18kn and 24kn in head seas in sea state 6. The statistical results of the significant values for the motions of the two ships are shown in Table 1.16.6, and the statistical results of green sea on deck are shown in Table 1.16.7.

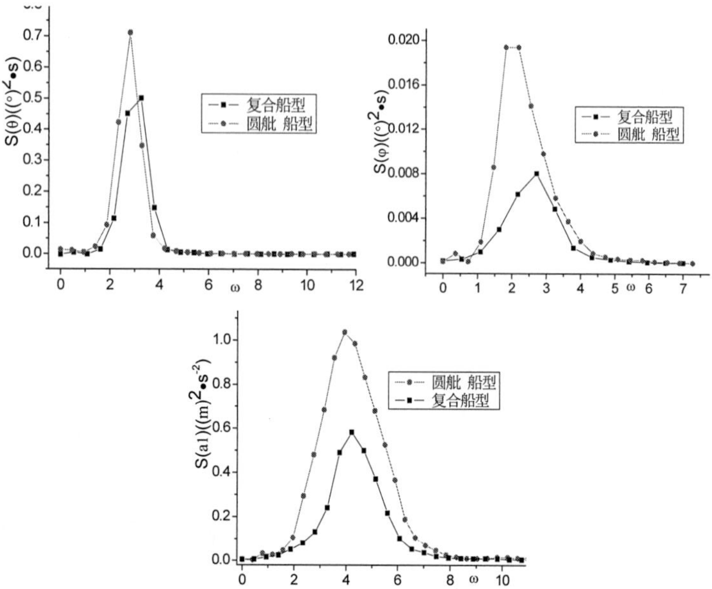

(a) Spectrum of pitch
(b) Spectrum of roll
(c) Spectrum of bow acceleration

Fig. 1.16.10. Large-scale model test results

Table 1.16.6. Significant values for motions of two models in the 6th sea state

Ship model	Speed/kn	Pitch angle/deg.	Bow acceleration/m/s^2
S_0	18	1.547	2.535
	24	1.6306	3.649
V_1	18	1.154	1.825
	24	1.18	2.35

Table 1.16.7. Green water frequency of two models

Speed/kn	Frequency/times/min	
	S_0	V_1
18	2.6	0
24	6.1	0.4
30	10.0	1.3

It can be seen from the large-scale model test results in actual sea conditions that the significant values of the pitching motion of the composite ship are at least 30% smaller than those of the round-bilge ship, and the green-sea frequency on deck has significantly decreased. It is thus clear that the calm-water resistance performance of the monohull deep-V ship based on the built-up stabilizing appendage is basically the same as that of the round-bilge ship, while its sea-keeping capacity has also been greatly improved in comparison with the round-bilge ship.

1.16.4 Conclusion

On the basis of the research above, the following conclusions can be drawn:

1. As shown from the research results of the tank model test of a 400t-class round-bilge ship fitted with a bow appendage and from sea trials, the bow appendage has few effects on the speed-ability and maneuverability of the original ship so that its sea-keeping capacity has been significantly improved, the speed loss of the waves has been greatly reduced, and the impact and vibration on the hull have been substantially reduced.
2. Through sea-keeping capacity tests in towing tanks and large-scale models in actual sea conditions, 1,000t-class and 4,000t-class deep-V monohull composite ships based on bow built-up appendage technologies and composite ship technologies show the sea-keeping capacities of these composite ships have been significantly improved with few effects on calm-water speeds.

It is thus evident that built-up appendages can dramatically improve the sea-keeping performance of round-bilge and deep-V ships under conditions that have little influence on the performance of the original ships, such as speed-ability, maneuverability, and so on.

Acknowledgments

This work is supported by the National Natural Science Fund of China (Grant No. 51209054), the National Defense Foundation (9140A140 30811CB01), and the Basic Research Foundation of HEU (HEUCFR1201).

References

[1-16-1] Cai Xingong, "Research on Hydrodynamics of Stabilizing Built-Up Appendage," Harbin: Harbin Engineering University, 2004.

[1-16-2] Liu Ya-dong, LI Ji-de, & Wang Qing, "New Concept Research on Mono-Hull Compounded Ship Form with High Sea-Keeping Performance," *Journal of Marine Science and Application* 2 (1), June 2003.

[1-16-3] Cai Xingong, Li Jide, Wang Jianfang, et al., "Experimental Study on Built-Up Stabilizing Appendage," *Shipbuilding of China* 44 (3): pp. 50–57, 2003.

[1-16-4] Cai Xingong, Li Jide, Wang Jianfang, et al., "Comparative Sea Trial of a High-Speed Ship with Built-Up Stabilizing Appendage," *Shipbuilding of China* 44 (4): pp. 13–17, 2003.

[1-16-5] Wang Xujie & Sun Shuzheng, "Research on Model Test of Thousand-Tons Class High Seakeeping Performance Hybrid Monohull," *Journal of Ship Mechanics* 15 (4): pp. 180–87, 2011.

[1-16-6] Sun Shuzheng, Li Jide, Zhao Xiaodong, et al., "Research on Longitudinal Motion Performance of Deep-V Hybrid Mono-hull," *Ship & Ocean Engineering* 38 (5): pp. 54–60, 2009.

[1-16-7] Li Jide, Zhang Heng, & Zhao Xiaodong, "Experimental Study on 4000t Class Deep-V Hybrid Monohull," *Journal of Ship Mechanics* 12 (5): pp. 709–15, 2008.

[1-16-8] Sun Shuzheng, "Research on Ship-Form and Hydrodynamics of Deep-V Hybrid Monohull," Harbin: Harbin Engineering University, pp. 69–76, 2010.

1.17

SEA-KEEPING RESEARCH ON AN AXE BOW SHIP WITH AN APPENDAGE (2014)

SHI JUN,[1] WU QINGWEI,[2] GAO XIAOPENG[2] AND DONG ZUSHUN[2]

[1] Shenzhen Hispeed Boat Tech & Dev Co. Ltd
[2] Naval University of Engineering

1.17.1. Overview

Since Delf University Daman Boatyard in Holland developed the axe bow ship, its range has expanded from hundreds of tons to one thousand tons, even thousands of tons, from monohulls to catamarans and trimarans.

The writers conducted a tank test on drag and sea-keeping comparison. The experiment was conducted at the tank of the Special Aircraft Research Institute in Jingmen, Hubei Province. The tank is 510m long, 6.5m wide, and 5m deep. The speed range of towing is 0.1–22m/s. The wave maker is equipped at the end.

The main particulars of the test vessels:

Table 1.17.1. Axe bow ship and deep-V hull main particulars

	Displacement Δ/t	Length (L_{WL})/m	Beam (B_{WL}) B_s/m	Draft FP T_f/m	Draft Aft T_a/m	$L_{cg}X_g$/m	$V_{cg}Z_g$/m
Deep-V hull	18	14.6	3.1	175	178	5.0	1.6
	22	14.6	3.1	175	185	5.5	1.6
	24	14.6	3.1	187	187	6.0	1.6
Axe bow ship	26.34	23	3.1	178	178	8.5	3.0
	28.29	23	3.1	178	190	9.0	3.0

The deep-V hull is a boat currently in service with a good track record of performance. The axe bow ship is based on the original beam of a deep-V hull, with its length extended from 14.6m to 23m (57.5% longer). Its displacement is correspondingly increased from 18t, 22t, and 24t up to 26.34t and 28.29t, an increase of 10.0–57%. Being small vessels, the shell plate is thicker than normal rules require out of consideration of corrosion. There is no significant weight increase following the length increasing. So the weight increase on structural members is small. But for a middle-size vessel, the weight increase is dramatic when the length increase. For a big vessel, it is even higher. So we chose a wide-range displacement to compare different sizes of axe bow ship.

1. Resistance contrast

In Fig. 1.17.1, which shows the resistance curves of five loading conditions, x is the Froude number and y is the resistance of unit displacement. It shows that at speed range Fr=0.4–0.9, the axe bow ship has low resistance. Especially at Fr=0.7, there is about 40% less resistance. If F_r is over 1.0, the deep-V hull will start planing. Resistance gets less. In conclusion: compared with a deep-V hull, an axe bow ship is suitable for Fr=0.4–0.9.[1-17-1]

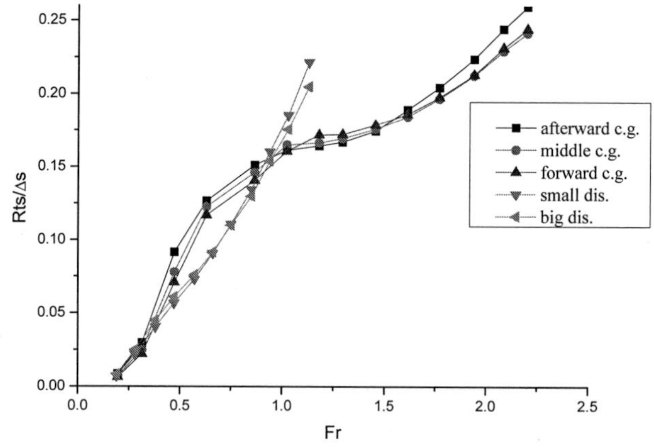

Fig. 1.17.1. Axe bow ship and deep-V hull resistance contrast

2. Contrast on sea-keeping

(1) The contrast between an axe bow ship and a deep-V hull:

Loading conditions are in middle range:

Deep-V hull displacement: Δ=22t, $L_{cg}X_g$=5.5m

Axe bow ship displacement: Δ=27t, $L_{cg}X_g$=8.5m

For performance prediction of sea-keeping based on tank test, see Table 1.17.2.

Table 1.17.2. Axe bow ship and deep-V hull sea-keeping performance

		Deep V hull				Axe bow ship			
Vs/kn	$H_{1/3}$/m	Θ	A_f	A_m	ΔR	Θ	A_f	A_m	ΔR
24	1.2	4.396	2.725	1.425	6.948	3.973	4.441	0.802	6.654
	1.6	6.399	2.532	1.009	7.035	5.673	4.081	0.742	6.665
	2.0	7.673	3.087	1.450	6.990	6.604	2.717	0.587	6.571
26	1.2	4.749	2.060	1.052	7.140	3.953	3.575	0.633	6.522
	1.6	6.887	3.235	1.630	7.217	5.687	3.691	0.698	4.466
	2.0	8.346	2.859	1.702	7.180	6.725	3.600	0.766	5.212

Table 1.17.2 shows that the pitch, ACG at amidships, and added resistance of an axe bow ship is less than that of a deep-V hull. The pitch is less at 10–20%, and the percentage increases with raises in speed and wave height. Added resistance follows the same rule. The range is between 4–38%, especially as the speed goes up. But an axe bow ship's ACG at the bow is higher than that of a deep-V hull, by about 15–80%. The ACG at amidships is less than that of the deep-V hull, by about 30–60%. The details are listed in Table 1.16.3. This is a special point about axe bow ships. Due to the small volume of the bow, the trim moment of axe bow ships is less than that of a deep-V hull and the trim lever is longer than for a deep-V. The pitch movement of an axe bow ship is smaller than that of a deep-V hull, and also adds resistance. The ACG at the bow is higher. There are two reasons for this. One is that the distance of the bow from the bow to the CG is longer than that of a deep-V hull. Another reason is that for an axe bow ship the bow always touches the water at all speed ranges. A deep-V hull goes to semi-planing when the speed reaches 24–26kn. The sensors for the ACG are out of the water and therefore the data is not accurate. The ACG at the amidships of an axe bow ship is less than that of a deep-V.

Table 1.17.3. Axe bow ship and deep-V hull added magnitude in waves

		Axe bow ship & deep-V hull			
Vs/kn	$H_{1/3}$/m	θ	A_f	A_m	ΔR
24	1.2	-9.6%	+63.0%	-43.7%	-4%
	1.6	-11.3%	+61.2%	-26.5%	-5%
	2.0	-14.0%	+12.0%	-59.5%	-6%
26	1.2	-16.8%	+79.0%	-39.8%	-8.6%
	1.6	-17.4%	+14.1%	-57.2%	-38%
	2.0	-19.4%	+25.9%	-55.0%	-27.4%

1.17.2. Axe bow ship performance advantages and further improvements

The test has proved that an axe bow ship has an advantage in resistance and sea-keeping. The reasons for this are as follows:

1. Keeping the beam and extending the waterline reduces wave-making resistance and improves sea-keeping.
2. The entry angle is reduced. Reducing the wave-making drag also reduces the changing rate of the trim moment. This creates less pitch movement.
3. The Keel line at the bow is going down. This means that the buoyancy center of the bow is also going down. The pitch is less and the change is smaller.

But axe bow ships also have some disadvantages:

1. As the length is longer, the displacement is bigger. Thus, an axe bow ship is more suitable for small craft. The shell thickness of small craft is normally bigger than rules require out of consideration for corrosion. When vessels get bigger, displacement increases slightly. But in middle-size vessels, detailed investigation must be conducted on both advantages and disadvantages.
2. It is difficult to keep on course because of the upwards bow, in particular for vessels that are driven by propellers in shallow water.

Thus, a skeg is introduced. But this makes more drag. Vessels with a water jet are unnecessary to add a skeg in deep water.

3. The axe bow ship has the same entry angle at a different draft. This makes bow movement bigger at the same time as it makes changing the rate of buoyancy smaller. It also makes a strong splash at the bow. So all axe bow ships add spray rails at the bow beneath the waterline—some also add one to two rails above the waterline.

Therefore, to make axe bow ships better, there are three ways to improve: (1) reduce drag at the same time as extending length; (2) reduce bow movement; (3) keep on course.

The normal way to keep on course is to add a skeg. We discuss this in the next chapter. To keep on course well, the size and position of the skeg can be adjusted. This will not add more drag. Let us now discuss reduced drag and bow movement.

1.17.3. Axe bow ship with an appendage

This research is on axe bow ships with an appendage to improve performance and sea-keeping. The test was conducted in the same tank as before. The normal way to do this is first to add a skeg and spray rails, then to add trim tabs to control the trim angle.

1. Axe bow ship

For convenience of adding an appendage, we made two other models to test—that is, a deep-V vessel with axe bow. The waterline lengths are 25m and 28m; Tλ=1/8. Fig. 1.17.2 shows one of them.

Table 1.17.4. Axe bow ship

Items	K	length (L_m)/m	Beam (B_m)/m	depth (H_m)/m	Displace-ment (D_m) /kg	LCG x_g /mm	z_g /mm
Axe-A	1:8	3.125	0.7424	0.35	87.9	1187.5	287.5
Axe-B	1:8	3.500	0.7421	0.35	97.7	1375	287.5

Fig. 1.17.2. Axe bow ship model

2. Added skeg
The skeg size is 400mm × 70mm. It is shown in Fig. 1.17.3.

Fig. 1.17.3. Skeg

3. Added trim tabs
The added trim tabs are 50–70mm long and extend downward, between 2 inches at a downwards angle of 2–5°, as shown in Fig. 1.17.4.

Fig. 1.17.4. Trim tables on a model

4. Added bow spray rails
There are two spray rails, 8mm wide and from 0# to 6# long (total 20#). The lower rail is above 100mm waterline at bow, 50mm at 6#. The Higher rails are 100mm above the lower one (Fig. 1.17.5).

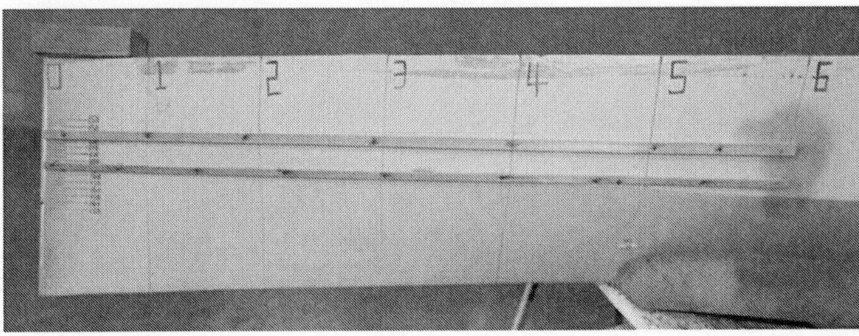

Fig. 1.17.5. Spray rails at bow

5. Added bow submerged body
The submerged body test involved two shapes: the first is a streamlined body with ellipse section. The second is the prior one, where a pair of triangular fins have been added. See Figs 1.17.6 and 1.17.7.

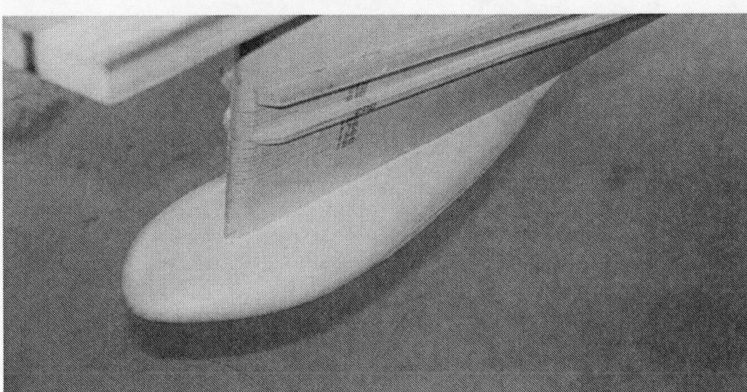

Fig. 1.17.6. Axe Bow ship submerged body

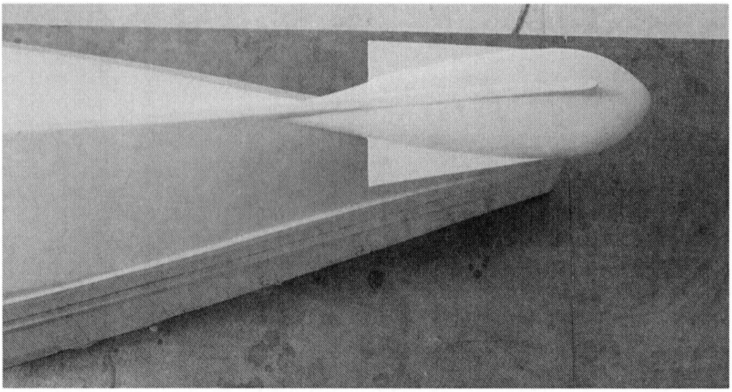

Fig. 1.17.7. Axe Bow ship submerged body with fin

6. Added wings—after adding the submerged body
The wing size is 170mm × 150mm, with a thickness of 10%, at a downward angle of 15°, and with an attack angle of 0°–2° (see Fig. 1.17.8).

Fig. 1.16.8. Axe bow ship with submerged body and wings

1.17.4. Tank test and result

For the conditions above, resistance and head waves were tested, and resistance data and movement data for head waves were generated.

1. Resistance

(1) The influence of spray rails, skeg, and trim tabs on resistance.

Fig. 1.17.9 shows the resistance curve for model Axe-A at three resistance statuses for trim tabs only, trim tabs plus skeg, and trim tabs with spray rails plus skeg.

The figure shows that at the speed range V_m=3m/s–5.8m/s, which is equal to Fn=0.54–1.04, adding an appendage as in the previous can reduce drag. The same is true for model Axe-B. The result is shown in Fig 1.17.10.

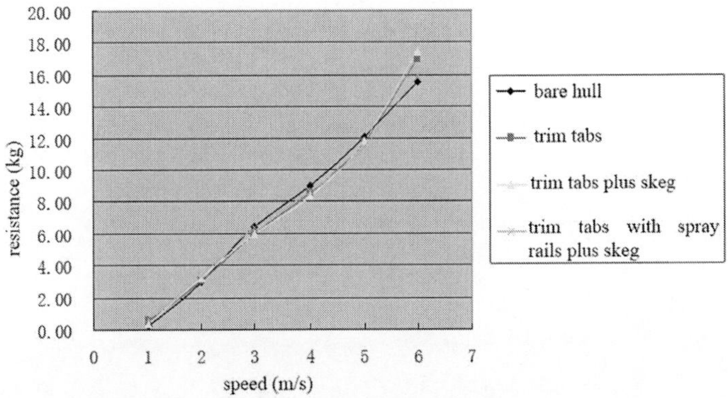

Fig. 1.17.9. Model Axe-A resistance curve with appendage

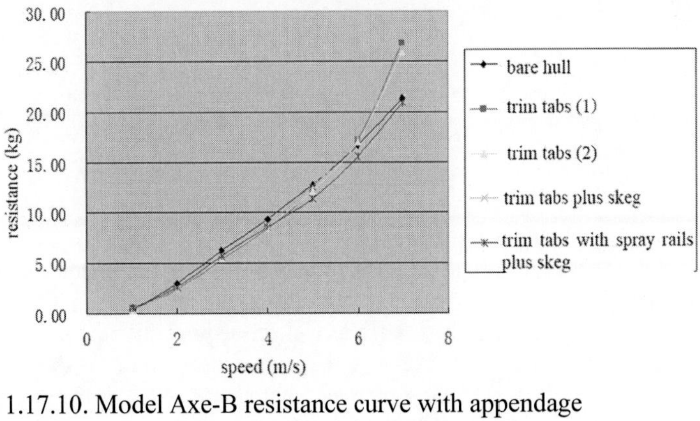

Fig. 1.17.10. Model Axe-B resistance curve with appendage

Tables 1.17.5 and 1.17.6 show the different appendages for model Axe-A and Axe-B and the resistance contrast of four combinations, as labeled.

The magnitude of the reduced drag is determined by the combinations. Obviously, it can be reduced by controlling the trim angle and trim tabs angle as well.

Table 1.17.5. Axe-A drag reduced for different combinations

speed/m/s	F_n	①	③	④	⑤	①-⑤/①
1	0.18	0.22	0.50	0.41	0.67	+206%
2	0.361	2.99	3.06	3.27	3.34	+11.7%
3	0.542	6.42	5.97	6.01	6.35	-1.1%
4	0.722	9.00	8.53	8.36	8.64	-4.0%
5	0.903	12.08	11.74	11.79	11.47	-5.0%
6	1.084	15.56	17.00	17.46	15.90	+2.2%

Table 1.17.6. Axe-B drag reduced for different combinations

speed/m/s	F_n	□	□	□	□	□	□-□/□
1	0.171	0.36	0.41	0.39	0.36	0.62	+72.2%
2	0.341	2.94	2.35	2.47	2.47	2.65	-9.7%
3	0.512	6.24	5.32	5.43	5.66	5.65	-9.5%
4	0.683	9.28	8.42	8.49	8.86	8.52	-8.2%
5	0.853	12.77	12.13	12.15	12.62	11.42	-10.6%
6	1.024	16.72	17.12	17.05	17.53	15.62	-6.6%
7	1.195	21.34	26.77	25.91	26.29	20.96	-2.2%

(2) Influence of adding a submerged body

Table 1.17.7 shows model A's resistance curve for different combinations of the submerged body, as well as for with a fin and with wings. It also compares an added appendage, such as a skeg, a spray rail, and trim tabs.

It can be seen that, for Axe-A, the speed is more than $F_n \geq 0.9$ and the resistance is reduced. It is the same when wings are added; however, it has reduced the drag. In the speed range F_n=0.5–0.8, resistance is increased by about 5–10%. But compared with the previous one, the drag increase only happens at Fn=0.5, and only at about 3%. For model Axe-B, trim tabs are better at reducing drag. This proved that changing the trim angle can make less drag.

Table 1.17.7. Axe-A added resistance submerged body and with wings

Speed V/m/s		1	2	3	4	5	6
F_n		0.180	0.361	0.542	0.722	0.903	1.084
Body with fin	ΔR/Ro%	+26.9%	+10.5%	+4.9%	+3.3%	-0.1%	-4.9%
	θ	1.25	1.30	1.72	1.68	1.47	1.07
Body with fin plus wings	ΔR/Ro%	+29.9%	+12.2%	+4.3%	+4.2%	-0.4%	-10.1%
	θ	0.66	0.76	1.56	1.87	1.82	1.77
Body plus wings	ΔR/Ro%	+32.8%	+17.7%	+5.2%	+5.0%	-0.9%	-10.0%
	θ	0.65	0.73	1.48	1.81	1.72	1.73

2. Sea-keeping
Sea-keeping was done in four different combinations for Axe-A:

1. Basic model with skeg, spray rails, and trim tabs
2. Submerged body with fin
3. Body with fin plus wings; four bodies plus wings

3. Encounter frequency
The encounter frequency curve is shown in Fig. 1.17.11–15.

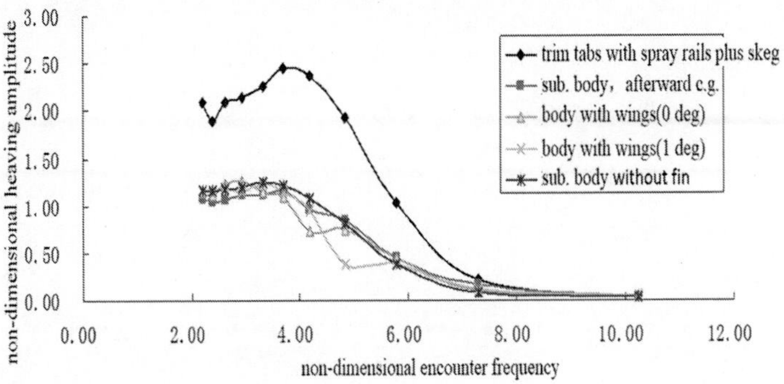

Fig. 1.17.11. Axe-A heave frequency

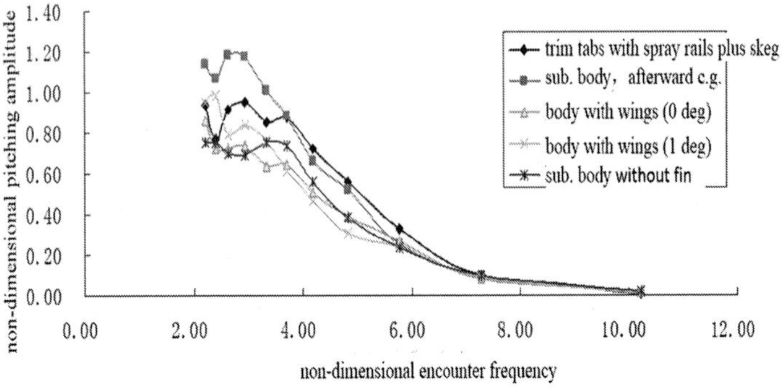

Fig. 1.17.12. Axe-A trim

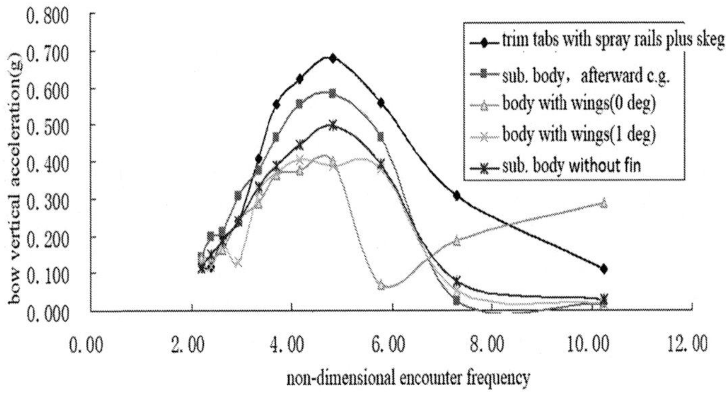

Fig. 1.17.13. Axe-A ACG at bow

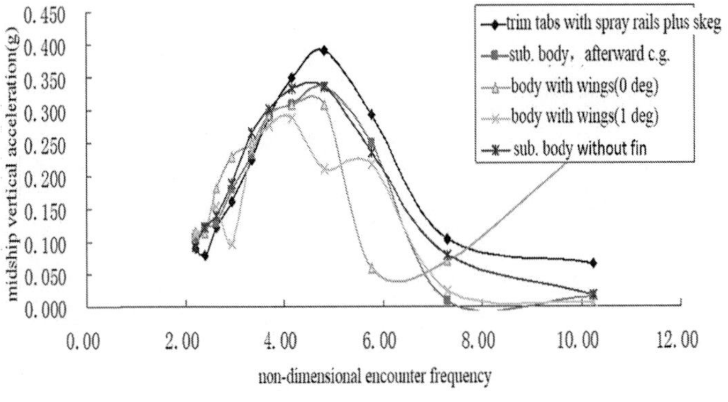

Fig. 1.17.14. Axe-A ACG at amidships

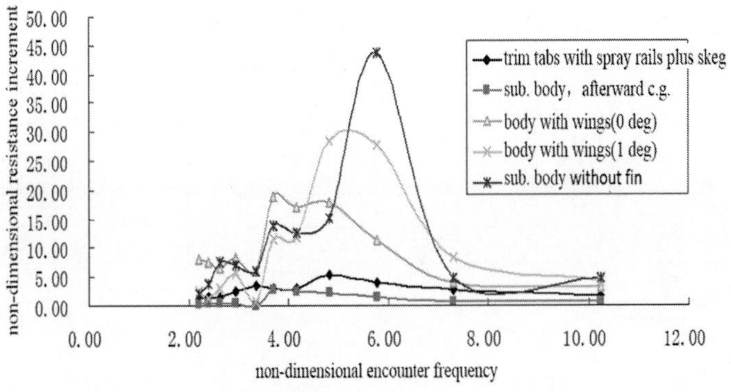

Fig. 1.17.15. Axe-A added resistance

In comparison with the previous model with skeg, spray rails, and trim tabs, these figures show:

1. All appendages lead to better heave movement. The body with fin plus wings is the best.
2. With only a submerged body with fin, there is little effect on pitch. But after adding wings, the pitch is reduced dramatically. The body with wings also has a good performance. This shows that a submerged body does not help the vessel's pitch; instead, wings do.
3. All help reduce the ACG at the bow, especially the submerged body with wings. For the ACG at amidships, a fined body with wings also helps, but not obviously.

4. For added resistance, only the one with the fined body helps. All others made it worse. As the added resistance is small, we can omit it.

4. Prediction

Table 1.17.8 lists prediction for wave height at 2m.

Table 1.17.8. Axe-A $H_{1/3}$=2.0m prediction

Combi-nations	Speed Vs/kn	Heave		Pitch		ACG at bow		ACG at amidships	
		$Z_{1/3}$ /m	%	$\theta_{1/3}$ /deg.	%	$A_{f1/3}$ /g	%	$A_{m1/3}$ /g	%
No body	22	1.6	0	6.73	0	1.62	0	0.91	0
	27.5	1.59	0	5.36	0	1.63	0	1.02	0
Subm-erged body	22	1.57	-1.9	7.42	+10.3	1.47	-9.3	0.87	-4.4
	27.5	1.57	-1.3	6.32	+17.9	1.55	-4.9	1.02	0
body with fin plus 0° wings	22	1.43	-10.6	5.08	-24.5	0.98	-39.5	0.82	-8.8
	27.5	1.53	-3.4	4.47	-16.6	1.11	-31.9	0.99	-2.9
body with plus 1° wings	22	1.64	+2.5	5.38	-20.1	1.07	-34.0	0.76	-16.5
	27.5	1.64	+3.1	5.41	+0.9	1.47	-9.8	0.99	-2.9
Body with wings	22	1.71	+6.9	5.33	-8.9	1.22	-24.7	0.90	-1.1
	27.5	1.67	+5.0	5.62	+4.8	1.37	-16.0	1.11	+8.8

Table 1.17.8 shows:

1. Compared with a status with no appendages, the body with fin only reduced the ACG at the bow by 5–10%.
2. The submerged body with wings can improve the pitch and ACG; nevertheless the heave gets worse as a result.
3. The body with 0° wings is best, leading to a reduction in heave of 5–10%, in pitch by 15–25%, and in the ACG at the bow by 30%–40%; the ACG at amidships also got smaller.
4. The appendage influence depends on speed.

1.17.5. Conclusion and analysis

1. Adding a skeg is necessary to keep on course; however, it adds drag. Adding spray rails and trim tabs can reduce resistance. It must be the correct trim angle for the vessel. The trim angle varies

depending on loading conditions. Spray rails and trim tabs also aid sea-keeping.

2. To improve sea-keeping further, adding a submerged body is beneficial. The purpose is to add damp and mass for the bow's vertical movement.
3. The research shows that adding only a submerged body or a submerged body with fin does not significantly aid sea-keeping. But this research was only tested on a submerged body. Further tests will be conduct for different body sizes and shapes.
4. Adding a submerged body with wings can improve sea-keeping in all aspects, especially in the pitch and ACG at the bow. These two factors are key points for the shipping industry.
5. The influence of a submerged body varies depending on speed. The right size and position of the submerged body will depend on the speed required. This is very important for axe bow ship design.
6. When designing a submerged body, it is necessary to think how bow spray rails and trim tabs are integrated. This will guarantee that the axe bow ship will have the best performance in trim angles, less drag, and good sea-keeping at cruising speed. In brief, when designing an axe bow ship, appendages should be integrated with the line plan.

References

[1-17-1] J. A. Kouning, J. Pinkster, & F. Van Walree, "Further Investigation into the Hydrodynamic Performance of the Axe Bow Concept," Proceedings of the WEGEMT Conference on High Performance Marine Vehicles, September 2002, Ischia Italy.

[1-17-2] Wei Song, Zeng Qing, Gao Xiaopeng, & Dong Zushun, "The Model Experiments Investigation on Resistance of Axe Bow Ship," China International Bow Show & High Performance Vessel (HPMV) Conference, Shanghai, 2011.

[1-17-3] Wu Qingwei, Zeng Qing, Gao Xiaopeng, & Dong Zushun, "Good Seakeeping of Axe Bow Ship," China International Bow Show & High Performance Vessel (HPMV) Conference, Shanghai, 2012.

CHAPTER 2

MULTIHULL SHIPS

2.1

Introduction

Since multihull ships overcome the limitations presented by the hull forms of conventional monohull ships, the seaworthiness, stability, and width of the deck area, as well as the combined hydrodynamic performance and so on, of the multihull has incomparable advantages over conventional monohull ships. At present, most research on multihull ships focuses on trimarans, SLICE craft, pentamarans, and so on. In 2002, *JANE's Fast Sea Transportation Yearbook* reported that "Since the emergence of steel ships, trimaran is the greatest progress in 140 years" (*sic*). The test vessel *RV Triton*, constructed in the UK, was launched in May 2000, and an extensive series of trails were carried out. The fast ferry *Benchijigua Express*, built in the Austal Shipyard, Australia, was also delivered to Olsen for applications. The first Independence-class Littoral Combat Ship, constructed in the United States, was launched in April 2008 and delivered to the US Navy at the end of 2009. Further similar trimarans, such as the Littoral Combat Ship (LCS), have continued to be constructed in the United States. The first search and salvage trimaran, *North Save 143*, was constructed in China and completed and commissioned in 2012. With respect to the research and investigation of trimarans, a large number of theoretical investigations, model tests, and full-scale ship tests have been carried out; compared with conventional monohulls, it has been demonstrated that the hydrodynamic performance, structural behavior, and ship layout of multihull ships is rather different from conventional monohull ships, which has led to many new techniques to resolve problems and hot spots in theoretical prediction methods, test methods, and ship and ship construction, and so on. Now, researchers are conducting correlative studies, and through practical research, design, and construction of trimarans, the technology behind and knowledge of multihull ships is accumulating and deepening, and technology, design, and construction levels are also gradually improving.

This chapter mainly describes the trimaran, stressing its hydrodynamics, performance, technology, and applications. We select some representative papers, mentioned above, which were published in the proceedings of the International Annual Conference on High Performance Marine Vessels, held in Shanghai, China, every spring, in recent years. The aim of the collected works is to introduce the research, design, construction, and applications of trimarans around the world in previous decades, and to reveal some of the critical performances and characteristics of multihull ships, so as to give readers a comprehensive understanding of up-to-date and correlative research on trimarans, pentamarans, and SLICE craft.

The first part of this chapter (sections 2.1–10) mainly introduces particular research, such as theoretical prediction methods for the hydrodynamic performance of trimarans, characteristics, hull form configuration, design technology, and proper navigational zone for the trimaran's zone, and so on. Sections 2.2 to 2.5 are based on potential flow theory; theoretical methods are introduced by Huang Debo, Sun Yumin, Lu Xiaoping, and Zhan Jinlin to predict the characteristics of resistance, sea-keeping quality, and maneuverability for different trimaran hull forms with different side-hull positions in terms of numerical calculation. In section 2.6, based on viscous theory, the application of overset grid techniques in predicting the resistance and flow field around the craft and the scaling technology of multihull ships are described by Shen Zhirong. In section 2.7 and section 2.8, selected research and investigations about innovative trimaran hull forms are introduced by Gao Lei and Zong Zhi, including semi-planing fore-body trimarans and trimarans with asymmetric side hulls. In section 2.9 and section 2.10, a full-scale trimaran car ferry is taken as an example by Tan Mei and N. A. Armstrong to introduce hull form characteristics, the design process, and the general arrangement of full-scale trimarans; in addition, their operational zone is also discussed.

The second part of this chapter (sections 2.11–13) mainly introduces SLICE craft, pentamarans, and so on. By comparing the performance of different hull forms and analyzing the operational area of full scale ships, the research status and development of the above-mentioned hull forms are discussed. In section 2.11, the wave interference effect between the main and side hulls as well as the speed characteristics of monohulls, trimarans, and pentamarans with the same displacement were compared with one another by E. Begovic; the author also analyzes and discusses the advantages and disadvantages of these hull forms. In section 2.12, Tricores, SLICE, and their characteristics, such as resistance and sea-keeping, are compared and analyzed by Dong Zushun. In section 2.13, the application, stability, structure, propeller, and maneuverability of SLICE are comprehensively discussed and summarized by John Kecsmar.

2.2

Calculations of the Resistance, Sea-Keeping Characteristics, and Optimization of High-Speed Trimarans (2006)

Huang Debo, Duan Wen Yang, Zhou Guangli, Li Yunbo, Han Kaijia and Li Jianghua

Harbin Engineering University, China

Abstract

There is an urgent demand for greater speed in marine transportation. Trimarans have become a heated research topic and it has turned out that they have a promising displacement hull form due to the various merits in hydrodynamic performance and other aspects over conventional hull forms, such as speed, sea-keeping, general arrangement, and other attractive characteristics. This paper presents some of our research on theoretical evaluation methods of the trimaran's resistance, sea-keeping, hull form, and configuration optimization. Related model tests show that the methods presented are effective. The conclusions confirmed that the trimaran, if designed properly, is a type of ship with an advantageous displacement.

2.2.1. Introduction

Peoples of all periods have attempted to design vessels with greater speed and better sea-keeping performance. The development of international world trade and water transportation requires faster vessels to compete with air shipping and trucking services. For example, containers are to be speeded up to 40–60kn, as opposed to today's usual 20–30kn. Some naval vessels follow the same trend. Since the end of the last century, many countries have developed fast-ship development projects. Through hundreds of years of effort and with great progress in construction and ship hydrodynamics, and so on, and the advent of more advanced and powerful computers and related numerical techniques, the development of shipbuilding has reached such an extent that conventional hull forms have little room left for improvement—a speed increase of a very small number of percentage points at the same power would be considered a remarkable success. This is due to the limitations of the natural rules for conventional types of hull form, which must be wide and full enough for adequate stability, space, and displacement. Among other requirements, unacceptably great power is demanded for vessels at high speeds. Unconventional hull forms are a solution in this regard. Various types of novel designs have appeared in recent decades. New techniques of planing or semi-planing, air cushions, hydrofoil-assisted craft, and catamarans and multihull ships, as well as hybrid versions of the above, burgeoned in recent years, but some are unsuitable to the high seas of the ocean.

Among displacement-type ships, the trimaran or multihull form has been paid increased attention. Though the concept of the trimaran can be traced back to the outrigger-assisted canoes of ancient Pacific islanders, and some research could be found in the 1960s, more serious development appeared only a few years ago, and the great advances have made it a heated topic. The many advantages of trimarans over monohulls and catamarans include excellent speed characteristics, high performance in sea-keeping, larger deck area, and many others; these advantages are attracting the eyes of both commercial and navy ship owners.

Trimarans such as QinetiQ's *RV Triton* and Austel's high-speed ferry *Benchijigua Express* have set a new horizon for displacement-type ships. More trimarans will come into being in a few years. Multihulls are even considered to be the "future type of ship" and a "breakthrough in water transportation." The United States, the United Kingdom, France, Russia, Italy, Australia, and various other countries have made efforts to develop this hull form.

We have studied trimarans since 1996, and have developed methods of resistance and sea-keeping performance, have optimized hull form and configuration, and have carried out some model tests, obtaining encouraging results.

2.2.2. Theoretical methods

The method of wave resistance developed by Huang Debo and Li Yunbo[2.2.11] is based on Noblesse's new slender ship theory,[2.2.10] and the method of wave steepness restriction is generalized for trimarans.

A ship of length L is assumed to move at velocity U in calm water. In the far field $\xi\to\infty$, which has the Green function

$$G=\frac{1}{\pi F_n^2}\int_{-\infty}^{\infty}\operatorname{Im} e^{kL(\xi+z)+ikL(\xi-x)\cos\theta+(\eta-y)\sin\theta}\,dt \quad (2.2.1)$$

From the Green theorem, the potential can be derived as

$$\phi(\vec{\xi})=\int_h(Gn_x-\phi G_n)da+F_n^2\int_c\left[G\left(n_x^2+t_x\phi_l-n_zt_y\phi_d\right)-\phi G_x\right]t_y dl$$

(2.2.2)

where h and c are the average wetted surface and its intersection line with the mean free surface, respectively, n and t are the hull's normal and tangential vectors respectively , and $d=n\times t$.

After some manipulation, we have the wave resistance

(2.2.3)

$$C_W=R/(0.5\rho U^2 L^2)=1/\pi\int_{-\infty}^{\infty}|K|^2\sqrt{(1+t^2)}dt$$

where R is the dimensional wave resistance, and K is the Kochin function.

$$K(t)=\frac{1}{F_n^2}\int_h(En_x-\phi E_n)da+\int_c\left[E\left(n_x^2+t_x\phi_l-n_zt_y\phi_d\right)-\phi E_x\right]t_y dl$$

(2.2.4)

and

$$E(x,\ y,\ z)\equiv e^{kLz-ikL(x\cos\theta+y\sin\theta)}=e^{kLz-ik_0L\left(\sqrt{1+t^2}x+t\sqrt{1+t^2}y\right)} \quad (2.2.5)$$

$$\left(k=k_0\sec^2\theta=k_0\left(1+t^2\right),\ k_0=g/U^2\right)$$

On the other hand, C_w can be written as

$$C_W=4\pi^3F^4\int_{-\frac{\pi}{2}}^{\frac{\pi}{2}}|A(\theta)/\lambda(\theta)|^2\cos^7\theta d\theta \quad (2.2.6)$$

where $A(\theta)$ is the wave amplitude and *l* is wave length of the elementary wave. As stated in 2.2.11, in the framework of potential theory, the discrepancy of numerical and test results could be considered from the over-estimation of the elementary wave steepness $A(\theta)/\lambda(\theta)$ for some θ, which is due to the neglecting of nonlinearities and viscosity. To compensate these, A/λ could be restricted somehow. A technique was introduced, and shown to be effective. From the relationship between *A* and *K*, the restriction can be applied to *K* instead of A/λ, and the expression of the integrand in 2.2.3 was expanded in special functions for more effective calculation, resulting in an effective and fast prediction method. This method was then extended to the case of the trimaran.

2.2.3. Optimization of hull form and configuration of trimarans

By use of thin ship theory, the relation between wave resistance C_w and half-breadth function $f(x,z)$ could be obtained. Further, the latter could be expressed in terms of tent functions, and C_w became a quadratic form of the changing-permissible offsets $y_{ij} = f\ (x_i, z_j)$. Then a nonlinear programming was used to minimize C_w, subjected to some prescribed constraints, such as block coefficient, midship section, and water surface coefficients, and so on. In the optimization process to prevent the hull from distortion, which occurs frequently in some hull optimization techniques, some penalty methods were introduced and satisfactory effects were shown. In our method, a "mixed resistance coefficient" C'_w involving the frictional resistance effect was used as the objective function instead of C_w itself. And a revision of this optimization method was created for the catamaran case. These two versions were applied respectively to the center hull and the outriggers, and the optimized components were combined with the configuration optimization to find the optimal trimaran. The prediction of C_w was completed by the better theoretical method, as stated in section 2.2.1, rather than by the thin ship theory.

As to the configuration optimization of the trimaran—the forms, sizes, and locations of the three components have a significant impact on resistance (for a certain range of speed). Some researchers use a serious and a great number of tests in various combinations to find the optimal configuration, and then produced practical charts for design use. We used both enumeration and a genetic algorithm to search for the optimal combination and configuration of the three hulls, which were made possible by our effective and fast resistance prediction method. The results showed that, in the low range of Froude number F_n, the location of the outriggers affected sensitively on C_w, but was not significant in magnitude; whereas in the higher F_n range resistance was found smaller when the outriggers were located far from the midship and toward the two ends of the center hull. It was found that the length of the outrigger should not exceed one third of that of the main hull, to avoid worsening the total resistance due to greater frictional resistance.

2.2.4. Sea-keeping performance prediction

A two-and-a-half-dimensional method was developed to predict the sea-keeping characteristics of the trimaran.[2.2.12] Assuming that the speed is high and the conditions required for the strip theory are met, $\phi_j (j = 1 \sim 7)$ satisfies

$$\begin{cases} \dfrac{\partial^2 \phi_j}{\partial y^2} + \dfrac{\partial^2 \phi_j}{\partial z^2} = 0 \quad (\text{in } D) \\ \left[(i\omega - U\partial / \partial x)^2 + g\partial / \partial z \right] \phi_j = 0 \quad (z = 0) \\ \dfrac{\partial \phi_j}{\partial n} = \begin{cases} i\omega n_j + U m_j & (j = 2 \sim 6) \\ -\partial \phi_0 / \partial n & (j = 7) \end{cases} \quad (\text{on } S) \\ \phi_j = \partial \phi_j / \partial x = 0 \quad (x > x_0) \end{cases} \tag{2.2.7}$$

with X_0 being the bow position.

Introducing the transforms $t(x) = (-x + x_0)/U$ and $\psi_j(t) = e^{i\omega t} \phi_j(t)$, one has

$$\begin{cases} \partial^2 \psi_j / \partial y^2 + \partial^2 \psi_j / \partial z^2 = 0 \\ \partial^2 \psi_j / \partial t^2 + g \partial \psi_j / \partial z = 0 \quad (z = 0) \\ \dfrac{\partial \psi_j}{\partial n} = \begin{cases} (i\omega n_j + U m_j) e^{i\omega t} & (j = 1 \sim 6) \\ -\partial \psi_0 / \partial n e^{i\omega t} & (j = 7) \end{cases} \quad (\text{on } S(t)) \\ \psi_j = \partial \psi_j / \partial t = 0 \quad (t = 0) \end{cases} \tag{2.2.8}$$

The integral equation of the source distribution function can be derived from the Green theorem by satisfying the boundary condition.

$$2\pi\psi_j(p,t) = -\int_{s(t)} \sigma(q,t)\ln(r/r')ds_q + \int_0^t d\tau\{\int_{s(\tau)} \sigma(q,\tau)\widetilde{G}ds_q -$$

$$-\frac{1}{g}\sigma\widetilde{G}n_y\Big|_{\eta=A,\zeta=0}\left[\dot{A}(\tau)\right]^2 + \frac{1}{g}\sigma\widetilde{G}n_y\Big|_{\eta=B,\zeta=0}\left[\dot{B}(\tau)\right]^2\}$$

(2.2.9)

and

$$-\pi\sigma(p,t) + \int_{s(t)} \sigma(q,t)\partial/\partial n_p \ln(r/r')ds_q = -2\pi\partial\psi_j/\partial n_p +$$

$$+\int_0^t d\tau\{\int_{s(\tau)} \sigma(q,\tau)\frac{\partial\widetilde{G}}{\partial n_p}ds_q - \frac{1}{g}\sigma\frac{\partial\widetilde{G}}{\partial n_p}n_y\left[\dot{A}(\tau)\right]^2 + \frac{1}{g}\sigma\frac{\partial\widetilde{G}}{\partial n_p}n_y\left[\dot{B}(\tau)\right]^2\}$$

(2.2.10)

where A and B are the intersection coordinates of the hull with the free surfaces port and the starboard, respectively. Solving these for σ and using the Bernoulli equation produces the following:

$$p_j(x,y,z) = -\rho(i\omega\phi_j - U\partial\phi_j/\partial x) = -\rho e^{-i\omega t}\partial\psi_j/\partial t$$

the hydrodynamic forces upon the hull will be

$$T_{ij} = \iint_S p_j n_i ds = -\rho\int_L e^{-i\omega t(x)}dx\left[\int_{S[x(t)]}\partial\psi_j/\partial t dl\right] \quad (2.2.11)$$

The diffraction force can be obtained likewise. And hence

$$\begin{cases}[(M+A_{33})(-\omega^2)+C_{33}+i\omega B_{33}]\xi_3 + [(A_{35})(-\omega^2)+C_{35}+i\omega B_{35}]\xi_5 = F_3\\ [(A_{53})(-\omega^2)+C_{53}+i\omega B_{53}]\xi_3 + [(I_{22}+A_{55})(-\omega^2)+C_{55}+i\omega B_{55}]\xi_5 = F_5\end{cases}$$

(2.2.12)

where $\eta_3(t) = \mathrm{Re}[\xi_3 e^{i\omega t}]$ and $\eta_5(t) = \mathrm{Re}[\xi_5 e^{i\omega t}]$ are heave and pitch displacements, respectively, and

$$T_{ij} = \omega^2 A_{ij} - i\omega B_{ij} \quad (2.2.13)$$

The STF and the two-and-a-half methods were used for added masses, damping coefficients, and diffraction force F_j^D. With these having been acquired, the equations (2.2.12) can be solved for motion response.

2.2.5. Calculations, optimization, and model tests

The numerical methods mentioned above were validated by various calculations, and the confirmations showed that they are reliable. The optimization method was used to obtain several optimum trimarans, for example, FA_1 and FA_2. Relevant models were made, and some different configurations were arranged for both resistance and sea-keeping tests. There are trimaran models FA_1-A, FA_1-B, FA_2-A, FA_2-B, and so on, with -A indicating the outriggers near the midships and -B indicating those near the sterns.

A serious of comparisons between prediction and test results showed satisfactory confirmation, see Figs 2.2.2–6, indicating that the hull optimization and performance prediction methods developed by us are effective, reliable, and have very little CPU time in calculation. The confirmation in total resistance between prediction and tests in Fig. 2.2.2 is remarkable (where the last higher test point should be discarded).

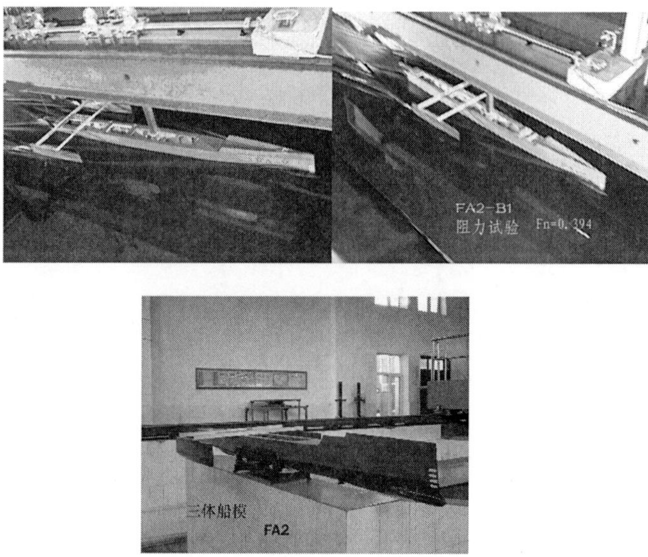

Fig. 2.2.1. Model test of optimized trimarans FA_2-A and FA_2-B

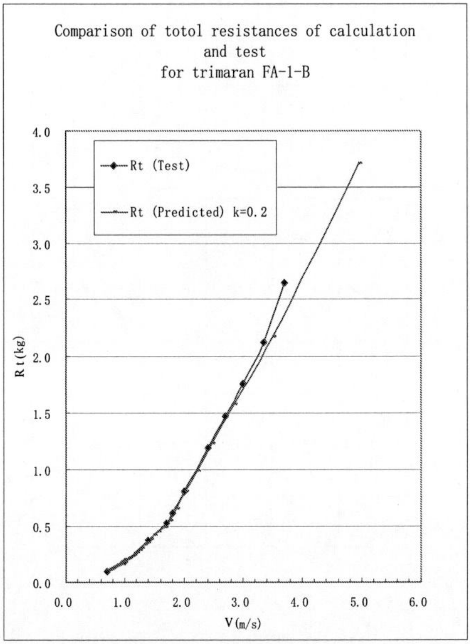

Fig. 2.2.2. Comparison between predicted and measured (shorter curve) total resistance of trimaran model FA_1-B

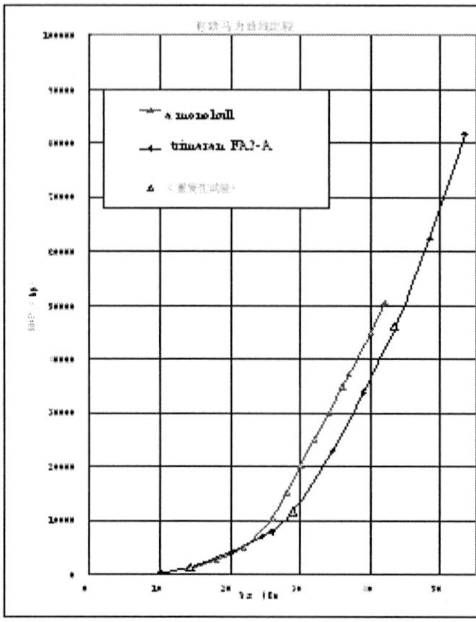

Fig. 2.2.3 A typical comparison between the EHP of a monohull (shorter curve) and trimaran FA_2-A, having the same displacement

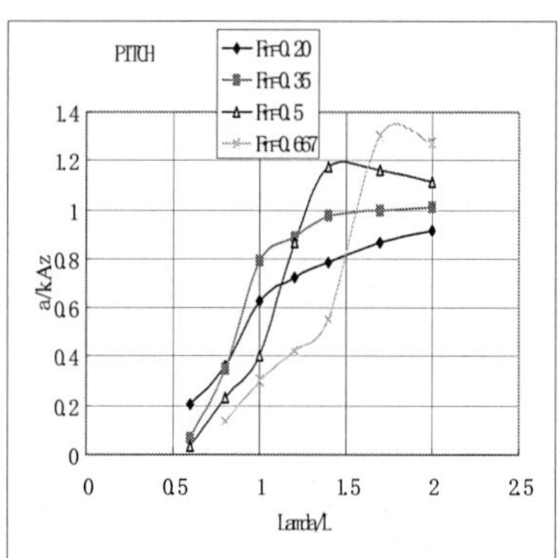

Fig. 2.2.4. ROA of pitch of trimaran model FA_I-B Fig.

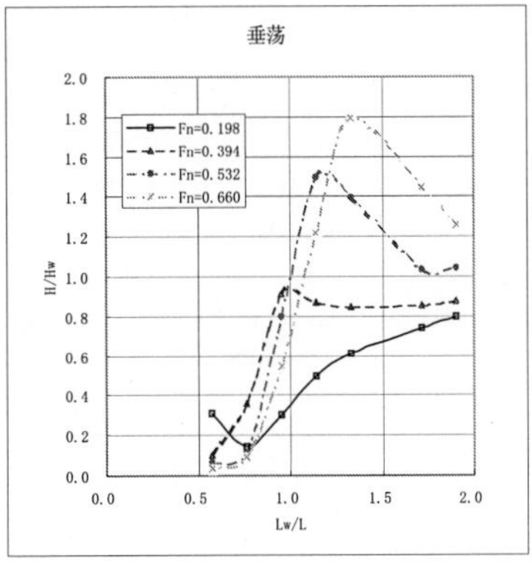

2.2.5. ROA of heave of trimaran model FA_2-B

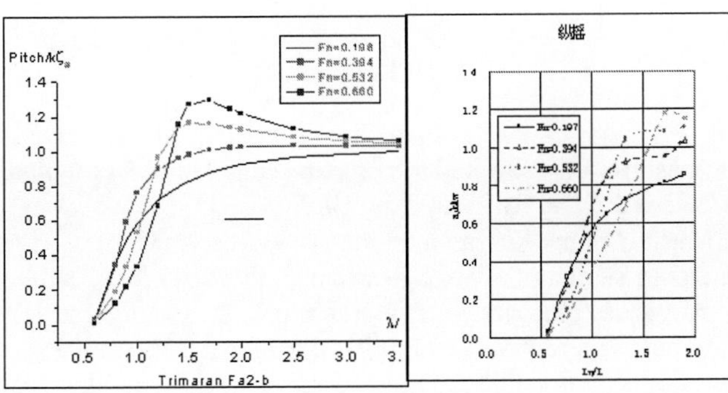

(a) (b)

Fig. 2.2.6. Comparison between calculation (a) and test (b) results of pitching of trimaran model FA_2-B in regular waves

The resistance tests for the optimized trimaran models showed that they have very low resistance at high speeds, with 15–20% less effective horsepower than the monohull form with the same displacement. But at lower speeds, they appear no better than the latter, mainly due to their larger wetted surface area. As to sea-keeping—for example, pitch—the trimaran showed an attractive performance, especially at higher speeds (see Fig. 2.2.4).

2.2.6 Conclusion

The authors' theoretical and experimental studies on calculations of resistance, sea-keeping, the optimization of hull form, and the configuration of trimarans showed that the methods we developed are effective and reliable, and that trimarans, if designed properly, will have very good hydrodynamic performances. In addition, as shown by other researchers, the trimaran has many other attractive features, which make it a highly promising type of displacement vessel with high speeds and very good sea-keeping performance.

In addition to this research, our team has completed various investigations into flow calculations with some viscous hydrodynamic solvers, and more are ongoing. They will be reported later.

We believe that, in the near future, the trimaran will become one of the most popular ship forms; and, in regard to oceangoing fast displacement vessels, it will be "a dominant ship of the new generation."

References

[2.2.1] B. Ackers, et al., "An Investigation of the Resistance Characteristics of Powered Trimaran Side-Hull Configurations," *SNAME Transactions* 105, 1997, pp. 349.

[2.2.2] S. Narita, "Some Research on the Wave Resistance of a Trimaran," International Seminar on Wave Resistance, Japan, 1976, p. 381.

[2.2.3] D. Andrews & J. Zhang, "Trimaran Ships, the Configuration for the Frigate of the Future," *Naval Engineering Journal*, May 1995.

[2.2.4] N. Takarada, et al. "R & D of a Displacement Type Ship (Part 2)," FAST 1993.

[2.2.5] J. Linstrom et al., <Super-Slender Mono-Hull with Outriggers," FAST 1995, p. 295.

[2.2.6] H. Langenberg, "Fast Displacement Ships—An Economical Option for High Speed Transportation," FAST 1995.

[2.2.7] N. Takarada et al., "R & D of a Displacement-Type High Speed Ship (part 1: A Conceptual Design Study)," *Proceedings of the Society of Naval Architects of Japan* 170.

[2.2.8] K. Suzuki & M. Ikehata, "Studies on Minimization of Wave Making Resistance for High Speed Trimaran with Small Outriggers," *Proceedings of the Society of Naval Architects of Japan* 177.

[2.2.9] Huang DeBo, Han KaiJia, & Zhao Wei, "A Study of Trimaran Ovoid Type Ship," 3rd International Conference for High Performance Marine Vehicles, 2000, Shanghai.

[2.2.10] F. Noblesse, "A Slender-Ship Theory of Wave Resistance," *J. S. R.* 1, March 1983.

[2.2.11] Huang DeBo & Li YunBBo, "Ship Resistance Based on Noblesse's Slender Ship Theory and Wave Steepness Restriction," *S. T. R.* 44, 1997.

[2.2.12] Duan Wenyang, Huang Debo, D. Hudson, & G. Price, "Comparison of Two Sea-Keeping Prediction Methods for High Speed Multi-hull Vehicles," FAST 2001, September 4–6, 2001, Southampton, UK.

[2.2.13] C. Hsiung, "Optimal Ship Forms for Minimum Resistance," *J. S. R.* 25 (2), 1981.

[2.2.14] Li Jianghua & Huang Debo, "Study of the Relation between Configuration of Trimaran and its Resistance," *Journal of Harbin Engineering University* 4 (23), 2002 (in Chinese).

[2.2.15] Huang DeBo, Han KaiJia, & Zhao Wei. "A Study of Trimaran Ovoid Type Ship," 3rd International Conference for Performance Vehicles, November 2000, Shanghai.

[2.2.16] Huang Debo, "Recent Development of High Speed Performance Displacement Vessels," 7th National Conference on Hydrodynamics, September 2005, Harbin (in Chinese).

2.3

On Resistance and Initial Transverse Stability of a High-Speed Transport Trimaran (2012)

Sun Yuming, Lu Xiaoping and Wang Zhong

Naval University of Engineering, China

Abstract

A target high-speed transport trimaran was designed using the Australian high-speed trimaran *Benchijigua Express* as the parent type. The resistance of this vessel was forecast on the basis of the linear wave-making thin-ship theory. It is shown that the speed from the resistance theoretical calculation under the main machine power MCR is close to that from *Benchijigua Express*'s full scale test results, and that the linear wave-making theory can be employed to predict the resistance of the practical high-speed trimaran. The wave-form diagrams of the target high-speed trimaran were calculated with the potential nonlinear panel methods presented as discrimination measures of the wave interference state among the piece hulls and as the design guide for the outrigger location. Monohull transverse stability calculation methods are extended to high-speed trimarans, based on which the transverse metacentric height and the stability arm curve initial segment of the target high-speed trimaran are calculated and analyzed.

Introduction

Since the late 1990s, the research and development of high-speed trimarans with small outriggers has become the mainstream trimaran hull configuration. It has received increasing interest from China and around the world, and several full scale high-speed trimarans have been put into operation already. This shows that the new high-speed trimaran hull form is about to enter a stage of commercial and military applications, which in turn will promote further study of the hull form design and hydrodynamic performance of high-speed trimarans in order to reinforce the lack of research in this area. However, research on high-speed trimarans from a practical engineering background is rare in China. On the basis of the high-speed trimaran wave resistance theoretical prediction method and a former study of trimaran stability, the present paper carried out theoretical calculations and feature analysis on the resistance and initial transverse stability of an engineering practical high-speed transport trimaran.

Australia built the world's largest aluminum car-passenger freight trimaran, *Benchijigua Express*, in 2005; the main hull is 126.96m, with a speed of 40kn at 500t dead weight, and 45kn at 4m significant wave height, which is of important business value. A target high-speed trimaran was designed with the Australian high-speed trimaran *Benchijigua Express* as the parent hull form. This was done to promote high-speed transport trimaran engineering applications in China. The resistance and transverse metacentric height of the target high-speed transport trimaran was predicted based on wave-making theory and stability calculation methods for trimarans, which show that the wave-making theory can be applied to the resistance prediction of engineering practical high-speed trimaran. Moreover, the wave-form diagrams from the potential nonlinear panel methods can be used as the discrimination measures of the wave interference state among the piece hulls and as the design guidance of the outrigger location. In addition, the monohull transverse stability calculation methods were extended to the high-speed trimarans, based on which the transverse metacentric height and the stability arm curve initial segment of the target high-speed trimaran were calculated and analyzed.

The high-speed trimarans in the present paper have practical engineering application prospects, and the jobs studied in the present paper provide technical support and foundation for the application of high-speed trimarans in China.

2.3.1. Calculation and analysis of high-speed trimaran resistance

2.3.1.1. General hull characteristics and geometry configuration of the target trimaran

On the basis of the plan lines and the principal dimensions of the car-passenger-freight trimaran *Benchijigua Express* given in the literature,[2.3.1] a new high-speed trimaran was designed, as shown in Figure 2.3.1. Its parent hull form plan line and configuration analysis and principal hull characteristics are listed in table 2.3.1. The center of the outriggers is located 14.627m beside and 29.4m behind the center of the main hull.

Table 2.3.1. General hull characteristics

	L/m	B/m	D/m	C_B	Δ/m^3	S/m^2
Main	127	9.6	4	0.565	2573	1526.658
Outrig	56	1.64	1.146	0.621	65.3	301.628

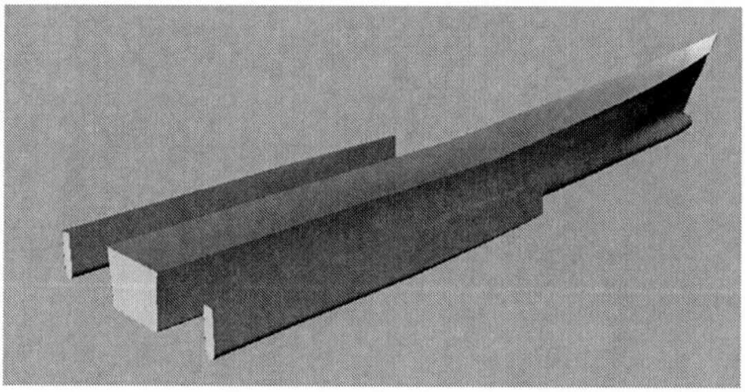

Fig. 2.3.1. Surface geometry

2.3.1.2. Linear wave-making thin-ship theory for high-speed trimarans

Linear wave-making thin-ship theory for high-speed trimarans is conducted from the linear superposition principle of the Kochine function and linear wave-making theory for monohulls.[2.3.2] The formula is expressed as:

$$R_W = \frac{8\rho g^2}{\pi U^2}\left(\frac{BT}{2}\right)^2 \int_0^\infty [(I_0 + I_1 + I_2)^2 + (J_0 + J_1 + J_2)^2] \frac{(u^2+1)^2}{\sqrt{u^2+2}} du \tag{2.3.1}$$

where: I_0, J_0, I_1, J_1, I_2, and J_2 denote Kochine functions of the piece hulls, and could be written as

$$I_0 = \int_{-1}^{0}\int_{-1}^{1} \frac{\partial \eta}{\partial \xi} \exp(G\zeta)\cos(R\xi)\mathrm{d}\xi\mathrm{d}\zeta \tag{2.3.2}$$

$$J_0 = \int_{-1}^{0}\int_{-1}^{1} \frac{\partial \eta}{\partial \xi} \exp(G\zeta)\sin(R\xi)\mathrm{d}\xi\mathrm{d}\zeta$$

(2.3.3)

$$I_1 = \int_{-1}^{0}\int_{-1}^{1} (\frac{B_1 T_1}{BT}) \frac{\partial \eta}{\partial \xi} \exp(G_1\zeta)\cos\{R_1[(\xi+\overline{a})+u\sqrt{u^2+2}\cdot\overline{p}]\}\mathrm{d}\xi\mathrm{d}\zeta$$

(2.3.4)

$$J_1 = \int_{-1}^{0}\int_{-1}^{1} (\frac{B_1 T_1}{BT}) \frac{\partial \eta}{\partial \xi} \exp(G_1\zeta)\sin\{R_1[(\xi+\overline{a})+u\sqrt{u^2+2}\cdot\overline{p}]\}\mathrm{d}\xi\mathrm{d}\zeta$$

(2.3.5)

$$I_2 = \int_{-1}^{0}\int_{-1}^{1} (\frac{B_1 T_1}{BT}) \frac{\partial \eta}{\partial \xi} \exp(G_1\zeta)\cos\{R_1[(\xi+\overline{a})-u\sqrt{u^2+2}\cdot\overline{p}]\}\mathrm{d}\xi\mathrm{d}\zeta$$

(2.3.6)

$$J_2 = \int_{-1}^{0}\int_{-1}^{1} (\frac{B_1 T_1}{BT}) \frac{\partial \eta}{\partial \xi} \exp(G_1\zeta)\sin\{R_1[(\xi+\overline{a})-u\sqrt{u^2+2}\cdot\overline{p}]\}\mathrm{d}\xi\mathrm{d}\zeta$$

(2.3.7)

The following expressions have been introduced in the above formulas:

$$\xi = \frac{x}{l},\ \eta = \frac{y}{B/2},\ \zeta = \frac{z}{T},\ \overline{a} = \frac{a}{l_1},\ \overline{p} = \frac{p}{l_1}, L = 2l$$

$$K_0 T = \frac{1}{F_n^2}(\frac{T}{L}) \quad , \quad K_0 T_1 = \frac{1}{F_n^2}(\frac{T_1}{L}) \quad , \quad K_0 l = \frac{1}{2F_n^2} \quad ,$$

$$K_0 l_1 = \frac{1}{2F_n^2}(\frac{l_1}{l});$$

$$G = (\frac{T}{L})(\frac{1}{F_n^2})(u^2+1)^2,\ G_1 = (\frac{T_1}{L})(\frac{1}{F_n^2})(u^2+1)^2;$$

$$R = (\frac{1}{2F_n^2})(u^2+1),\ R_1 = (\frac{l_1}{l})(\frac{1}{2F_n^2})(u^2+1)$$

2.3.1.3. Panel methods of the potential nonlinear wave-making theory

On the basis of the literature,[2.3.3] [2.3.4] [2.3.5] the theoretical model and numerical method of the nonlinear potential wave-making theory panel methods for high-speed trimarans are summarized here. Trimaran speed U is assumed along the x-axis direction. In the Cartesian coordinate system *o-xyz* on board, the *xy*-plane is coincident with the calm water surface, the *x*-axis points to the bow, the *y*-axis points to the starboard, and the *z*-axis upwards. Following the potential wave-making theory, the flow field satisfies the following equations and boundary conditions:

$$\nabla^2\Phi = 0\ ,\ z\leq \zeta(x,y) \tag{2.3.8}$$

where: ζ denotes wave height.

The kinematic boundary condition on the free surface $z=\zeta(x,y)$ is as

$$\zeta_t + \Phi_x\zeta_x + \Phi_y\zeta_y = \Phi_z \tag{2.3.9}$$

The dynamic boundary condition on the free surface is as

$$\Phi_t + \frac{1}{2}\nabla\Phi\cdot\nabla\Phi + g\zeta = \frac{1}{2}U^2 \tag{2.3.10}$$

The velocity potential is as

$$\Phi = -Ux + \phi \tag{2.3.11}$$

The hull's surface condition is as

$$\nabla\phi\cdot\mathbf{n} = Un_x \tag{2.3.12}$$

where n denotes the unit normal vector of the hull surface, and n_x denotes the x component of n. Put formula 2.3.11 into formula 2.3.9 and 2.3.10 and move all the linear parts to the left of the equation, and the numerical calculation procedures can be expressed as

$$\frac{1}{\Delta t}\zeta^{n+1} - U\zeta_x^{n+1} - \phi_z^{n+1} = \frac{1}{\Delta t}\zeta^n - \frac{1}{2}\{(\phi_x^n\zeta_x^n - \phi_y^n\zeta_y^n) + (\phi_x^{n-1}\zeta_x^{n-1} - \phi_y^{n-1}\zeta_y^{n-1})\} \tag{2.3.13}$$

$$\frac{1}{\Delta t}\phi^{n+1} - U\phi_x^{n+1} + g\zeta^{n+1} = \frac{1}{\Delta t}\phi^n - \frac{1}{2}(\frac{1}{2}\nabla\phi^n\cdot\nabla\phi^n + \frac{1}{2}\nabla\phi^{n-1}\cdot\nabla\phi^{n-1}) \tag{2.3.14}$$

Make the time interval infinity, $\Delta t \to \infty$, and the iterative procedures of the kinematic boundary and dynamic boundary condition are transformed into

$$-U\zeta_x^{n+1} - \phi_z^{n+1} = -\frac{1}{2}\{(\phi_x^n \zeta_x^n + \phi_y^n \zeta_y^n) + (\phi_x^{n-1}\zeta_x^{n-1} + \phi_y^{n-1}\zeta_y^{n-1})\} \quad (2.3.15)$$

$$-U\phi_x^{n+1} + g\zeta^{n+1} = -\frac{1}{2}\{\frac{1}{2}\nabla\phi^n \cdot \nabla\phi^n + \frac{1}{2}\nabla\phi^{n-1} \cdot \nabla\phi^{n-1})\} \quad (2.3.16)$$

where $n-1$ and n, $n+1$ denote iteration steps.

2.3.1.4. Calculation methods of frictional resistance and form resistance

The frictional resistance formula of high-speed trimarans is $R_f = 1/2\rho U^2 SC_f$, where S denotes the wet surface area of the trimaran, S_1, S_2 denotes the wet surface area of the main hull and the outrigger, and $S = S_1 + 2S_2$. C_f could be calculated with the following formula:

$$C_f = \frac{C_{f1}S_1 + 2C_{f2}S_2}{S_1 + 2S_2} \quad (2.3.17)$$

where C_{f1}, C_{f2} denotes the frictional resistance coefficient of the main hull and the outrigger, and C_{f1}, C_{f2} by ITTC-57 Formula. The form factor *K* for the form resistance can be determined according to empirical methods.

2.3.1.5. Calculation results and validation

The resistance target of the high-speed trimaran is predicted on the basis of the methods mentioned above. The wave-resistance prediction results from the linear wave-making thin-ship theory are shown in Figure 2.3.2 (a), and the whole trend is rather reasonable. The target is a high-speed trimaran with transoms. According to the former studies,[2.3.4] [2.3.5] the theoretical prediction of wave resistance for transom trimarans is much more difficult than that of ordinary monohulls, so that the calculations should be verified by the information produced by the experiment. Related model tests are carried out to validate the results of the full nonlinear wave-resistance prediction for the target high-speed trimarans. To verify the effectiveness of the present preliminary theoretical results, the target trimaran speed under the main machine power MCR was predicted according to the above theoretical resistance curve. The main machine power is 4 × 8,200kW, according to the parent trimaran, *Benchijigua Express*, and it is assumed that P.C. = 0.65. The target trimaran speed is obtained as V = 37.6kn, shown in Fig. 2.3.2 (b), which is close to the service speed of the parent trimaran, V = 38kn, and the linear wave-making thin-ship theoretical calculation results are validated.

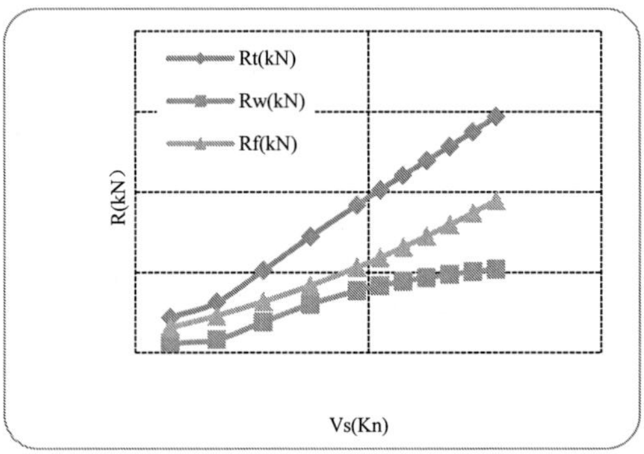

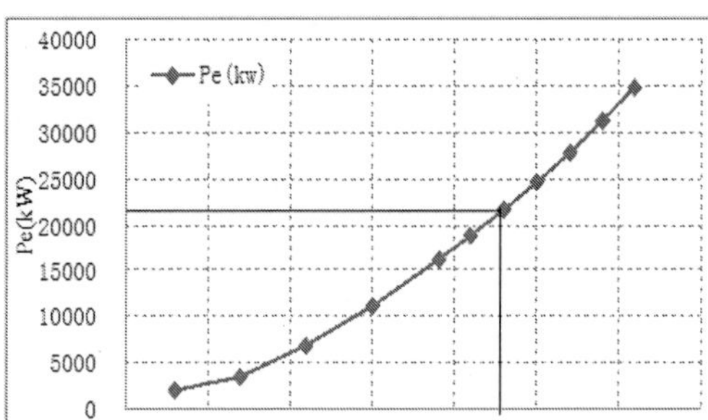

(a) Resistance calculated by linear method
(b) Speed prediction

Fig. 2.3.2.

To describe the wave interference state between the main hull and the outriggers, the wave-form diagram of the target high-speed trimaran was calculated by the nonlinear potential wave-making theory panel method at different speeds, shown in Fig. 2.3.3. The computational field was extended to 0.5*L* upstream, 0.7*L* sideward, and 1*L* downstream, while 1,750 mesh grids and 1,998 mesh grids are created on the trimaran hull surface and free water surface, respectively. The wave-form diagram shows the wave crest and the wave trough. It is shown from the previous investigation that the favorable wave interference emerged when the outrigger bow in the wave trough was produced by the main hull. As shown in Fig. 2.3.3, favorable wave interference emerged when the target high-speed trimaran was at speeds of 22–30kn, and even more favorable wave interference can be realized by moving the outriggers further backward.

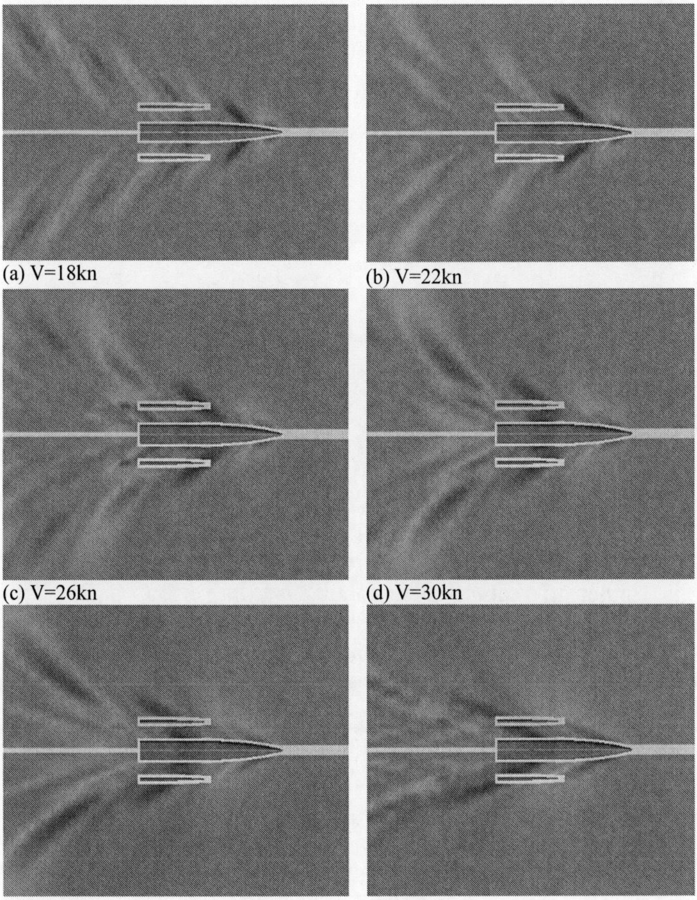

(a) V=18kn (b) V=22kn

(c) V=26kn (d) V=30kn

(e) V=34kn (f) V=44kn

Fig. 2.3.3. Wave-form diagram

2.3.2. Calculation and analysis of initial transverse stability of the target trimaran

2.3.2.1. Initial transverse stability calculation formula

The transverse stability of high-speed trimarans is much better than that of monohulls, due to the great water plane moment of inertia provided by the outriggers. However, the metacentric height of catamarans is not so big and it doesn't incur the fast roll. Moreover, the metacentric height of trimarans is highly affected by the clearance between the main hull and the outriggers and by the water plane area of the outriggers, so that we can adjust its metacentric height by varying the outrigger clearance to realize the multi-objective optimization for the resistance reduction, stability improvement, and sea-keeping performance reinforcement. Therefore, it is important to calculate and analyze the stability performance of present high-speed trimarans.

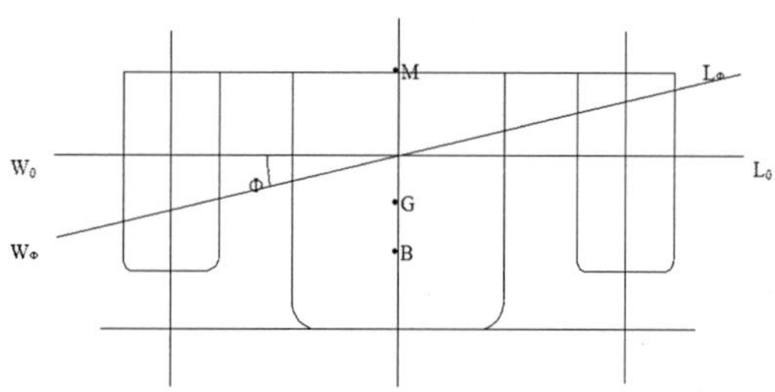

Fig. 2.3.4. Static transverse stability of trimarans

The calculation principle of a trimaran's transverse metacentric height (*h*) is similar to that of a monohull, but the effect of the outrigger water-plane area should be taken into account to allow the formula for the monohull's initial stability calculation to be suitable to be extended to the high-speed trimaran situation. As shown in Fig. 2.3.4, take W_0OW_Φ and L_0OL_Φ as the wedges with vacancy, and the formula for the trimaran initial stability calculation procedures can be expressed as [2.3.6]

$$BM = [I_0 + 2(I_1 + A_{WS}b^2)]/\nabla \qquad (2.3.18)$$

$$GM = z_b + BM - z_g \qquad (2.3.19)$$

where I_0 and I_1 denote the water-plane inertia moments of the main hull and outriggers around their own longitudinal central axes; A_{WS}, the water plane area of the outriggers; b, the horizontal clearance between main hull and outriggers; ∇ the total displacement, where $\nabla = \nabla_0 + 2\nabla_1$, and ∇_0,

∇_1 the displacement of main hull and each outrigger; z_b, z_g the height of buoyancy center and gravity center, respectively, and

$z_b = (\nabla_0 z_{b0} + 2\nabla_1 z_{b1})/\nabla$ (2.3.20)

where z_{b0}, z_{b1} are the height of the buoyancy centers of main hull and the outriggers, respectively.

2.3.2.2. Upright side-wall formula for high-speed trimaran stability

The upright side-wall formula for calculation of the transverse stability contains the water-plane inertia moment, displacement, gravity center height, and so on. According to the upright side-wall assumption, the straight side-wall formula is suitable for the high-speed trimaran stability calculation in wide-angle range, so that the straight side-wall formula of the monohull can be extended to the high-speed trimaran stability arm calculation, the calculation procedures of which can be expressed as

$$GZ = \sin\theta(GM + \frac{1}{2}BM\tan^2\theta) \qquad (2.3.21)$$

where Gm and BM are conducted from the related parameters of high-speed trimarans.

2.3.2.3. Calculation results and analysis

For estimation, z_b and z_g of the target high-speed trimaran are 2.67m and 9.6m, respectively, and the transverse metacentric height calculation results of the present high-speed trimaran is shown in Table 2.3.2. The transverse metacentric height of it is much bigger than for monohulls, but it is not so big just, like a catamarans. In addition, the transverse metacentric height of those trimarans similar to the present target can be controlled between 3m and 10m by adjusting the clearance between the main hull and the outriggers or adjusting the z_g of the high-speed trimarans. The stability arm curves of target high-speed trimaran calculated by the straight side-wall formula and initial stability formula are drawn in Fig. 2.3.5, which shows that the initial stability formula is suitable for the target high-speed trimaran in a smaller heel-angle range than that of an ordinary monohull. In other words, the straight side-wall formula is suited to a larger heel-angle range than the initial stability formula in the target high-speed trimaran. A comparison of the stability arm curves for the *Benchijigua Express*, the target trimaran, and a catamaran is shown in Fig. 2.3.6, where the transverse metacentric height of *Benchijigua Express* is about 4m, surveyed from the related curve in the figure. As shown in Fig. 2.3.6, *Benchijigua Express* and the target trimaran hull all meet a satisfactory stability status, fulfilling the initial stability and large-angle stability requirements, with its initial stability that is much better than a monohull, while not being so big as to incur a fast roll like a catamaran.

Table 2.3.2. Transverse metacentric height calculation

Projects	Water plane area /m^2	I /m^4	Displacement /m^3	H /m
Results	1033.38	21749.88	2703.6	16.24732

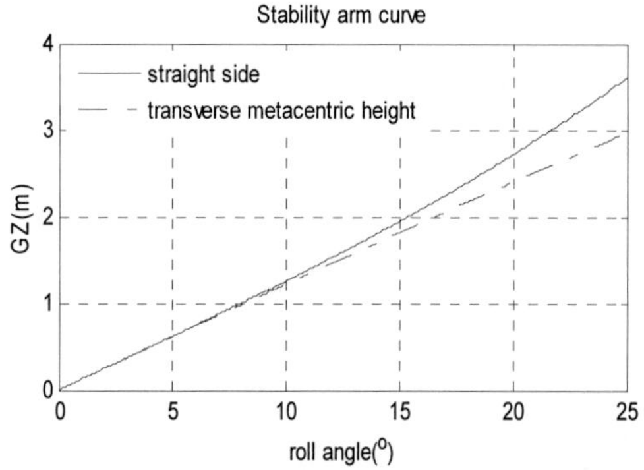

Fig. 2.3.5. Stability arm curves by different calculations

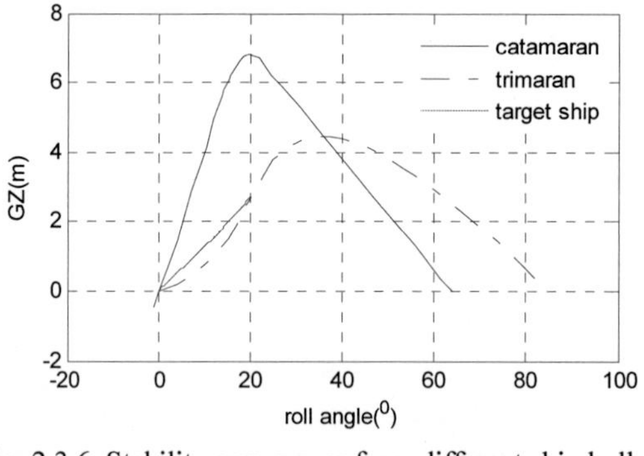

Fig. 2.3.6. Stability arm curves from different ship hulls

2.3.3. Summary

A new high-speed transportation trimaran, taking the Australian high-speed trimaran *Benchijigua Express* as the prototype, was designed. The linear wave-making thin-ship theory, the nonlinear potential wave-making theory panel method and the initial stability calculation methods for the high-speed trimaran are presented. The theoretical calculations of the resistance and the metacentric height were carried out based on those methods. The main conclusions from the present jobs are summarized as follows:

1. The linear wave-making thin-ship theory predicts the high-speed trimaran resistance performance reasonably, which is validated with the service speed information of the parent high-speed trimaran, *Benchijigua Express*, and is applicable for the quick prediction of resistance in the practical engineering design of high-speed transportation trimaran hulls.
2. The wave-form diagrams of the target trimaran can be calculated by the nonlinear potential wave-making theory panel method at different speeds and can be used as the discrimination measures of the wave interference state among the piece hulls and for design guidance for the outrigger location.
3. The initial transverse stability calculation formula is suitable for high-speed trimarans in a smaller heel-angle range than for ordinary mononhulls, and the straight side-wall formula is more suitable in a larger heel angle than the initial stability formula for high-speed trimarans.
4. The transverse stability of high-speed trimarans is much better than for ordinary monohulls, and they don't roll as fast, like catamarans, and can be adjusted in a large range by the outrigger position design. This allows the realization of the multi-objective optimization for resistance reduction, stability improvement, and sea-keeping performance reinforcement.

References

[2.3.1] T. Armstrong et al., "A New Generation of Large Fast Ferry—From Concept to Contract Reality," Proceedings of the 8th International Conference on Fast Sea Transportation, 2003, Ischia, BERTRAM, V, pp. 75–84.

[2.3.2] C. W. Dawson, “Calculations with the XYZ Free Surface Program for Five Ship Models,” Proceedings of the Workshop on Ship Wave Resistance Computation, DTNSRDC, Bechesda, MD, November 1979: pp. 232–55.

[2.3.3] D. C. Wyatt, “Development and Assessment of a Nonlinear Wave Prediction Methodology for Surface Vessels,” *Journal of Ship Research* 44 (2): pp. 96–107, 2000.

[2.3.4] Wang Zhong, Lu Xiaoping, & Wang Wei, “A Solution to Wave Making Problem Using Full Nonlinear Free-surface Boundary Condition,” *Journal of Huazhong University of Science and Technology* (Nature Science Edition) 37 (8): pp. 112–15, 2009.

[2.3.5] Wang Zhong, “Nonlinear Wave Resistance Numerical Research and Resistance Characteristics Analysis and Resistance Reduction for Trimarans,” Wuhan, Naval University of Engineering, 2010.

[2.3.6] Cheng Hangtao. “The Trade-off Analysis of Resistance and Stability of Trimaran Design,” Wuhan, Naval University of Engineering, 2010.

2.4

An Investigation into the Wave Resistance of High-Speed Trimarans (2003)

Lu Xiaoping and Pan Yucun

Dept. of Naval Architecture & Ocean Eng., Naval Univ. of Engineering, Wuhan, China

Abstract

On the basis of the linear wave-resistance principle, a wave-resistance formula was established for high-speed trimarans with a main center hull and two small side hulls. Following the present wave-resistance formula, the regulation of the wave interference effect from each of the piece hulls and the trend of the wave-resistance curve were investigated. In addition, the method for optimizing the side hull position from the viewpoint of reducing wave resistance was studied, incorporating the contour graph method and the simple enumeration method.

2.4.1. Introduction

The underwater piece hulls of a special type of high-speed trimaran are constructed with a main center hull and two small outer hulls, which could be called small outer hull trimarans (SOHT). This paper discusses special high-speed trimarans (SOHTs), which provide spacious deck surfaces as one of their main advantages. Because of their particular construction, high-speed trimarans (SOHTs) have lower resistance, excellent sea-keeping qualities, and obvious virtue in general arrangement and stealth compared with single hull ships or catamarans. In particular, high-speed SOHTs cancel the waves by the wave interference effect from each of the piece hulls so as to reduce wave resistance; they also alleviate "diagonal shake" and "fast roll," which often occurs in catamarans (damping, pitching, and heaving); they shield noise from underwater ship hulls; and they incorporate better RCS. In present high-speed battleship designs, designers pursue satisfactory comprehensive performance in contrast to the former design ideas, which mainly included a consideration of speedy performance. Now, for example, more attention is paid to sea-keeping quality and stealth. Therefore, in recent years unconventional ship forms have been studied and applied more than ever, and their sizes have tended to increase. High-speed trimarans manifest obvious virtue in these aspects, so that high-speed trimarans are a desirable platform for the frigates of the future.

In the later 1970s, the concept of a trimaran was proposed abroad and a monograph on this aspect was published, which investigated the resistance of trimarans in shallow water and deep water in detail as one of its main contents. Soon afterwards, various papers were published that studied the hydrodynamic performance of trimarans theoretically, performed model tests, or carried out analysis of ship types. The excellent speed and sea-keeping of trimarans were validated in these initial studies. This type of ship made a favorable impression on some ship firms for its outstanding speed and sea-keeping performance, and a few trimarans were put into service, for example, *Adventurer*, a trimaran made in Ireland in the later 1990s. In recent years, high-speed trimarans gained the attention of the military, for example, that of the United States, Great Britain, and some other NATO countries. The British have invested a large fund for pre-research into trimarans. A test trimaran for warship application was built by the British military in 2000. This information shows the broad prospects for trimarans in the military. However, references to trimaran research and design are deficient.

In China, Harbin Engineering University engaged in some tentative theoretical studies and model tests on the wave resistance of the trimarans from the later 1990s,[2.4.1] [2.4.2] which proved the advantages of the trimaran initially.

On the basis of the linear wave-resistance principle and a mathematic (Wigley) ship form, the wave resistance on the special high-speed trimarans (SOHTs) was studied to establish the wave-resistance formula. From the present wave-resistance formula, the regulation of the wave interference effect from each of the piece hulls and the trend of the wave-resistance curve were investigated; in addition, the method for the outer hull position optimization from the viewpoint of reducing wave resistance was studied, incorporating the contour graph method and the simple enumeration method.

2.4.2. The computation method of the wave resistance on trimarans

From the linear wave-resistance theory, the wave resistance of the single hull ships could be written as:

$$R_w = \frac{8\ \rho g^2}{\pi\ U^2} \int_0^{\infty} (I^2 + J^2) \frac{(u^2+1)^2}{\sqrt{u^2+2}} du \qquad (2.4.1)$$

where I and J express the cosine portion and the sine portion of the generalized Kochine function, respectively, written as:

$$I = \int_{-T}^{0} \int_{-l}^{l} f_x \exp[k_0 z (u^2+1)^2] \cos[k_0 (u^2+1)(x + u\sqrt{u^2+2}y)] dxdz$$

$$J = \int_{-T}^{0} \int_{-l}^{l} f_x \exp[k_0 z (u^2+1)^2] \sin[k_0 (u^2+1)(x + u\sqrt{u^2+2}y)] dxdz$$

(2.4.2)

$$k_0 = \frac{g}{U^2}$$

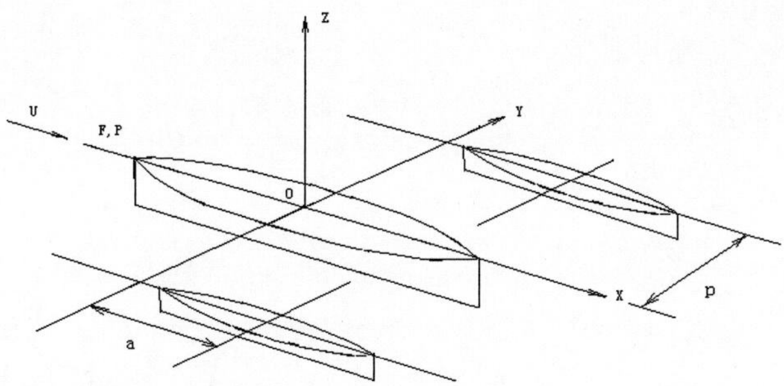

Fig. 2.4.1. Cartesian coordinate

Following the linear superposition principle of the Kochine functions, formulas 2.4.1 and 2.4.2 are able to extend to the linear wave-resistance formula of trimarans. If the origin of the Cartesian coordinate is put on the center hull, shown as Fig. 2.4.1, where the ordinate and abscissa of the outer hull center points are (a, p), (a, -p), the cosine portion and the sine portion of the generalized Kochine function for trimarans were able to be rewritten as:

$$I = I_0 + I_1 + I_2 \quad I_0, I_1, I_2;$$
$$J = J_0 + J_1 + J_2$$

J_0, J_1, J_2 refer to the cosine portion and the sine portion of the generalized Kochine function of the center hull and the outer hull of the trimarans, respectively. Thus, the wave-resistance formula of trimarans was able to be reformed as follows:

$$Rw = \frac{8\rho g^2}{\pi U^2}(\frac{BT}{2})^2 \int_0^\infty [(I_0 + I_1 + I_2)^2 + (J_0 + J_1 + J_2)^2] \frac{(u^2+1)^2}{\sqrt{u^2+2}} du$$

$$= Rw_0 + 2Rw_1 + Rw_{012} + Rw_{12}$$

(2.4.3)

$$Rw_0 = \frac{8\rho g^2}{\pi U^2}(\frac{BT}{2})^2 \int_0^\infty (\overline{I_0}^2 + \overline{J_0}^2) \frac{(u^2+1)^2}{\sqrt{u^2+2}} du$$

(2.4.4)

$$Rw_1 = \frac{8\rho g^2}{\pi U^2}(\frac{BT}{2})^2 \int_0^\infty (\overline{\overline{I_1}}^2 + \overline{\overline{J_1}}^2) \frac{(u^2+1)^2}{\sqrt{u^2+2}} du$$

(2.4.5)

$$Rw_{012} = \frac{8\rho g^2}{\pi U^2}(\frac{BT}{2})^2 \times 2\int_0^\infty ([2\cos(R_1 \overline{a})\cos(R_1 u\sqrt{u^2+2}\,\overline{p})][\overline{J_0}\overline{\overline{J_1}} + \overline{I_0}\overline{\overline{I_1}}]$$

$$[2\sin(R_1 \overline{a})\cos(R_1 u\sqrt{u^2+2}\,\overline{p})][\overline{J_0}\overline{\overline{J_1}} - \overline{I_0}\overline{\overline{J_1}}] \frac{(u^2+1)^2}{\sqrt{u^2+2}} du$$

(2.4.6)

$$Rw_{12} = \frac{8\rho g^2}{\pi U^2}(\frac{BT}{2})^2 \times 2\int_0^\infty \cos(2\ u\sqrt{u^2+2}\,\overline{p}\ )[\ \overline{\overline{I_1}}^2 + \overline{\overline{J_1}}^2\ ]$$

$$\frac{(u^2+1)^2}{\sqrt{u^2+2}}du \quad (2.4.7)$$

where $\overline{I_0}$, $\overline{J_0}$, $\overline{\overline{I_1}}$, $\overline{\overline{J_1}}$ refer to the non-dimensional generalized Kochine function, written as:

$$\overline{I_0} = \int_{-1}^{0}\int_{-1}^{1} \frac{\partial \eta}{\partial \xi}\exp(G\zeta)\cos(R\xi)d\xi d\zeta$$

$$\overline{J_0} = \int_{-1}^{0}\int_{-1}^{1} \frac{\partial \eta}{\partial \xi}\exp(G\zeta)\sin(R\xi)d\xi d\zeta$$

$$\overline{\overline{I_1}} = \int_{-1}^{0}\int_{-1}^{1} (\frac{B_1T_1}{BT})\frac{\partial \eta}{\partial \xi}\exp(G_1\zeta)\cos(R_1\xi)d\xi d\zeta$$

$$\overline{\overline{J_1}} = \int_{-1}^{0}\int_{-1}^{1} (\frac{B_1T_1}{BT})\frac{\partial \eta}{\partial \xi}\exp(G_1\zeta)\sin(R_1\xi)d\xi d\zeta$$

In the proceeding deductions, the following expressions were introduced:

$$k_0T = \frac{1}{F_n^2}(\frac{T}{L}), k_0T_1 = \frac{1}{F_n^2}(\frac{T_1}{L}), k_0l = \frac{1}{2F_n^2}, k_0l_1 = \frac{1}{2F_n^2}(\frac{l_1}{L})$$

$$G = (\frac{T}{L})(\frac{1}{F_n^2})(u^2+1)^2, G_1 = (\frac{T_1}{L})(\frac{1}{F_n^2})(u^2+1)^2$$

$$R = (\frac{1}{2F_n^2})(u^2+1), R_1 = (\frac{l_1}{l})(\frac{1}{2F_n^2})(u^2+1)$$

2.4.3. The computation example

The piece hulls of the trimarans are generally slender. At the stage of the regular analysis and the investigation into the arrangement of the piece hulls, for simplicity without losing significant results, the piece hulls were taken as the mathematic ship form, the non-dimensional ship form function of which is $\eta(\xi,\zeta)=(1-\xi^2)\times(1-\zeta^2)$, and the piece hulls were assumed to be similar to each other. Thus,

$$L_c=(\sqrt[3]{\frac{\nabla}{\nabla_0}})L_z,\qquad B_c=(\sqrt[3]{\frac{\nabla}{\nabla_0}})B_z,\qquad T_c=(\sqrt[3]{\frac{\nabla}{\nabla_0}})T_z\quad,$$

$$S_c=(\sqrt[3]{\frac{\nabla}{\nabla_0}})^2 S_z$$

Let the wave-resistance coefficient be $Cw=\dfrac{Rw}{\frac{1}{2}\rho U^2 S}$, where S is the sum of the entire piece-hull surface area. Subscription C denotes two outer piece hulls and Subscription Z denotes the center piece hull. For the mathematic piece hull that was adopted, the wave-resistance formula of the trimarans was simplified as

$$Rw_0=\frac{8\rho g^2}{\pi U^2}(\frac{BT}{2})^2\int_0^\infty [(-2)(\frac{-2\cos(R)}{R}+\frac{2\sin(R)}{R^2})$$

$$(\frac{e^{-G}(2+2G)-(2-G^2)}{G^3})]^2\frac{(u^2+1)^2}{\sqrt{u^2+2}}du \tag{2.4.8}$$

$$Rw_1=Rw_2=\frac{8\rho g^2}{\pi U^2}(\frac{BT}{2})^2\int_0^\infty (\frac{B_1T_1}{BT})^2(\frac{e^{-G_1}(2+2G_1)-(2-G_1^{\,2})}{G_1^{\,3}})^2$$

$$[(-2)(\frac{-2\cos(R_1)}{R_1}+\frac{2\sin(R_1)}{R_1^{\ 2}})]^2\frac{(u^2+1)^2}{\sqrt{u^2+2}}du$$

(2.4.9)

$$Rw_{012}=\frac{8\rho g^2}{\pi U^2}(\frac{BT}{2})^2\times 2\int_0^\infty(\frac{B_1T_1}{BT})[2\cos(R_1\bar{a})\cos(R_1u\sqrt{u^2+2}\bar{p})]$$

$$[(\frac{e^{-G_1}(2+2G_1)-(2-G_1^{\ 2})}{G_1^{\ 3}})$$

$$(-2)(\frac{-2\cos(R_1)}{R_1}+\frac{2\sin(R_1)}{R_1^{\ 2}})\ (-2)(\frac{-2\cos(R)}{R}+\frac{2\sin(R)}{R^2})$$

$$(\frac{e^{-G}(2+2G)-(2-G^2)}{G^3})]\frac{(u^2+1)^2}{\sqrt{u^2+2}}du$$

(2.4.10)

$$Rw_{12}=\frac{8\rho g^2}{\pi U^2}(\frac{BT}{2})^2\times 2\int_0^\infty \cos(2\ R_1\ u\sqrt{u^2+2}\bar{p}\ )(\frac{B_1T_1}{BT})^2$$

$$(\frac{e^{-G_1}(2+2G_1)-(2-G_1^{\ 2})}{G_1^{\ 3}})^2\ [(-2)(\frac{-2\cos(R_1)}{R_1}+\frac{2\sin(R_1)}{R_1^{\ 2}})]^2$$

$$\frac{(u^2+1)^2}{\sqrt{u^2+2}}du \qquad (2.4.11)$$

Combining the non-dimensional mathematical ship form and the main ship sizes, the ship form could be described as:

1. The water line length of the center ship hull equals 100m; the water line broad of that equals 8m; the draft of that equals 3.555m; the block coefficient of that equals 0.44444; the displacement of that equals 1264m^3, the wet surface area of that equals 945.17m^2

2. The proportion of the single outer hull and the main hull $\frac{\nabla}{\nabla_0}=\frac{1}{20}$, the water line length of the outer hull equals 36.8403m; the broad of that equals 2.947m; the draft equals 1.3098m; the block coefficient equals 0.44444; the displacement equals 63.2m^3; the wet surface area equals 128.279m^2.

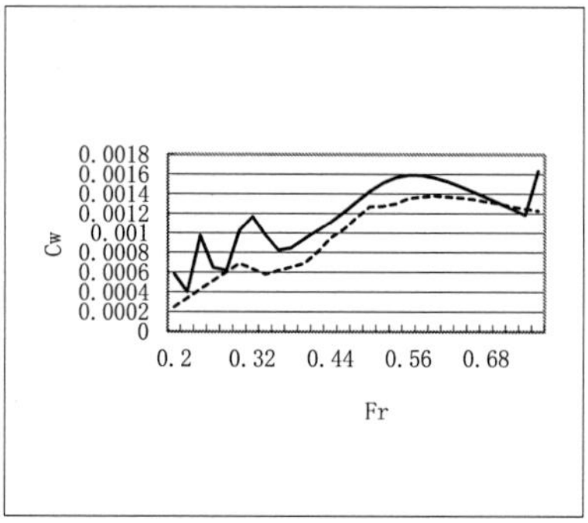

Fig. 2.4.2. The wave-resistance curves of the trimarans

3. The centers of the outer hulls, expressed as (a,p), are (40m, 15m) (40m–15m), respectively.

The wave-resistance curve of computation (solid) is illustrated in Fig. 2.4.2, the regular of which is consistent with that of the relevant reference[2.4.3] (dotted). The result gives a validation to the propriety of the wave-resistance formula for the trimarans, which was applicable for the investigation of the optimum problems of the piece-hull arrangement and the regular wave-resistance of the trimarans.

2.4.4. The problems of piece-hull position optimization

2.4.4.1 The tentative program for (*a,p*) selection by a simple enumeration algorithm

The problems of the piece hull arrangement of the trimarans are rather complex. The decisive factors about the optimum outer hull arrangement include the position of the outer hulls, the size and ship form of the outer hulls, and the distribution of the displacement on the outer hulls and the center hulls. As the first step, the wave-resistance formula for the mathematic piece-hull form built in the last section was introduced; the computation programs for this were made with some slight changes to develop a simple enumeration program for the selection of the best outer piece-hull arrangement. In the programs, the enumerated parameter *a* changes from -50 m to 50m and *p* changes within 8–40m, when the minimal value of Cw and the corresponding (*a, p*) are primarily obtained in the proceeding. The change steps of *a* and *p* are 1m, letting *F*r = 0.3, 0.5, 0.7. Moreover, let $\frac{\nabla}{\nabla_0} = \frac{1}{20}, \frac{1}{40}, \frac{1}{60}$ for each Froude number. Then there are nine groups of selected $(a, p)_{opt}$ and corresponding Cw. Because the objective ships optimized in the present paper are trimaran escorts, the main sizes of the considered trimarans are determined after summarizing some reference of the escorts and researching the characters of the escorts. After detailed consideration, the main sizes of the trimarans were determined, and the outer hull sizes were taken as the former. The determined main sizes and the outer hull position are listed in table 2.4.1.

Table 2.4.1. The main sizes and the outer hull position of the considered trimarans

	center hull	outer hulls ($\frac{\nabla}{\nabla_0}=\frac{1}{20}$)	outer hulls ($\frac{\nabla}{\nabla_0}=\frac{1}{40}$)	outer hulls ($\frac{\nabla}{\nabla_0}=\frac{1}{60}$)
waterline length	100*m*	36.8403*m*	29.2402*m*	25.5436*m*
waterline broad	8*m*	2.947*m*	2.3392*m*	2.0435*m*
draft	3.555*m*	1.3098*m*	1.0396*m*	0.9082*m*
block coefficient	0.44444	0.44444	0.44444	0.44444
displacement	1264 m^3	63.2 m^3	31.6 m^3	21.06666 m^3
wet surface area	945.17 m^2	128.279 m^2	80.81 m^2	61.6755 m^2

Table 2.4.2. $(a,p)_{opt}$ selected by enumeration

	$(a, p)_{opt}$	Cw corresponding $(a, p)_{opt}$
Fr=0.3,1/20	（0，8）	1.96668E-04
Fr=0.3,1/40	(0,12)	2.95110E-04
Fr=0.3,1/60	(0,12)	3.77285E-04
Fr=0.5,1/20	(50, 16)	1.29686E-03
Fr=0.5,1/40	(50,15)	1.43921E-03
Fr=0.5,1/60	(50, 15)	1.50909E-03
Fr=0.7,1/20	(50, 8)	1.04282E-03
Fr=0.7,1/40	(50, 8)	1.09989E-03
Fr=0.7,1/60	(50, 8)	1.13025E-03

Tentatively the range of *a* was covered from -50m to 50m. If the range was extended further, it is possible that the center points of the outer hulls would extend the stern (or the bow) of the main hull, so that the total length of the trimarans would be too long, or the ship structure could not be rationally designed for application. The range of *p* was covered from 8m to 40m, so that the blockage of the water follow could be avoided when the space is less than 8m, and the favorable interference of the ship waves was kept when the space is less than 40m. On the other hand, if the space *p* is too large, there will also be problems with the structure.

2.4.4.2. Optimum location of the piece hulls determined by incorporating the contour graph method to the simple enumeration method

The optimum (*a*, *p*) were roughly known in the last section; however, simple enumeration is not sufficient to determine the optimum outer hull position. This method cannot provide information on the wave-resistance change tendency caused by the varying of (*a*, *p*). Furthermore, there is no idea about the regular of the best–better–worse outer hull arrangement and the influence on the wave resistance of that. To overcome these shortages, the contour graph method is incorporated to the simple enumeration method to investigate the influence from the outer hull arrangement.

Again some slight changes to the computation program were able to produce the data of *C*w vs. (*a*, *p*). According to these data the counter graphs were able to be plotted, which took *C*w as the contour value. In the proceeding Froude number *Fr*=0.3, 0.5, 0.7, the volume of the outer hulls was as follows: $\dfrac{\nabla}{\nabla_0}=\dfrac{1}{20},\dfrac{1}{40},\dfrac{1}{60}$. As to the results, there are nine groups of counter graphs, shown in Figs 2.4.3–11.

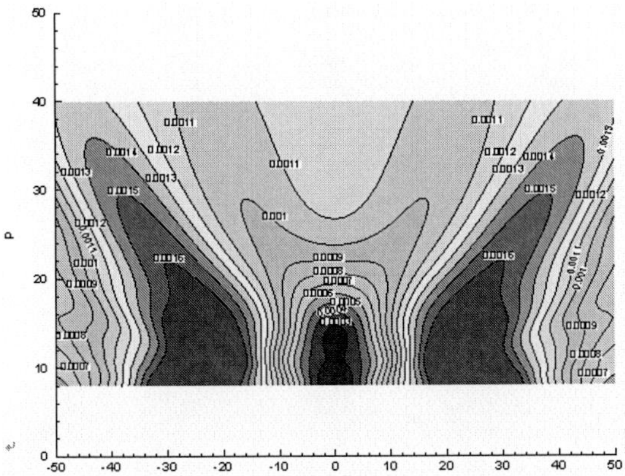

Fig. 2.4.3. Fr=0.3, $\dfrac{\nabla}{\nabla_0}=\dfrac{1}{20}$,C_W contour graph

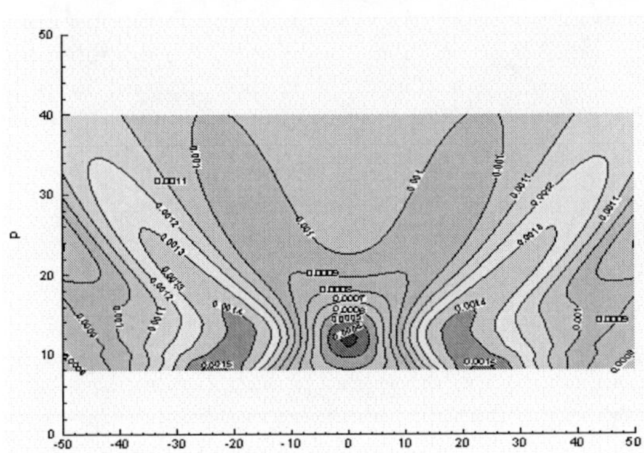

Fig. 2.4.4. Fr=0.3, $\dfrac{\nabla}{\nabla_0}=\dfrac{1}{40}$ C_W contour graph

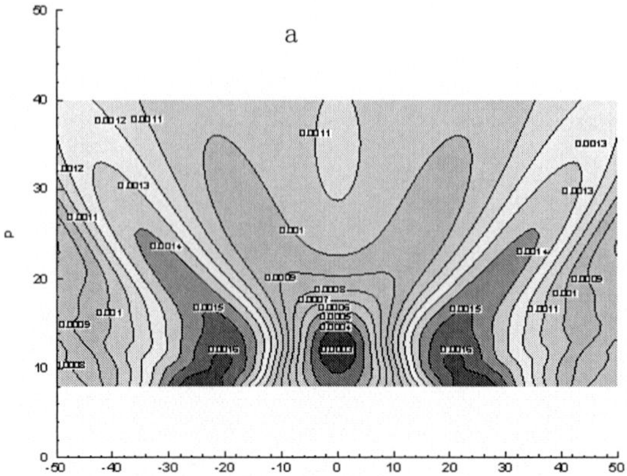

Fig 2.4.5. Fr=0.3, $\frac{\nabla}{\nabla_0} = \frac{1}{60}$,C_w contour graph

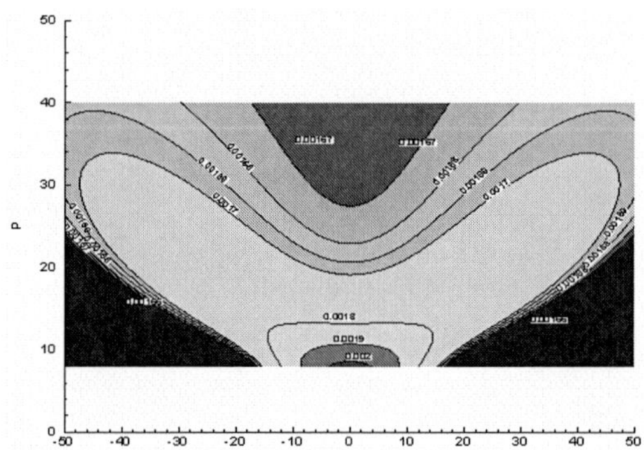

Fig. 2.4.6. Fr=0.5, $\frac{\nabla}{\nabla_0} = \frac{1}{20}$,C_w contour graph a

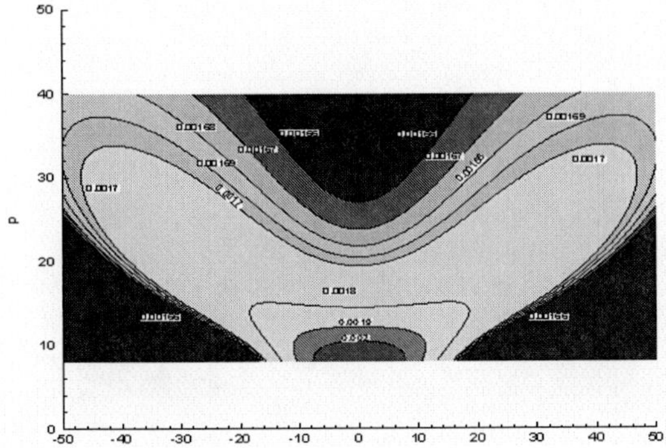

Fig. 2.4.7. Fr=0.5, $\frac{\nabla}{\nabla_0} = \frac{1}{40}$ C_w contour graph

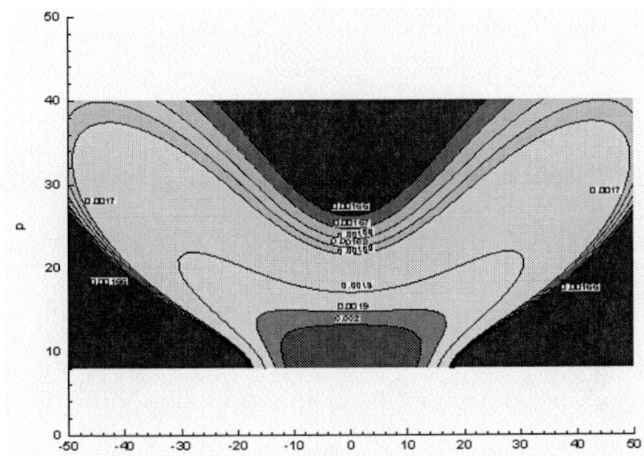

Fig. 2.4.8. Fr=0.5, $\frac{\nabla}{\nabla_0} = \frac{1}{60}$, C_w contour graph

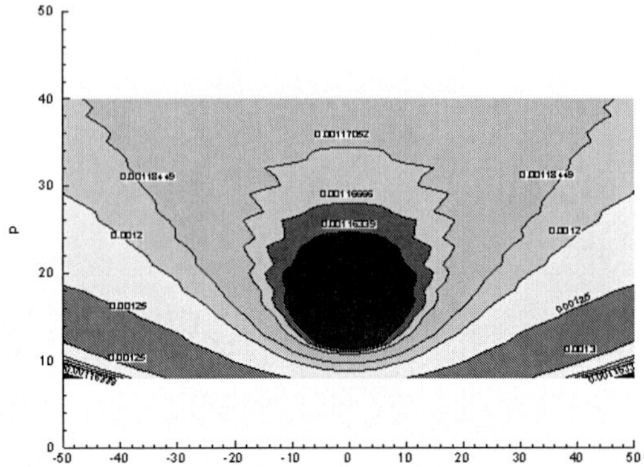

Fig. 2.4.9. Fr=0.7, $\frac{\nabla}{\nabla_0} = \frac{1}{20}$, C_w contour graph

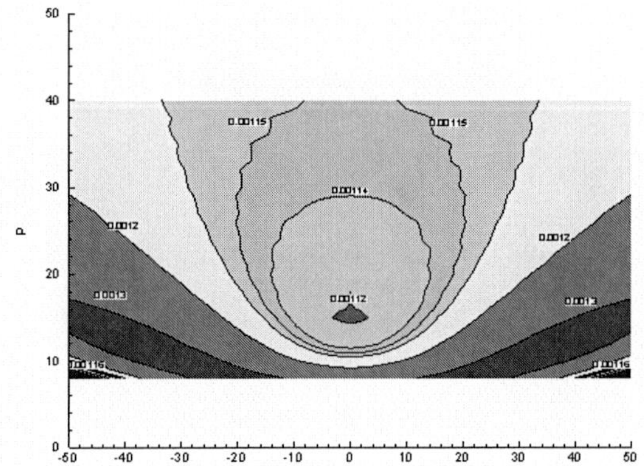

Fig. 2.4.10. Fr=0.7, $\frac{\nabla}{\nabla_0} = \frac{1}{40}$ C_w contour graph

We could sum up that as follows:

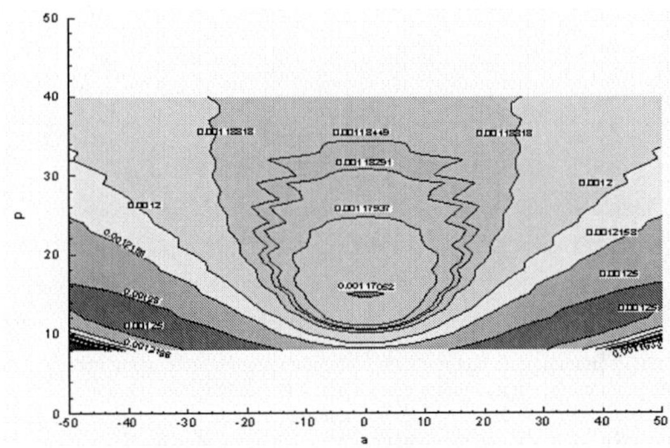

Fig. 2.4.11. Fr=0.7, $\frac{\nabla}{\nabla_0}=\frac{1}{60}$, C_w contour graph

1. When Fr=0.3, there is a district, a=0, $p\approx$10m, where C_w reaches a minute value for the three outer hull sizes selected.
2. When Fr=0.5, there are two districts, a∈30m–50m, p∈8m–20m, a∈50m–30m, p∈8m–20m, where C_w reaches the minute value; and one district, a∈–10m–10m, p∈30m–40m, where C_w is able to obtain lesser values for the three outer hull sizes selected.
3. When Fr=0.7, there is a district, a∈45m–50m, p∈8m–9m, where C_w reaches a minute value; and two districts, a∈–50m–45m, p∈8m–9m, a≈0, p≈16m, where C_w is able to obtain a lesser value for the three outer hull sizes selected.

These optimum values of a, p optimized from the C_w contour graphs are consistent with the one selected by enumeration. The contour graphs are more marginal and direct, and the contour graphs provide more information and are more convenient to use.

2.4.4.3. Analysis of the optimization results

Now the wave-resistance curves corresponding to the different outer hull positions were investigated by contrast. Take the outer hulls with $\frac{\nabla}{\nabla_0}=\frac{1}{40}$ as an example: the curves of C_w vs Fr were plotted, as shown in Fig. 2.4.12. The curves in the figure correspond to the situation that *Fr*=0.3, 0.5, 0.7, respectively.

From Fig. 2.4.12 we could sum up as follows:

1. Outer hull positions optimized at different speeds are very different, so that it is difficult or impossible to obtain the desirable outer hull arrangement for every speed of application. From the viewpoint of ship resistance, the outer hull position should be determined by the speed often used, such as the design speed or the cruising speed.
2. Generally speaking, the peak value of the wave resistance on the trimarans occurs when *Fr*=0.4–0.5. Corresponding to these Froude numbers, the outer hull position after optimization is located at the afterbody of the center ship.
3. The resistance of the trimarans is small in the low speed range (*Fr*=0.1–0.3), their resistance is large in the high speed range (*Fr*>0.4). But the resistance of another vessel is small in the high speed range; its resistance is not large in the low speed range. From this viewpoint, in the theoretical design procedure, the outer hull position should be determined by the optimization results at high speed.

The design speed of the high-speed escort in concept is greater than 30kn, and the optimal Froude number *Fr* equals 0.4–0.52. According to the optimization results (*Fr*=0.5), some references being incorporated, an outer hull-position arrangement was proposed in the present paper, which is that a=40m, p=12m, $\dfrac{\nabla}{\nabla_0} = \dfrac{1}{20}$

The results of optimization are analyzed as follows:

(1) Optimization effect

Assuming that a single hull ship is similar to the proposed trimaran, and both are equal in displacement, curves of their wave resistance of unit displacement are plotted in Fig. 2.4.13. The figure shows that the wave resistance of unit displacement on the trimaran is less than that of the single hull ship (*L*/*B*=12.5) in the high speed range (13m/s–19m/s, or 25–35kn). At the point of the maximal resistance reduction ratio, the wave resistance is able to decrease more than 10%. The single hull ship by comparison is rather slender; therefore, if the trimaran is compared with the ordinary single-hull ships, the trimaran is more advantageous in resistance performance.

(2) Wave interference effect

To investigate the wave interference effect of the trimaran, the wave resistance in the favorable wave interference of the trimaran is compared with the sum of the resistance from each single hull constituting the trimaran. The wave resistances of unit displacement in both situations are plotted in Fig. 2.4.14. It is shown in the figure that in the middle- and high speed range (*Fr*=0.35–0.6, or 22–35kn), the favorable wave interference of the trimaran is able obviously to decrease the wave resistance. At the point of the maximal resistance reduction the wave resistance is decreased by 20%.

It is regarded that for the concept plan of the high speed trimaran, SHOT, *a*=40m, *p*=12m, $\frac{\nabla}{\nabla_0}=\frac{1}{20}$, which is rather favorable in resistance performance.

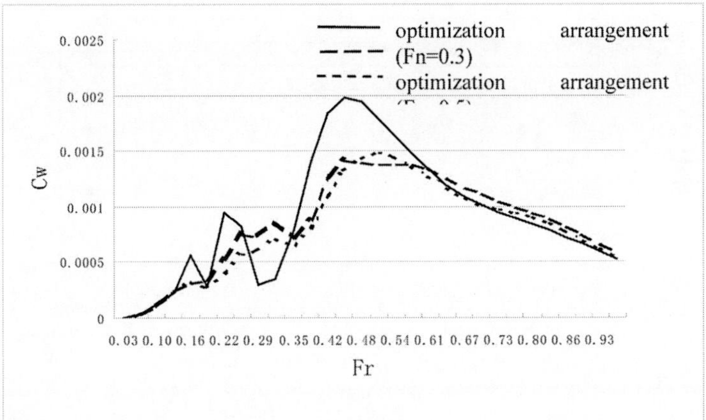

Fig. 2.4.12. Wave-resistance curves corresponding to optimal arrangement when *F*r=0.3,0.5,0.7

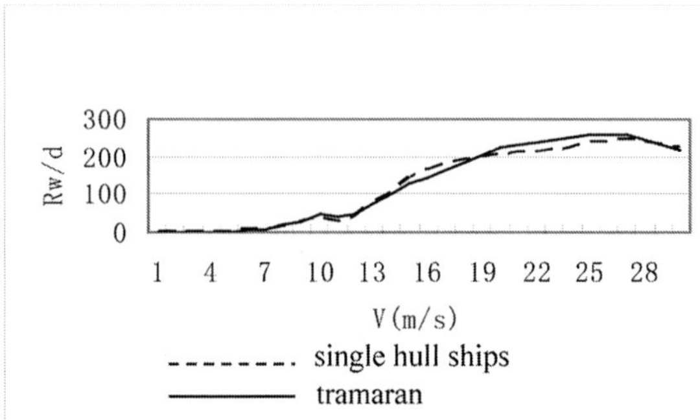

Fig. 2.4.13. Wave resistance of unit displacement on a single-hull ship and the trimaran

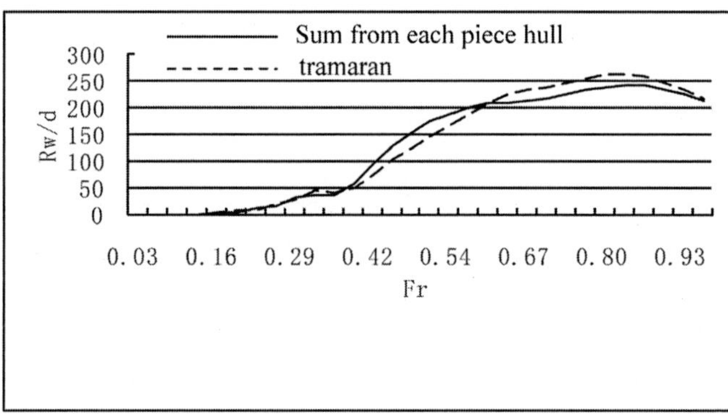

Fig. 2.4.14. Wave resistance (of unit displacement) in the favorable wave interference of the trim ran and the sum of wave resistance (of unit displacement) from each single hull

2.4.5. Conclusions

1. The computation method of the wave resistance on the trimarans is applicable to outer hull optimization, and is suitable to approximate computation of the wave resistance of the trimarans.
2. The contour graph method incorporating the simple enumeration as proposed by the present paper is able to give reliable optimization results about the outer hull position of the trimarans. It is practical for purposes of engineering.

3. The piece-hull arrangement plan of the trimaran proposed by the present paper is obviously favorable in the resistance performance compared with regular single-hull ships.

References

[2.4.1] Han Kaijia & Huang Debo, "The Computation Method of the Wave Resistance on the Trimarans," *Journal of Harbin Engineering University* 21 (1), February 2000: pp. 6–10, Harbin (in Chinese).

[2.4.2] Zhao Lianen, He Yi, Li Jide, et al., "An Investigation of Wave Piercing Mutihull Ships," *Shipbuilding of China* 1997 (4): pp. 20–28 (in Chinese).

[2.4.3] Ba Xin A. M., "Ship Dynamics in the Shallow Water," 1976 (in Russian).

2.5

Calculation and Analysis of High-Speed Trimaran Maneuvering Motion Based on Potential Flow Theory (2013)

Zhan Jinlin,[1,2] Lu Xiaoping,[2] Zhang Shulong[2] and Chen Yang[2]

[1] Bengbu Naval Petty Officer Academy
[2] Naval University of Engineering, China

Abstract

Predicting the maneuvering motion performance of high-speed trimarans on the basis of potential flow theory is important to the selection, design, and evaluation of propulsion-maneuvering devices. This paper carried out research into the calculation of maneuvering hydrodynamic forces with the three-dimensional panel method of the potential flow theory, on the basis of which maneuvering motion characters were numerically calculated and analyzed by solving the maneuvering motion equation. Additive masses related to the trimaran maneuvering motion were calculated by the three-dimensional panel method of the potential flow theory without lift, and the counterparts of the positional hydrodynamic derivatives and rotary hydrodynamic derivatives were calculated with the surface vortex lattice method of the potential flow theory with lift. According to present methods, maneuvering hydrodynamic derivatives and the turning radius were calculated for a high-speed sample trimaran with transom sterns, and the calculation results were validated by model test results. Furthermore, the zigzag experimental curves and turning trajectory of the trimaran were captured, and the influence of the outrigger position arrangement on the maneuvering turning radius and the turning trajectory of the trimaran were analyzed through solving the related maneuvering motion parameters of the sample trimaran with the different outrigger position arrangement. The present results of the trimaran maneuvering hydrodynamic derivatives and maneuvering motion characters are reasonable and rather accurate, which shows that the calculation of the maneuvering hydrodynamic force by the three-dimensional panel method of the potential flow and solution of related maneuvering motion equation in this paper have excellent application prospects in the prediction, analysis, and design of the maneuvering performance of high-speed trimarans.

Introduction

The high-speed trimaran is configured of a central hull and two outriggers, all of which are slender piece hulls. The new ship hull form has excellent stability, a spacious deck area, and, as a warship platform, is suitable for stealth performance design and modularized construction, and is favorable to weapon system dispersion arrangements and helicopter operation. Furthermore, in terms of the hydrodynamic performance aspect, it creates lower resistance at high speed and improves the sea-keeping quality. The hydrodynamic performance of high-speed trimarans has been investigated for more than 20 years at home and abroad, mainly concentrating on wave-resistance calculation and the optimization of outrigger position for the reduction of wave resistance,[2.5.1–4] and, more recently, in the calculation and analysis of sea-keeping performance and so on. In contrast, the research on the maneuvering performance of the trimaran was scarcely discussed, [2.5.5–9] which also covers the main hydrodynamic performances of the new ship-hull form. Therefore, there is an urgent need to investigate the maneuvering performance of high-speed trimarans, as the hydrodynamic feature and the foundation of the configuration and design of the propulsion-maneuvering device.

In the field of the maneuvering performance of the trimaran there exist some particular hydrodynamic problems, such as hydrodynamic interference from the side hull in the theoretical prediction of maneuvering performance, which is not referred to in the situation of conventional single-hull ships. There are two categories of research tasks in the investigational field of maneuvering performance of high-speed trimaran, with the prediction and analysis of maneuvering performance as one category, and the design of propulsion-maneuvering devices from the viewpoint of the hydrodynamic feature as another category. [2.5.10] This paper carried out research in the first category, with the trimaran maneuvering hydrodynamic performance calculated with the three-dimensional panel method of potential flow theory. On this basis, the maneuvering motion characters, such as the zigzag steering response curve and the turning trajectory curve were numerically simulated and analyzed according to the maneuvering motion equations. Moreover, a comparison of the turning diameter data of the sample trimaran trajectory from the theoretical calculation with those from the model test was presented to validate the theoretical calculation. Finally, according to the above work, the impact of the outrigger position arrangement on the turning trajectory diameter of the high-speed trimaran was discussed.

2.5.1. Hydrodynamic performance calculation for high-speed trimarans

2.5.1.1. Coordinate system

The coordinate system is shown in Fig. 2.5.1, with the *O* representing the coordinate origin point, located at the center position of the main hull, the *X*-axis pointing aft, and the *Y*-axis pointing starboard and the *Z*-axis pointing upwards, perpendicular to the free surface. Here, the outriggers are symmetrically arranged on both sides of the main hull, with the symbol *a* expressing the stagger of the outrigger center point to the main hull center point, and *p* expressing the clear of the outrigger center point to the main hull center position, and the inflow aft, i.e., the direction along the *X*-axis as a positive.

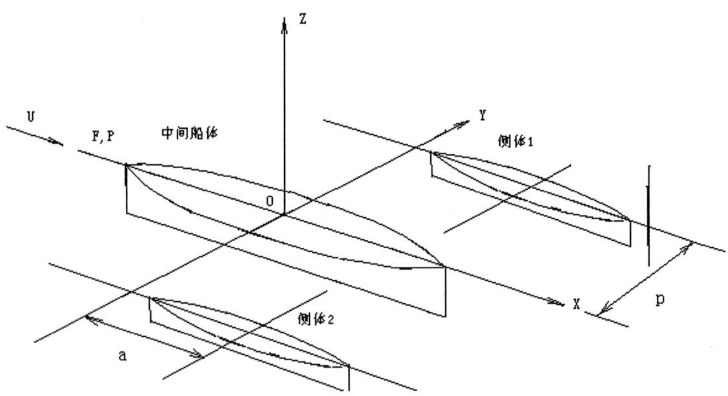

Fig. 2.5.1. Coordinate system for the trimaran

2.5.1.2. Mathematical model for the theoretical calculation

It is assumed that the uniform stream velocity is represented with V_0, while its direction has an angle β with the longitudinal section of the main hull. Therefore, V_0 could be expressed as

$$\boldsymbol{V}_0 = V_{0x} \cdot \boldsymbol{i} + V_{0y} \cdot \boldsymbol{j} + V_{0z} \cdot \boldsymbol{k} \quad (2.5.1)$$

where V_{0x}, V_{0y}, V_{0z} are the fluid velocity components in each axis direction. On the basis of the theory of potential flow, the flow field has velocity potential Φ, which involves the potential of uniform flow velocity and the potential of perturbation velocity φ, and is expressed as follows:

$$\Phi = x \cdot V_{0x} + y \cdot V_{0y} + z \cdot V_{0z} + \varphi \quad (2.5.2)$$

As the influence of free surface wave making is not taken into account for the time being, the perturbation velocity potential φ should satisfy the following control equation and boundary conditions:

$$\nabla^2 \varphi = 0 \quad \text{(In flow field)} \qquad (2.5.3)$$

$$\frac{\partial \varphi}{\partial n} = -\boldsymbol{V}_0 \cdot \boldsymbol{n} \quad \text{(On the object surface)} \qquad (2.5.4)$$

$$\frac{\partial \varphi}{\partial n} = 0 \quad \text{(On the free surface)} \qquad (2.5.5)$$

$$\varphi \to 0 \quad \text{(Infinity)} \qquad (2.5.6)$$

For the theoretical model with lift, the Kutta condition at the hull trailing edge should be satisfied for the water following, which could be expressed as

$$p_{TE}^{+} = p_{TE}^{-} \qquad (2.5.7)$$

where p_{TE}^{+}, p_{TE}^{-} are the vortex surface pressures at the trailing edges.

2.5.1.3. Numerical methods for perturbation velocity potential solution and hydrodynamic coefficient calculation

(1) Theoretical calculation model for potential flow without lift

The present paper adopted the theoretical model of unbounded potential flow without lift to calculate the additional mass m_{11}, m_{22}, and m_{66} for maneuvering performance of the trimaran, which can be expressed as follows:

$$m_{ji} = \rho \iint_{sb} \varphi_i \frac{\partial \varphi_j}{\partial n} ds \quad (i, j = 1, 2 \ldots 6) \qquad (2.5.8)$$

where φ_i is the unit perturbation velocity potential of the flow field around the hull for the motion model i, which can be obtained from the following formula:

$$\varphi(p) = \iint_s \frac{\sigma(q)}{r} ds_q \quad (2.5.9)$$

In formula 2.5.9, point q is referred to as the source point on the trimaran hull surface; point p is the flow field point; r is the distance between point p and q ; s is the hull surface area domain; and $\sigma(q)$ is the distribution source density on the hull surface area.

As can be seen from the above equation, the key step to solving the perturbation velocity potential, and, furthermore, the additional mass m_{11}, m_{22}, and m_{66}, is to solve the distribution source density $\sigma(q)$. According to the method in the literature,[2.5.11] the quadrilateral ΔQ_j is approximately taken as the integral panels approaching the initial curved surface panels, and as long as the panel segmentation is not too rough, the σ can be considered as the constant on each panel ΔQ_j , indicated by σ_j . Therefore, the integration on the curved surface s can be approximated by the sum of the integrations on the N square planar panels as follows:

$$\iint_s \sigma(q) \frac{\partial}{\partial n_p} \left(\frac{1}{r_{pq}}\right) ds_q \approx \sum_{j=1}^{N} \sigma_j \iint_{\Delta Q_j} \frac{\partial}{\partial n_p} \left(\frac{1}{r_{pq}}\right) ds_q$$

Incorporating with the hull surface condition (i.e., the formula 2.5.4), we can further develop linear equations to solve the distribution source density σ_j .

$$\sum_{j=1}^{N} a_{ij} \sigma_j = b_i \quad (i = 1, 2, \ldots, N) \qquad (2.5.10)$$

In the above formula (2.5.10), a_{ij} can be expressed as

$$a_{ij} = \begin{cases} \iint_{\Delta Q_j} \frac{\partial}{\partial n_{P_i}} \left(\frac{1}{r_{P_i q}}\right) ds_q \cdots\cdots i \neq j \\ 2\pi \cdots\cdots\cdots\cdots\cdots\cdots\cdots i = j \end{cases} \qquad (2.5.11)$$

$$b_i = -\vec{V}0 \cdot \vec{n}_{P_i} \qquad (2.5.12)$$

With the distribution source density obtained according to equation 2.5.10, the velocity potential and the additional masses can be further resolved.

(2) Lift potential flow theoretical calculation model

On the basis of the lifting potential flow theory and panel method of the Green function, the perturbation velocity potential is caused by the singularity system, usually assumed to be the dipole or vortex, distributed on the surface of the hull and the trailing vortex area. In the present paper, the method was conducted by distributing the vortex lattice on the hull surface and the trailing vortex surface. According to the method in the literature,[2.5.12] the surface of the hull and its trailing vortex area are divided into many panel elements, and as long as the surface panel division is small enough, the induced velocity on the panel element control point can be expressed as the velocity induced by the vortex lattices as

$$v(p) = -\frac{1}{4\pi}\sum_{j=1}^{N} m_j \oint_{l_j} \frac{\boldsymbol{r}\times d\boldsymbol{l}}{r^3} - \frac{1}{4\pi}\sum_{k=1}^{L} m_k \oint_{l_k} \frac{\boldsymbol{r}\times d\boldsymbol{l}}{r^3} \quad (2.5.13)$$

where the first item is the perturbation velocity at the point p induced by the vortex lattices on the hull surface panels; the second item, that induced by the vortex lattices on the trailing vortex area; and m_j and m_k, the strength of the vortex lattice l_j and l_k, respectively, which represent the perimeter of the vortex lattice on the hull surface panels and the tail vortex surface panels. As the flow involved in the present paper is steady, the Meng Ruinuo Kutta condition is adopted at the trailing edge as

$m_k = m_1 - m_2$ (2.5.14).

In formula 2.5.14, m_k express the strength of the horseshoe vortex, while m_1 and m_2 express the vortex strength of the panel element on the upper and lower surface at the trailing edge, respectively. With the hull surface divided into N panel elements, as well as the perturbation velocity expression (2.5.13) and the Kutta condition formula (2.5.14) incorporated into the hull surface condition represented by equation 2.5.4, the hull surface boundary condition can be converted into the discrete form as the following linear equations:

$$v_n(p_i) = \sum_{j=1}^{N} m_j \boldsymbol{n}_{p_i} \cdot \boldsymbol{C}_j(p) = -\boldsymbol{n}_{p_i} \cdot \boldsymbol{V}_0,\ i = 1,2,\ldots N \quad (2.5.15)$$

where the influence coefficient of the panel elements not adjacent to the trailing edge can be expressed as:

$$C_j(p) = -\frac{1}{4\pi}\oint_{l_j} \frac{\boldsymbol{r}(p,q)}{r^3(p,q)} \times d\boldsymbol{l} \quad (2.5.16)$$

The influence of the wake horseshoe vortex is included in the influence coefficient of the panel elements adjacent to the trailing edge, expressed as:

$$C_j(p) = -\frac{1}{4\pi}\oint_{l_j} \frac{\boldsymbol{r}(p,q)}{r^3(p,q)} \times d\boldsymbol{l} \pm \boldsymbol{H}(p) \qquad (2.5.17)$$

where the term H represents the velocity induced by the horseshoe vortex per unit strength, and the positive sign is for the upper surface panel, and the negative sign is for the lower. According to the vortex lattice strength solved by the formula (2.5.15), we can calculate the flow velocity, and thus obtain the pressure P_i at each control point. Then the lateral hydrodynamic force and corresponding yaw moment can be calculated as formula 2.5.18 and 2.5.19:

$$\boldsymbol{Y}_{\boldsymbol{h}} = \sum_{i=1}^{N} P_i S_i \boldsymbol{n}_{yi} \qquad (2.5.18)$$

$$\boldsymbol{N}_{\boldsymbol{h}} = \sum_{i=1}^{N} S_i \boldsymbol{p}_i \times (P_i \boldsymbol{n}_{xi} + P_i \boldsymbol{n}_{yi}) \quad (2.5.19)$$

2.5.2. Mathematical model for maneuvering motion

Once the hydrodynamic force related to the maneuvering performance was obtained, the differential equations of the maneuvering motion can be derived from the fundamental theorem of Newtonian mechanics as

$$\left.\begin{aligned} m(\dot{u} - vr) &= X \\ m(\dot{v} + ur) &= Y \\ I_z \dot{r} &= N \end{aligned}\right\} \qquad (2.5.20)$$

In differential equations (2.5.20) for maneuvering motion, the ship, propeller and rudder can be treated as an integer or as separate individual units, which was called the integration model or the separation model of the maneuvering motion, also known as the Abkowitz model and MMG (Maneuvering Model Group) model, respectively. The main features of the MMG maneuvering motion mathematical model is that the whole hydrodynamic force on the research objects can be decomposed into independent components such as the ship hull hydrodynamic force component, the propeller hydrodynamic force component and the rudder hydrodynamic force component, or the propulsion-maneuvering device hydrodynamic force component, and so on, and the interference effect of the above hydrodynamic force components. According to the MMG maneuvering-motion mathematical model, the ship hull and the propulsion-maneuvering device (such as the propeller-rudder system, the waterjet propulsion system, or the pod propulsion device) are considered as the separate objects for the hydrodynamic analysis of the maneuvering performance. Thus, the MMG mathematical model offers great conveniences to the maneuvering performance investigation in cases where the propulsion-maneuvering device is not determined in detail; for, in the MMG mathematical model, the process of solving the maneuvering hydrodynamic and motion problem is separated from the propulsion-maneuvering device design problem, just as in the case of the high-speed trimaran maneuvering performance investigation where the propulsion-maneuvering device was not determined in detail.

In accordance with the MMG mathematical model, the hydrodynamic forces and moments on the trimarans can be deconstructed as:

$$\left.\begin{array}{l} X = X_H + X_P + X_R \\ Y = Y_H + Y_P + Y_R \\ N = N_H + N_P + N_R \end{array}\right\} \qquad (2.5.21)$$

where the subscripts H, P, R, of X, Y, and N denote the hull, propeller and rudder (or other propulsion-maneuvering device), which can be expressed as formula 2.5.22, with the nonlinear hydrodynamic forces denoted by Y_{NL}, N_{NL}:

$$\left.\begin{array}{l} X = -\lambda_{11}\dot{u} + \lambda_{22}vr + X(u) + X_{R0} + (1-t)T(J_p) - F_N \sin\delta \\ Y = -\lambda_{22}\dot{v} + Y_v v - \lambda_{11}ur + Y_r r + Y_{NL} + (1+\alpha_H)F_N \cos\delta \\ N = -\lambda_{66}\dot{r} + N_v v + N_r r + N_{NL} + x_\Theta Y_H + (x_H + \alpha_H x_H)F_N \cos\delta \end{array}\right\} \qquad (2.5.22)$$

Incorporating formula 2.5.22 into equation 2.5.20, the maneuvering motion differential equations could be conducted in more detail with the nonlinear hydrodynamic forces retained. When ignoring the terms of the nonlinear hydrodynamic forces and involving the trajectory differential equations, the maneuvering motion differential equations can be linearized as equation 2.5.23:

$$\left.\begin{array}{l} (m+\lambda_{11})\dot{u} - X_u \Delta u = 0 \\ (m+\lambda_{22})\dot{v} + (m+\lambda_{11})ur - Y_v v - Y_r r = Y_\delta \delta \\ (I_z + \lambda_{66})\dot{r} - N_v v - N_r r = N_\delta \delta \\ \dot{\psi} = r \\ \dot{x}_0 = u\cos\psi - v\sin\psi \\ \dot{y}_0 = v\cos\psi + u\sin\psi \end{array}\right\} \qquad (2.5.23)$$

According to the MMG mathematical model, in formula 2.5.23 the first equation is independent of the others. Furthermore, for maneuvering motion trajectory calculation, the first equation can be obliterated while only retaining the later five equations to be taken into consideration. Thus, the problems of the propulsion-maneuvering device design and the equilibrium between thrust and resistance are separated from the maneuvering motion trajectory calculation problems, which clearly presents the unique advantages of the MMG mathematical model for present high-speed trimaran maneuvering performance prediction and analysis.

When the hydrodynamic force features are determined according to the hydrodynamic calculation methods based on the theoretical potential flow models, described in section 1, the trimaran maneuvering motion features—including the maneuvering motion trajectory and the tangential velocities of the gravitation center or the rotating velocity of the trimaran hull—can be further solved according to the later five equations in formula 2.5.23.

2.5.3. Validation and analysis of the calculation results for a sample trimaran

A high-speed trimaran of the piece hull with the transom stern was presented as the calculation sample, the principal dimensions of which were shown in Table 2.5.1. For more detailed information, the lateral clear span of the outrigger was presented as p=0.673m, and the longitudinal stagger distance as a=2.25m; the calculation mesh of the sample trimaran is shown in Fig. 2.5.2.

Table 2.5.1. Principal dimensions of the sample trimaran

	L	B	d	C_B
Main hull/m	7	0.56	0.28	0.47
Outrigger/m	2.5	0.12	0.1	0.50

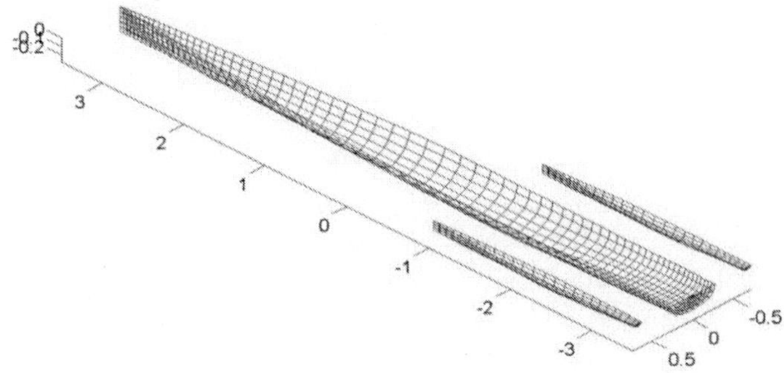

Fig. 2.5.2. Calculation mesh of the sample trimaran

Self-propulsion model tests of the gyration motion were carried out for the sample trimaran model, which was fitted with a propeller-rudder system as the propulsion-maneuvering device in the model test procedure for the time being. To facilitate a comparison between the calculation results and the model test results for validation, the propeller-rudder system is also introduced into the calculation sample as the propulsion-maneuvering device following the test sample model. As mentioned above for the MMG maneuvering motion mathematical model, the maneuvering performance calculation results don't depend on the detailed form of the propulsion-maneuvering device, so that the results and the calculation method in the present section can be generalized to practical cases with other forms of propulsion-maneuvering devices, such as a waterjet propulsion system or pod propulsion device.

Theoretical calculation results of the hydrodynamic parameters for the maneuvering performance are listed in Table 2.5.2. For the theoretical calculation values, additional masses m_{11}, m_{22}, and $m_{66,}$ position derivatives Y_v and N_v, and rotating derivatives Y_r and N_r were calculated by potential flow theory models and numerical methods, with and without lift, and presented in section 1, while rudder derivatives Y_δ and N_δ were calculated by lifting panel theory, which were not listed in Table 2.5.2, for their values varied compared with the rudder angle.

Table 2.5.2. Hydrodynamic parameters for maneuvering performance

Parameters	Theoretical calculation values
$M_{11/}$ kg	10.85
$M_{22/}$ kg	713.37
$M_{66/}$ kg·m	1394.20
Y_v	-464.69
N_v	-676.46
Y_r	68.78
N_r	-2901.51

With the hydrodynamic parameters as shown in Table 2.5.2, the gyration radius of the trimaran maneuvering motion, R_0, can be further calculated as in the following formula:

$$R_0 = \frac{Y_v N_r + N_v (mV_0 - Y_r)}{N_v Y_\delta - Y_v N_\delta} \frac{V_0}{\delta} \quad (2.5.24)$$

where δ represents the rudder angle; and $V_{0,}$ represents the tangential velocities in the steady gyration stage, and, taken as V_0=3.9m/s, following the model test procedure.

A comparison of the numerical calculation values with experimental values of R_0/L was shown in Table 2.5.3.

Table 2.5.3. Calculation values of R_0/L vs. trial values

Rudder angle /deg.	15	25	35
R_0/L calculation values	8.075	5.271	4.296
R_0/L model test values	7.965	5.692	4.719

The comparison of R_0/L listed in Table 2.5.3 shows that the calculation values agreed well with the model test ones, with 8.9% of the maximum relative error, and 1.36% of the minimum one only, which validated the calculation results from the present procedures for the calculation and analysis of the trimaran maneuvering performance.

2.5.4. Trimaran maneuvering motion feature analysis based on numerical calculation results

2.5.4.1 Numerical simulation of the zigzag steering response curve

To facilitate a comparison of numerical simulation results with model test results, both in the present paper and in research in the near future, the calculation and analysis in section 4 were carried out based on the information of the sample trimaran model presented in section 2.5.3. The zigzag steering response curves of the object trimaran were simulated by present theory models and numerical methods mentioned above, where the maximum rudder angle was taken as 10° with steering speed 2°/s, and with the outrigger position arrangement as *a*=2.5m, 0m, 2.5m, *p*=0.5m, 1m. The numerical simulation results of the zigzag steering response curves were shown in Fig. 2.5.3 (a), (b), and (c), and Fig. 2.5.4 (a), (b), and (c).

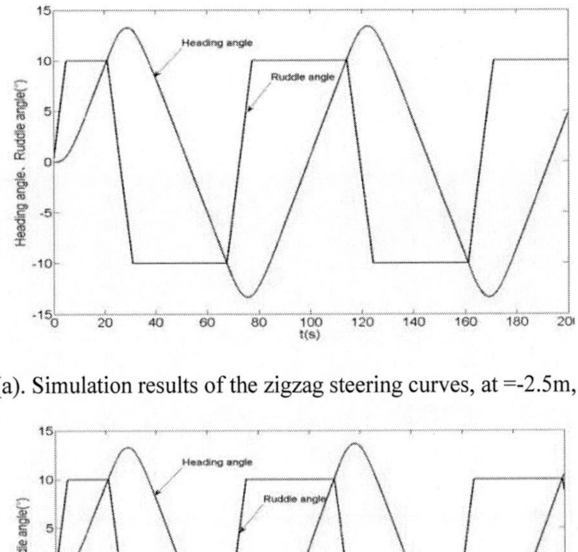

Fig. 2.5.3(a). Simulation results of the zigzag steering curves, at =-2.5m, *p*=0.5m

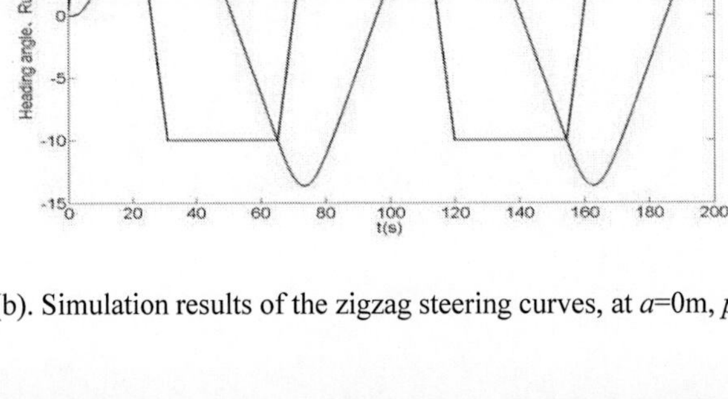

Fig. 2.5.3(b). Simulation results of the zigzag steering curves, at *a*=0m, *p*=0.5m

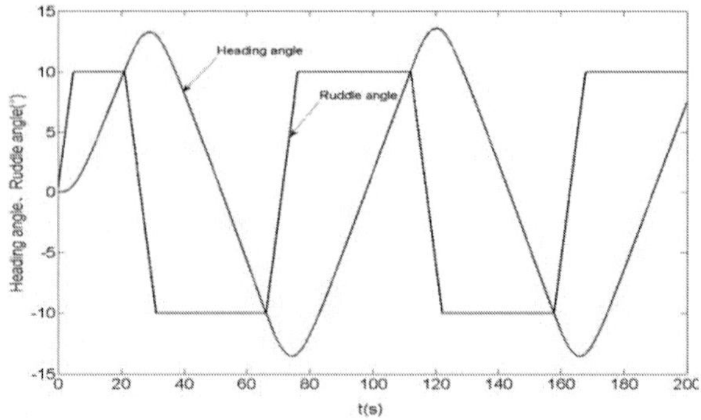

Fig. 2.5.3(c). Simulation results of the zigzag steering curves, at a=2.5m, p=0.5m

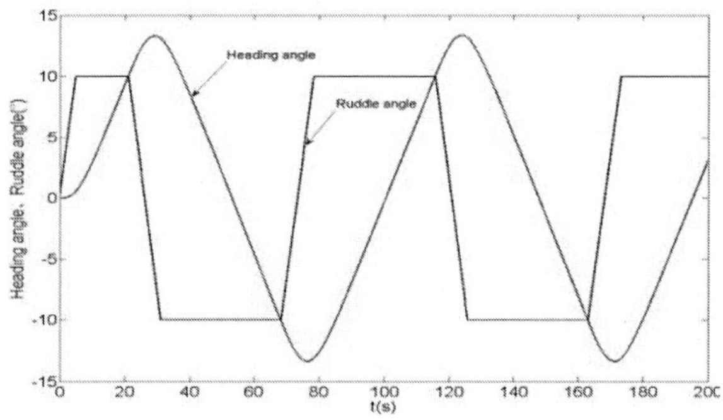

Fig. 2.5.4(a). Simulation results of the zigzag steering curves, at a=-2.5m, p=1m

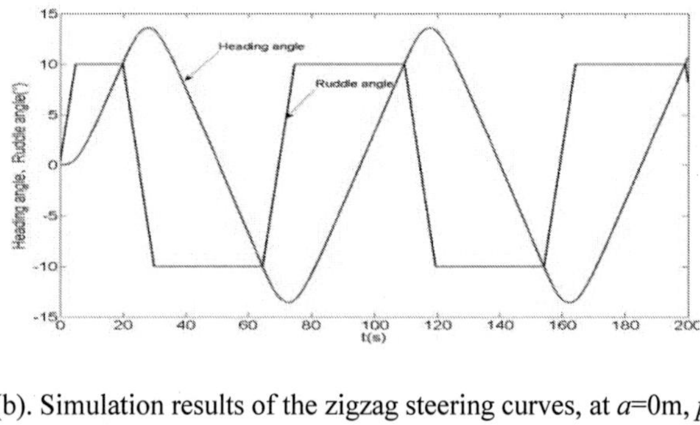

Fig. 2.5.4(b). Simulation results of the zigzag steering curves, at a=0m, p=1m

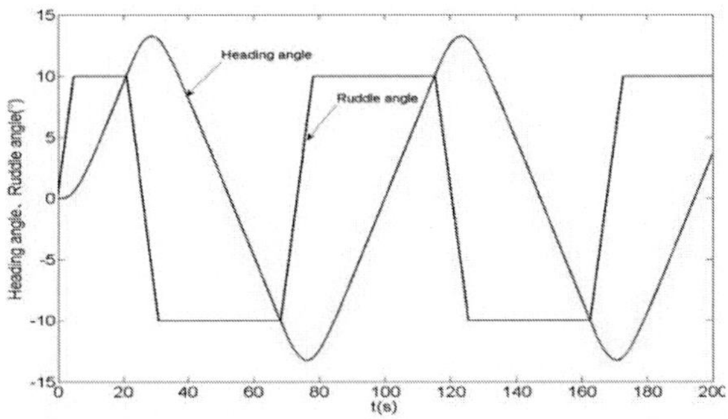

Fig. 2.5.4(c). Simulation results of the zigzag steering curves, at a=2.5m, p=1m

2.5.4.2. Numerical simulation of turning trajectory

The turning motion trajectory numerical simulation of the object trimaran was simulated by the present theory models and the numerical methods mentioned above, where the rudder angle was taken as 35°. The turning trajectory numerical simulation results are shown in Fig. 2.5.5 (a) and (b).

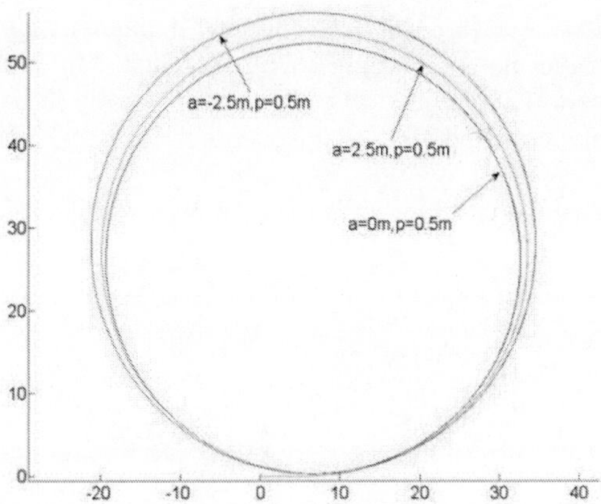

Fig. 2.5.5(a). Gyration motion trajectory, p=0.5m

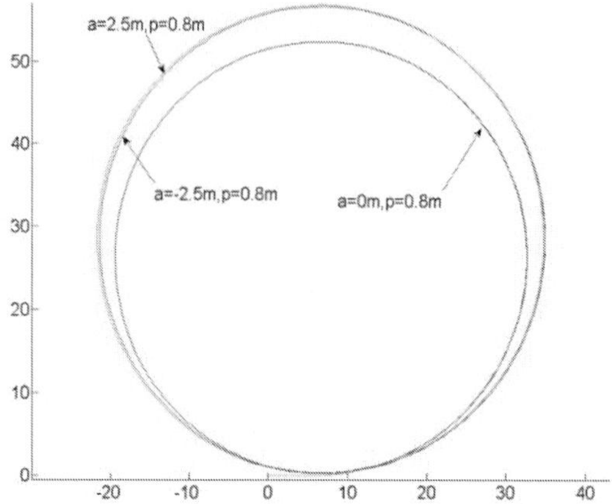

Fig. 2.5.5(b). Gyration motion trajectory, p=0.8m

2.5.4.3 Influence of the outrigger position arrangement on the zigzag steering test curve

The initial turning time t_a, overswing time t_s, and overswing angle φ_s of the trimaran zigzag steering test curve in the numerical simulation under different outrigger position arrangements were listed in Table 2.5.4, where three data sets were presented in every grid, from top to bottom, for t_a, t_s, φ_s, respectively, corresponding to the curves in Fig. 2.5.3 and Fig. 2.5.4.

Table 2.5.4. Maneuvering parameters from the results of the zigzag steering test curves

a/m \ p/m	-2.5	0	2.5
0.5	20.9 8.1 3.372	19.8 8.1 3.600	20.4 8.2 3.567
0.8	21.1 8.1 3.330	19.9 8.1 3.592	21.0 8.0 3.276

As shown in Table 2.5.4, with the same lateral clear span of the outrigger, the smallest initial turning time of t_a (i.e., favorable initial turning performance) can be obtained, when the outrigger is located in the middle region of the main hull (i.e., a=0m); in contrast, the initial turning performance was slightly better for the outrigger located aft than that located fore, when the outrigger arranged the longitudinal stagger distance asymmetrically. On the other hand, with the same longitudinal position of the outrigger, the greater lateral clear span corresponds to the greater initial turning time of t_a. Comprehensively, arranging the outrigger near the main hull middle region longitudinally and close to the main hull laterally yields a better initial turning performance.

Also as shown in Table 2.5.4, the outrigger position arrangement has little influence on the overswing time t_s.

Finally, Table 2.5.4 shows that, with the same lateral clear span, the nearer to the middle of the outrigger the greater the overswing angle φ_s. With the same longitudinal stagger distance, the greater the lateral clear span and the smaller the overswing angle φ_s. Clearly, then, arranging the outrigger aft longitudinally and close to the main hull laterally yields better lagging performance.

2.5.4.4. Influence of the outrigger position arrangement on the gyration motion

It can be seen from Fig. 2.5.5(a) and Fig. 2.5.5(b) that the outrigger position arrangement has little influence on the gyration motion trajectory in the initial stage, but has greater influence on the gyration diameter in the steady-motion stage.

In Fig. 2.5.5(a), with a smaller outrigger lateral clear span, when the outrigger is arranged fore longitudinally, the gyration motion performance gets worse; when the outrigger is located in the middle region of the main hull longitudinally, the gyration motion performance improves; when the outrigger is arranged aft longitudinally, the gyration motion performance is neither better nor worse. In Fig. 2.5.5(b), with the greater outrigger lateral clear span, the gyration motion performance gets worse when the outrigger is arranged fore longitudinally; while, in contrast, the gyration motion performance improves when the outrigger is arranged aft longitudinally, and approximately symmetrical longitudinally to the state of the outrigger being arranged fore.

For comprehensively reflecting the influence of the outrigger position arrangement on the gyration diameter in the steady motion stage, the authors calculated the steady gyration radius for the object trimaran presented in section 2.5.4.1–3 under different outrigger position arrangements. According to formula 2.5.24 and assuming a rudder angle of δ=15°, the results are shown in a cloud figure (see Fig. 2.5.6).

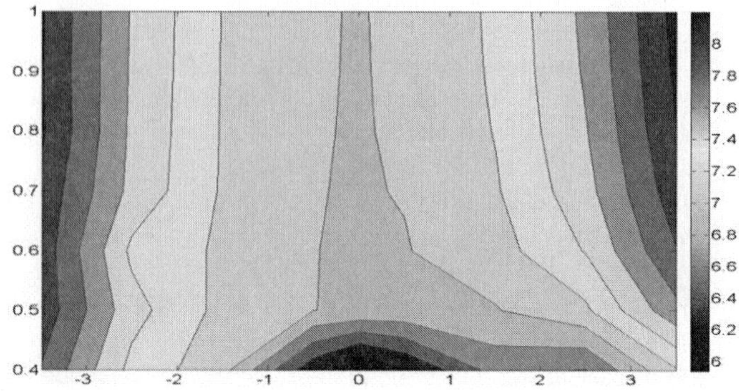

Fig. 2.5.6. R_0/Lvs. outrigger position arrangement

Fig. 2.5.6 shows that the outrigger position arrangement has a significant influence on the steady gyration radius. Generally speaking, the longitudinal stagger distance has greater influence than the lateral clear span. In greater detail, Fig. 2.5.6 shows: (1) locating the outriggers both fore and aft longitudinally yields poor steady gyration performance; (2) when the outrigger is close to the middle region of the main hull, the lateral clear span has some influence on the steady gyration radius, which declines rapidly and tends to disappear with the lateral clear span increased; (3) as a particular case, when a=0m and p=0.4m, corresponding to the arrangement very close to the middle region of the main hull, the steady gyration radius is only six times that of the main hull waterline, which is approximate to that for a conventional monohull. As the general regulation, detailed performance, or particular case, all the above mentioned results are consistent with fundamental physical concepts and actual situations.

2.5.5. Conclusion

The mathematical model and numerical method based on potential flow theory for the maneuvering performance of trimarans, involving hydrodynamics and motion, were conducted and established. According to these tests, the numerical calculation, simulation, and analysis of the maneuvering performance were carried out for the sample and object trimaran hull models. Our conclusions are as follows:

1. The proposed mathematical model and numerical method based on potential flow theory for the trimaran maneuvering performance have satisfied accuracy and reliability, which lays the foundation for predicting and analyzing trimaran maneuvering performance with great values and broad prospects in engineering.
2. The outrigger position arrangement has significant influence on the steady turning radius, with the influence of longitudinal stagger greater than the lateral clear span. Arranging the position of the outrigger close to the middle region of the main hull will yield better gyration performance.

The main intention of the present research is to establish a mathematical model and numerical method for prediction and analysis of the trimaran maneuvering performance, and carry out the calculation for sample and object trimaran models with the propeller-rudder assumed as the propulsion-maneuvering device. Nevertheless the results and methods can be generalized to practical cases with other forms of propulsion-maneuvering devices, such as a waterjet propulsion system or a pod propulsion device.

In the near future, based on the research in the present paper, the authors will carry out an investigation of the design of propulsion-maneuvering devices and the impact of wave making on maneuvering performance for high-speed trimarans.

References

[2.5.1] Han Kaijia & Huang Debo, "Wave Resistance Calculation of Trimaran," *Journal of Harbin Engineering University* 2000 (01): pp. 8–12.

[2.5.2] Lu Xiaoping, "Research of High-Speed Trimaran's Wave Resistance and Outriggers' Displacement Optimization," *Journal of Hydrodynamics* Ser. A, 2004 (03): pp. 347–54.

[2.5.3] Chen Kang, "CFD Application in Trimaran Resistance Research," *Journal of Harbin Engineering University* 2006 (03): pp. 68–75.

[2.5.4] Lu Xiaoping, "Research Summary of High Speed Trimaran," *Journal of Nav University*, 17 (2): pp. 89–96, 2005.

[2.5.5] Lu Xiaoping, "Trimaran's Maneuverability Calculation and Characteristic Analysis," *Journal of Naval University of Engineering* 21 (1): pp. 47–53, 2009.

[2.5.6] Zhan Jinlin, "Calculation of Trimaran's Maneuverability Hydrodynamics by the Potential Flow Theory," *Journal of Harbin Engineering University*, 2012 (05): pp. 642–47.

[2.5.7] Zhan Jinlin, "Hydrodynamic Analysis of Ship Maneuverability by the Potential Flow Theory," *Chinese Journal of Ship Research*, 6 (5): pp. 41–45, 2011.

[2.5.8] Yao Di, "Trimaran's Hydrodynamics Coefficient Calculation and Maneuvering Emulation Research," *Journal of Naval University of Engineering* 2008, 12.

[2.5.9] Yao Di, "Trimaran Hydrodynamic Derivative Calculation and Maneuvering Characteristics Simulation Research," Wuhan: Naval University of Engineering, 2008.

[2.5.10] Wu Xiuheng, *Ship Maneuverability*, Beijing: National Defense Industrial Press, 2004.

[2.5.11] Dai Yishan, *Potential Flow Theory of Ship Motion in Waves in Frequency and Time Domain*, Beijing: National Defense Industrial Press, 1998.

[2.5.12] Wang Xianfu, *Ship Computational Fluid Dynamics*, Shanghai: Shanghai Jiaotong University Press, 1992.

2.6

Numerical Analysis of Resistance Characteristics of and Flow Fields around Multihull Ships Based on Overset Grid Techniques (2011)

Shen Zhirong, Ma Juan and Wan Decheng

Shanghai Jiao Tong University

Abstract

Since multihull ships have good resistance and sea-keeping performances in sea waves as well as a bigger deck area compared with ordinary monohull ships, they have been applied not only for civil use but also for military use. In the present work, numerical analysis of resistance characteristics and flow fields around two common types, the catamaran and the trimaran, based on RANS method coupled with overset grid technique and level set method, is presented. The free-surface is taken into account and treated by the level-set method. The overset grid technique is employed to embed the body-fit hull grids into an orthogonal background grid. The distance gap between ship hulls can be adjusted easily by the overset grid technique without regenerating the mesh, so that the optimal distance gap between ship hulls can be figured out. And for each distance gap, a serial of calculations with different Froude numbers is carried out to predict the resistance characteristics and find out the optimal distance gap and ship hull layout. Complex wave patterns produced by multihull ships are also included in the analysis.

2.6.1. Introduction

In recent decades, the multihull ship has developed rapidly not only for civil use but also for military applications owing to its good rapidity and sea-keeping. There are many advantages of the multihull ship compared with the traditional monohull ship. First, since the multihull ship has a larger deck area than the monohull ship, it has good capacity; therefore, it is quite economical for transportation. Second, the multihull ship has good rapidity with less wave resistance. The main hull and side hull of the multihull ship are usually more slender. So the larger length–width ratio can reduce wave resistance. Third, since the multihull ship has a larger transverse section area, it has better stability in rolling than the monohull ship. Therefore, it is quite suitable for passenger ships, cruise ships, and marine combat ships, which require good sea-keeping.

Traditional multihulls include catamarans, trimarans, pentamarans, small waterplane area twin hulls (SWATH), and so on. In this paper, the resistance performance of catamarans and trimarans are the main research objects.

Hydrodynamic research on catamarans and trimarans mainly depends on experiments and numerical calculation methods. For the aspect of experimental method, C. Y. Wang[2.6.1] analyzed the resistance performance of catamarans with different hull forms, hull separated distances, and Froude numbers based on experimental results. X. Q. Wang[2.6.2] found out the best ratio of hull distance to ship length (K/L) of high-speed catamaran through analyzing experimental dates. P. Y. Li[2.6.3] took a series of experiments on trimaran ships to analyze the influence of side hulls on the main hull and the wave disturbance of each hull. On the basis of the experimental results of Series Taylor and Series 64, R. Saunders[2.6.4] analyzed the influence of side-hull displacements. In the area of numerical methods, many methods have been applied to the region of high-performance multihull ships. Z. Elcin[2.6.5] applied the three-dimension Rankie Panel method to research wave-making resistance characteristics of trimaran hulls. J. P. Chen et al.[2.6.6] developed a numerical code based on an improved Dawson method, which is suitable not only for catamaran hulls but also for trimaran hulls.

However, almost all the numerical methods are based on potential flow, and viscous flow is rarely considered. This is probably because of the limitation of computation at that time. However, with the development of computation science, many methods based on viscous flow are developed to deal with ship hydrodynamics problems.[2.6.7–11] In this paper, a widely used method, Reynolds-Averaged Navier-Stokes (RANS), is applied to solve the resistance characteristics of catamaran and trimaran ships. Beside, different positions of side bodies or demihulls with different ship speeds are taken into account. The overset grid method is also applied in this paper. In the overset method, one or more body fitted grids, which wrap up the objects, can be embedded into a Cartesian background grid. And, thus, the body fitted grids can translate and rotate in any direction without changing the background mesh. The overset method has been successfully applied to ship sea-keeping and maneuverability problems.[2.6.7] [2.6.8]

In this paper, the resistance characteristics of multihull ships, such as catamarans and trimarans, are presented. By calculating the resistance prediction of multihull ships with different hull distances or side-body positions, an optimal ratio of hull distance to ship length (*K*/*L*) for catamarans and an optimal position of the side body for trimarans is found.

2.6.2. Numerical method

2.6.2.1. Governing equation

In this paper, the viscous flow is solved through continuity equations and RANS momentum equations:

$$\nabla \cdot \mathbf{U} = 0 \quad (2.\ 6.1.)$$

$$\frac{\partial \rho \mathbf{U}}{\partial t} + \nabla \cdot (\rho \mathbf{U}\mathbf{U}) = -\nabla p + \rho \cdot \mathbf{g} + \nabla \cdot \left[\mu \left(\nabla \mathbf{U} + (\nabla \mathbf{U})^T \right) \right] - \nabla \cdot \left(\rho \overline{\mathbf{U}'\mathbf{U}'} \right) + f_s \quad (2.6.2.)$$

where $\mathbf{U}$ is velocity, p is pressure, μ is dynamic viscosity, $\mathbf{g}$ is gravity acceleration, f_s is the source term, $\overline{\mathbf{U}'\mathbf{U}'}$ is Reynolds stress tensor, which satisfies Boussinesq approximation.

$$-\overline{\mathbf{U}'\mathbf{U}'} = \nu_t \left(\nabla \mathbf{U} + (\nabla \mathbf{U})^T \right) + \frac{2}{3} k\mathbf{I} \quad (2.6.3)$$

For the turbulence model, a blended $k-\varepsilon / k-\omega$ model, shear-stress transport (SST), which was introduced by Menter,[2.6.12] is used to solve the turbulent viscosity.

2.6.2.2. Level-set method

The level-set method, which was first introduced by Osher,[2.6.13] is applied to capture the surface interface. In this method, a distance function ϕ is introduced to denote the distance from each mesh node to the surface interface. $\phi = 0$ indicates the surface interface. $\phi > 0$ and $\phi < 0$ represent water and air, respectively. The distance function ϕ is solved by a transport equation:

$$\frac{\partial \phi}{\partial t} + \mathbf{U} \cdot \nabla \phi = \mathbf{0} \quad (2.6.4)$$

However, the distance function cannot remain the distance function during the advection. Therefore, a reinitialization step is required to make it into a distance function. The details are available in reference 2.6.14.

2.6.2.3. Pressure-velocity solution procedure

The Navier-Stokes solution and continuity equation is obtained by applying the pressure-implicit split-operator (PISO) algorithm, which was first developed by Issa.[2.6.15] The method uses a predictor-corrector method to solve the Navier-Stokes equations by applying the continuity equation.

2.6.2.4. Overset grid method

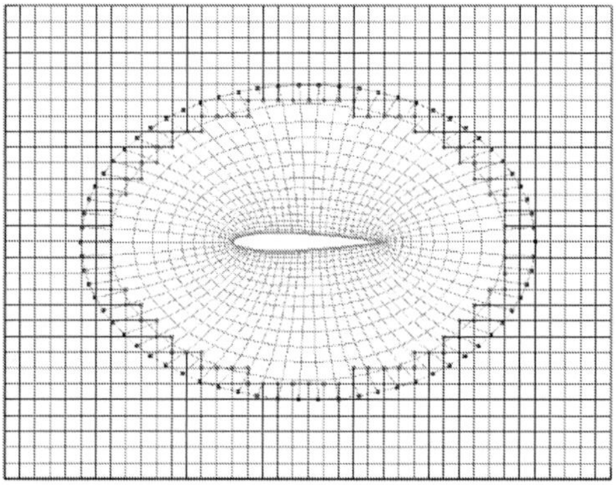

Fig. 2.6.1. Demonstration of overset grid

For multihull ships, each hull can be wrapped up by a body fitted grid, which can be embedded into a Cartesian background grid by applying the overset grid method. And thus each hull can move in any direction without restriction. Therefore, it is convenient to move the hull or side body from one position to another without regenerating meshes.

The main procedure of the overset grid method is the domain of connectivity, which consists of three steps: projection, hole cutting, and fringe-point interpolation.[2.6.16] The projection process shifts the surface points to remove the effect of surface-discretization errors, which often occur in more complex conditions, such as surface interpolation. However, in this paper, since the hull shape is simple and no surface interpolation is required, the procedure is ignored. The hole cutting is to remove the unnecessary grid points from a gird that is inside another grid. The interpolation step is to search stencil points that provide interpolation information for fringe points (the remaining boundary points after hole cutting). The interpolation coefficients are calculated in this step. And thus, the fringe points can receive information from the other meshes though the interpolation of stencil points.

$$\phi_{\text{int}} = \sum_{i=1}^{8} \xi_i \phi_i + \eta_i \phi_i + \zeta_i \phi_i \tag{2.6.5}$$

where ϕ_{int} is the interpolated fringe point; ϕ_i is any computational variables in the domain, such as velocity, pressure, and so on; and ξ_i, η_i, ζ_i represents the interpolation coefficients in three directions for each stencil point.

For example, Fig. 2.6.1 demonstrates the overset grids, which consist of two grids: one is a body fitted airfoil grid, the other is a Cartesian background grid. The body fitted airfoil grid is embedded in the background grid. The grid nodes residing inside the airfoil mesh are removed, but two layers of overlapping nodes remain. These remaining nodes are prepared for interpolation.

2.6.3. Ship geometry and calculation condition

2.6.3.1. Catamarans

The catamaran ship is composed of two simple Wigley hulls with a ship length of 1m. As shown in Fig. 2.6.2, *L* is length and *B* is the width of each hull; *K* is the transverse distance between the two hulls. In this paper, the optimal ratio of the transverse distance to ship length (*K*/*L*) is to be found. And therefore, several *K*/*L* are taken into account, which are 0.2, 0.3, 0.4, 0.6, and 1.0. Beside, five Froude numbers need to be calculated, which are 0.25, 0.316, 0.4, 0.5, and 0.6, respectively.

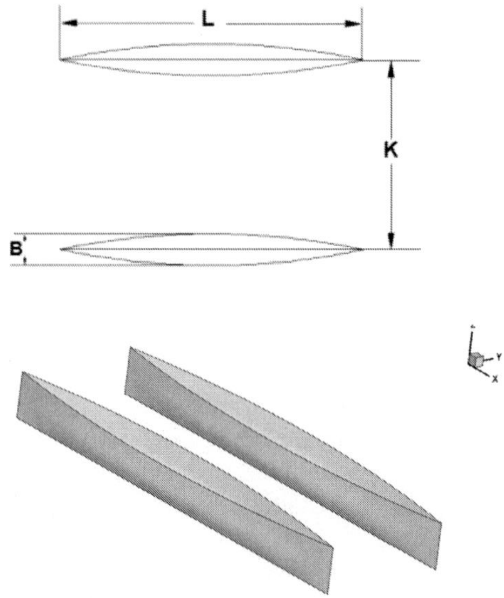

Fig. 2.6.2. Geometry of a catamaran ship

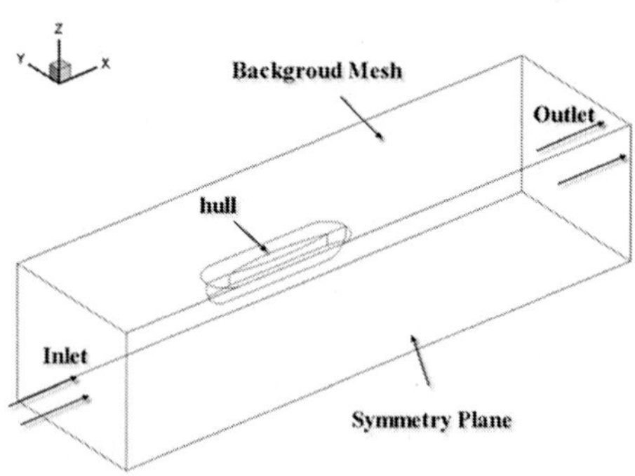

(a) Computation domain

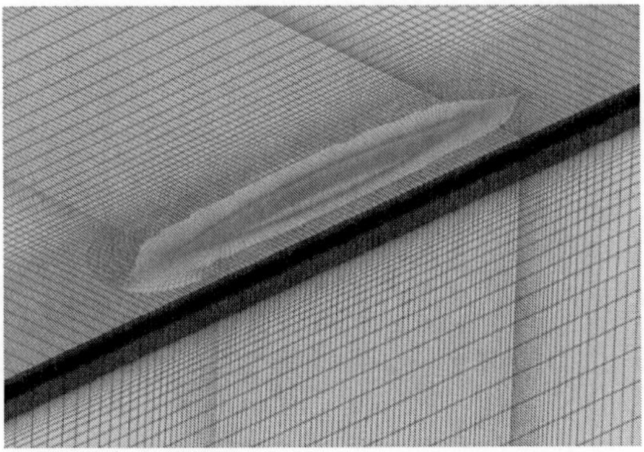

(b) Overset grid (after hole cutting)

Fig. 2.6.3. Computational domain and overset grid

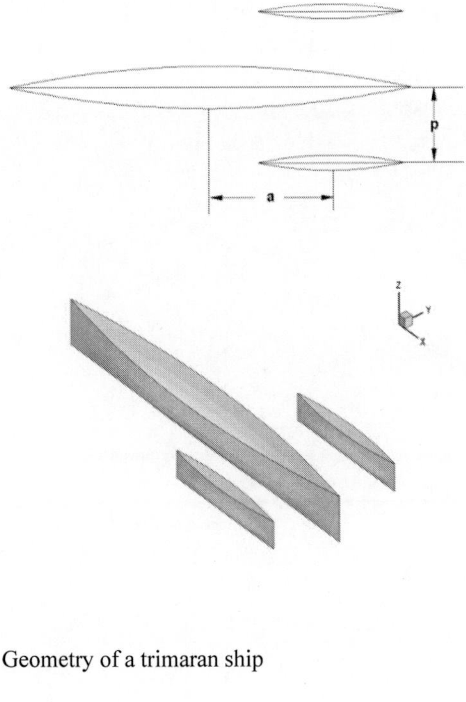

Fig. 2.6.4. Geometry of a trimaran ship

The computation domain is shown in Fig. 2.6.3(a). Since the Wigley catamaran is symmetrical about the *y* axis, only half of the domain is presented, and a symmetry boundary condition is applied at the symmetry plane. The domain ranges from -1.2*L* to 4*L* in the *x* direction, and the width and depth are both about 1.5*L*. The overset grid is shown in Fig. 2.6.3(b). The blue mesh, which is the body fitted hull grid, is embedded into the Cartesian background grid (black grid). After hole cutting, the grid points, which reside inside the blue grid are removed, and the remaining boundary fringe points of each grid are ready to be interpolated to receive information from the other grid. And thus, the connectivity between the two independent grids is established.

2.6.3.2. Trimarans

As shown in Fig. 2.6.4, the trimaran ship is composed of three Wigley hulls, one for the main body, and two small ones for the side bodies. The length of the main body is equal to one meter, and for the side bodies the length is 0.36 meters. To find out the optimal position of side bodies, two key parameters are taken into account. One is the longitudinal distance between the centers of the main body and the side bodies. Another is the transverse distance between the main body and the side ones. The two parameters are represented by symbol *a* and *p*, respectively, as shown in Fig. 2.6.4. Besides, the Froude numbers, which are considered in this section, ranged from 0.4 to 0.7.

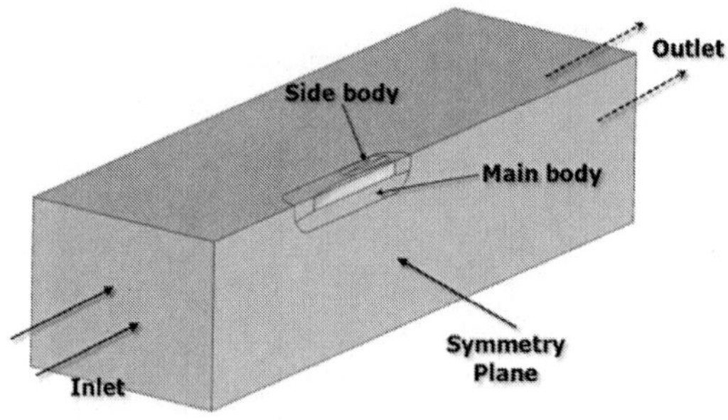

(a) Computation domain

(b) Overset grid (after hole cutting)

Fig. 2.6.5. Computational domain and overset grid

Being similar to a catamaran, the trimaran is symmetrical about the longitudinal center plane of the main body. And, therefore, only half of the computational domains need to be presented in this case, as shown in Fig. 2.6.5(a). The computation domain is the same as that of the catamaran, but two body fitted grids are embedded into the background mesh. One is half of the main body and another is one of the side bodies. The overset grid of the trimaran ship is shown in Fig. 2.6.5(b). The main body grid is fixed, but the side body grid will be moved for different parameters "a" and "p." The transverse section of the overset grid is shown in Fig. 2.6.5(c). The blue grids represent the side bodies and the red one is the main body. Each hull grid is embedded into the black background grid and some layers of overlapping points remain to build the connectivity between each grid.

2.6.4. Results and discussion

2.6.4.1. Catamaran ships

The results of resistance coefficients are shown in Fig. 2.6.6. C_t represents the total resistance coefficients, which consist of a frictional (viscous) force coefficient and a pressure force coefficient. The results are compared with those of a monohull ship, shown by the black solid line. The results indicate that the interference between the two hulls can increase the resistance of a catamaran ship. For the smallest hull separation distance, K/L=0.2, the interference is most severe, as shown with the dash line. The resistance is significantly greater than that of a monohull ship. However, with the increase of K/L, the interference is reduced. For K/L>0.6, such as K/L=1.0, the interference of each hull is so small that it can be neglected. Beside, the resistance coefficient reaches peak value at *Fn*=0.5 for all *K*/*L*.

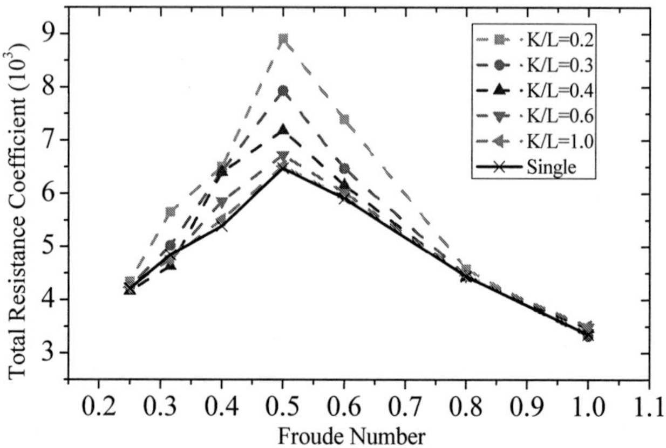

Fig. 2.6.6. Summary of total resistance coefficient for each *K*/*L*

However, if the Froude number is larger than 0.8, the results are quite different. As shown in Table 2.6.1, the resistances are close to that of a mono Wigley hull, compared with the results with low Froude numbers (*Fn*<0.8). The resistance increment due to the wave interference can be neglected. When *Fn* is equal to 1.0, the resistance reduces with the decrease of *K*/*L*, and reaches the minimum value at *K*/*L*=0.3, which is even smaller than that of the mono Wigley hull. Therefore, the beneficial interference between the two hull occurs at *Fn*>0.8.

Table 2.6.1. Results of total resistance coefficients with a high Froude number ($\times 10^3$)

Fn	*K*/*L*=0.2	*K*/*L*=0.3	*K*/*L*=0.4	*K*/*L*=0.6	*K*/*L*=1.0	Single Hull
0.8	4.57	4.45	4.43	4.46	4.46	4.44
1.0	3.37	3.32	3.46	3.49	3.51	3.35

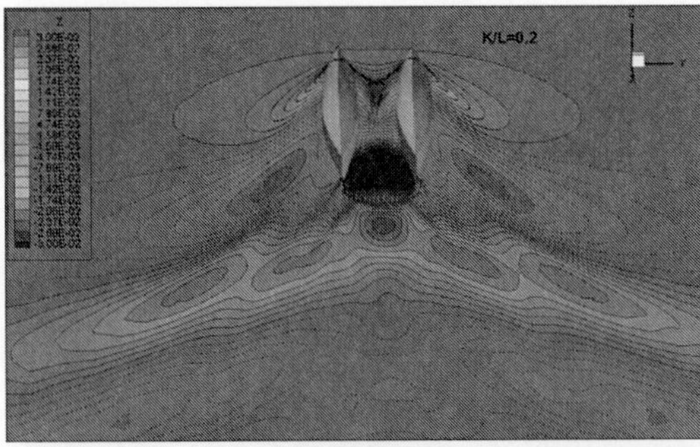

(a) *K*/*L*=0.2

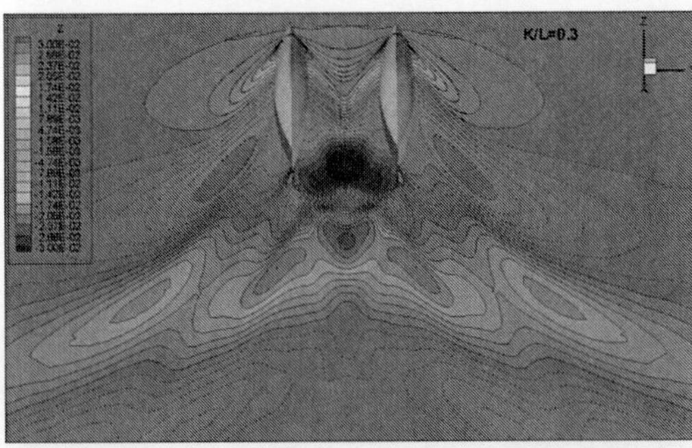

(b) *K*/*L*=0.3

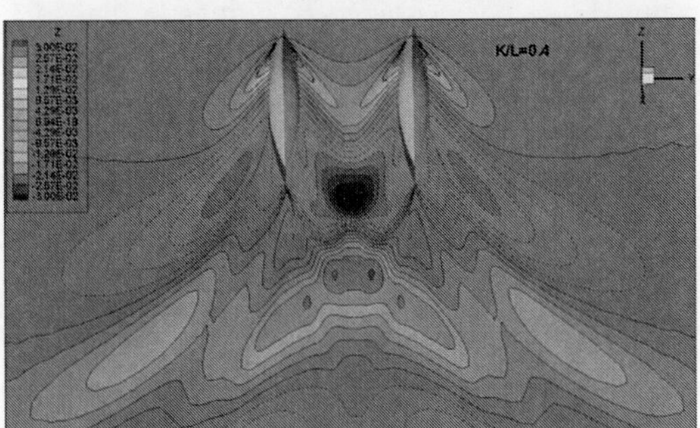

(c) *K*/*L*=0.4

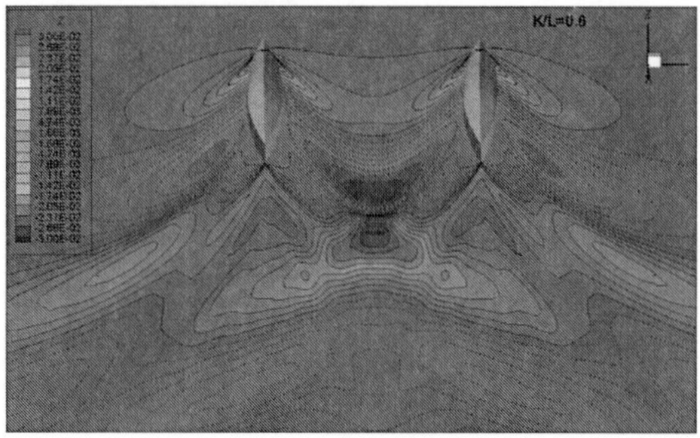

(d) *K*/*L*=0.6

Fig. 2.6.7. Wave patterns of *Fn*=0.6

Fig. 2.6.7 shows the wave patterns with *Fn*=0.6. For the case of *K*/*L*=0.2, for which the interference is most severe, the wave patterns interference is most significant. Both bow waves and stern waves generated by each hull overlap each other. For the bow waves, two overlapped wave patterns create an extremely high wave elevation. So do the stern waves, which cause a wave trough. Because of the combined effects of wave peak and wave trough, large wave resistance force is caused. However, with the increase of *K*/*L*, the overlapped wave height has been reduced. For the case of *K*/*L*=0.6, there is little stern-wave overlap, but no bow-wave interaction. In this case, the resistance is similar to that of a single hull, which indicates that the interference between the two hulls is negligible.

2.6.4.2. Trimaran ship

For the trimaran ship, an optimal position is to be found. And two parameters are introduced: one is the longitudinal distance between the centers of the main body and the side bodies, represented by symbol "a"; another is the transverse distance between the main body and the side ones, represented by symbol "p," as shown in Fig. 2.6.4.

First, the longitudinal position of the side bodies is taken into account. We fix the parameter *p*/*L* and calculate the cases of different *a*/*L* but with the same *p*/*L*, as shown in Fig. 2.6.8. The results indicate that for each transverse position *p*/*L*, the resistance for *a*/*L*=0.3 is always the smallest one. With the increase of *a*/*L*, the resistance continuously reduces. Therefore, *a*/*L*=0.3 is the optimal distance for longitudinal position.

Second, we fix *a*/*L*=0.3 and consider the resistance with different *p*/*L*. The results are shown in Fig. 2.6.9. From the results, it can be found that, for *Fn*>0.50, resistance with *p*/*L*=0.12 is always optimal, and for *Fn*<0.50, *p*/*L*=0.18 is always smallest. However, when considering the speed of a high-performance ship, the Froude number is usually greater than 0.5, so it can be concluded that *p*/*L*=0.12 is the optimal distance for transverse distance.

In conclusion, the case with *a*/*L*=0.3 and *p*/*L*=0.12 is the optimal position for the side bodies.

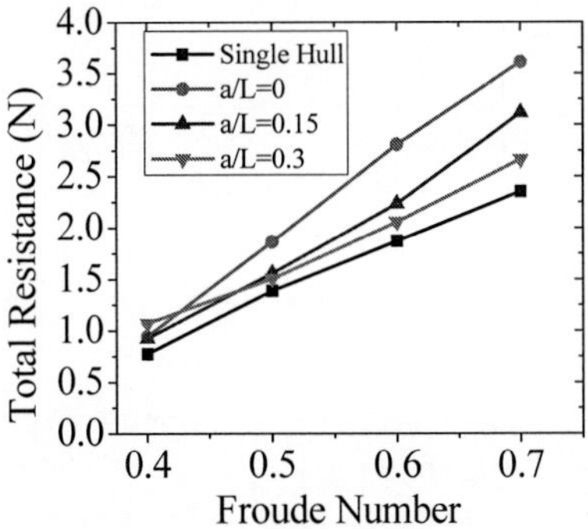

(a) *p*/*L*=0.12

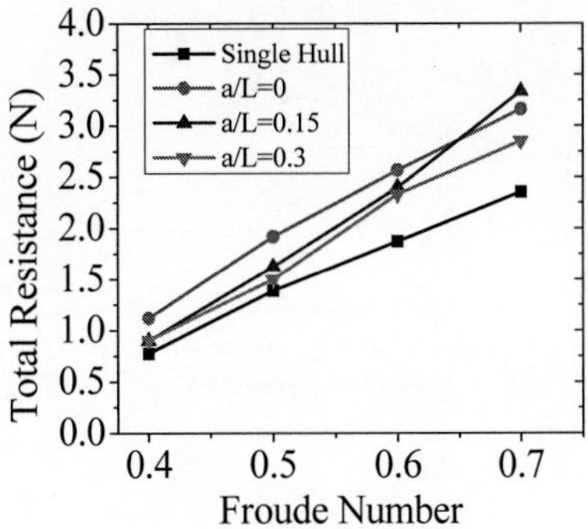

(b) *p*/*L*=0.15

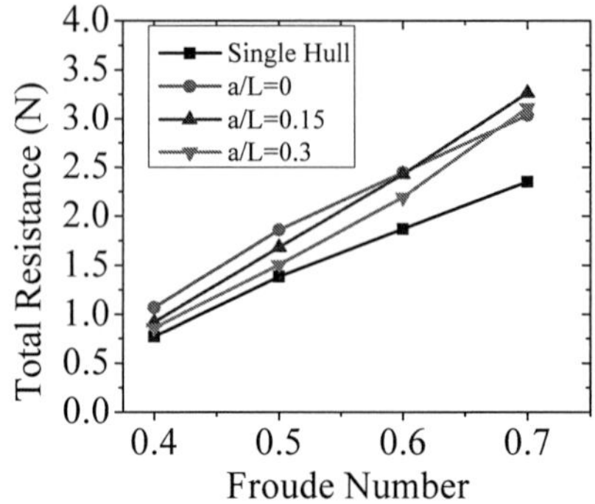

(c) *p/L*=0.18

Fig. 2.6.8. Summary of resistance with different *p/L*

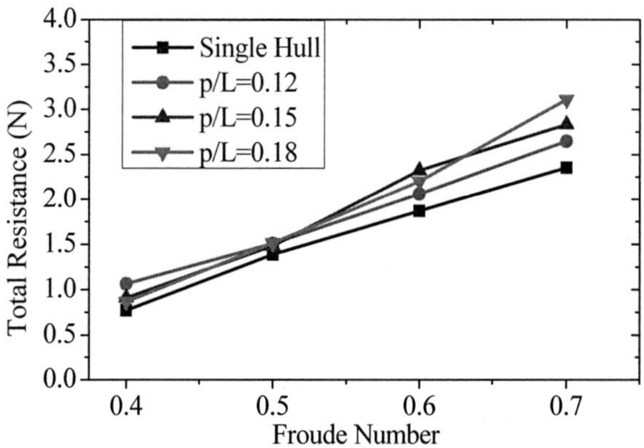

Fig. 2.6.9. Resistance with *a/L*=0.30

The wave patterns with different *a/L* and *p/L* are shown in Fig. 2.6.10. In Fig. 2.6.10(a), the bow of the side bodies resides in the bow waves produced by the main body, and the two kinds of bow waves overlap and create a large wave height, as do the stern waves. And, therefore, the wave resistance in this case is the largest. However, with the side bodies moving to the stern of the main body, they leave the bow waves region produced by the main body, as shown in Fig. 2.6.10(b) and (c). In these conditions, since the wave height is reduced, the wave resistance reduces to a minimum. But if the transverse distance is increased, as shown in Fig.

2.6.10(d), the side bodies return to the bow wave region again. And, therefore, the resistance increases again. So from the above discussion, we can conclude that the condition shown in Fig. 2.6.10(c) is the optimal condition for trimaran ships.

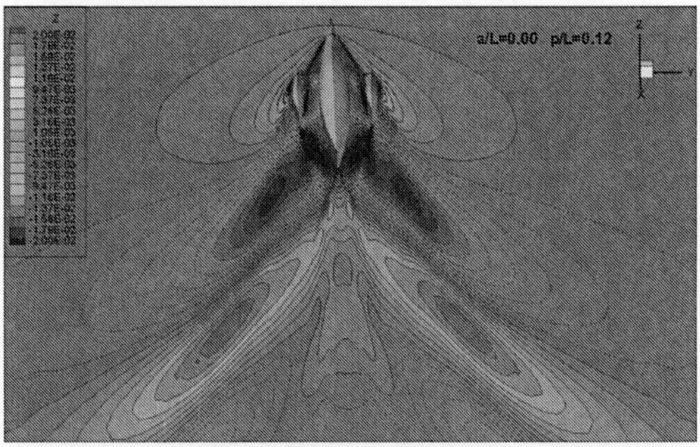

(a) *a*/*L*=0.00, *p*/*L*=0.12

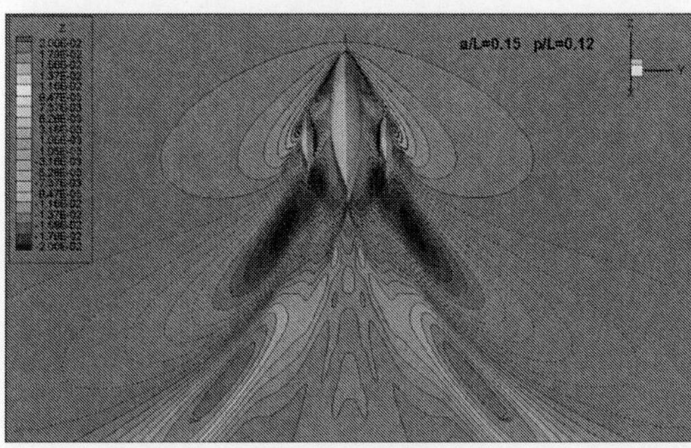

(b) *a*/*L*=0.15, *p*/*L*=0.12

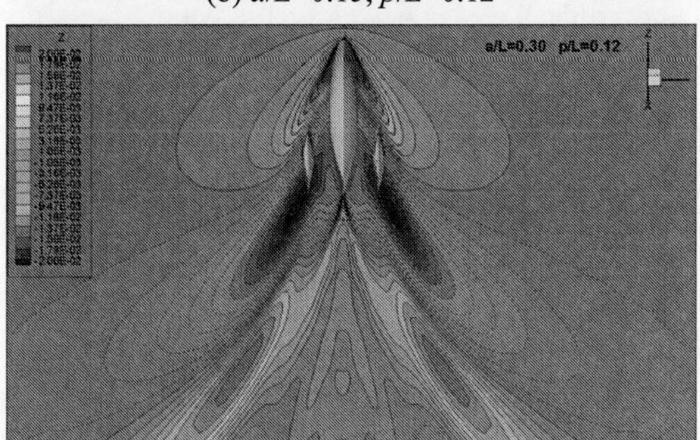

(c) *a*/*L*=0.30, *p*/*L*=0.12

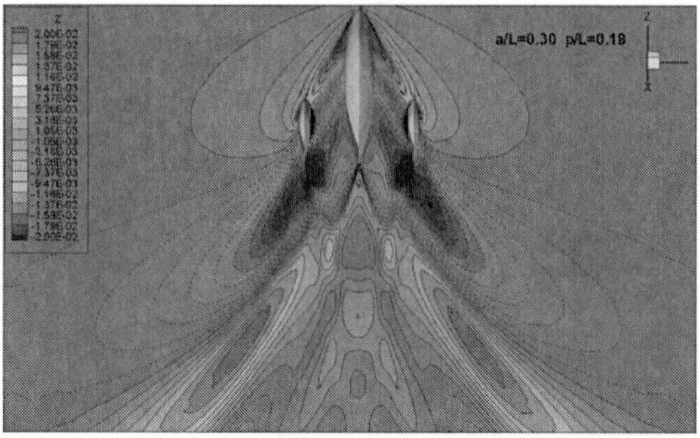

(d) a/L=0.30, p/L=0.18

Fig. 2.6.10. Wave patterns with different positions of side bodies

2.6.5. Conclusions

In this paper, RANS equations coupled with an overset grid mesh are applied to solve the resistance characteristics of multihull ships—catamarans and trimarans. The surface is solved by the level-set surface capturing method. To simplify the problems, both catamarans and trimarans are composed of a Wigley hull.

For the catamaran ship, the transverse distance between the two hulls are taken into account. The results show that, for Froude numbers from 0.25 to 0.6, the interference between the two hulls increases the resistance of the catamaran ship. With a decrease in transverse distance, the interference increases. For the smallest distance, the resistance force is much larger than that for a single-hull ship. However, the beneficial interference between the two hulls occurs when the *Fn* is larger than 0.8. The results show that the resistances of the catamaran ship with different K/L are close to that of a mono Wigley hull. And when the *Fn* is equal to 1.0, the benefit is most significant. The resistance reduces with the decrease of K/L and reaches its minimum at K/L=0.3.

For the trimaran ship, the longitudinal and transverse positions of the side bodies are considered, with Froude numbers ranged from 0.4 to 0.7. Two steps are taken to investigate the resistance characteristics of the trimaran ship. The first step is to analyze different longitudinal positions with a fixed transverse distance. The second step is to calculate each transverse hull separation distance with the optimal longitudinal position concluded from the first step. According to the calculation results, it can be concluded that the position with a longitudinal position of a/L=0.3 and a transverse distance of p/L=0.12 is the optimal position for the side bodies.

This is because in other positions the side bodies may reside in the peak of the waves produced by the main bodies, which can cause a large wave of significant height that leads to an increase in wave-resistance force.

Acknowledgements

This work is supported by National Natural Science Foundation of China (grant nos. 51379125, 51490675, 11432009, 51411130131), the High Technology of Marine Research Project of The Ministry of Industry and Information Technology of China, Chang Jiang Scholars Program (Grant No. T2014099), and the Program for Professors of Special Appointment (Eastern Scholar) at Shanghai Institutions of Higher Learning (Grant No. 2013022), to which the authors are most grateful.

References

[2.6.1] C. Y. Wang, "On Resistance and Interference of Wave Systems of High-speed Catamaran," *Shipbuilding of China*, 1989 (in Chinese).

[2.6.2]. X. Q. Wang, "Resistance Property of High Speed Twin Hull Ship and Determination of k/l and Ship Speed Limits," *Journal of Wuhan Transportation University*, 1997 (in Chinese).

[2.6.3] P. Y. Li, Y. M. Qiu, M. T. Gu, et al., "Experimental Investigation on Resistance of Trimaran," *Ship Building of China*, 43 (4), 2002 (in Chinese).

[2.6.4] R. Saunders Jr, "An Investigation of the Resistance Properties of a Modern Trimaran Combatant Ship based on Taylor Standard Series and Series 64," Master's thesis, Naval Postgraduate School, Monterey, CA, 1995.

[2.6.5] Z. Elcin, "Wave Making Resistance Characteristics of Trimaran Hulls," Naval Postgraduate School, Monterey, CA, 2003.

[2.6.6] J. P. Chen, D. X. Zhu, S. L. He, "Research on Numerical Prediction Method for Wavemaking Resistance of Catamaran/Trimaran," *Journal of Ship Mechanics*, 10 (2): pp. 23–29, 2006 (in Chinese).

[2.6.7] P. M. Carrica, R. V. Wilson, R. W. Noack, et al., "Ship Motions using Single-phase Level set with Dynamic Overset Grids," *Computers and Fluids*, 36 (9): pp. 1415–33, 2007.

[2.6.8] P. M. Carrica, K. J. Paik, H. S. Hosseini, et al., "Urans Analysis of a Broaching Event in Irregular Quartering Seas," *J Mar Sci Technol*, 13 (4): pp. 395–407, 2008.

[2.6.9] N. Sakamoto, "Urans and Des Simulations of Static and Dynamic Maneuvering for Surface Combatant," 2009.

[2.6.10] R. V. Wilson, P. M. Carrica, & F. Stern, "Unsteady Rans Method for Ship Motions with Application to Roll for a Surface Combatant," *Computers & Fluids* 35: pp. 501–24, 2006.

[2.6.11] S. Rhee & F. Stern, "Unsteady Rans Method for Surface Ship Boundary Layer and Wake and Wave Field," *Int. J. Numer. Meth. Fl.* 37 (4): pp. 445–78, 2001.

[2.6.12] F. Menter, "Two-Equation Eddy-Viscosity Turbulence Models for Engineering Applications," *AIAA J* 32 (8): pp. 1598–605, 1994.

[2.6.13] S. Osher & J. Sethian, "Fronts Propagating with Curvature-Dependent Speed: Algorithms based on Hamilton-Jacobi Formulations," *J. Comput. Phys.* 79 (1): pp. 12–49, 1988.

[2.6.14]. P. M. Carrica, R. V. Wilson, & F. Stern, "An Unsteady Single-Phase Level Set Method for Viscous Free Surface Flows," *Int. J. Numer. Meth. Fl.*, 53 (2): pp. 229–56, 2007.

[2.6.15]. R. Issa, "Solution of the Implicitly Discretised Fluid Flow Equations by Operator-Splitting," *J. Comput. Phys*. 62 (1): pp. 40–65, 1986.

[2.6.16]. W. Chan, R. Gomez, S. Rogers, et al., "Best Practices in Overset Grid Generation," AIAA paper 2002-3191, 2002.

2.7

Research on Asymmetric Outrigger Hull Form Design and Resistance of Trimarans (2014)

Gao Lei, Chang Liang and Qiu Gengyao

China Ship Scientific Research Center

Abstract

Everyone knows that the monohull form is symmetrical to aid maneuvering stability and to gain as little resistance as possible. Asymmetrical monohull resistance is usually much higher than for symmetrical hulls. Trimaran hull form design is developed from that of a monohull and the outrigger hull form is usually symmetrical.

During the research on trimaran hull form and resistance, it was found that the proportion of outrigger resistance of a trimaran is much higher than that of the displacement or wetted area. Further research has been carried out and it has been found that the flow around symmetrical outriggers is asymmetrical as the result of hydrodynamic interaction between multihulls. Transverse flow occurs and is considered to be a component of the total resistance from the energy view. A new kind of asymmetrical outrigger hull form design has been developed and asymmetrical coefficient is applied to describe the character of the hull design. The research is applied to a new strategy for trimaran hull form design.

2.7.1. Introduction of a trimaran model test

Researchers at the deep-water towing tank of the China Ship Scientific Research Center found an interesting phenomenon in model resistance tests of a trimaran in 2011, which they had undertaken as a commercial task. The model principals are shown in Table 2.7.1. Special attention should be paid to the asymmetric outrigger hull form. The outriggers, each appearing in half of a conventional deep-V mono hull, were lying on two sides of the main hull symmetrically and the vertical broadside was outboard.

It is observed that when the Froude number was larger than 0.79, a huge amount of spray was produced in the design draught condition during the model resistance tests. The spray was so large that it washed the cross-deck. As this speed was higher than the design speed, the designer considered that it was not necessary to improve this.

Further work was later carried out in the deep-water towing tank by the researchers. Two side hulls were replaced. The starboard outrigger was moved to the port side and the port outrigger was moved starboard. The vertical broadside became inboard. The clearance and stagger remained the same. The trimaran with inboard vertical broadside of outriggers had lower resistance than the trimaran with the outboard vertical broadside of outriggers. The decrease in the residuary resistance coefficient was 9.1% (from 2.697 to 2.453) when the Froude number was 0.705. And no huge spray was produced.

A new idea was applied by the model tests. The asymmetrical hull with outrigger would affect the total resistance of the trimaran.

Table 2.7.1. Main dimensions of the deep-V hull form trimaran

Principals	Symbols/Units	Trimaran (model scale)	
Displacement	∇/m^3	0.1611	
Beam	B/m	1.1111	
Area of wetted surface	S/m^2	2.6894	
Clearance	Cl/m	0.5556	
Stagger	St/m	0.7781	
-		Main hull	Outrigger
Displacement by %	$\%\nabla$	86.34%	6.83%
Length of waterline	L_{WL}/m/	4.8000	1.6800
Maximum mould breadth at the waterline	B_{WL}/m	0.3842	0.0641
Draught	T/m	0.1782	0.0454

Stagger=longitudinal distance between outrigger transom and the main hull transom
Clearance=Transversal distance between the external side of the outrigger and the main hull symmetry plane

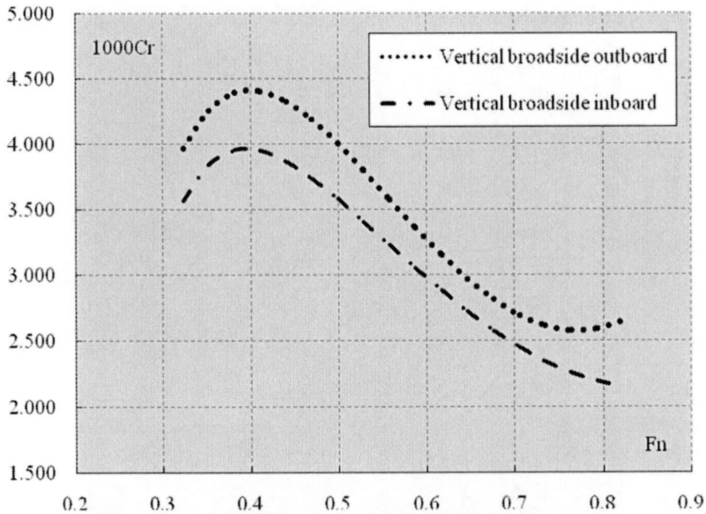

Fig. 2.7.1. Residuary resistance coefficient for two trimarans

Fig. 2.7.2. Photos of model tests (Fn=0.80, inboard vertical broadside outriggers on the left side and outboard vertical broadside outriggers on the right)

2.7.2. Research on asymmetric outrigger hull form design

2.7.2.1. Similar research

In 2001, Antonio Cardo started a project on fast trimaran ferries supported by the Italy Ministry of Education and several universities and research institutes, among them the Universities of Genova, Napoli, and Trieste. Asymmetrical outriggers were applied to one trimaran with the vertical broadside outboard, shown in Fig. 2.7.3. During the model tests, huge spray was generated at Froude number 0.65, shown in Fig. 2.7.4. Spray rails were added to the outriggers but the effect was negligible. Then the round bilge outriggers were applied, shown in Fig. 2.7.3. There was no large spray at Froude number 0.65.

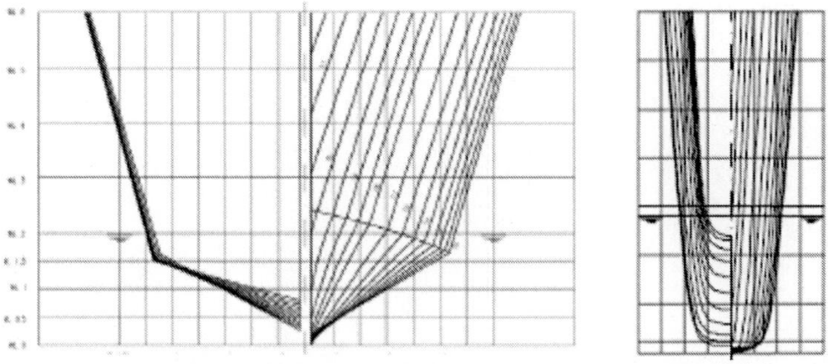

Fig. 2.7.3. Body plans of the asymmetric outrigger (top) and the round bilge outrigger (bottom)

Fig. 2.7.4. Photos of the deep-V asymmetric outrigger (Fn=0.65)

2.7.2.2. Asymmetrical outrigger hull design

A research project was started on trimaran hull form design by the China Ship Scientific Research Center (CSSRC) in 2012, focused on the spray of outriggers and the total resistance of trimarans. The trimaran M_1 was composed of a round-bilge main hull and two symmetric round-bilge outriggers. The model principals are shown in Table 2.7.2. Then an asymmetrical outrigger hull form was developed from the symmetric one with the same vertical central plane. The difference of the displacement, wetted surface area, and L/B ratio were less than 0.5%. The hull forms of the two outriggers are shown in Fig. 2.7.5 and the principals are shown in Table. 2.7.3 for the outrigger model scale.

Table 2.7.2. Main dimensions of the round bilge hull form trimaran M_1

Principals	Symbols/Units	Trimaran (model scale)
Displacement	∇/m^3	0.1767
Beam	B/m	1.2731
Area of wetted surface	S/m^2	2.9065
Clearance	Cl/m	0.8750
Stagger	St/m	0.0000

Table 2.7.3. Main dimensions of the two outriggers

Principals	Symbols/Units	Symmetric outrigger	Asymmetric outrigger
Length of waterline	L_{WL}/m	2.5000	2.5000
Maximum mould breadth at the waterline	B_{WL}/m	0.1531	0.1525
Draught	T/m	0.1500	0.1500
Displacement	∇/m^3	2.3668×10^{-2}	2.3554×10^{-2}
Area of wetted surface	S/m^2	0.6725	0.6742

The asymmetric outriggers were positioned as shown in Fig. 2.7.6, and thus the M_2 and M_3 trimarans were developed. The M_2 trimaran could be regarded as thin inboard and fat outboard. The M_3 trimaran could be regarded as fat inboard and thin outboard.

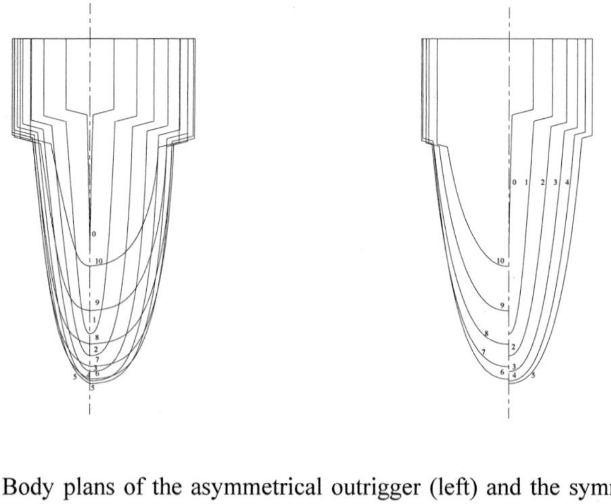

Fig. 2.7.5. Body plans of the asymmetrical outrigger (left) and the symmetrical outrigger (right)

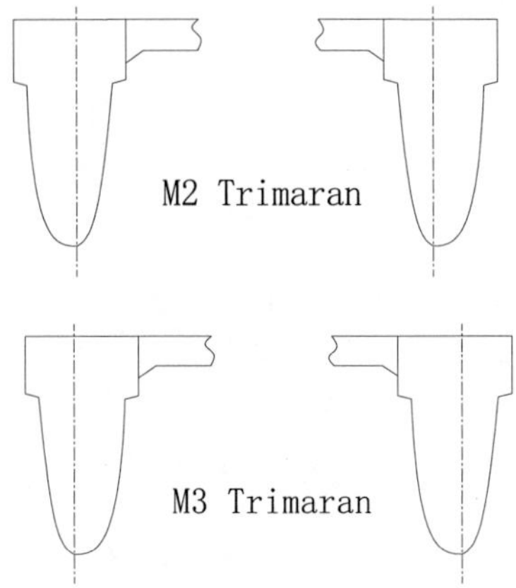

Fig. 2.7.6. Different positions of asymmetrical outriggers for different trimarans

2.7.2.3. Research on the resistances of different trimarans with asymmetrical outriggers

CFD calculations were carried out on the resistance of three trimarans. The flow of the outrigger was asymmetrical even for the symmetrical hull form as the result of the hydrodynamic influence of the main hull. Here there would be lateral flow around the outriggers. Their direction was from port to starboard, or opposite. Then the lateral forces impacted on the outriggers were also calculated. The result is shown in Fig. 2.7.7.

The contour lines of the trimaran waves near the outrigger are shown in Fig. 2.7.8. The wave pattern was not symmetrical with the center plane of the symmetrical outrigger. For different outrigger hull forms, the wave pattern was different but the wave patterns near the same hull forms of different trimarans did not change much. The effect of the main hull on the outrigger was large, but the effect of the outrigger on the main hull was small.

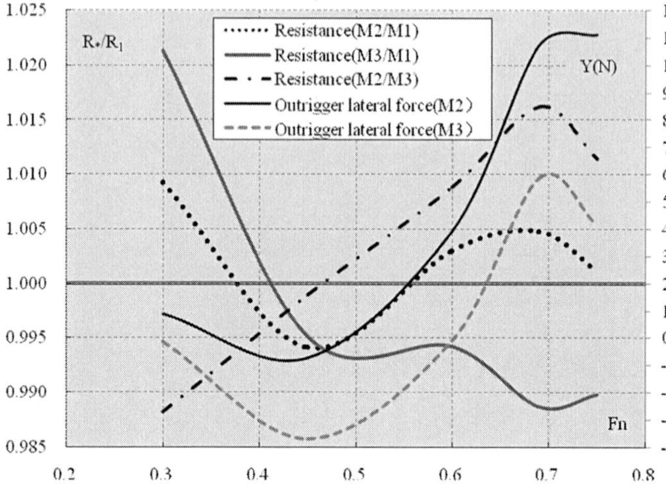

Fig. 2.7.7. Total resistances and outrigger lateral forces

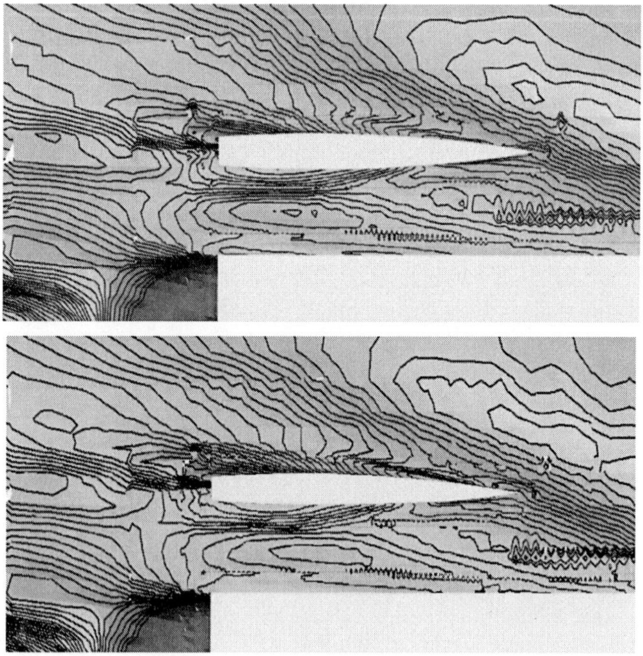

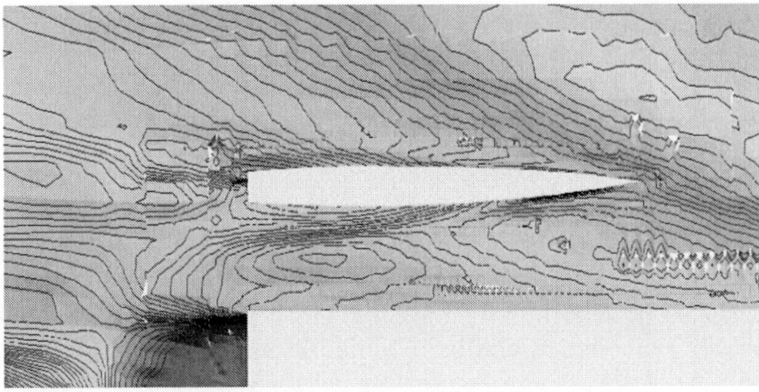

Fig. 2.7.8. Contour lines of the trimaran wave near the outrigger (Fn=0.69, from top to bottom, M_1, M_2, M_3)

2.7.2.4. Analysis of the calculations

First, the resistance was compared and analyzed.

1. When the Froude number was below 0.37, the resistances of M_2 and M_3 were larger than that of M_1.
2. When the Froude number was between 0.37 and 0.56, the resistance of M_2 was less than that of M_1.
3. When the Froude number was above 0.56, the resistance of M_2 was larger than that of M_1.
4. When the Froude number was above 0.42, the resistance of M_3 was less than that of M_1.
5. When the Froude number was above 0.46, the resistance of M3 was less than that of M_2.

The analysis of resistance shows the complex relation between symmetrical and asymmetrical outrigger hull forms. And it is different for different speeds.

It could be concluded that if the Froude number of a trimaran is larger than 0.46, the asymmetrical outrigger regarded as a fat inboard and thin outboard would be the best hull form choice from the resistance viewpoint. And if the Froude number of a trimaran is less than 0.37, the symmetrical outrigger would be the best hull form choice, like the Triton (Fn=0.36).

Next, the outrigger lateral force was compared and analyzed.

(1) When the Froude number was below 0.56, the direction of lateral force of M_2 and M_3 was outboard.

(2) When the Froude number was above 0.56, the direction of lateral force of M_2 was inboard and when the Froude number was above 0.63, the direction of the lateral force of M_3 was inboard.
(3) The lateral force of M_2 was less than that of M_3 only near the Froude number 0.45.

It would be interesting to discover whether the lateral force would be zero for some speeds. In other words, the impact of the water flow near the outrigger could be zero in the transverse plane, which should be received by the cross-deck and as the load of its structure. For M_2, the Froude number was 0.35 and 0.50 when the lateral force became zero and for M_3 it was 0.30 and 0.60.

2.7.3. Further research on the asymmetrical outriggers

2.7.3.1. Definition of asymmetrical coefficient
A new coefficient was applied here to define the asymmetrical outrigger, C_{AS}, a symmetrical coefficient. It should be calculated as follows:

$$C_{AS} = \frac{2}{\pi} \arctan \frac{\nabla_{OUT}}{\nabla_{IN}} \quad (2.7.1)$$

Define the vertical plane that runs through the bow as a bow vertical plane. Then the outrigger will be cut into two parts. The displacement of the outboard volume is ∇_{OUT} and the displacement of the inboard volume is ∇_{IN}. The C_{AS} of M_1 should be 0.5, M_2, 0.324, and M_3, 0.673. The C_{AS} of the deep-V trimaran should be 1.0 for the vertical broadside outboard and 0 for the vertical broadside inboard.

2.7.3.2. Relationship between trimaran resistance and asymmetrical coefficient
The relationship between trimaran resistance and the asymmetrical coefficient is shown in Fig. 2.7.9. The C_{AS} should be larger to gain lower resistance or better power performance when the Froude number is larger than 0.60.

When the Froude number is 0.30, the C_{AS} should be 0.50 on the resistance view. In other words, the symmetrical outrigger hull-form design should be the best choice.

The C_{AS} of the deep-V trimaran should be 1.0 for the vertical broadside outboard and 0 for the vertical broadside inboard.

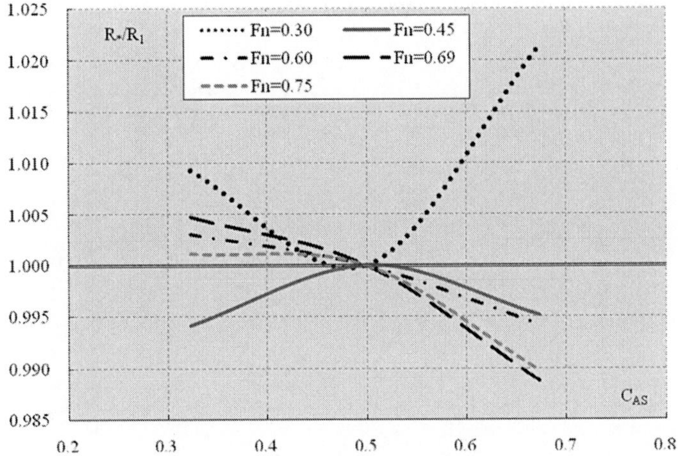

Fig. 2.7.9. Resistance and asymmetrical coefficient curve

2.7.3.3. Relationship between lateral force and asymmetrical coefficient

The lateral force coefficient C_Y was applied here. It should be calculated as follows:

$$C_Y = \frac{Y}{0.5\rho V^2 S} \quad (2.7.2)$$

where Y=the outrigger lateral force, it is above 0 when the direction is to the central plane of the main hull. S=area of single outrigger wetted surface.

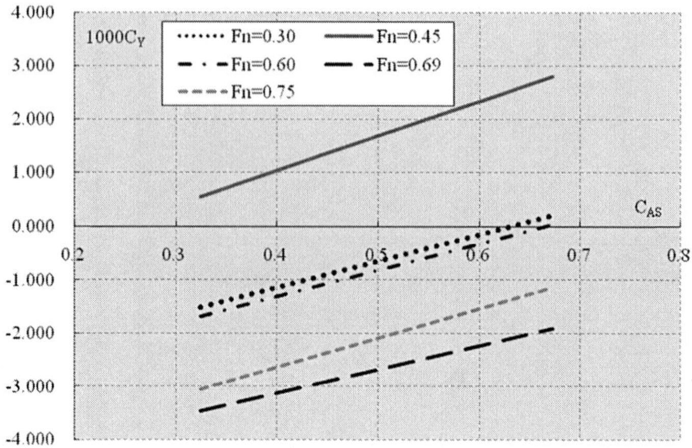

Fig. 2.7.10. Lateral force coefficient curves

Taken as the sign of the lateral force value, it could be concluded that the lateral force increased as the asymmetrical coefficient increased. This factor could be used to decrease the load on the outrigger and optimize cross-deck structure design.

This conclusion was made from the result in calm water. There would be wave load on the outrigger and the cross-deck in wave-making conditions. The relationship between the wave load and outrigger asymmetrical coefficient would be an important part of the research of next stage.

It is difficult to find a very clear relationship between the resistance and lateral forces. However, it is possible to find a better choice when the resistance and lateral forces were both taken into consideration. For example, when the Froude number was 0.60, the asymmetrical outrigger with C_{AS}=0.65, had the latest resistance and its lateral force was almost 0.

Then to the deep-V trimaran, whose model tests were carried out in CSSRC in 2011. The result is shown in Fig. 2.7.10. The C_{AS} of the trimaran with the outrigger vertical broadside outboard should be 1.0 and the C_{AS} of the trimaran with the outrigger vertical broadside inboard should be 0. When the Froude number was larger than 0.60, the resistance of the lower C_{AS} trimaran was lower. It seemed to be contrary to the conclusion and research on the asymmetrical round-bilge outrigger trimaran. It showed the complex relationship between resistance and C_{AS}.

The C_{AS} should be enlarged to a larger range in the research at the next stage.

2.7.4. Further research on the direction of outrigger lateral force

The use of a reflex model in wind tunnels or CFD calculation provided an approximate means of directly measuring or calculating the viscous drag of the model, without generation of waves. So the reflex model of the M_1 trimaran was made to the calculation.

The result was shown in Fig. 2.7.11. And the result of M_1 with free water surface was compared with the reflex model calculation, which had already been carried out at the early stage.

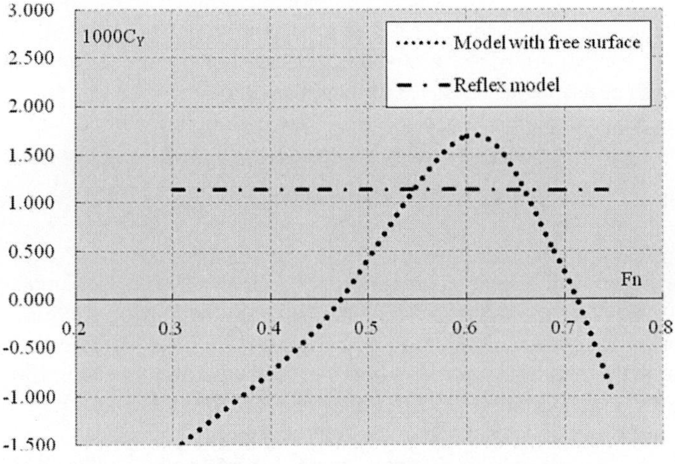

Fig. 2.7.11. Lateral force coefficient curves (the reflex model and the model with the free surface)

The outrigger lateral force was positive from the reflex model calculation. In other words, the direction of the outrigger lateral force was toward the central plane of the main hull or inboard. The reflex model eliminated the influence of waves generated by the trimaran. The comparisons showed that the direction of the outrigger lateral force was effected by the wave generation of the main hull to a large degree.

It is suggested that further research is carried out on this subject.

2.7.5. Further research plan

1. The C_{AS} should be enlarged to encompass a larger range; it could even be minus. If the waterline shape of the outrigger is like the double arch wing, its C_{AS} could be minus. The outboard and inboard volumes are on the same side of the head vertical plane.
2. The research on the relationship between the lateral force and C_{AS} should be carried out in wave conditions.
3. The model tests should be carried out to prove the calculations. Now the balance and trimaran model have been designed and the model tests will be carried out in the middle of 2014.

2.7.6. Summary

A new kind of asymmetrical hull-form design of an outrigger has been developed.

The research on asymmetrical and symmetrical outrigger hull form has been carried out in the first stage. The asymmetrical coefficient was applied to describe the character of the hull design.

The research applies a new strategy for trimaran hull-form design.

References

[2.7.1] A. Cardo, M. Ferrando, et al., "Influence of Hull Shape on the Resistance of a Fast Trimaran Vessel," International Conference on Fast Sea Transportation, FAST 2005, June, St Petersburg, Russia.

[2.7.2] S. Brizzolara, M. Capasso, M. Ferrando, et al., "Effect of Hull Form Variation on Hydrodynamic Performance of a Trimaran Ship for Fast Transportation," International Conference on Fast Sea Transportation, FAST 2005, June, St Petersburg, Russia.

[2.7.3] M. G. Morabito, "Experimental Investigation of the Lift and Interference of Asymmetric Planing Catamaran Demi-hulls," 11th International Conference on Fast Sea Transportation FAST 2011, September, Honolulu, Hawaii, USA.

[2.7.4] C. Lugni, A. Colagrossi, M. Landrini, et al., "Experimental and Numerical Study of Semi-displacement Mono-Hull and Catamaran in Calm Water and Incident Waves," 25th Symposium on Naval Hydrodynamics, 8–13 August 2004, St. John's, Canada.

2.8

MODEL TESTING OF TRIMARAN RESISTANCES AND THE SEMI-PLANING FPOT (FORWARD PLACED OUTRIGGER TRIMARAN) CONCEPT

ZONGZHI[1] AND JIA JINGPEI[2]

[1] Dalian University of Technology
[2] Dalian Ocean Universit

Abstract

Model testing methodology is used in this paper to study the influence of side-hull positions on the resistance of trimaran hull forms. Experimental results show that the longitudinal positions of side hulls have significant influence on the resistance. At semi-planing state, the configurations with side hulls placed forward will improve resistance performance greatly. This observation leads us to propose the concept of the Semi-Planing FPOT (Forward Placed Outrigger Trimaran).

2.8.1. Introduction

Recent years have seen rapid development in multihull vessel technologies. Among them is the trimaran. A trimaran hull form consists of three single hulls, one main hull and two side hulls. Generally, the two side hulls are placed symmetrically on the two sides of the main hull. A trimaran hull form is superior to the traditional monohull or catamaran hull form due to its large deck area, good transverse performance, low wave-making resistance, and improved sea-keeping.

In the past two decades, there have been many publications on experimental and numerical works concerning trimaran hydrodynamics. Ackers et al.[2.8.1] experimentally investigated the effects of side-hull position, symmetry, displacement, and angle-of-attack on the resistance characteristics of trimarans. They proposed a useful design tool for quantifying the resistance characteristics of different configurations. Li et al.[2.8.2] presented an experimental investigation of the resistance characteristics in calm water of a round-bilged trimaran over a Froude number range of 0.3 to 0.8. Gu et al.[2.8.3] presented a new type of trimaran. The center hull of the trimaran is of a small waterplane area. A series of model tests were carried out to study the resistance characteristics. Degiuli et al.[2.8.4] experimentally studied the wave-pattern resistance of the trimaran configuration, which consists of standard Wigley's models. Oh et al.[2.8.5] presented a model test for the resistance of a particular trimaran for hull design. The influence of the spacing between the mainhull and the outriggers and the longitudinal locations of the outriggers has been examined to find the optimal configuration. The roll motion of a trimaran ship composed of three Wigley hulls in beam waves has been studied on an experimental basis by Francescutto et al.[2.8.6] Li et al.[2.8.7] conducted the experiments with a supper slender trimaran model to investigate sea-keeping performance in regular waves. In addition, extensive theoretical and numerical studies of trimaran hydrodynamics have been developed.[2.8.8–14] Moreover, the results from numerical and theoretical studies have been verified against the related experimental data.

A peculiar unsolved problem concerning trimarans is that the arrangement of the side hulls may have a very important influence on hydrodynamic performance. In the present paper, the results of a trimaran model experiment are described. On the basis of the experimental data, the influence exerted by the longitudinal position of the side hulls on trimaran resistance was analyzed. In addition, a new trimaran configuration is proposed to improve the overall performance of the trimaran hull form.

2.8.2. Trimaran model tests

2.8.2.1. Model design

The tested trimaran model consists of a single main hull and two scaled-down side hulls. All the hulls follow a knuckle-line plan with a transom stern, as shown in Fig. 2.8.1. The main dimensions of both the main hull and the side hulls are listed in Table 2.8.1.

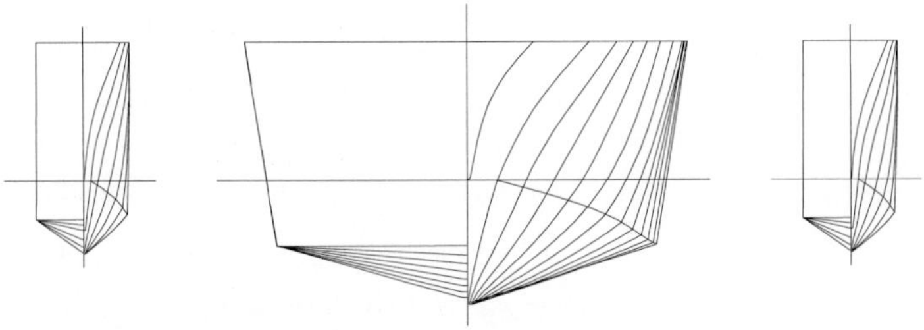

Fig. 2.8.1. The cross section plan of the model

Table 2.8.1. Particulars of the trimaran models

Item	Symbol	Value
Waterline length of main hull/m	Lm	4.00
Waterline breadth of main hull/m	Bm	0.358
Draft of main hull/m	Dm	0.17
Displacement of main hull/kg	Δm	129.0
Wetted surface area of main hull/m2	Sm	1.899
Block coefficient of main hull	Cbm	0.530
Waterline length of side hull/m	Ls	1.00
Waterline breadth of side hull/m	Bs	0.085
Draft of side hull/m	Ds	0.10
Displacement of side hull/kg	Δs	4.5
Wetted surface area of side hull/m2	Ss	0.201
Block coefficient of side hull	Cbs	0.524
Longitudinal position of side hull/m	a	-1.3,-1. 0,-0.7,0 .1,0
Transverse position of Side hull/m	b	0.7

To investigate the influence of the positions of the side hulls, different trimaran configurations were designed. The origin of the coordinate system is at the center of the main hull. The trimaran configurations are defined by parameters a and b, as can be seen in Figs 2.8.1 and 2.8.2.

The parameters a and b denote the longitudinal and transverse position of the side hulls, respectively. The value of a is positive when the midship section of the side hull is in front of that of the main hull, while the value of b is always positive. In the present experiment, the value of b was taken as a fixed value, that is, $b = 0.7$. However, the parameter a is a variable. Their particular values are listed in Table 2.8.1. It is can be seen that, in the experiments, the waterline length of the trimaran is identical to that of the main hull. The configuration is schematically shown in Fig. 2.8.2.

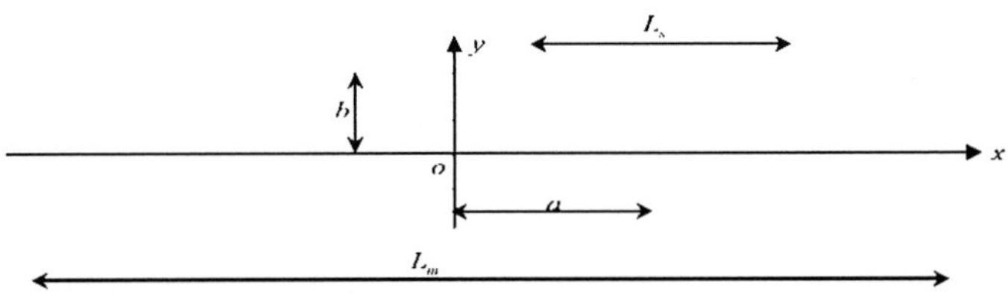

Fig. 2.8.2. Schematic of the trimaran configuration

2.8.2.2. Test design series

Model tests were carried out to study the influence of the side hull position on trimaran capability, including resistance and sea-keeping at various speeds. The resistance experiments were conducted in calm water, while the sea-keeping experiments were undertaken in regular waves. Five configurations were considered and tested in the present work. The trimaran models were built in the towing tank lab at Dalian University of Technology. The towing tank of the lab is 160m long, 7m wide, and 3.7m deep. The trimaran model tests were carried out in the towing tank, as shown in Table 2.8.2.

Table 2.8.2. Trimaran configuration parameters

Models	Model 1	Model 2	Model 3	Model 4	Model 5
a /m	-1.3	-1.0	-0.7	0.0	1.0
b /m	0.6	0.7	0.7	0.7	0.7

The trimaran's high speed is regarded as one of its outstanding advantages. Thus, the presented experiments focus on high speed. The total resistance in calm water was measured over a range of Froude numbers, varying from 0.6 to 1.0. The heave measurements were recorded during the pitch test for each configuration. Both pitch and roll tests were performed in regular waves. The wave length range is from 2m to 12m. The wave height is in the range of 35–40mm in the model tests. The pitch tests were carried out in head seas at Fn=0.701, while roll tests were carried out in beam seas at Fn=0.0. The towing point was taken as the center of floatation for all configurations in the model experiments.

2.8.3. Trimaran resistance in calm water

The resistance measured from the tests is the total resistance in calm water as shown in Fig. 2.8.3.

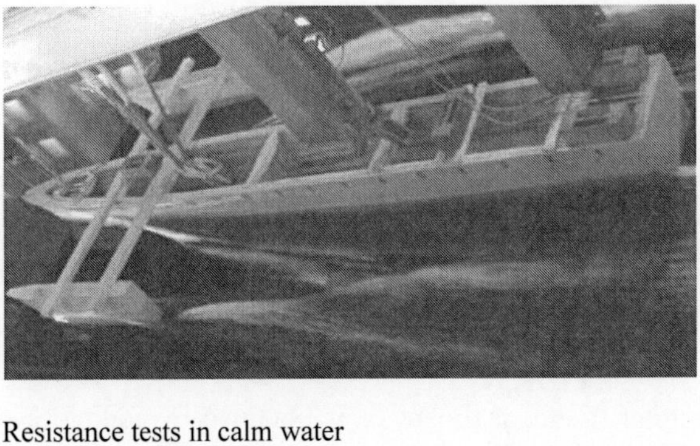

Fig 2.8.3. Resistance tests in calm water

On the basis of the Froude assumption, the total resistance can be divided into two parts: frictional resistance and residuary resistance. Thus, the residuary resistance can be derived by subtracting the frictional resistance from the total resistance, that is $R_t = R_f + R_r$. In terms of dimensionless quantities, the coefficient of total resistance is the sum of the friction resistance coefficient and the residual resistance coefficient in the form of $C_t = C_f + C_r$.

The frictional resistance coefficient can be calculated according to the ITTC-57 formula:

$$C_f = \frac{0.075}{(\log_{10} \mathrm{Re} - 2)^2} \quad (2.8.1)$$

where $\mathrm{Re} = uL/\nu$

u —the ship velocity;

L —the ship length;

ν * the viscosity coefficient—it is assumed to be an constant: $\nu = 0.000001$.

And then the frictional resistance R_f can be obtained from the following formula:

$$R_f = C_f \times \frac{1}{2} \rho u^2 S \quad (2.8.2)$$

where

ρ —the water density;

S —the wetted surface area of the ship hull.

Because of the difference between lengths of the main hull and side hulls, it is necessary to compute the variables C_f and S respectively. Thus, residuary resistance R_r can be derived from:

$$R_r = R_t - R_f \quad (2.8.3)$$

Finally, the total frictional resistance is obtained by adding the frictional resistances together.

Similarly, we can obtain the residuary resistance coefficient C_r with the formula:

$$C_r = R_r \Big/ \frac{1}{2} \rho u^2 S \quad (2.8.4)$$

It has to be noted that here the parameter S is the total wetted surface area of the trimaran model. The sinkages at the stem and stern of the main hull during resistance tests were recorded and plotted in Figs 2.8.4 and 2.8.5, respectively. The parameter Sk denotes the sinkage of the designed waterline (above the water surface is positive). The influence by trim on the wetted surface area is negligible. Thus, the total wetted surface area was taken as the surface area below the designed waterline.

The test results are given in Fig. 2.8.4. Comparison of the results presented in Fig. 2.8.4 shows that the optimum positions of the side hulls are dependent on Froude numbers. In the interval Fn=01–0.2, the residual resistance coefficients of different configurations do not show a big difference for inside the interval, and frictional resistance is dominant over wave-making resistance. In the interval Fn=0.2–1.0, the residual resistance coefficients of different configurations are significantly different for the wave-making resistance, which becomes important. The interference effects among different configurations produce different wave-making resistances.

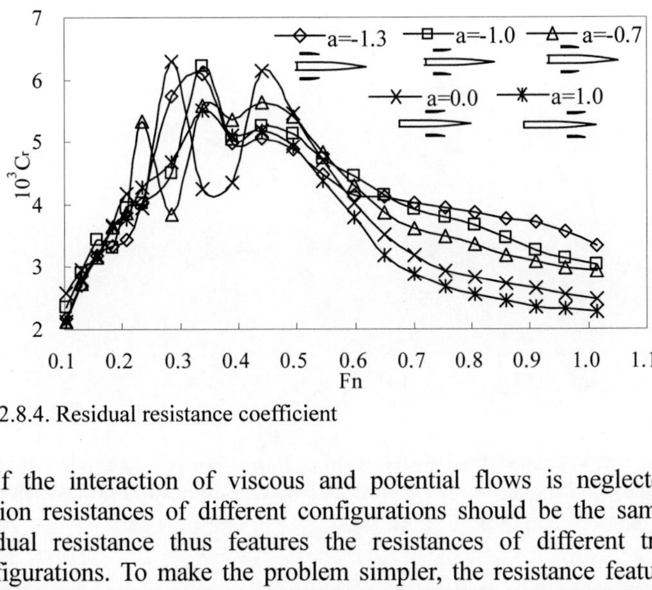

Fig. 2.8.4. Residual resistance coefficient

If the interaction of viscous and potential flows is neglected, the friction resistances of different configurations should be the same. The residual resistance thus features the resistances of different trimaran configurations. To make the problem simpler, the resistance features are discussed in three intervals defined by Fn=0.0–0.5, Fn=0.5–0.65, and Fn=0.65–1.0.

Fig. 2.8.5 shows the residual resistance coefficients in the interval of Fn=0.2–0.5. Large fluctuations can be observed for all configurations. The biggest ones are a=-0.7 and a=0.0. This results from the strong interferences among the main and side hulls and is dependent on Froude number. As beneficial interferences are produced between the main and side hulls, wave-making resistance is greatly reduced leading to lower residual resistance coefficients. If the interference is unbeneficial, the opposite effects are observed.

In the interval Fn=0.27–0.32, the configuration a=-0.7 produces the minimum resistances among all configurations. At Fn=0.29, its residual resistance coefficient is about 39% smaller than the residual resistance of the configuration a=0.0. At Fn=0.32–0.41, the residual resistance coefficient of the configuration a=0.0 is much smaller than others: about a 30% reduction compared with that of the configuration a=-1.0 at Fn=0.36. In the interval Fn=0.41–0.50, the two configurations a=0.0 and a=-0.7 produce larger residual resistances, but with minor differences between the remaining three configurations.

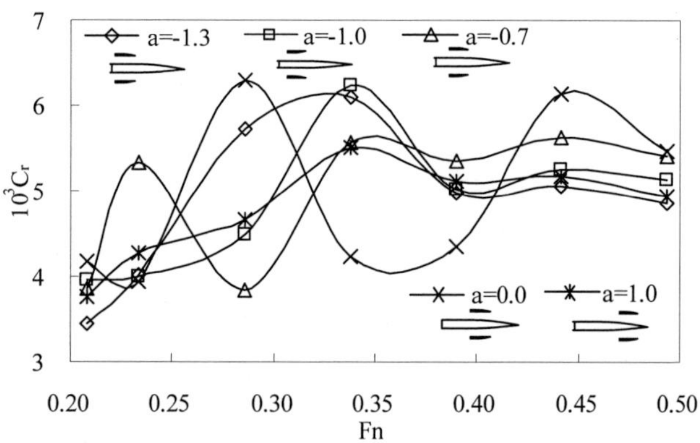

Fig. 2.8.5. Residual resistance coefficient in the interval Fn=0.2–0.5

Fig. 2.8.6 shows the results for Fn=0.5–0.65. The smallest residual resistance falls to the configuration a=1.0 with side hulls placed forward. The resistance reduction is about 10%.

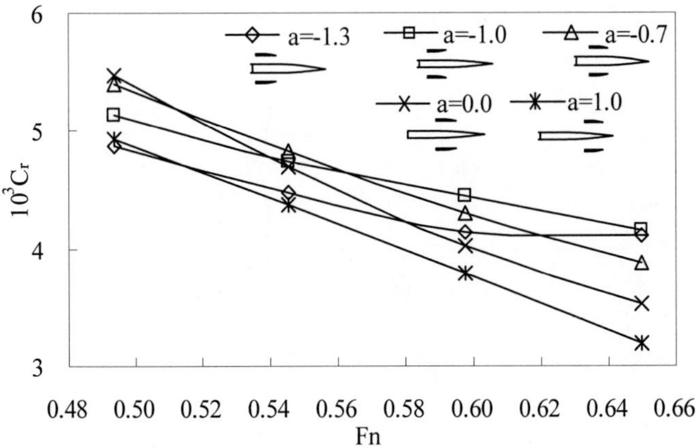

Fig. 2.8.6. Residual resistance coefficient for Fn=0.5–0.65

Fig. 2.8.7 shows the residual resistance for all configurations between Fn=0.65–1.0 The five curves are nearly parallel without any intersection. The fives, from higher to lower, correspond to a=1.0, a=0.0, a=-0.7, a=-1.0, and a=-1.3; that is, the residual resistance decreases as the side hulls are moved from the aft to the forward part. Within the experimental range, the residual resistance of the configuration a=1.0 is at a minimum compared with the maximum resistance of the configuration a=-1.3. About a 30% reduction in resistance is achieved.

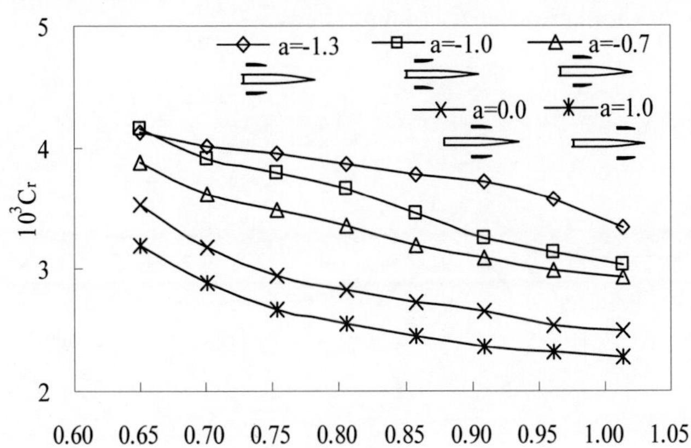

Fig. 2.8.7. Residual resistance for Fn=0.65–1.0

2.8.4. Resistance in waves

We use R_{tw} (unit N) to denote resistance in waves. In the model tests, the wave lengths fall between 2–12m with wave height between 35mm–40mm. Figs 2.8.8 to 2.8.10 are the total resistance of all configurations in head sea with frequency for Fn=0.234, Fn=0.494, and Fn=0.701.

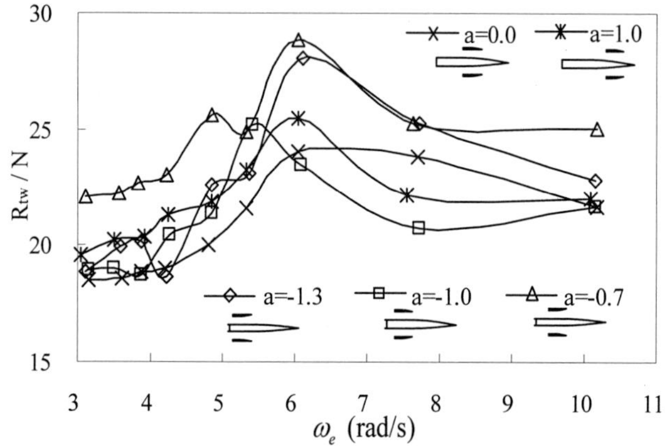

Fig. 2.8.8. Total resistance in regular waves at Fn=0.234

Fig. 2.8.9 shows that the configuration a=1.0 with side hulls placed forward produces the minimum resistances at Fn=0.494, about 7% smaller than the maximum resistance configuration a=-0.7.

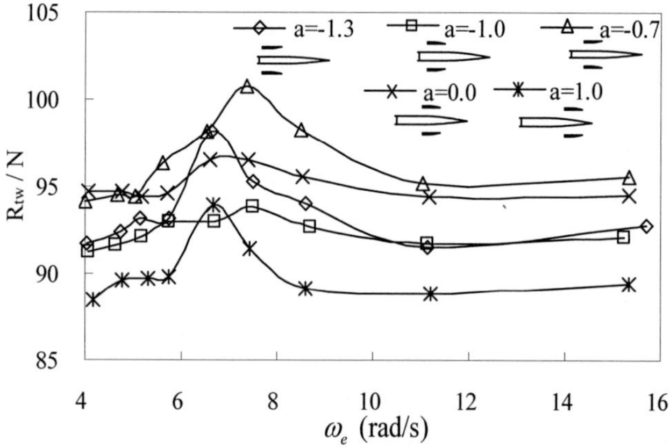

Fig. 2.8.9. Total resistance in regular waves at Fn=0.494

Fig. 2.8.10 shows that at Fn=0.701, the configuration with side hulls (a=1.0) placed forward produces minimum resistance in regular waves among all configurations, with a 16% reduction in resistance compared with one of maximum resistance a=-1.0. Another observation from the figure is that the resistances for the two cases of a=1.0 with forward side hulls and a=0.0 with central side hulls are nearly the same with narrow difference 1%. Thus at Fn=0.701 a trimaran form with forward or central side hulls are advantageous in resistance performance.

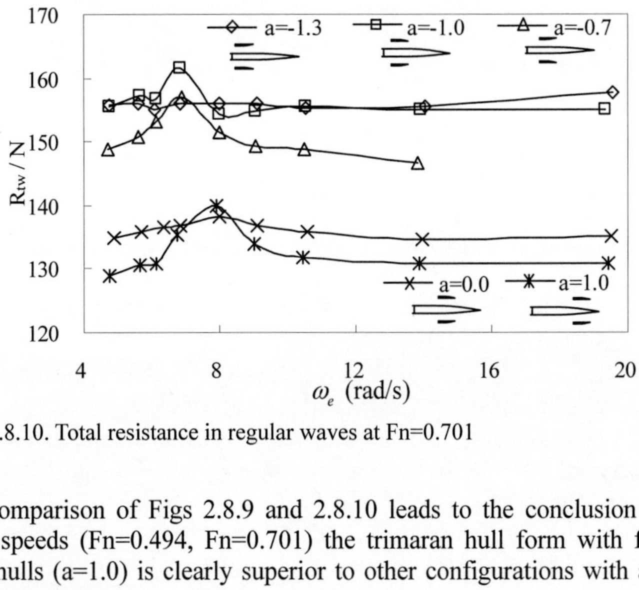

Fig. 2.8.10. Total resistance in regular waves at Fn=0.701

Comparison of Figs 2.8.9 and 2.8.10 leads to the conclusion that at high speeds (Fn=0.494, Fn=0.701) the trimaran hull form with forward side hulls (a=1.0) is clearly superior to other configurations with aft side hulls in resistances.

2.8.5. Semi-planing fore-placed outrigger trimaran (FPOT) hull form

The trimaran hull form studied in this paper has a volume Froude number $Fr_{\nabla}=1.4$ at Froude number Fn=0.50. This shows that the hull is in the semi-planing state at Fn>0.50. The results obtained in this paper shows that once the Froude number is greater than 0.5, the resistance performance will be greatly improved if the side hulls are placed forward. Based on this observation, we thus propose a new concept Semi-planing

Forward-Placed Outrigger Trimaran (FPOT), which is schematically shown in Fig. 2.8.11.

Fig. 2.8.11. Schematic of FPOT

2.8.6 Conclusions

The experimental results in this paper lead to the following conclusions:

- In between 0<Fn<0.2, friction resistance is dominant with negligible configuration influence.
- In between Fn=0.2–0.5 there exist strong interferences among hulls dependent on configurations and Froude numbers.
- At semi-planing state Fn>0.50, the trimaran configuration with forward side hulls has superior resistance over those with aft configurations.
- At semi-planing state Fn>0.65, residual resistance shows strong dependence on side hull positions and Froude number. As side hulls are further moved forward, residual resistance decreases.
- The experimental results in this paper lead to the concept of semi-planing Forward-Placed Outrigger Trimaran (FPOT) hull form.

Acknowledgements

This project is supported by National Natural Science Foundation of China under Grant No. 51279030.

References

[2.8.1] B. B. Ackers, T. J. Michael, & O. W. Tredennick, "Investigation of the Resistance Characteristics of Powered Trimaran Side-Hull Configurations," *Society of Naval Architects and Marine Engineers* 105, pp. 349–73, 1998.

[2.8.2] P. Y. Li, Y. M. Qiu, M. T. Gu, & W. F. Wang, "Experimental Investigation on Resistance of Trimaran," *Shipbuilding of China* 43 (4), pp. 6–12, 2002.

[2.8.3] M. T. Gu, F. Zheng, & Y. M. Qiu, "Research on the Resistance Test of Trimaran Small Waterplane Area Center Hull," *Journal of Shanghai Jiaotong University* 37 (8), pp. 1222–25, 2003.

[2.8.4] N. Degiuli, A. Werner, & Z. Doliner, "Experimental Determination of Wave Pattern Resistance of a Trimaran," *Proceedings of XVII IMEKO World Congress*, pp. 1396–401, 2003.

[2.8.5] S. M. Oh, S. H. Lee, & Y. G. Lee, "Hull Form Optimization of a Small Trimaran by Model Testing," *Ship & Ocean Technology* 9 (3), pp. 14–22, 2005.

[2.8.6] A. Francescutto, "On the Roll Motion of a Trimaran in Beam Waves," *Proceedings of the Eleventh International Offshore and Polar Engineering Conference*, pp. 321–25, 2001.

[2.8.7] P. Y. Li, Y. M. Qiu, M. T. Gu, & W. F. Wang, "Super Slender Trimaran Model Experiments," *Ocean Engineering* 20 (4), pp. 1–4, 2002.

[2.8.8] G. A. Li and H. K. Ye, "Wave Making Computation in Time Domain for Multi-Hull Ships, *Journal of Hydrodynamics*, Ser. B, 18 (2), pp. 225–30, 2006.

[2.8.9] P. Y. Li, Y. M. Qiu, & M. T. Gu, "Study of Trimaran Wave-Making Resistance with Numerical Calculation and Experiments," *Journal of Hydrodynamics*, Ser. B, 2, pp. 99–105, 2002.

[2.8.10] M. Igor, A. Eduard, & C. Leonard, "High-Speed Trimaran Drag: Numerical Analysis and Model Tests," *Journal of Ship Research* 48, pp. 248–59, 2004.

[2.8.11] K. Hebblewhite, P. K. Sahoo, & L. J. Doctors, "A Case Study: the Oretical and Experimental Analysis of Motion Characteristics of a Trimaran Hull Form," *Ships and Offshore Structures* 2 (2), pp. 149–56, 2007.

[2.8.12] S. Brizzolara and D. Bruzzone, "Hydrodynamic Optimisation of High-Speed Trimaran Hull Forms," *Proceedings of the 18th 2008 International Offshore and Polar Engineering Conference*, pp. 547–54, 2008.

[2.8.13] M. R. Javanmardi, E. Jahanbakhsh, M. S. Seif, & H. Sayyaadi, "Hydrodynamic Analysis of Trimaran Vessels," *Polish Maritime Research* 15 (55), pp. 11–18, 2008.
[2.8.14] I. Mizine & G. Karafiath, "Wave Interference in Design of Large Trimaran Ship," *Transactions of the Royal Institution of Naval Architects Part A: International Journal of Maritime Engineering* 150 (4), pp. 57–72, 2008.

2.9

ON THE HYDRODYNAMIC CHARACTERISTICS OF TRIMARANS AND THEIR APPLICATION (2012)

TAN MEI, XIONG FEI AND WANG ZHONGFU

CIMC Ship & Offshore Research Institute

Abstract

The hydrodynamic performance of trimaran ships compared with that of conventional ships is presented in this paper. The application fields of trimarans are also studied, and then the main characteristics of trimarans are analyzed based on the concept design of a trimaran ferry ship that is suitable to operate in the Taiwan Strait. The result can be used as a reference for trimaran ship design in future.

Introduction

Trimarans are a new kind of high-performance ship. Since the concept was proposed, it has attracted great worldwide attention. Even the United States and Britain think that trimaran technology will change the twenty-first-century navy. Over the past decade, the United Kingdom's *Triton* test ship, the Australian ferry *Benchijigua Express*, and the United States Navy's *Independence* have been launched successively. In China, Harbin Engineering University, MARIC, and other relevant units have also carried out a large amount of theoretical and experimental study. However, work in the application field has been sparse. On the basis of the analysis of the trimaran's characteristics, this paper explores the initial design of this kind of ferry ship.

2.9.1. Overall performance comparison research

There are some contradictions between the resistance, sea-keeping, and stability performance requirements of conventional ships, which limit overall performance improvement and their applications. Trimarans differ significantly from conventional ships in their hydrodynamic aspects; as a result, their applications are greatly improved compared with conventional ships.[2.9.1]

2.9.1.1. Resistance comparison

Round-bilge ships, deep-V-shaped ships, wave-piercing catamarans, and trimarans all have the advantage of high speed. But there exists a big difference at different Fr. All the ships are displacement vessels, and in a transitional state from a practical use perspective. So these high-performance ships' resistance performance, suitable speed range, and navigation conditions can be clear by comparison.

Generally, the indicator of resistance performance can be shown as the total resistance per displacement (Rt/D). On the basis of the model test results in reference 2.9.2, the total resistance comparison results in displacement and length equivalency are given for 1,000t of a round-bilge ship, a deep-V ship, a wave-piercing catamaran, and a trimaran, shown in Figs 2.9.1 and 2.9.2.

As shown in Fig. 2.9.1, almost across the entire speed range, trimarans have the best resistance performance. In the low or medium speed range, a round-bilge ship is better than a deep-V ship or a wave-piercing catamaran. But in the high speed range, wave-piercing catamaran is better than a round-bilge ship or a deep-V ship; however, the round-bilge ship and deep-V ship are very similar. As in the ship-length equivalent view, Fig. 2.9.2 shows that trimarans are best in the $Fn<0.8$ range and wave-piercing catamarans are best in the $Fn>0.8$ range. In the high speed range, the resistance of round-bilge ships and deep-V ships resistance are very similar. In the low or medium speed range, the round-bilge ship is better than deep V-shaped ship. In the high speed range, the resistance performance of trimarans and wave-piercing catamarans is better than that of the other two.

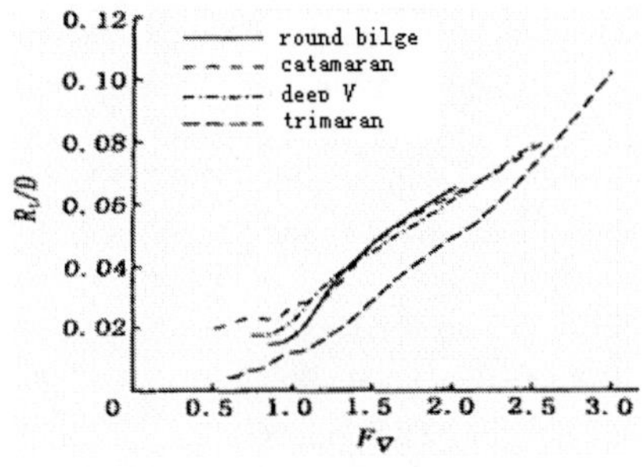

Fig. 2.9.1. Total R comparison of equal displacement

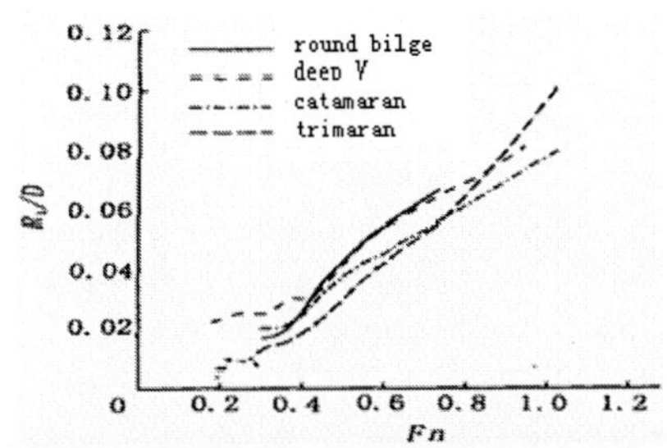

Fig. 2.9.2. Total R comparison of equal length

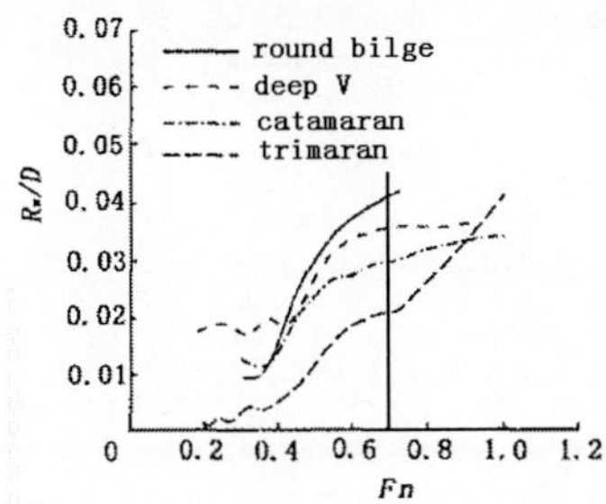

Fig. 2.9.3. Comparison of wave-making resistance of different vessels with the same displacement

Generally, trimarans sail at about 40kn. If the ship length is 90m, the corresponding Fn is about 0.7. Wave-making resistance will be a large ratio of the total resistance during navigation. So the above high-performance ships' wave-making resistance study is essential. Fig.

2.9.3 shows a comparison of the resistance of the displacement equivalent for round-bilge ships, deep-V ships, wave-piercing catamarans, and trimarans in waves. The figure shows that the trimaran's resistance advantage is obvious when the Fn is about 0.7. This further explained the trimaran's optimal resistance performance reasons in the high or medium speed range.

On the basis of the above analysis, deep-V ships and wave-piercing catamarans are the most suitable in the high speed range and transition state; trimarans are adaptable to a wide range of speeds, from the transition state to a medium or high speed range. Round-bilge ships are suitable for transition state except in the high speed range. Therefore, the speed performance of trimarans and wave-piercing catamarans is better for considering in terms of resistance only.

2.9.1.2. Sea-keeping comparison

One reference book[2.9.9] compared the sea-keeping performance of a round-bilge ship, a deep-V ship, and a wave-piercing catamaran at 18kn and 30kn and sea states four (wave height H=2.0m) and five (wave height H=3.5m). Next, the sea-keeping performances of wave-piercing catamarans and trimarans are analyzed respectively.

Table 2.9.1. Sea-keeping comparison (H=2.0m, Vs=18kn)

	Pitching /deg.	heave/m	stem acceleration/g	stern acceleration/ g	Rw enhance/ kN
Catamaran	2.48	0.85	0.60	0.24	0.95
Deep-V	2.17	0.43	0.41	0.24	2.48
Round bilge	2.75	0.65	0.62	0.30	2.63

Table 2.9.2. Sea-keeping comparison (H=2.0m, Vs=30kn)

	Pitching /deg.	heave /m	stem acceleration/g	stern acceleration/ g	Rw enhance/kN
Catamaran	1.44	0.73	0.45	0.30	1.51
Deep-V	2.10	0.57	0.58	0.37	3.06
Round Bilge	2.56	0.83	0.74	0.42	2.92

Table 2.9.3. Sea-keeping comparison (H=3.5m, Vs=18kn)

	Pitching /deg.	heave /m	stem acceleration/g	stern acceleration/g	Rw enhance/kN
Catamaran	5.25	1.84	0.83	0.37	3.31
Deep-V	4.33	1.14	0.64	0.39	5.43
Round bilge	5.13	1.46	0.83	0.45	5.77

Table 2.9.4. Sea-keeping comparison (H=3.5m, Vs=30kn)

	Pitching /deg.	heave/m	stem acceleration/ g	stern acceleration/g	Rw enhance/kN
Catamaran	3.87	2.06	0.82	0.57	7.30
Deep-V	4.54	1.47	1.03	0.65	9.08
Round bilge	5.33	1.92	1.33	0.74	9.23

From the above tables, the sea-keeping quality of the wave-piercing catamaran is close to two other kinds of ship in the low speed range; its resistance in waves increased to a smaller degree. In the high speed range, the wave-piercing catamaran is superior to the round-bilge ship and deep-V ship. For the design speed of 40kn, the wave-piercing catamaran is relatively good; however, in the process of comparing the resistance, we found that the wave-piercing catamaran and high-speed trimaran are both the optimal selection. Therefore, the 800t class's wave-piercing catamaran and trimaran-motion performance comparison is carried out separately in Table 2.9.5.

Table 2.9.5. Sea-keeping comparison of catamarans and trimarans

Vs/ kn	24/2.5		24/3.5		42/2.5		42/3.5	
H / m	Cat	Tri	Cat	Tri	Cat	Tri	Cat	Tri
pitching /deg.	2.37	1.39	3.82	2.63	2.32	2.04	3.90	3.51
heave/m	1.34	1.37	2.21	2.25	0.69	0.74	1.42	1.45
stem acceleration /g	0.29	0.25	0.45	0.40	0.36	0.31	0.56	0.48
stern acceleration /g	0.31	0.31	0.40	0.40	0.32	0.33	0.54	0.52

The result shows that the trimaran's performance is better than the wave-piercing catamaran in many indicators, especially for pitching. In low speed ranges, the two ships' heave bow and stern accelerations are roughly equal. In high speed ranges, both at a wave height of 2.5m or 3.5m, the trimaran's pitching, heave, bow, and stern accelerations are better than that of the wave-piercing catamaran. At the same time, the wave-piercing catamaran is prone to synchronism, but the trimaran can effectively avoid this shortcoming.

2.9.2. Application exploration

The world's developed countries have created high-speed ship development programs, such as Japan's high-tech ship plan (50 kn, 1,000t deadweight), the Europe Express Plan (40–60kn for high-speed ferries and container ships), the United States' 21st-Century LCS, the DDX program and SEALIFT high-speed assistance plan, and the United Kingdom's future surface combatant FSC plan. In all these programs, trimarans are the focus.[2.9.3] [2.9.4]

The British Royal Navy launched a trimaran test ship in 2000 that was 90m long and had a displacement of 1,200t. On the basis of this test ship, UCL proposed the trimaran concept for offshore patrol boats, anti-submarine frigates, corvettes, air defense destroyers, and aircraft carriers. The US Navy began a trimaran Littoral Combat Ship (LCS-2) construction plan in 2005. This ship has the maximum payload, tank arrangement capacity, stern spaciousness, and regularity of deck area with a speed of 50kn.

In the field of civilian vessels, the Austal Shipyard launched the 126m trimaran high-speed passenger ship *Benchijigua Express* in 2003. The vessel was designed according to SOLAS and the *International Code of Safety for High-Speed Craft 2000*. Practical applications proved its outstanding speed and sea-keeping performance, and it was thought to establish a new standard for high-speed ferries. In addition to the above example, the international current plans to develop high-speed trimarans are:

1. France's plan to build the *Sea Union* trimaran ferry BGV (120m length, 75 kn)
2. Norway's hydrofoil trimaran ro-type (50kn)
3. Italy's container ships (40kn with a displacement of 28,000t)
4. Russia's SWATH trimaran light aircraft, with a better sea-keeping performance than a 100,000t conventional vessel

5. America's VHST ultra-high-speed transportation trimaran projects (1,500t and 42kn)

On the basis of the above analysis, although there are no other example ships, the trimaran's advantages in resistance, sea-keeping and stability coordination, and abundant arrangement area are significantly obvious. Thus, the trimaran has good application prospects for those who demand high speed, high sea-keeping and comfort, but a low load factor, such as light surface navy ships, yachts, ferries, marine research vessels, patrol boats, the guardian ships of deep-sea oil bases, supply, fire at sea, anti-risk rescue, diving support ships, and some other special projects, and so on. Of course, initial investment costs and the complex factors of the construction process should be considered for civilian ships.

2.9.3. Trimaran ferry application study

2.9.3.1. Main technical requirements

The basic design requirements of the concept ship are as follows:

1. The intended operation field is the Taiwan Strait. Because of the terrain effects, the ship should be suitable to cross transverse and oblique waves.
2. The single voyage should be completed in no more than four hours. On the basis of range calculations, the service speed should be 38kn. Taking the high waves into account, the maximum continuous speed should be 40kn at full loaded displacement with a maximum range of more than 400 nautical miles.
3. Normal operation is in sea state 5 ($H_{1/3}\leq 3.5$ m); safe operation is in SS6 ($H_{1/3} < 6.0$ m).
4. The anti-wind capacity should be wind state 10 according to GB/T 3526-94 requirements.
5. All loading conditions for intact stability and damage stability should meet the *International Code of Safety for High-Speed Craft 2000*'s multihull ship requirements.
6. The motion seasickness index (MSI) should be below 10% (exposure time less than two hours) according to ISO2631-1.
7. The cabin noise level during operation should be less than 65dB, public spaces less than 75dB.
8. The safety requirements issued by International Maritime Organization (IMO) for high speed craft should be met.
9. There will be a crew of 20.

10. The deadweight will not be less than 320t: this is designed for 140 small family cars, or 10 heavy duty trucks and 110 small family cars, with a passenger capacity of more than 700 people, with a specific roll loading device.

2.9.3.2. Principal dimension

It is a typical car passenger ferry ship. Thus, the principal dimensions related mainly to its deck area should meet personnel and vehicle layout needs and the main engine's arrangement demands.

The main deck's length and width depend on the vehicle cabin's features. On the basis of the vehicle layout requirements, the main hull and the side hulls' width depend on the maximum width of the main deck and the space between the lateral side hulls. According to the design index, the final preliminary principal dimension is given as follows:

Table 2.9.6. Principal parameters of trimarans worldwide

Projects	*Benchijgua Express*	*Benchijgua* 102	Trimaran ferry
L/m	127.00	102.00	91
B/m	30.40	26.80	25.4
D/m	8.20	7.60	8
Design draft/m	4.00	4.50	4
Passenger no.	1,291	950–1,165	746
Vehicle no.	341	245	150

Referring to the actual operation of worldwide trimarans in Table 2.9.6, three main engines with 9,100kW each, and a waterjet with 2x125 S□ and 1x120 B□ were chosen, respectively, as the main propulsion installations. Through initial speed estimation calculations, the ferry can meet the requirements of maximum continuous speed of 40kn.

Table 2.9.7. Propulsion characteristic of trimarans worldwide

Projects	Triton	LCS $_2$	*Benchijigua Express*	Trimaran ferry
Mainhull	2×12VP185 1×FPP	2×LM2500 2×MTU20V8000 4×Wartsila (WJ)	4×MTU20V8000M71L 2× Kamewa 125 S II 1× Kamewa 180 B II	3×MTU20V8000M71L 2× Kamewa 125 SIII 1× Kamewa 120 B II
Side	2×Motors 2×Waterjets	No	No	No
L/m	98	127.1	126.7	91

2.9.3.3. General arrangement characteristics

The arrangement of the decks is as follows: main deck, length 7m; to serve as the vehicle deck, length 9.5m; middle deck to serve as the vehicle deck, length 12m; upper deck to serve as passenger deck, length 15m; navigation deck to serve as crew and VIP deck and compass deck length 18m.

The main deck is successive in longitudinal dimensions. There are passenger embarkation doors from the vehicle deck to the passenger compartment. The vehicle cabin has a middle deck with both sides fixed and a movable middle part. When only small cars are fully loaded, some of the vehicles can be lifted through the movable deck to the side decks. The maximum load capacity is 150 small cars. When the movable mid deck is used, 10 large trucks and 12 small cars can be loaded using this space and 106 small cars can be loaded in the other space. The two loading conditions all meet the capacity requirements.

The passenger deck is above the vehicle deck with part of the aft-end weather deck. As the main escape space, there are two bulkheads dividing the room into three parts with 416 passengers accommodated in the aft cabin, 176 in the middle cabin, and 134 in the front cabins. The general arrangement is as follows:

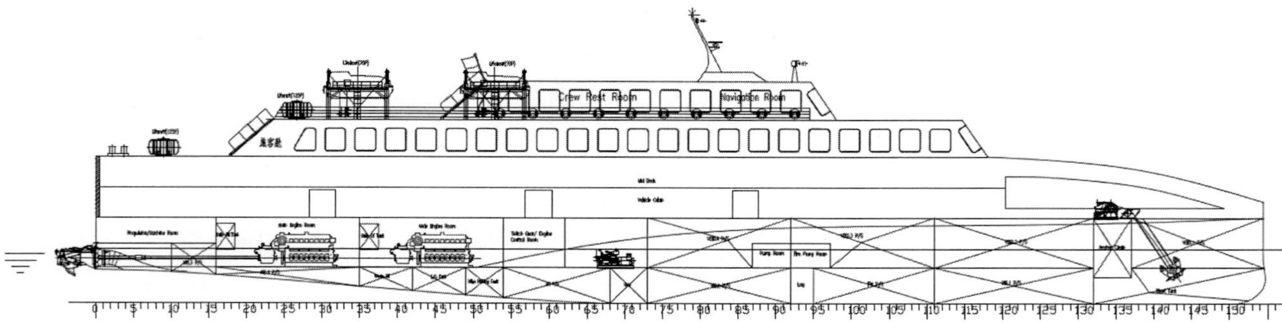

Fig. 2.9.4. Side view

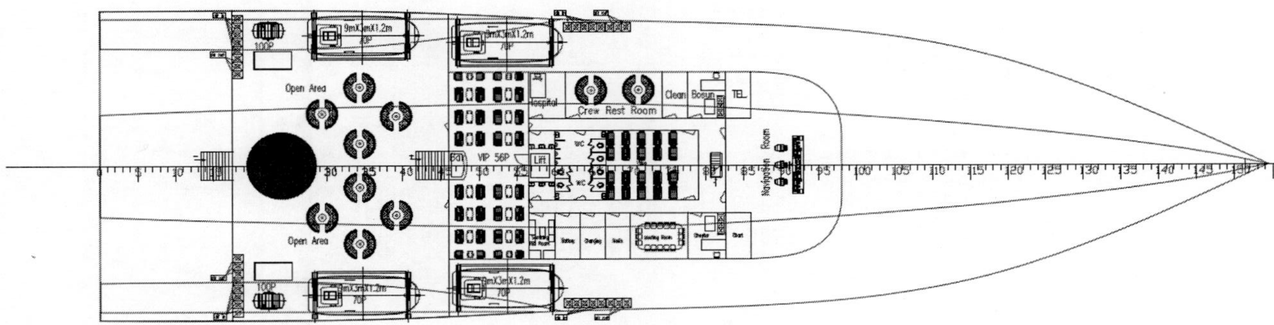

Fig. 2.9.5. Bridge deck

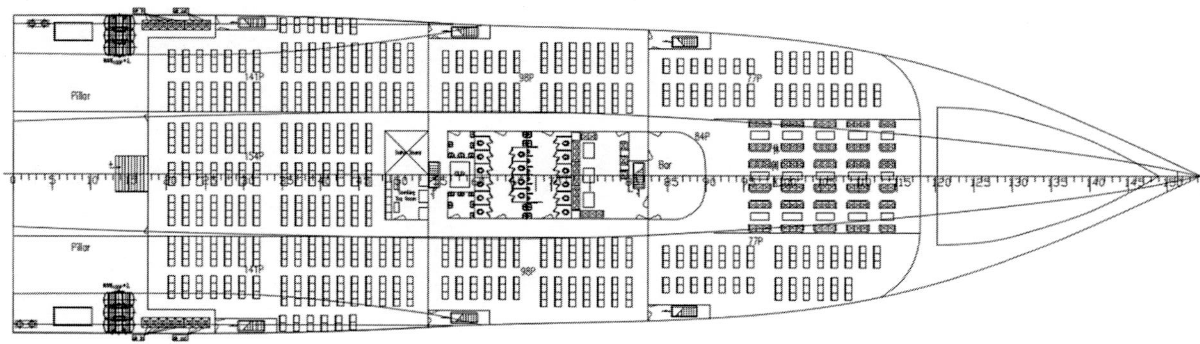

Fig. 2.9.6. Passenger deck

2.9.3.4. Intact transverse stability analysis

The trimaran's stability criteria should comply with the *International Code of Safety for High-Speed Craft 2000*. Its appendix 7 provides the criteria for the stability of multihull ships. The stability calculation principle is similar to that of conventional monohull vessels[2.9.5] [2.9.6] as shown in Fig. 2.9.9. The transverse stability characteristics of a trimaran with a transverse recovery moment versus a large heeling angle are shown in Fig. 2.9.10:

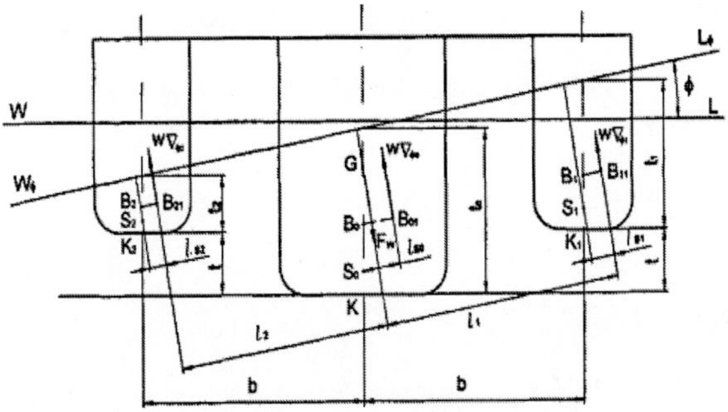

Fig. 2.9.9. Trimaran stability calculation theoretical diagram

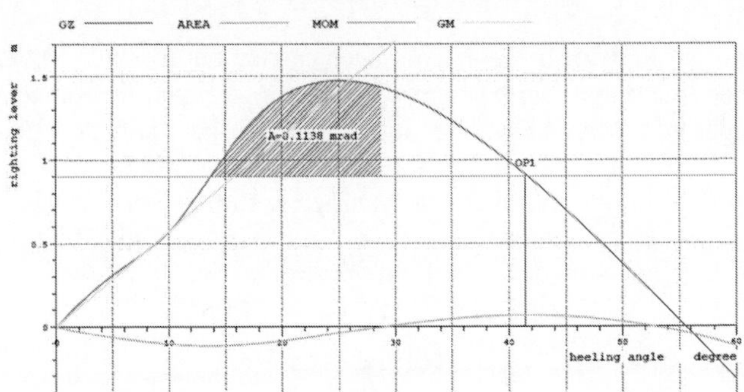

Fig. 2.9.10. GZ curve and environment moment

Referring to the rule's corresponding requirements, the intact transverse stability criterion covers:[2.9.9]

1. The area under the GZ curve up to the given angle
2. The maximum GZ value shall occur at an angle of at least 10°

3. Heeling due to wind
4. Heeling due to passenger crowding or high-speed turning
5. Rolling in waves

According to the ferry's design requirements, the checking results show that the ship satisfies the rule's stability demands.

Table 2.9.8. Intact stability criteria check

Item	Demand value	Calculated value	Checked state
GZ area	0.066 m.rad	0.351 m.rad	√
Maximum GZ angle/deg.	10	24.95	√
Wind heeling angle/deg.	10	7.363	√
Crowding heeling angle/deg.	-	6.769	-
Turning heeling angle/deg.	-	8.425	-
Wind plus turning heeling	0.028m.rad	0.114 m.rad	√

2.9.4. Conclusions

Although the study of trimarans has been carried out for over 20 years, the completed ships that can be referred to are few. Some theoretical results obtained from research investigations remain to be completed before the actual test. This paper compares the performance of trimarans with that of conventional ships, and then takes a ferry ship concept design as an example for investigation. Hopefully, it can be of some help to the development, research, and design of trimarans.

References

[2.9.1] Zhao Lianen, Xie Yonghe, *Theory and Design of High-Performance Ship*, National Defence Industry Press, 2009.
[2.9.2] Zhao Lianen, *Hydrodynamics Theory and Design of High-Performance Ship*, Harbin Engineering University Press, 2000.
[2.9.3] Xu Tongquan. "Development of Multi-Hull Ship Technology," *Journal of SSSRI* 23 (2): pp. 135–43, 2000.

[2.9.4] Lu Xiaoping, Li Yun, & Dong Zu Shun, "Research Summary of High Speed Trimaran," *Journal of Naval University of Engineering* 17 (2): 43–48, 2005.

[2.9.5] *International Code of Safety for High-Speed Craft 2000.*

[2.9.6] Li Peiyong, "Research of Concept Design for Fast Trimaran," *Journal of Shanghai Jiaotong University* 38 (11), 2004.

2.10

THE PRACTICAL DESIGN OF A 102M TRIMARAN FERRY FOR THE TAIWAN STRAIT (2010)

N. A. ARMSTRONG AND V. MORETTI

Austal Ships, Australia

Summary

This paper describes the development of a 102m-long trimaran high-speed vehicular ferry designed for operation across the Taiwan Strait, which was recently completed at the Austal shipyard in Australia. The design methodology is revealed, together with some of the performance characteristics compared with an equivalent catamaran; the operational behavior in the Taiwan Strait is also discussed.

Nomenclature

GM_T	Metacentric height
KG	Vertical center of gravity above base
LCG	Longitudinal center of gravity from transom
MSI	Motion sickness index[2.10.5]

2.10.1. Introduction

A commonly-adopted approach to the design of a craft to operate at high-speed is to make the hull as long and as thin as practical. The long-and-thin approach provides low wave-making resistance and superior ship motions, but it then becomes essential to solve the resultant stability issues, as long-and-thin hulls do not generally have sufficient stability to meet the various regulations, particularly those for passenger craft.

One solution to the stability issue is to duplicate the hull into a twin-hull arrangement—the catamaran. Another solution is to add two small outer hulls to provide additional stability—the stabilized monohull or trimaran. The outer hulls are termed "amahs."

The widely separated hulls of a catamaran provide plenty of stability; but, as a result, catamarans can be uncomfortable in beam seas because of the high levels of acceleration resulting from the excessive metacentric height GM_T.

Another limitation of the catamaran when operated in a seaway is the wave height in which the cross-deck structure may impact with the waves, commonly called slamming. The trimaran has no cross structure at the forward part of the craft and therefore slamming is not generally a limitation as it may be for catamarans.

2.10.2. Background

The first large trimaran ferry was the 127m-long mv *Benchijigua Express*, completed in 2005 and operated since that date on a regular daily service between the various islands in the Canary Islands in the North Atlantic Ocean. Details of this craft and its development are given in References 2.10.1 and 2.10.2.

As this boat was the first of its type, it was subjected to many trials, and instrumentation remained on board for the first year of operation. The results of these measurements were used to identify areas where the performance of the concept could be improved.

As a result of the development of this trimaran, and its subsequent successful operation, the hull form was used as the basis for a 127m-long warship for the US Navy developed as a Littoral Combat Ship (LCS_2), as illustrated in Fig. 2.10.1. Two of these craft have been ordered to date, and trials on the first craft have confirmed the superior behavior of the trimaran concept, particularly when compared to alternative craft designed for the same tasks.

Fig. 2.10.1. *USS Independence*, a 40kn 126m trimaran designed and built by Austal

2.10.3. A new generation of trimarans

2.10.3.1. Selection of roll performance

Taking advantage of the novel capabilities of trimaran shapes, *Benchijigua Express* had been deliberately designed with a low metacentric height GM_T in order to provide a slow roll characteristic and thereby enhance passenger comfort. However, for the next stage of development it was decided to increase the GM_T value in order to reduce the magnitude of the roll. A new ride control system was also designed specifically to suit the trimaran layout, comprising an active T-foil mounted at the bow of the main hull and at the transom of each amah. The hull form and associated ride control arrangement was modeled in a sea-keeping simulation, a variety of sea conditions and wave headings using the software Veres from Marintek. Several underwater amah shapes were developed to provide varying GM_T values; from this data, reproduced in Fig. 2.10.2, a choice was made of the design GM_T, based on the lowest value at which roll was minimized at the worst heading. A design GM_T of about 6 meters was selected.

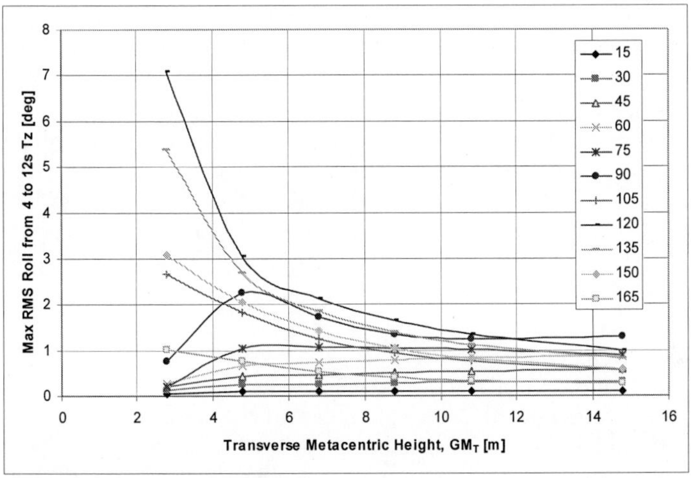

Fig. 2.10.2. Roll angle resulting from varying GM_T values at various wave headings (head seas = 0°)

2.10.3.2. Basis of design for a trimaran

The design of the 102m trimaran was based on various broad steps. The specification for the craft was based on our analysis of likely trade on the selected route, the most efficient operating speed and the minimum vessel size necessary to provide a comfortable ride in the wind and wave conditions of the Taiwan Strait. The vessel was primarily sized to suit an economical purchase price while offering a substantial carrying capability. The capability requirements were as follows:

Passenger numbers	>1000
Vehicle numbers	>245
Truck capacity	190 lane meters
Design speed at 450t (dwt)	39kn
Maximum deadweight	700t

The car capacity dictated the overall ship beam (26.8m) which is related to the standard car size of 4.5m × 2.35m. The vehicle numbers also dictated the minimum ship length (102m).

The required speed and outline ship dimensions indicated the necessary propulsive power (3 × 9,100 kW), and the dimensions of the most likely engines with the required power and their associated waterjets determined the minimum width of the main hull.

In this way the main dimensions were fixed, and an initial general arrangement could be drawn. The passenger cabin was flexibly mounted on isolation mounts above the vehicle-carrying space.

An estimate was made of the *LCG* and *KG* values. A hull form having the necessary general characteristics outlined above was then developed, which also had the stability characteristics of GM_T and *GZ* as detailed above. The hull form was then subject to numerical analysis to determine the resistance and motions, and the shape was manipulated in order to provide the minimum resistance and best motion characteristics.

2.10.3.3. Ship motions and the ride control system

An active ride control system for the new trimaran was developed based on three T-foils located at the extremities of the craft, being the bow and the after end of the amahs. The bow T-foil was a proven design having a hydraulically-actuated flap to reduce pitching motions. The aft-mounted T-foils were a novel design being hydraulically activated swinging foils mounted on the transom of the amahs, and used to minimize roll. A new control algorithm was developed to suit the slow roll characteristics of the craft, and to automate the system so that no input was required from the operator. The influences of various sizes of T-foils on the behavior of the hull were analyzed in a realistic sea condition in order to help choose the T-foil size.

2.10.3.4. Predicted performance

(1) Speed in calm water

Speed in calm water was initially predicted by using the panel Code Shipflow. This was also used to evaluate the effect of small changes to the hull form in order to reduce resistance and to examine the effect on dynamic trim and hence resistance of fitting various wedges and transom interceptors. Tank testing was then conducted on the selected hull form, again with a variation in the wedge size at the after end of the main hull. The resistance was found to be in line with estimation developed from the performance of the existing larger trimarans and their model tests, as shown in Fig. 2.10.3.

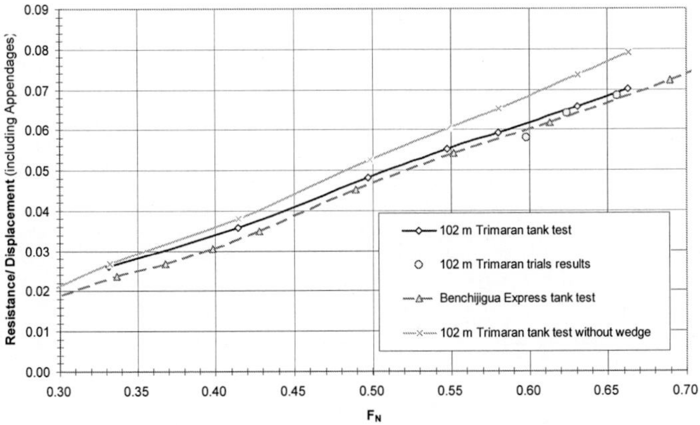

Fig. 2.10.3. Tank testing results of three hull forms scaled to the same length, and three points calculated from full-scale trials results

The speed was measured during ship trials and correlated against that calculated from resistance tank tests for the trimaran together with data on the thrust provided by the waterjets. The same procedure was carried out for an equivalent 88m catamaran, both crafts having a deadweight of 400t, and assuming full power. It was found that both the trimaran and the catamaran had the same speed for an output power of 27,000kW, as illustrated in Fig. 2.10.4. As the speed was reduced, a small advantage became evident for the trimaran at the required power for a given speed. At higher deadweights there was a substantial reduction in power required for the trimaran compared to the catamaran, and at 29kn the trimaran offered a 21% fuel saving compared to the catamaran.

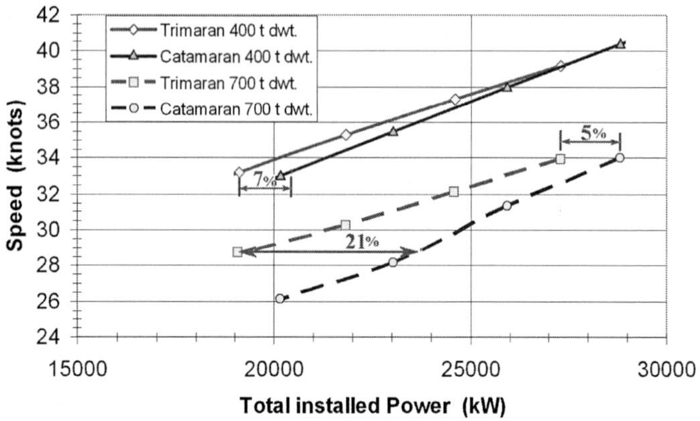

Fig. 2.10.4. Speed versus power for the 102m trimaran and an equivalent 88m catamaran, showing power savings

(2) Speed in waves

The speed loss in waves was calculated using the expression of Joosen described in reference 2.10.3. From an initial speed of 34kn in a 3m significant wave height at a deadweight of 550t, the speed loss for the trimaran was calculated as 3.7kn, and for the 88m catamaran it was 5.0kn. This relative advantage to the trimaran in maintaining speed in a seaway has also been reported by the operator of *Benchijigua Express*.

(3) Speed limitation

The 102m trimaran was designed to the requirements of the *HSC Code*,[2.10.4] in common with most high-speed craft. These regulations are based on the philosophy of trading off certain aspects of safety against practical limitations that permit the craft to be built to relatively light scantlings. A major limitation describes the worst intended conditions in which the vessel is permitted to operate, generally described by the wave environment, particularly the maximum significant wave height.

This permits the vessel structure to be designed to an acceleration level, and the structural loads can be limited through the observance of the speed–wave height curve. The 102m trimaran has been structurally designed to a limit of 40kn for operation in a significant wave height of 2.5m. In a 3m sea the maximum allowable speed is reduced to 31kn, and in a 4.5m sea to 18kn, with a maximum significant wave height in operation of 5.0m.

2.10.3.5. Maneuverability

Maneuverability relates to high-speed ability to maintain a course, as well as a low-speed ability to maneuver through small and congested waterways and to safely dock the craft. The course-keeping ability of the 102m trimaran at high-speed has proved to be excellent on trials, no doubt assisted by the struts associated with the aft T-foils on the amahs. The principal steering mechanism is by swiveling the waterjets, but in order to maintain speed when operating in a seaway, the waterjets are maintained in the ahead position and small heading adjustments are made using a rudder to maintain heading. The waterjets automatically take over when helm angles exceed a set value or when manual control is selected.

Two retractable bow thrusters are fitted to assist with docking at low speeds. The vessel is capable of docking in winds of about 24kn, as illustrated by the station-keeping diagram in Fig. 2.10.5.

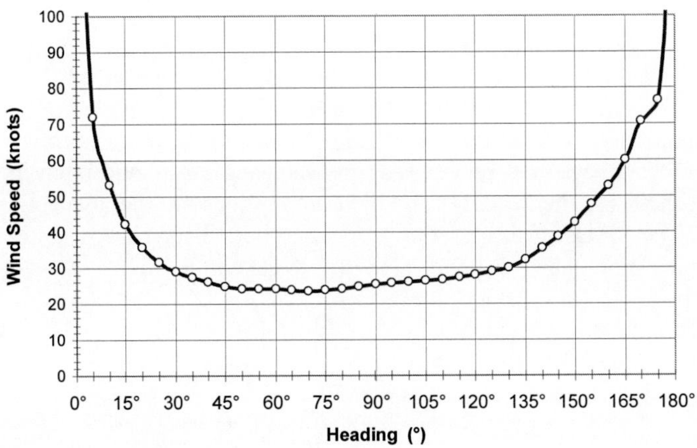

Fig. 2.10.5. Maximum wind speed for dynamic positioning, with two 44kN retractable bow thrusters (head wind=0°)

2.10.3.6. Ship motions

The vessel motions were calculated using Veres strip-theory software suitable for high-speed craft. The trimaran hull form is designed to operate with the ride control system, and therefore this system was included in all of the analyses, as indicated in Fig. 2.10.6.

	Vessels	102 m Tri	98m Cat	Existing ferry
Vessel data	Lightship	990	1060	470
	Analysed DWT	**340**	**340**	**240**
	Draught (DWT)	3.2	3.6	2.2
	Length WL	101.4	90.7	57.7
	Beam WL	27.23	26.45	16
	Analysed Speed	**37**	**37**	**31**
	LCG	35.7	37.64	24.43
	VCG	7.2	7.08	4
	Roll radii gyr	6.8	9.25	5.6
	Pitch radii gyr	24.45	26.3	16.7
	Yaw radii gyr	24.45	22.67	14.4
Passenger position	position description	midship at side	midship at side	midship at side
	Pax LCG	43.6	47	30
	Pax VCG	6.75	7	4
	Pax TCG	14.4	14.4	10.6
Ride control	RCS 1	main hull T-foil	central T-foil	no ride control
	Size	10 sqm	7.2sqm	
	RCS 2	amah T-foils	trim flaps	
	Size	2 x 2.7m^2	2 x 10m^2	
	RCS 3	interceptor		
	Size	100mm		

Fig. 2.10.6. Principal characteristics of three craft

The major ship motions, expressed as non-dimensional response amplitude operators RAOs, are illustrated in Fig. 2.10.7, including vertical acceleration at amidships for head seas and bow quartering seas, roll-in-beam seas, and bow-quartering seas, and heave and pitch in head seas. These are the headings at which the motions are the greatest for the craft. The RAOs of the existing ferry are omitted for clarity, as they are generally more than twice the value of the catamaran.

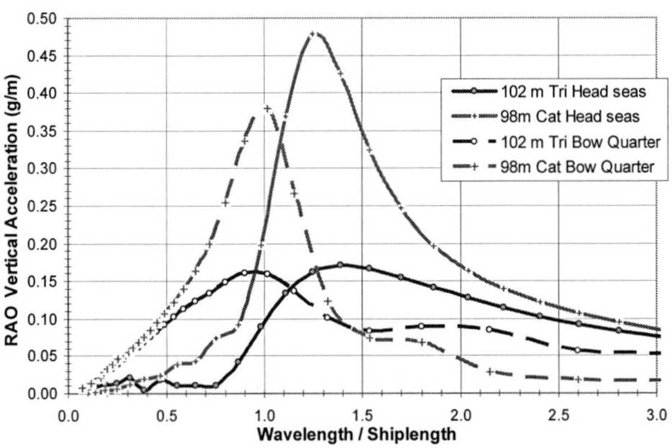

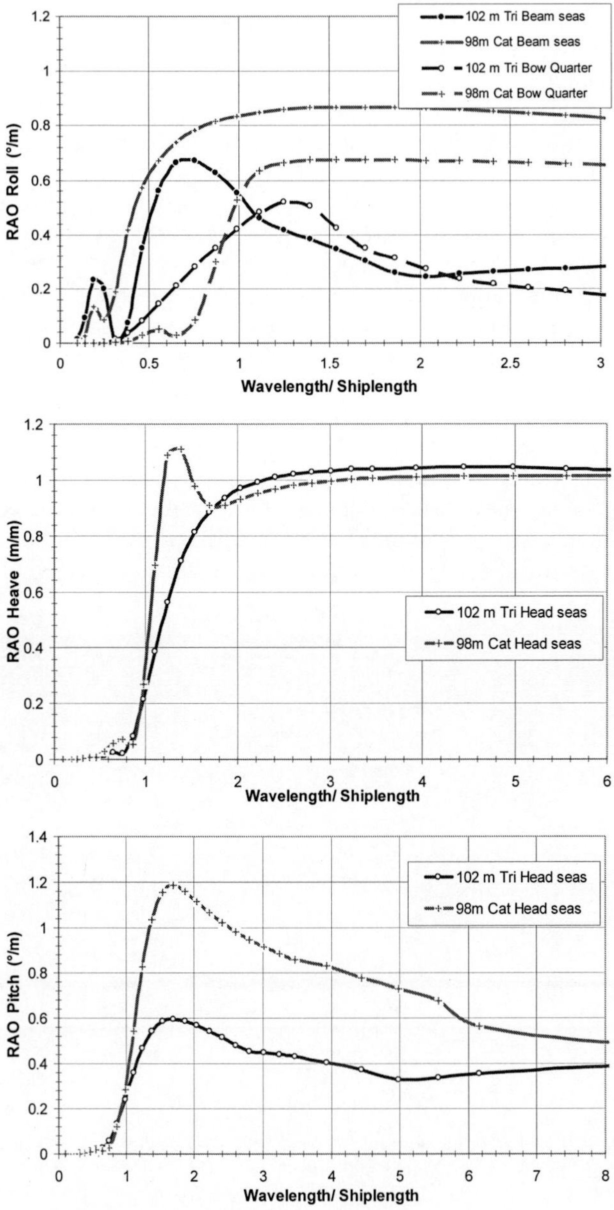

Fig. 2.10.7. RAOs for the major motions of two fast craft

It is apparent that the 102m trimaran has superior motion characteristics compared with the 98m catamaran, and for wavelengths greater than the ship length, the motion of the trimaran was approximately half that of the catamaran, except for heave, which is similar. The high level of vertical acceleration of the 98m catamaran in head seas is surprising.

While the RAOs are indicative of the behavior of the three craft, it is more relevant to compare the motions in a practical seaway. Therefore the behavior is calculated in a Pierson-Moscovitz spectrum, in long crested seas, with a significant wave height of 2.5m and for a variety of wave periods. The trimaran and the catamaran are investigated at a speed of 37kn, and the existing ferry at 31kn, which is the maximum speed of the craft. The rms values of vertical acceleration in head seas and lateral acceleration in beam seas are illustrated in Fig. 2.10.8(a), and the roll angles in beam seas in Fig. 2.10.8(b). The motions of the trimaran are found to be superior in all cases.

It is interesting to note that the maximum values of lateral and vertical acceleration appear to be very similar. However they should be compared at the same wave period, and for most practical wave periods (5s and above) the vertical acceleration levels are greater than the lateral ones.

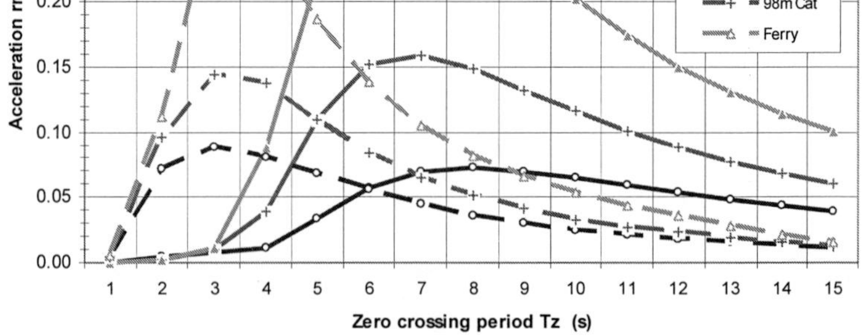

Fig. 2.10.8(a). Vertical acceleration in head seas, and lateral acceleration in beam seas, in a significant wave height of 2.5m

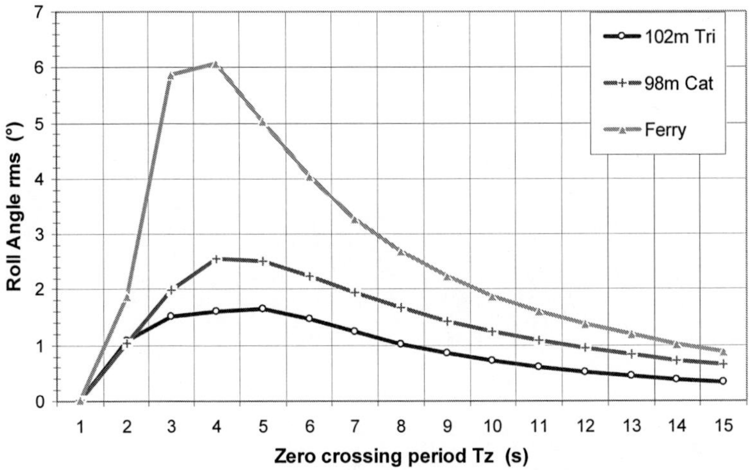

Fig. 2.10.8(b). Roll angles in various wave periods, in a significant wave height of 2.5m, beam seas

2.10.4. Operation in the Taiwan Strait

The Taiwan Strait sea area is characterized by a prevailing wave and wind direction from the north-east. Historical wave data are available from a number of sources, including from satellite data (www.waveclimate.com, also www.oceanweather.com), from an offshore platform (KK platform), and from various buoys from which data are collected and distributed by the Central Weather Bureau of Taiwan.

Several existing ferry routes operate in the area, some of which are indicated in Fig. 2.10.9. Two hypothetical routes are identified and the behavior of the 102m trimaran is further investigated in the sea conditions representative of the area. Wave data are taken from published information on the KK Platform as this is a less sheltered location than some of the other buoys, and is also mid-way between the two chosen routes.

An operability analysis is carried out based on the wave scatter diagram for the KK platform, reproduced in Fig. 2.10.10, for the 102m Trimaran and a 98m Catamaran, as well as an existing ferry operating in the area. Long-crested waves are assumed of a Pierson-Moscowitz spectrum.

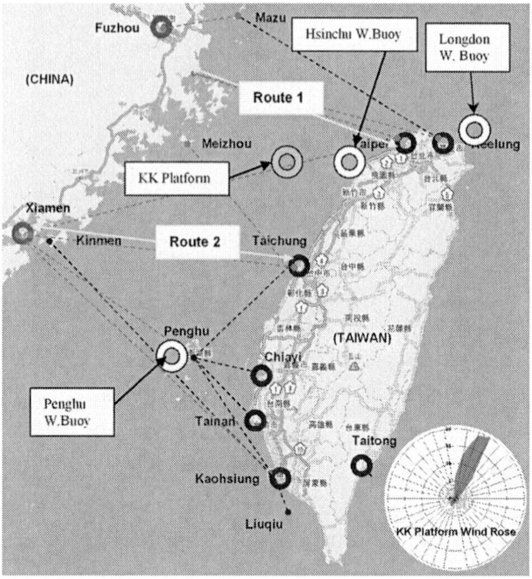

Fig. 2.10.9. Taiwan Strait, showing existing routes, two proposed routes, and existing wave data buoys

			Tz (s)					
Hs	4-5	5-6	6-7	7-8	8-9	9-10	10-11	11-12
7 - 10		0.2						
5 - 7							0.2	
4 - 5						0.3		
3.5 - 4.0					1.1	1.1		
3.0 - 3.5				0.6	4.3	0.6	0.3	0.2
2.5 - 3.0				3.7	3.1	0.6	0.0	1.0
2.0 - 2.5			1.3	7.7	1.9	0.8	1.1	
1.5 - 2.0		0.3	6.8	5.8	1.0	0.8		
1.0 - 1.5	1.3	8.7	11.4	3.1	0.6			
0.5 - 1.0	5.9	12.7	7.1	1.9	0.5			
0.0 - 0.5	0.5	1.3	0.2					

Fig. 2.10.10. Wave scatter diagram for the KK platform, Lat. 24-47'07"N, Long. 120-39'53"E.

An operability analysis indicates the amount of time that a vessel can be operated in the chosen sea area throughout the year, and remains within the stated criteria. This approach excludes typhoon events. Typical chosen criteria are MSI (<10% in 2h), vertical acceleration (<0.05g rms), lateral acceleration (<0.08g rms), roll angle (<4°), and pitch angle (<3°). MSI and vertical acceleration are related, as are roll angle and lateral acceleration. The most relevant criterion is probably MSI or vertical acceleration because these relate directly to passenger comfort. The results of the operability analysis using the Motion Sickness Indices (MSI) as defined in reference 2.10.5 is illustrated in Fig. 2.10.11, using a chosen criterion of 10% of the passengers becoming sick within a two hour crossing period. The trimaran outperforms the 98m catamaran and the existing ferry by a considerable margin.

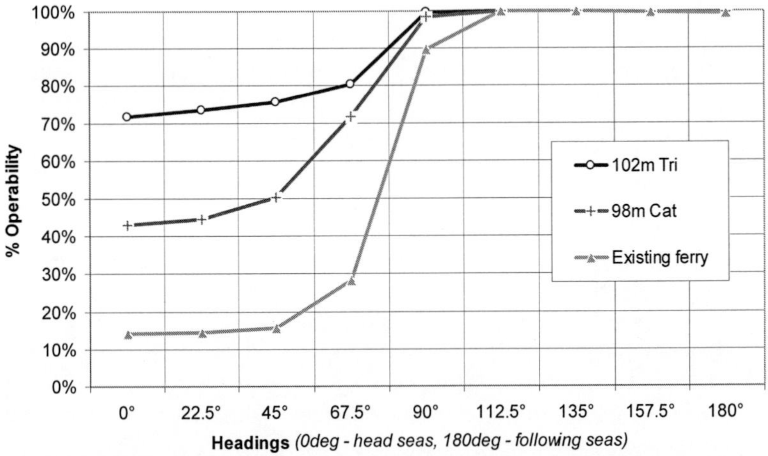

Fig. 2.10.11. Operability analysis based on sea data from the KK platform, for 10% MSI level within two hours

2.10.5. Conclusion

The approach to the design of this novel craft, outlined at the beginning of this paper, appears to have been validated by the performance of the craft as recently measured during ship trials. The motions of the ship are immediately obvious as being different to other high-speed craft, with low levels of acceleration being evident and as predicted by the use of numerical simulation.

Although the performance of the vessel with a bare hull (without ride control) are not given in this paper, the low levels of lateral acceleration and roll angle result from the combination of the novel hull and its associated ride control arrangement. This combination makes the vessel concept particularly suited to operate in areas where the prevailing wave direction combined with the vessel route results in predominantly beam seas, and is also enhanced when the area is generally exposed and subject to longer wave periods. A good example of such an area is the Taiwan Strait, and it is concluded that the trimaran would provide a stable and comfortable platform and be ideally suited to operate in the Taiwan Strait.

The results of a mathematical simulation of the vessel behavior using real sea data representative of this area indicates that the 102m trimaran could provide a 71% operability level throughout the year without exceeding the chosen criterion, compared with only 42% operability level for a 98m long catamaran, and compared with only 15% for an existing ferry operating in the area.

The above absolute values are affected by the choice of criteria values, which are set quite low. A choice of higher criteria values would result in higher levels of operability, although the relative difference between the three craft would remain.

References

[2.10.1] N. A. Armstrong & K. Holden, "A New Generation of Large Fast Ferry—From Concept to Contract Reality," 7th International Conference on Fast Sea Transportation, FAST'03, Ischia, Italy, October 2003.

[2.10.2] N. A. Armstrong, "Coming Soon to a Port Near You—The 126 m Austal Trimaran," International Conference on the Design & Operation of Trimaran Ships, RINA, London, April 2004.

[2.10.3] N. A. Armstrong & A. Schmieman, "On the Added Resistance of Catamarans in Waves," 8th International Conference on Fast Sea Transportation, FAST'05, St. Petersburg, Russia, June 2005.

[2.10.4] IMO, *International Code of Safety for High-Speed Craft*, 2nd ed., Resolution MSC.97 (73), International Maritime Organisation, London, 2008.

[2.10.5] J. F. O'Hanlon & M. E. McCauley, "Motion Sickness Incidence as a Function of the Frequency and Acceleration of Vertical Sinusoidal Motion," *Aerospace Medicine* 45, 1974.

2.11

High-Speed Multihull Craft for Medium-Distance Marine Transportation (2005)

E. Begovic, C. Bertorello, S. Caldarella and P. Cassella

Department of Naval Architecture, University of Naples "Federico II," Italy

Abstract

Ship owners' demand for medium-distance marine transportation is focused on HSC for passengers, cars, and trucks with L_{WL} around 110–120m and cruising speeds in the range of 35–40 kn.

In the last few years, designers and builders have developed proposals mainly based on preceding experiences from smaller ships, but the importance of this segment in the whole picture of marine transportation has called for further investigations, which recently suggested the development and the study of new hull forms.

Within this frame both pentamarans and trimarans suitable for medium-distance routes have been considered in a research program jointly carried out by Naples, Genoa, and Trieste Universities. Main hulls and outriggers are high-speed displacement slender hull forms. Intact stability prescriptions by IMO HSC code have been also verified.

Hydrodynamic characteristics and powering performances have been evaluated and optimized through experimental tests performed with different outriggers positions. The powering performances have been assessed and compared with those relative to a monohull ship with equivalent service capabilities.

Introduction

The increasing demand for fast sea transportation on medium range routes has led to an increasing interest in multihull ships, with the catamaran as the leading commercial type. Recently also other unconventional multihull craft have been proposed. The study and the development of such hull forms aims to reduce the high fuel consumption inevitably related to the requested high cruising speeds as well as achieving the advantages of large deck areas.

The trimaran and the pentamaran seem interesting possibilities due to the benefits given by their very slender hull forms compared to monohull ships at medium-high speeds as regards both sea-keeping and powering performances. At the same time they present simpler structural aspects in comparison with medium- or large-sized catamarans.

Nevertheless, trimarans have had a very limited use till now and there is very little information about pentamaran hull forms and configuration, although some proposals have been presented and prototype ships have been built. Besides, it is necessary to compare the performances and the safety among this new ship type and the other different high-speed vessels.

Within this frame, experimental research has been carried out at Naples University towing tank involving model tests of pentamaran and trimaran hull forms, both derived from the same main hull and suitable for a ship to be used on medium distance Mediterranean routes. The basic design has been developed to achieve service capabilities equivalent to operative monohulls apt for transportation of 800–1000 passengers and 280–300 cars at a cruising speed of 35–40 kn.

In previous work,[2.11.1] the pentamaran hull configuration has been experimentally studied. Transversal and longitudinal positions of after and forward outriggers have been varied with the aim of powering performance optimization. The draught of the pentamaran forward outriggers has been varied also and finally the best outrigger position able to get the minimum total resistance for Fn ranging from 0.5 to 0.6 has been identified. The resistance results have been compared with those relative to an equivalent operative monohull ship, showing the pentamaran to be very interesting in regard to power performance.

In this paper, the search for the optimum trimaran configuration derived from the same pentamaran main hull is presented, so that a fair comparison between these two new hull forms—both based on the "slender main hull plus outriggers" concept—can be performed.

A hull form from the NOVA series is used to compare multihull power performance with those of monohull ships presently in service. This hull form has been chosen due to its proven good performances in the considered Fn range.

Intact stability has been also verified to preliminarily assess the highest possible KG/D values complying with prescriptions by the IMO HSC Code. This has the aim of verifying whether such hull configurations are suitable for a realistic ship and also has the aim of comparing pentamaran and trimaran stability performances.

The aim of this work is to provide useful information on the performances of a realistic multihull based on a main-hull-plus-outriggers configuration.

2.11.1. Main hull, pentamarans, trimarans, and equivalent ship hull forms

2.11.1.1. Main hull

The design of the main hull to be used for both the pentamaran and the trimaran has been developed on the basis of a realistic general layout, with displacement volume equivalent to the monohull presently in service taken as a reference. The main hull is a slender round-bilge hull form from series 64. The main hull form is shown in Fig. 2.11.1 and its principal characteristics are reported in Table 2.11.2.

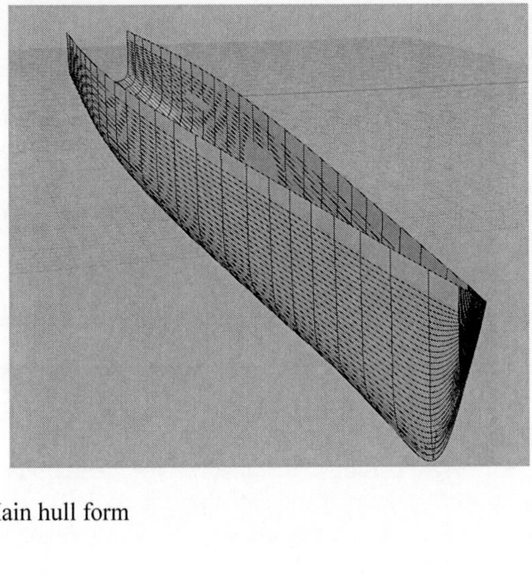

Fig. 2.11.1. Main hull form

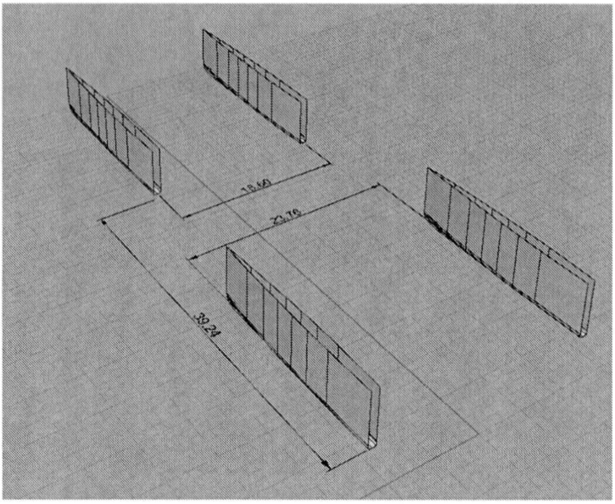

Fig. 2.11.2. Pentamaran outrigger hull form

2.11.1.2. Pentamaran hull

Hull design of the pentamaran has been developed on the basis of a realistic maximum beam allowing the largest possible deck area and at the same time with the position of the forward outriggers constrained by sea-keeping and structural aspects. The outrigger hull forms are from series 64 also, and are modified by geometrical affinity.

From the results of previous research about trimaran performance optimization, the outrigger displacement volume has been taken as the minimum realistic value to provide adequate low-angle stability. Both the forward and after outrigger length is 20% of the main hull length, and the longitudinal gap between outriggers is equal to their length. The outrigger hull form is shown in Fig. 2.11.2 and its principal characteristics are reported in Table 2.11.2.

In previous work,[2.11.1] the whole pentamaran hull at the best trim of the main hull was tested in the following different configurations:

1. After outriggers with transoms coincident with the main hull transom and three different drafts of the forward outriggers
2. Two longitudinal positions of the forward outriggers
3. Two transverse positions (distance of the centerplane of the side hull from the center plane of the main hull) of the forward outriggers, selected as percentage 1.05 and 1.22 of the main hull beam

4. Two transverse positions of the after outriggers, selected as percentage 1.22 and 1.35 of the main hull beam

From this large number of tests, the lowest resistance pentamaran configuration in the Fn range 0.5–0.6 has been identified and it is shown in Fig. 2.11.3. The pentamaran's main characteristics are reported in Table 2.11.2.

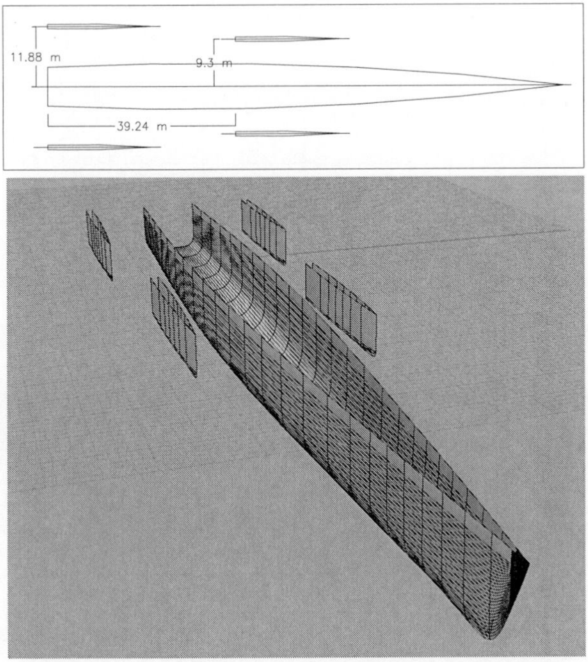

Fig. 2.11.3. Best pentamaran configuration in the Fn range 0.5–0.6

2.11.1.3. Trimaran hull

For the trimaran also, the outrigger displacement volume has been taken as a minimum realistic value of the total volume. The outrigger length has been fixed as 30% of the main hull length. This value is large enough to provide an adequate main deck surface and a feasible general layout and it is about the average value from the examples reported in the literature relative to medium-sized trimarans.

As regards the trimaran configuration, further tests have been necessary to identify the stagger and clearance allowing the lowest hydrodynamic resistance. To this aim, 12 hull configurations (three clearances and four staggers), as detailed in Table 2.11.1, have been considered and tested.

The outrigger hull form is from series 64 and is shown in Fig. 2.11.4. The main geometrical characteristics are reported in Table 2.11.2. The lowest resistance trimaran configurations identified are shown in Fig. 2.11.5.

Table 2.11.1. Trimaran staggers and clearances

Trimaran stagger				
Distance from main hull and outrigger transoms				
(m)	0.000	0.352	0.704	1.056
% L_{WL}	0	10	20	30
Code	S_0	S_1	S_2	S_3

Trimaran clearance			
Distance from main hull and outrigger centerplane			
(m)	0.348	0.390	0.471
% L_{WL}	9.9	11.10	13.40
Code	CL_1	CL_2	CL_3

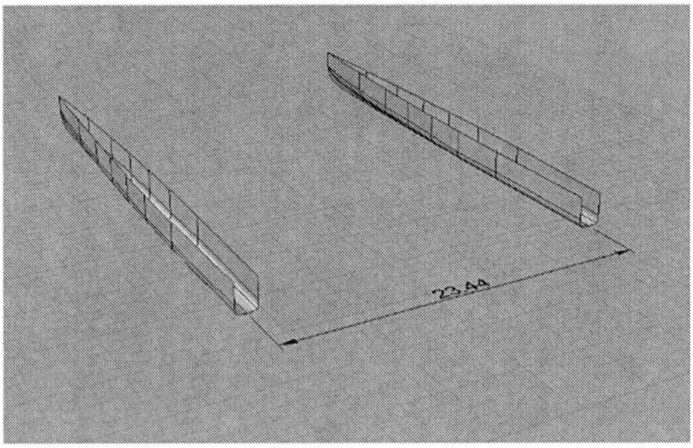

Fig. 2.11.4. Trimaran outrigger hull forms

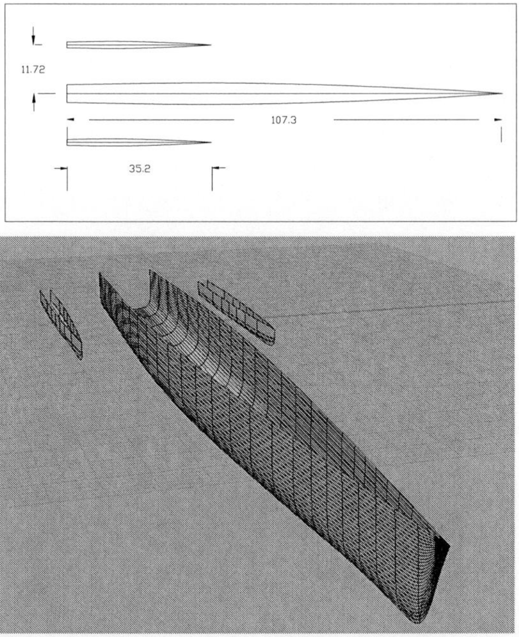

Fig. 2.11.5. Identified trimaran best configuration

2.11.1.4. Equivalent monohull

To identify the potential of the main-hull-plus-outriggers concept in respect to the monohull ships presently in service, a comparison has been arranged. To this aim, from the very beginning of this research the equivalence criterion to compare different hull forms has been identified in the service capability.

Ships to be used on medium distance routes—for example, Sardinia or Majorca land connections—must be adequate for transportation of 800–1,000 passengers and 280–300 cars at a cruising speed of 35–40 kn.

Such a criterion implies adequate deck areas and displaced volume as well as brake horsepower and a propulsion system able to obtain such a cruising speed. Furthermore, the main dimensions have constraints due to the port sizes. As regards this last characteristic, both the pentamaran and the trimaran have been limited to the maximum beam of 25m.

The basic choice to perform the comparison at the same full load displacement led to the pentamaran and trimaran having an overall length greater than the monohull has; this has been considered acceptable provided the increase would be limited to 15%.

The equivalent ship hull form is from the NOVA Series,[2.11.9] a further development of the well-known NPL series. This form is also a round bilge with transom but with considerably different L/B and B/T values in respect to the pentamaran/trimaran main hull. The fuller body allows reduced main dimensions for equivalent service capabilities. The wetted and total surface of the monohull are also smaller. The body plan is shown in Fig. 2.11.6. The main geometrical characteristics are reported in Table 2.11.2.

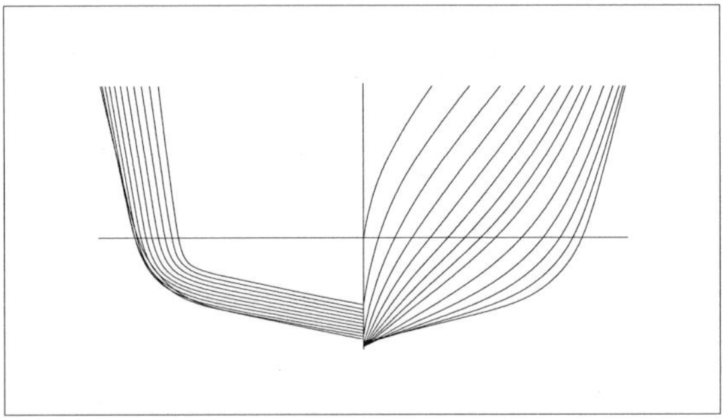

Fig. 2.11.6. Equivalent monohull body plan

Table 2.11.2. Principal characteristics of the considered hull forms in full scale

	Main hull	Pentourt	Trimourt	Pent.	Trim.	Monohull
L_{OA}/m	107.3	21.4	35.4	107.3	107.3	89.3
L_{WL}/m	105.6	21.1	34.8	105.6	105.6	86.5
T/m	4.416	0.417	0.680	4.412	4.416	3.600
B/m	8.832	1.092	1.621	24.796	25.138	15.90
DWT/t	2,322	3.1	13.9	2,334	2,350	2,242
WS/m^2	1,270	8.1	20.4	1,302	1,311	1,274
V/kn	36	36	36	36	36	36
C_b	0.550	0.350	0.350			0.440
L/B	11.95	21.50	21.50			5.44
B/T	2.000	2.356	2.356			4.410
Fn	0.576	1.288	1.003			0.635

2.11.2. Experimental set up and model tests

The resistance tests were conducted on the 1/30 pentamaran and trimaran scale models in the Naples University towing tank (140m × 9m × 4.20m).

The tests have benefitted from experiences performed in previous works about intrinsic aspects of towing tests of slender hull forms at medium high Fn that highlighted the opportunity of no turbulence stimulation[2.11.3] and the preferable model ship correlation.[2.11.6] The scale ratio used, which leads to a 3.570m L_{WL} main hull model, allows for the reliability of the results.

The isolated main hull, the isolated outriggers, and the whole pentamaran and trimaran hulls, in different configurations, have been tested in calm water.

In previous work,[2.11.1] the results relative to the main hull alone and to the pentamaran outriggers alone were reported. In the same paper the tests relative to several outrigger positions of the pentamaran hull are also reported and analyzed. From those results the best pentamaran configuration has been identified as that reported in Fig. 2.11.3. The rank has been performed according to the hydrodynamic total resistance at Froude numbers ranging from 0.5 to 0.6. In Fig. 2.11.7 the 1/30 model of the best performing pentamaran configuration towed at 2.3 m/s is shown.

Fig. 2.11.7. 1/30 pentamaran model towing test

2.11.3. Results and analysis

2.11.3.1. Pentamaran hull

As regards the pentamaran hull, analysis of the test data was performed by the ITTC '57 correlation methodology using the classical Froude procedure of the resistance components subdivision:

$C_{R(P)} = C_{T(P)} (F_n, R_n) - C_{F(P)} (R_n)$

$C_{F(P)} = C_{F(M)} . S_{(M)} / S_{(P)} + 2\, C_{F\,(OF)} . S_{(OF)} / S_{(P)} + 2\, C_{F(OA)} . S_{(OA)} / S_{(P)}$

where $C_{(TP)} = R_{T(P)} / 0{,}5\ \rho\ S_{T(P)}.V^2$, the pentamaran total resistance coefficient by the model tests.
$C_{R(P)}$—the pentamaran residual resistance coefficient.
C_F,—the frictional resistance coefficient according to the ITTC frictional line.
Suffix codes P, M, OF, and OA refer to the whole pentamaran, the main hull, the forward outriggers, and the after outriggers, respectively.

In Fig. 2.11.8, the curves of C_R and C_T versus Fn relative to the best performing pentamaran configuration are shown. To have complete information about the influence of the outriggers position, refer to reference 2.11.1.

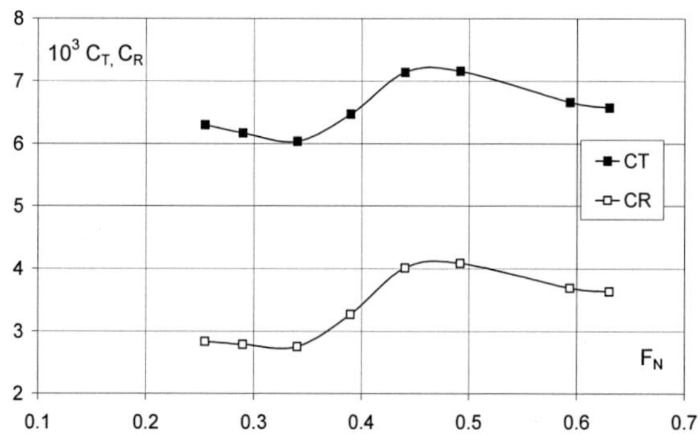

Fig. 2.11.8. C_R and C_T values for the best performing pentamaran configuration from reference 2.11.1

2.11.3.2. Trimaran hull

Also for the trimaran hull, analysis of the test data was performed by ITTC '57 correlation methodology using the Froude procedure for the resistance components subdivision:

$C_{R(T)} = C_{T(T)}(F_n, R_n) - C_{F(T)}(R_n)$ (2.11.1)

$C_{F(T)} = C_{F(M)}.S_{(M)} / S_{(T)} + 2\ C_{F(O)}\ S_{(O)} / S_{(T)}$ (2.11.2)

where $C_{T(T)} = R_{T(T)} / 0{,}5\ \rho\ S_{T(T)}.V^2$, the trimaran total resistance coefficient by the model tests.
$C_{R(T)}$—the trimaran residual resistance coefficient
C_F—the frictional resistance coefficient according to the ITTC frictional line. Suffix codes T, M, and O refer to the whole trimaran, the main hull, and the outriggers, respectively. In Figs 2.11.9 and 2.11.10, respectively,

the curves of C_R and C_T relative to the different staggers for the most interesting clearance (11.10 % L_{WL}) are shown.

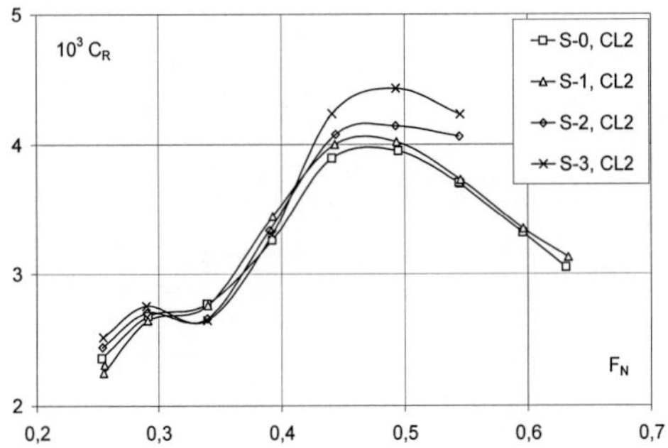

Fig. 2.11.9. C_R values for staggers 0, 10, 20, 30% L_{WL} at clearance 11.10% L_{WL}

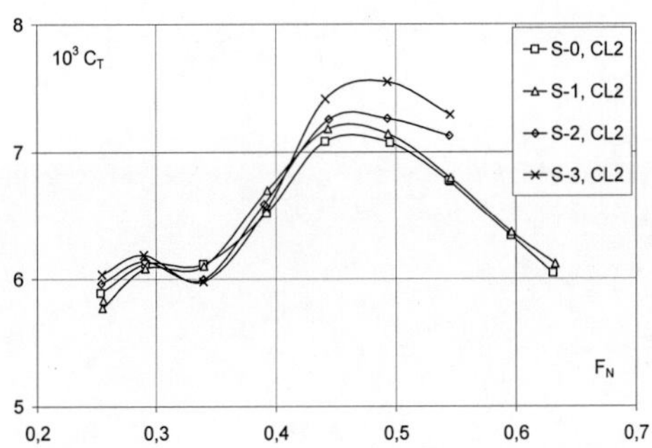

Fig. 2.11.10. C_T values for staggers 0, 10, 20, 30 % L_{WL} at clearance 11.10% L_{WL}

Figs 2.11.11 and 2.11.12 show the curves of C_R and C_T relative to the most interesting stagger (0% L_{WL}) relative to the three clearances CL_1 CL_2 CL_3, corresponding to 9.9%, 11.1%, 13.4% L_{WL}, respectively.

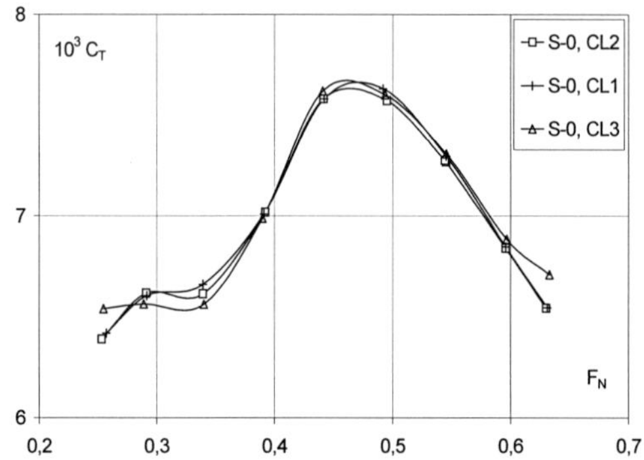

Fig. 2.11.11. C_R values for different clearances at stagger 0 % L_{WL}

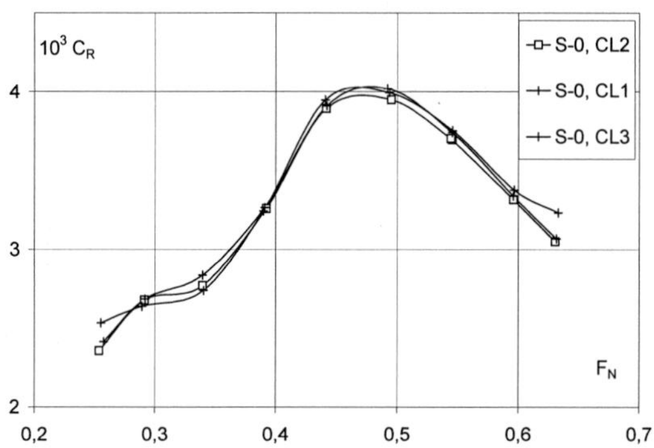

Fig. 2.11.12. C_T values for clearances at stagger 0 % L_{WL}

It is interesting to appreciate that the stagger seems greatly to influence the resistance values in respect to the clearance.

In Fig. 2.11.13 the curves of C_R and C_T versus Fn relative to the best-performing trimaran configuration (stagger 0, clearance 11.1% L_{WL}) are shown. The large gap between the curves highlights the importance of frictional resistance for this ship due to the very slender hull forms.

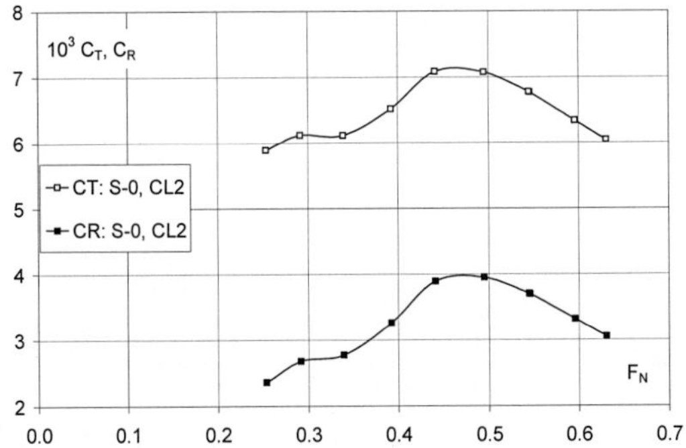

Fig. 2.11.13. C_R and C_T values for the best performing trimaran configuration

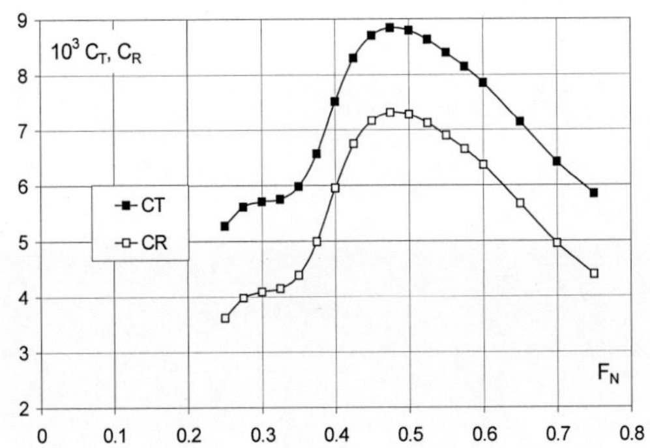

Fig. 2.11.14. C_R and C_T values for the equivalent monohull

2.11.3.3. Equivalent monohull

Fig. 2.11.14 shows C_R and C_T values for the monohull ship. Results are from the NOVA series data[2.11.9] analyzed through the Froude procedure, applying the ITTC 57 frictional line.

2.11.3.4. Results comparison

Figs 2.11.15 and 2.11.16 show the comparison of C_R and C_T values for the pentamaran, for the trimaran, and for the equivalent monohull.

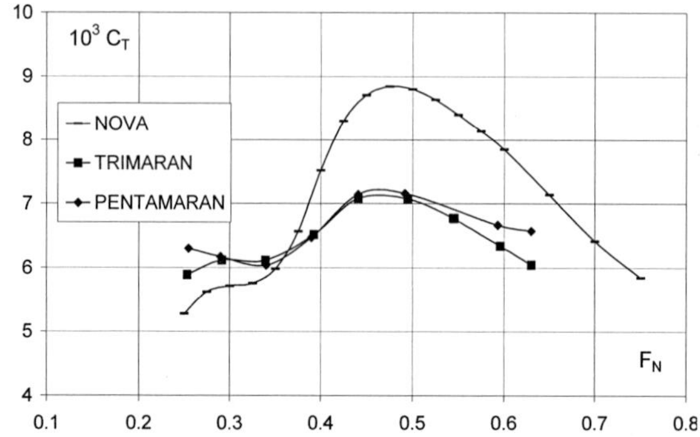

Fig. 2.11.15. C_T values for the pentamaran, for the trimaran, and for the equivalent monohull

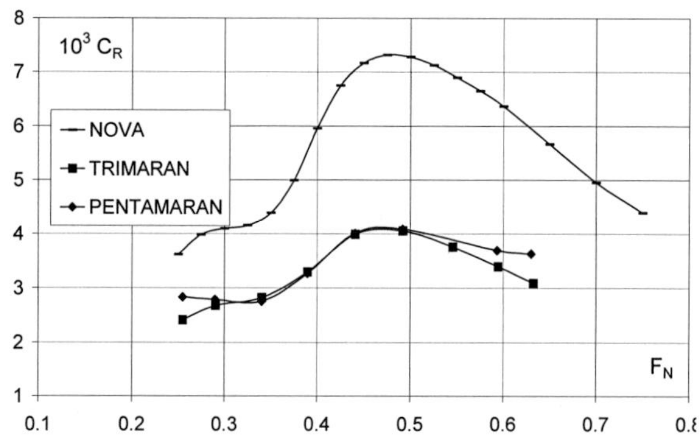

Fig. 2.11.16. C_R values for the trimaran and for the equivalent monohull

2.11.4. Interference phenomena

The available experimental data allows us to present preliminary information about the interference phenomenon always present in multihulls with narrow tunnels among the hulls. As shown in previous works relative to trimaran hulls for small size fast ferries[2.11.7] and to pentamaran hulls,[2.11.1] the interference is a very complex phenomenon that is due to viscous and wave interference among the hulls whose effects can be very different according to the relative speed.

In this paper, focused on a pentamaran–trimaran comparison, a preliminary attempt to identify the interference effects on these hull configurations can be done.

To such an extent, Fig. 2.11.18 reports the total resistance of the main hull R_{MH}, of the trimaran outrigger R_{TOU} times two, of pentamaran outrigger R_{POU} times four, and finally of the whole pentamaran R_{PW} and trimaran R_{TW}.

To assess the interference values, the same quantities for the given values of the Fn are reported in Table 2.11.3. All values are in N and are relative to the tested models.

Table 2.11.3. Interference preliminary assessment at Fn 0.3–0.6

R/N	Fn 0.3	Fn 0.4	Fn 0.5	Fn 0.6
R_{MH}	12.5	25.3	41.8	52.4
R_{TW}	13.9	26.7	44.3	57.0
R_{PW}	13.9	26.6	44.5	59.8
R_{TOU}	0.63	0.99	1.70	2.61
R_{POU}	0.30	0.53	0.87	1.32
No interference values	Fn 0.3	Fn 0.4	Fn 0.5	Fn 0.6
$R_{TW}* = R_{MH} + 2\,R_{TOU}$	13.8	27.3	45.3	57.7
$R_{PW}* = R_{MH} + 4\,R_{POU}$	13.8	27.5	45.3	57.6
Interference values	Fn 0.3	Fn 0.4	Fn 0.5	Fn 0.6
$R_{TW} - R_{TW}*$	.05	-.06	-1.0	-.07
$R_{PW} - R_{PW}*$	.05	-.09	-.07	2.52

From the obtained results it is possible to appreciate that interference values are very small and are only marginally beneficial for the trimaran at Fn 0.5. The larger unfavorable value obtained for the pentamaran at Fn 0.6 is due to spray and wave interference with the forward outrigger bracket.

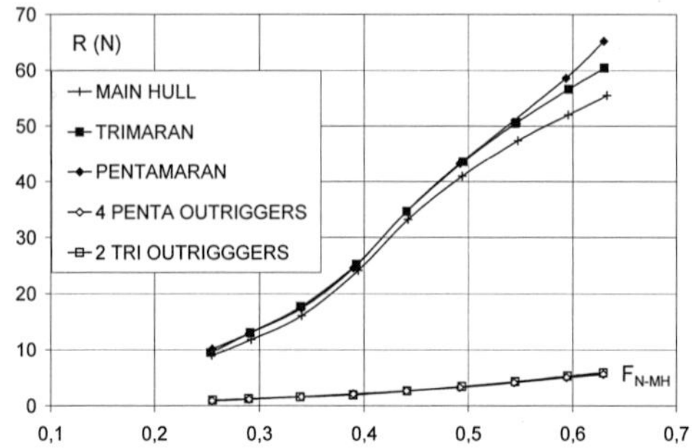

Fig. 2.11.18. Total resistance of the main hull, of the trimaran outrigger times two, of the pentamaran outrigger times four curves comparison, and of the whole pentamaran and trimaran

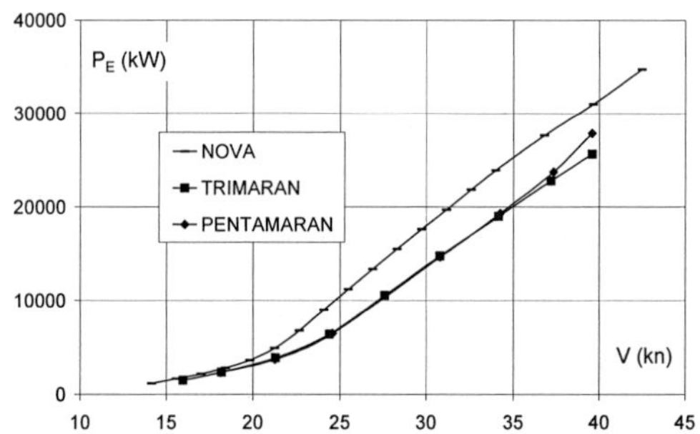

Fig. 2.11.19. Monohull, trimaran, and pentamaran effective power

2.11.5. Powering performances

Fig. 2.11.19 shows the comparison in the speed range V = 15–40kn between the effective horsepower of a pentamaran, a trimaran, and a monohull ship evaluated by the ITTC '57 methodology correlation and frictional line. This figure highlights that in the speed range from 25kn to 35kn, pentamaran and trimaran are equivalent. At higher speed, the trimaran presents a marginally better performance and the margin increases with the speed. It must be noticed that this difference at the highest considered speed is due, for the pentamaran, to unfavorable interference of the forward outriggers with the main hull bow wave. This inconvenience has been reduced through higher positioning of the forward outriggers.

The trimaran does not suffer this inconvenience due to the after considered stagger.

Anyway both pentamarans and trimarans need significant lower power in respect to the monohull in the speed range 30–40kn, being a percentage difference in the order of 20–30%.

2.11.6. Intact stability

To get a more complete picture of the potential of the pentamaran and trimaran hull configuration, their intact stability has been assessed. The ships considered have to be treated as HSC for cruising speeds over 26.2kn; thus, the criteria proposed by HSC IMO Code have been used. In this paper, the presented results are relative to just one criterion, i.e., the area

$A\alpha_{GZmax}$ below GZ curve up to max GZ angle α_{GZmax}.

For three different freeboards, the highest possible KG values complying with the above mentioned criterion have been determined for both the trimaran and the pentamaran, considering the outrigger position used for the previous performance comparison.

The corresponding KG/D values can give useful information to evaluate the suitability of these hull forms in respect to the typical layouts requested by light goods transportation.

Figs 2.11.20–22 show the curves of the stability arms GZ versus heel angles for the pentamaran and the trimaran at freeboards FB = 4.96m, 4.34m, and 3.72m, respectively. Table 2.11.4 reports the values of the stability indexes.

The aim of this part of the work is to furnish a preliminary evaluation of the stability characteristics in order to verify the feasibility of the main-hull-plus-outrigger configurations with realistic KG values, suitable to the layouts requested for passenger/car transportation. From this preliminary analysis limited to intact stability the limit values obtained for KG height for both the pentamaran and the trimaran seem in agreement with the results of preliminary weight analysis of structural and general layouts.

From the diagrams, the pentamaran configuration presents lower initial stability, but better global stability characteristics in respect to the trimaran. Both multihulls have obviously a quite different behavior in respect to the monohull presently in service. This last hull form is surely preferable for the safety at high heeling angles.

This preliminary investigation will be integrated with damage stability analysis.

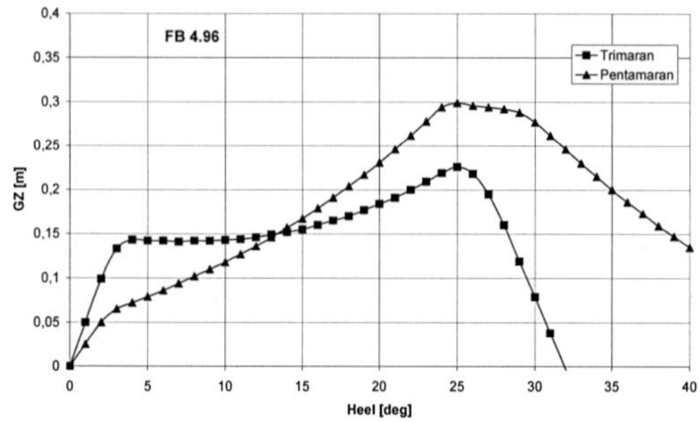

Fig. 2.11.20. Pentamaran and trimaran righting arms curves for FB=4.96

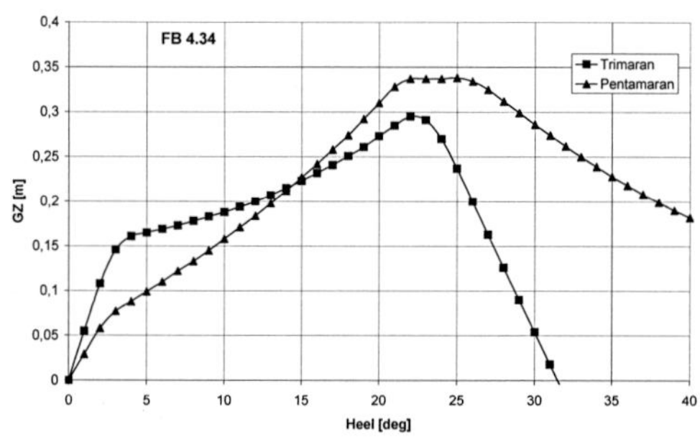

Fig. 2.11.21. Pentamaran and trimaran righting arms curves for FB=4.34

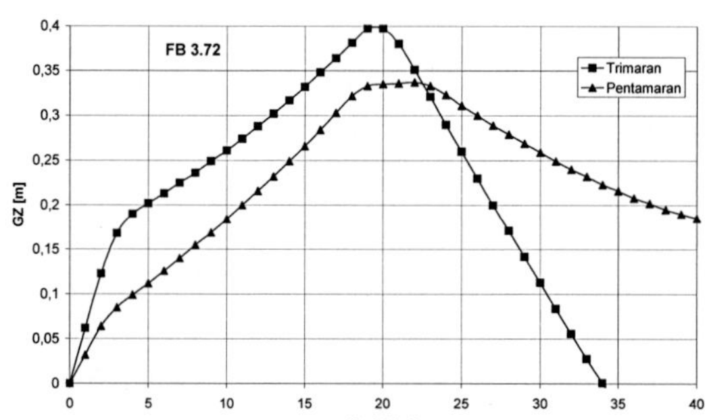

Fig. 2.11.22. Pentamaran and trimaran righting arms curves for FB=3.72

Table 2.11.4. Trimaran and pentamaran stability parameters

Case		Pentamaran	Trimaran
1	FB/m	3.72	3.72
	B/m	18.75	16.10
	D/m	8.16	8.16
	KG/m	5.11	6.10
	KG/D	0.62	0.75
	GM/m	1.94	3.67
	α_{GZmax}/deg.	21.8	19.5
	$A_{\alpha GZmax}$/m rad	0.0759	0.0846

Case		Pentamaran	Trimaran
2	FB/m	4.34	4.34
	B/m	18.75	16.10
	D/m	8.78	8.78
	KG/m	5.26	6.52
	KG/D	0.60	0.74
	GM/m	1.79	3.25
	α_{GZmax}/deg..	24.8	22.3
	$A_{\alpha GZmax}$/m rad.	0.0842	0.0759

Case		Pentamaran	Trimaran
3	FB/m	4.96	4.96
	B/m	18.75	16.10
	D/m	9.40	9.40
	KG/m	5.49	6.78
	KG/D	0.58	0.72
	GM/m	1.56	2.99
	α_{GZmax}/deg.	25.0	25.0
	$A_{\alpha GZmax}$/m rad.	0.0663	0.0666

2.11.7. Conclusions

The main-hull-plus-outriggers concept consists of a very slender unstable central hull stabilized by very small outriggers. Both the trimaran and the pentamaran, although presenting small differences between their relative powering performances, are definitively better than the equivalent monohull at medium high Fn.

The trimaran and the pentamaran seem interesting possibilities due to the benefits given by their very slender hull forms compared with monohull ships at medium-high speeds as regards both sea-keeping and powering performances. At the same time, they present simpler structural aspects in comparison with medium-large sized catamarans

These hull forms appear well suited to medium-size ships, such as those considered in this paper. Other advantages of the multihulls considered here are the large deck and the high intact stability. In particular, their GM values complying with IMO stability requirements are generally greater than that of a monohull. This feature and the large deck area allow a great amount of load to be carried on the main deck.

Furthermore the main hull plus outriggers concept appears easy to fit main dimensions constrains and easy to approach from the structural point of view, allowing to consider it also for larger ships.

It is difficult to give a defined preference for either the pentamaran or the trimaran configuration. The trimaran presents slightly better powering performances in the 0.5–0.6 Fn range, which is operative for the ship considered in this paper, but investigating other characteristics such as damage stability or sea-keeping performance could modify this rank.

On the other hand, the trimaran and the pentamaran seem, as the main dimensions increase, preferable also to the standard or wave-piercing catamaran configurations due to evidently simpler structural layouts and to the possibility of containing the maximum beam within imposed limits. A fair and detailed comparison between performances and characteristics of the two basic multihull concepts, twin hulls or main-hull plus outriggers, would be important to identify operational conditions and dimensional ranges most suited to each hull form.

The results of this study are applicable only to the considered hull form and configuration. The high-speed trimaran has to be seen as a possible competitor to the catamaran and to the monohull. However, for the selection of the optimum high-speed craft it is necessary to consider the relation between the resistance characteristics and other important design variables as the structural weight and the cost implications due to the overall length.

Acknowledgements

This work has been supported by the Italian Ministry of University and Scientific Research in the frame of the 2003–2004 research plan.

References

[2.11.1] E. Begovic et al., "Pentamaran Hull for Medium Size Fast Ferries," ICHD 2004 Conference Proceedings, 2004.

[2.11.2] C. Bertorello et al., "On Hydrodynamic Performance of High Speed Crafts," *Proceedings of the 7th International Conference on Fast Sea Transportation, FAST 2003*, vol. I, pp. 101–8, 2003.

[2.11.3] C. Bertorello et al., "From Model Scale to Full Size: Investigation on Turbulence Stimulation in Resistance," *Proceedings of the 7th International Conference on Fast Sea Transportation, FAST 2003*, vol. I, A1, pp. 1–9, 2003

[2.11.4] L. J. & R. J. Scarce, "The Optimisation of Trimaran Sidehull Position for Minimum Resistance," *Proceedings of the 7th International Conference on Fast Sea Transportation, FAST 2003*, vol. I, pp. 1–12 (keynote lecture), 2003.

[2.11.5] N. Gee & J. Roy, "The Effect of the Length on the Powering of Large Slender Hull Forms," *Proceedings of the 7th International Conference on Fast Sea Transportation FAST 2003*, vol. I, pp. 23–30 (Keynote Lecture), 2003.

[2.11.6] E. Begovic et al., "High Speed Trimarans Validation of Numerical Results by Geosim Tests," *Proceedings of the 6th International Conference on Fast Sea Transportation, FAST 2001*, vol. II, pp. 285–94, 2001.

[2.11.7] C. Bertorello et al., "Experimental Performance Investigation on Different High Speed Crafts: Monohull Catamaran and Trimaran," *Proceedings of HIPER 2001*, pp. 107–14, 2001.

[2.11.8] A. Kolmhoss et al., "Prediction of Global Wave Induced Response for the ADX Express High-Speed Pentamaran," *Proceedings of the 6th International Conference on Fast Sea Transportation, FAST 2001*, vol. II, pp. 71–79, 2001.

[2.11.9] E. Lahtiharju et al., "Resistance and Seakeeping Characteristics of Fast Transom Stern Hulls with Systematically Varied Form," *SNAME Transactions* 99, pp. 85–118, 1991.

[2.11.10] H. Y. H. Yeh, "Series 64 Resistance Experiments on High Speed Displacement Forms," *Marine Technology*, July, pp. 248–72, 1965.

2.12

THE EVOLUTION OF ADVANCED SLICE® TECHNOLOGY ADAPTED TO SATISFY THE HSC CODE AND COMMERCIAL REQUIREMENTS (2007)

JOHN KECSMAR,[1] NIGEL WARREN[2] AND JOHN MOORE[3]

[1] Ad Hoc Marine Designs (Japan)
[2] Ad Hoc Marine Designs (England)
[3] Lockheed Martin (USA)

Abstract

This paper describes a unique hull form patented by Lockheed Martin Corporation that realizes all of the sea-keeping advantages of a SWATH hull with the added benefit of increased speed capacity for the same horsepower. This increased speed is obtained by increasing the Froude Number of an equivalent SWATH by employing four shorter buoyancy hulls. This configuration is named SLICE®. The concept was successfully demonstrated in 1997 with the vessel SeaSLICE®, which achieved a sustained speed of 30kn. Two commercial, DNV-classed crew boats using this patented technology were recently designed, built, tested, and delivered to Mexico for deployment in the Gulf of Mexico oil fields using the SLICE® hull configuration. To maximize passenger comfort, a sophisticated active ride control system using active canards and unique rudder/stabilizers were incorporated into the design. The SLICE® technology poses unique structural, arrangement, and noise and vibration design considerations due to the hull form, size of the lower hulls, and machinery arrangements. This paper describes the evolution of SLICE® and the inherent benefits of this unique hull form.

2.12.1. Concept and background

SLICE® is a new, patented ship technology that enables small waterplane area twin hull (SWATH) ships to operate at higher speeds while retaining their characteristic low motions in a seaway.[2.12.1] SLICE® technology's key innovation is a reduction of wave-making drag, which is accomplished by the introduction of four short struts, four teardrop-shaped submerged hulls, and speed well beyond the "hump" on the Froude resistance curve. Combining increase speed with reduced motions in high seas, SLICE® opens up new commercial and military markets to SWATH technology (see Fig. 2.12.1).

Throughout history, ship designers have sought a hull form for ocean-going vessels that combines excellent sea-keeping in high sea states with high speed. For centuries, the prevailing design was the monohull. In the early 1900s with the successful integration of onboard power, high speed became achievable and demand for ever-faster ships increased. As available installed power reached its limit, however, designers began investigating ways to increase speed by reducing hull resistance.

Fig. 2.12.1. Original SLICE® configuration

Several advanced hull forms resulted from these efforts. Displacement hull variants like catamarans, which utilize buoyancy, and hull forms incorporating dynamic lift such as planing hulls, hydrofoils, and surface effect ships were designed and tested. Each of these innovations confirmed that design improvements could produce higher speeds.

Throughout this quest for speed, ship motions in high seas remained a critical issue. A light-weight high-speed vessel is prone to damage in unprotected water unless it can survive the ocean's unforgiving environment. As a naval officer once observed, "Sea State is a war stopper." Thus, for generations, a vessel that combines high speeds and excellent sea-keeping in high seas has been the holy grail of ship designers.

The challenge: high sea state performance

In high seas, ships must sacrifice either speed or sea-keeping ability, and neither can be achieved without size. As speed (Froude number, F_n) increases, wave resistance becomes a higher percentage of total resistance—until at the critical or "hump speed" wave resistance exceeds viscous resistance. This large increase occurs when $F_n = 0.4$ and is maximum at $F_n = 0.5$. Conventional ships operate at Froude numbers below this primary hump speed. To achieve high speed, naval architects design ships to operate below the $F_n = 0.4$ threshold by increasing their length. HSC vessels and naval ships with high installed-propulsion power can operate at a Froude number above 0.4.

To maintain high speed in an increasing sea state, conventional displacement ships must have a long waterline length. For example to reach a speed of 30kn with a reasonable power, a vessel must be at least 150m in length (see Fig. 2.12.2). [Froude number = 0.4 = velocity /square root (g × length).] Consequently, small conventional displacement ships are unable to consistently perform high-speed missions.

A ship's size also limits its ability to perform in a seaway.

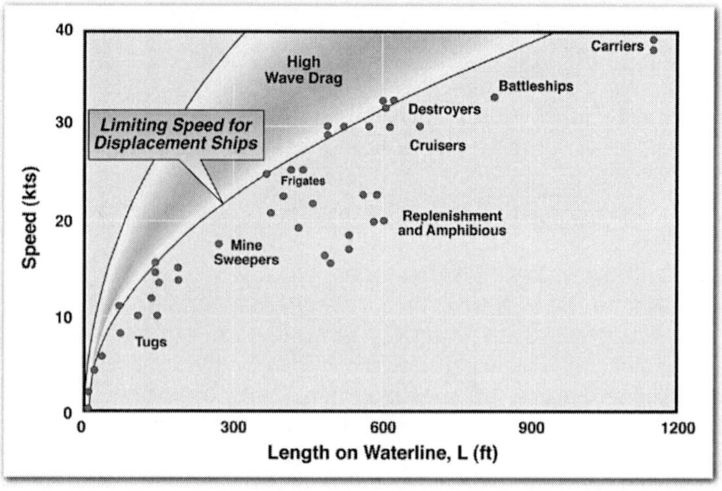

Fig. 2.12.2. Speed versus waterline length

For example, to be fully operational in a seaway of 4.5m seas, a vessel must be at least 160m in length (see Fig. 2.12.3). This sea-state limitation further emphasizes the unsuitability of small, conventional displacement ships for high-speed missions, especially in high sea states.

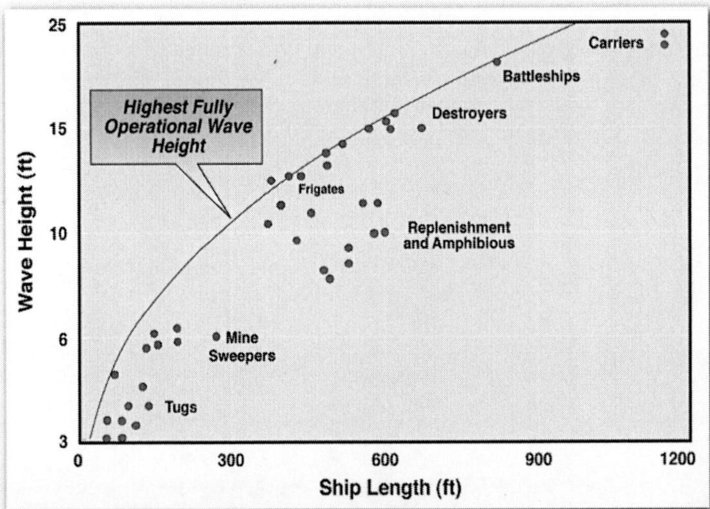

Fig. 2.12.3. Wave height versus ship length

The choice of a hull configuration to behave as a stable platform in high sea states dictates the design in operations like fast ferry services, offshore energy exploration and production, and piloting services that move further offshore. Conventional hull forms and even more advanced hull forms like catamarans, planing hulls, hydrofoils, and air cushion vessels must reduce speed to avoid passenger discomfort and crew fatigue to maintain an acceptable level of safety.

The Response: high speed SWATH technology

The quest to improve sea-keeping led to the development of the SWATH hull form, which utilizes submerged hulls, thin wave-piercing struts, and an elevated platform. The SWATH hull form has a low waterplane area that is less affected by waves than conventional vessels, resulting in reduced motions in high seas.

Lockheed Martin undertook an effort, supported by the US Navy's Office of Naval Research (ONR) to investigate designs that would provide the sea-keeping performance of a SWATH in high seas and still provide the high speeds required.

Lockheed Martin focused on rearranging the displacement hulls to increase the Froude number by reducing the effective length. This resulted in four lower hulls sized to provide the required displacement and buoyancy distribution. The minimum practical powering of SLICE®, for a given length over all, is determined by the power required to overcome the Froude "hump." At speeds beyond the Froude hump, the reduction in wave-making resistance resulted in an approximately 20% increase in speed over a similarly sized conventional SWATH (see Fig. 2.12.4).

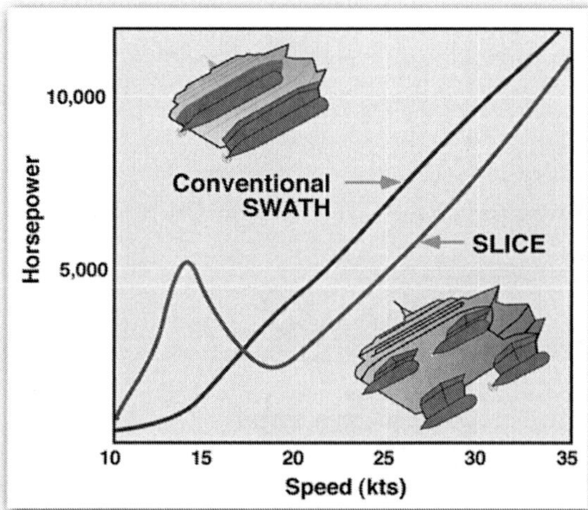

Fig. 2.12.4. Typical SWATH powering

SLICE® provides advantages in reduced wake and improved endurance owing to lower fuel consumption. Additional benefits include large open decks and multiple lower hulls, flexibility in propulsion arrangements, modular payload capability, and an unobstructed stern for loading and unloading.

The status: proven performance

In 1998 the first SLICE® was put to sea and met or exceeded every performance goal established by Lockheed Martin and ONR (see Fig. 2.12.5). Since then, SLICE® has been put to use demonstrating high-speed stable performance in Hawaii and from San Francisco to San Diego. She has been used in US Navy warfare demonstrations Fleet Battle Experiments Hotel and Juliet in 2000 and 2002, respectively. She has been used to demonstrate the technology to ferry operators, oil companies, and military officers from around the world. She is currently home-ported in San Diego, California, USA.

In 2004, based on SLICE® technology, Lockheed Martin was awarded a contract to supply a Crew Transport Vessel through HSP, a Mexican offshore operator, to PEMEX, the national oil company of Mexico. Two completed vessels are currently in Mexico and will be put into service in 2007 (see Fig. 2.12.6 & Fig. 2.12.7). Discussions are ongoing for using the same technology for ferry operations, pilotage service, and other offshore support requirements.

Fig. 2.12.5. *SeaSLICE*

Fig. 2.12.6. *Lider* at launch

Fig. 2.12.7. *Tenaz* at launch

2.12.2. General description

Lider and *Tenaz* are 29.3m crew boats carrying 150 workers and 6 crew at speeds of 20kn to oil rigs in the Mexican Gulf; the workers are transferred to the oil rig by basket transfer. The two vessels are of all-aluminum construction powered by twin diesels and controllable pitch propellers. There is crew accommodation onboard for occasional overnight stops consisting of double cabins plus single cabins for the master and engineer. A galley and mess are provided together with an office for the PEMEX representative. Full air conditioning is fitted aided by thermal insulation and double-glazed saloon windows.

The craft are built to DNV requirements specifically, +1A1 HSLC R2 Crew Boat. Essentially these requirements are according to the IMO HSC Code for fast passenger carrying vessels. Very few of the relaxations given in the DNV Rules for crew boats were in fact taken up in this design.

The hull form is most unusual as can be seen from the general arrangement drawing (see Fig. 2.12.8). The four airship-shaped pods form most of the buoyancy. The bottom of each pod is flat in order to reduce the draft. The forward pods are set inboard of the aft ones so that the propeller wash from the propellers at the aft end of the forward pods does not impinge on the after pods. Large haunches above water connect to the platform raft and provide buoyancy for damage stability and a good root connection to the raft and limit slamming on the underside of the raft.

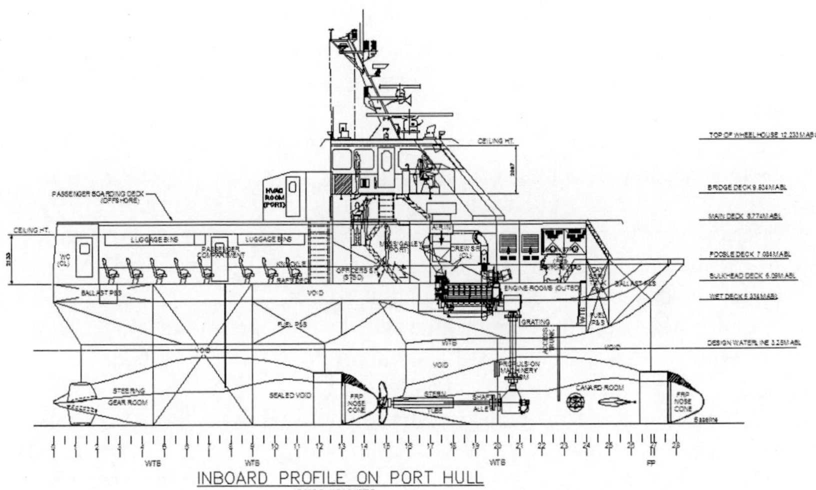

Fig. 2.12.8. G.A. profile

The principal particulars are as follows:

Length overall/m	29.3
Beam/m	16.2
Displacement full load/t	200
Draft at full load/m	3.25
Workers	150

Crew	6
Engines	Twin Diesel
Speed/kn	20
Fuel/t	21
Water/t	3.3

At the design waterline of 3.05m the waterplane area is 25m^2, which at 200t displacement gives a waterplane area/displacement ratio of 0.73. Comparing this with other SWATH craft shows the figure to be very low indicting an inherently steady platform in a seaway, especially when the large control surfaces are also taken into account.

The twin engines are sited within the haunches/raft while the engine room actually extends right up to the roof of the superstructure. Both engines drive via right angle gearboxes to a horizontal shaft line to CP propellers.

Steering and ride control functions are combined. Projecting inboard from the after pods are two large fins angled down to the horizontal (see Fig. 2.12.17). Projecting inboard from the forward pods are two smaller fins.

Two 175kW diesel generators supply 440V 3-phase 60Hz, each machine capable of coping with normal seagoing loads. Transformers also supply 220V 3-phase 60Hz and 127V 1-phase 60Hz. A bank of 24V DC batteries supply essential emergency loads. All engines are started electrically by dedicated 24V batteries. A separate switchboard room is sited in the deckhouse between the engine rooms.

Other systems in the boat include side thrusters set into the pods. There is one in the after starboard pod and one in the forward port pod. Each is rated at 1.7t and is driven hydraulically.

There is a powerful ballast system controlled from the wheelhouse. There are four ballast tanks sited at each corner of the raft totaling 34t. Being at each corner, the ballast has the best lever to counteract changes of trim and list. They are positioned high up near the vertical center of gravity of the craft so that changes in the amount of ballast do not affect VCG and hence GM. It also allows discharge of the tanks by gravity rather than pump.

A sewage holding tank and two sewage treatment plants are fitted for redundancy.

Lifesaving appliances, passive fire protection, gas drenching of the engine rooms, firefighting, and non-combustible or flame-retardant furnishings and minor bulkheads are fitted all in accordance with the HSC Code.

Similarly, the bilge system consisting of individual electric pumps meets the requirements of DNV. System services have to extend down from the raft into each of the four pods and the extent and weight of the piping and contents (hydraulic thrusters, SW, FW, and bilge and vents) is considerable. Four fuel tanks and two service tanks feed the four diesel engines.

A comprehensive navigation and communications suite if fitted all meeting the IMO and HSC Codes. In addition, a night vision system and CCTV system are fitted.

2.12.3. Stability

Designing a SWATH to meet intact and damage stability criteria presents some different challenges compared to monohulls. Like a catamaran but worse, when one hull is flooded large angles of heel can be expected. On the other hand, if the raft structure is made watertight clear of the side damage the sheer volume of intact buoyancy created is enough to ensure the craft is not in danger of sinking. The HSC 2000 Code increased the bottom damage scenario considerably as a result of grounding at speed of a number of passenger catamarans. The requirements now are that an orderly evacuation should be possible after bottom damage of 55% of the craft length from the fore end. Today many catamaran designs have double bottoms to cope with this requirement but this was not practical in the case of the SLICE or any horizontal watertight division.

The 55% criteria meant that all the forward pod/strut/haunch was flooded and parts of the after pod too. This lead to the idea of having buoyancy filled nose cones on each of the four pods. These are of GRP filled with foam and bolted to the aluminum structure. This arrangement was previously used on the PTC craft.[2.12.2] This arrangement also obviates the difficulties of creating the double curvature shape in aluminum. The large haunches, where intact, create a stabilizing influence above 5° or so of heel. Nevertheless, the angle of heel after such a large amount of flooding exceeds the 10° criteria.

The code allows this provided the heel can be corrected quickly. The master's first line of defense is to fill the high side ballast tanks and empty the low side ballast tanks from the control panel in the wheelhouse. His second line of defense is to action the counter flooding arrangement. In all four pods, 75mm diameter sea valves are fitted in the bottom of the pods. These may be opened remotely from the raft area. Flooding the opposite pods plus using the ballast system, effectively brings the craft down on its wide stable raft. Examples of the intact and damage stability are shown in Figs 2.12.9(a) and 2.12.9(b).

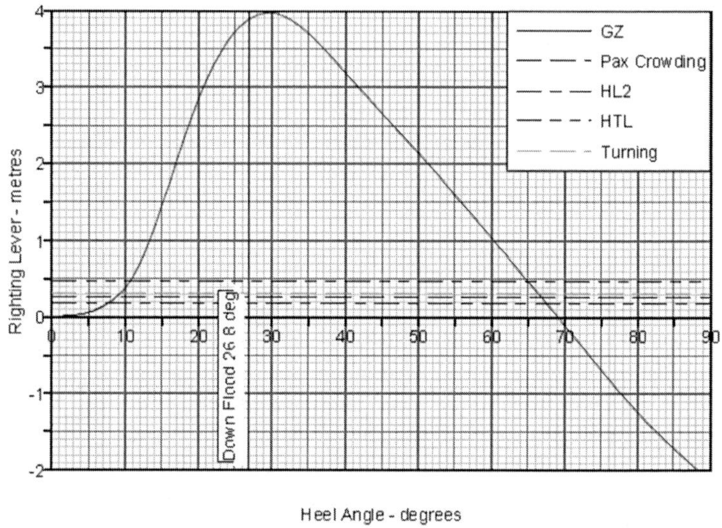

Fig. 2.12.9(a). Intact stability

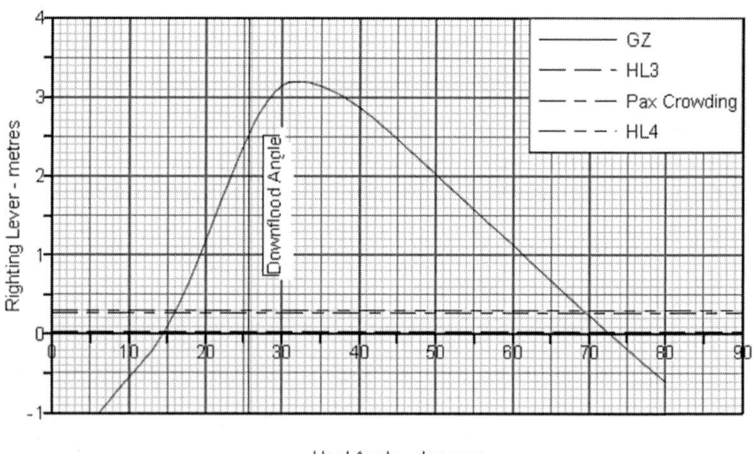

Fig. 2.12.9(b). Damage stability

2.12.4. Structural and vibration design

A vessel with such a unique geometry naturally poses problems in its analysis. Being a multihull and falling inside the limits to be classed as a HSC vessel under the IMO code, classification society rules were used as guidelines. However, the class rules, in this case DNV, have no rules regarding four separate hulls.[2.12.3] The classic hull torsional moments about the longitudinal and transverse axis are of no use, since the hull is very short compared to the overall length of the vessel and is compounded by the fact that each hull is also axially out of plane with the others (see Fig. 2.12.10).

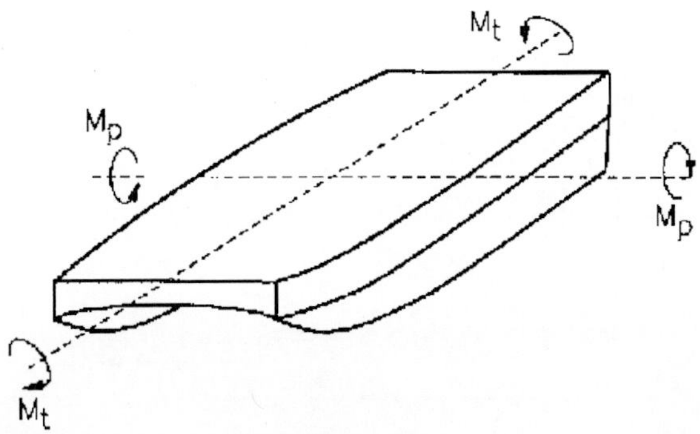

DNV Torsional moments

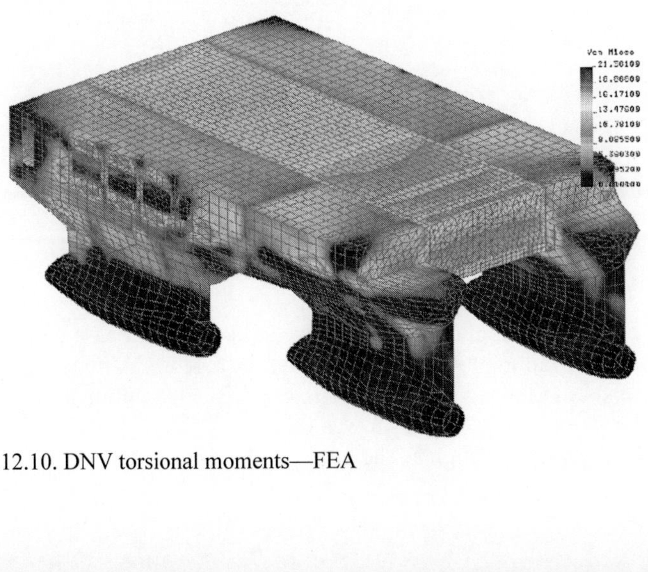

Fig. 2.12.10. DNV torsional moments—FEA

The prescriptive global torsional loads in DNV are concerned with the raft structure—that is, the platform to which the hulls are attached. Therefore, the load application to establish the torsional rigidity of the raft structure requires a different approach, or a modification of the existing rule. It was decided to use the DNV rule and apply the loads, as if the hulls were a continuous member, since this was seen to be a load case greater than the vessel could actually experience.

The haunch-strut-pod required a different approach from the conventional "secondary" loading condition (see Fig. 2.12.11). Applying dynamic loads was not possible owing to the deep draft. The draft is a limiting factor when applying the dynamic sea pressures from DNV rules. This meant defaulting to the standard hydrostatic sea pressures, as on conventional vessels.

Many lessons were learnt when designing the FBM SWATH *Patria* in the late 1980s, now called *Sea Flower 2* running between Korea and Japan. These lessons formed the basis of the design methodology in the FBM SWATH for the UK MoD, the Passenger Transfer Craft (PTC) in the late 1990s. The design principals applied have proven to be effective and satisfactory.

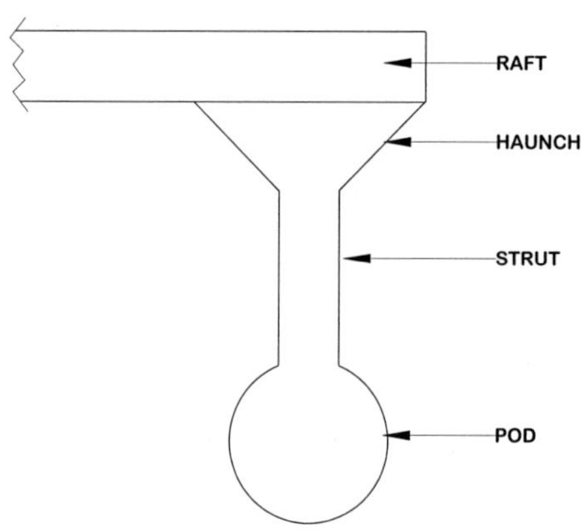

Fig. 2.12.11. Hull geometry

The PTC, to the authors' knowledge, has not suffered a single crack nor structural failure.[2.12.2] Owing to this, the UK MCA now surveys her biannually. All the load cases that were established, a nominal "1.5g" load applied horizontally at half draft was the most onerous (see Fig. 2.12.12). This load case highlights the low stiffness of the strut-pod and strut-haunch connection as being the weak link. During the design of the PTC, a global fatigue allowable design stress value was established: the value used was 40MPa. All critical regions were designed using this nominal value. However, when designing in aluminum, the stress levels are not always the driving factor, it is deflection.

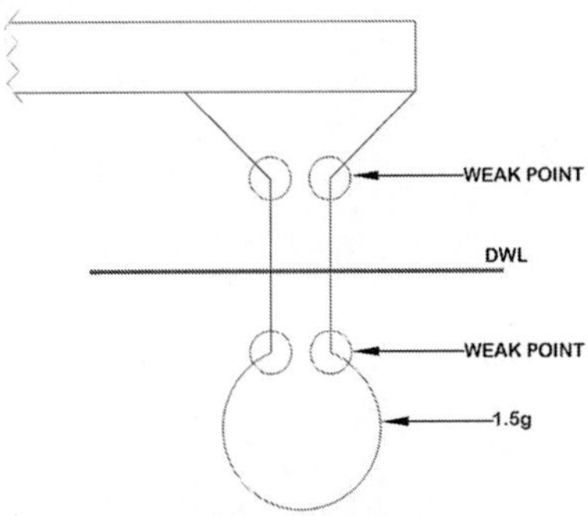

1.5*g* Load case

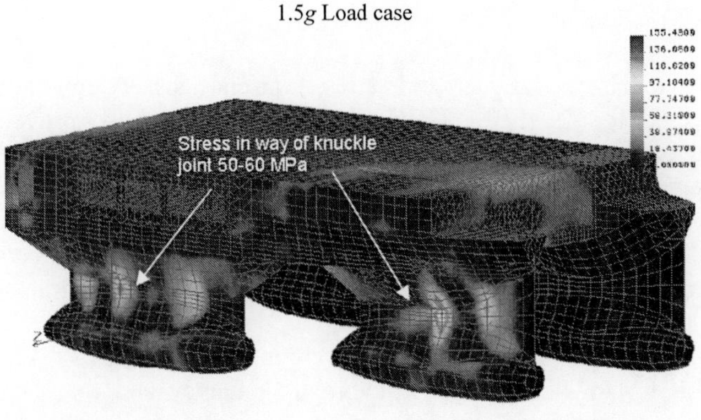

Fig. 2.12.12. FEA of 1.5*g* load case

When applying a 1.5g load to the pod, the stress exceeded the 40MPa global fatigue limit. To reduce the stress to 40MPa required an increase in plate thickness from 6mm to 8mm. However, to maintain the same deflection ratios that were used on the PTC, required a further increase in plate thickness to 12mm. This also had the positive effect of reducing the stress at this critical region even further. The strut section of the vessel is very thin, more so than a conventional SWATH. This inevitably raises issues with stiffness and the rate of change of stiffness within the strut joint. However, reducing the ratio of deflections between the aft and fwd pods was far more complex. This required a more detailed FEA analysis (see Fig. 2.12.13).

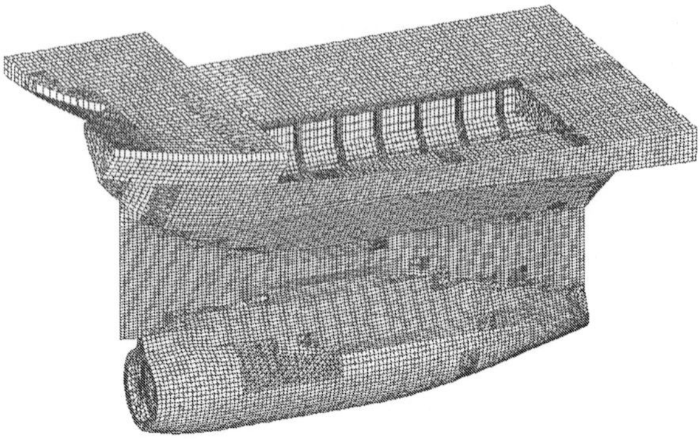

Fig. 2.12.13. FEM of SLICE®

For localized fatigue and stress raisers, the design guidelines established for the FBM TRICAT Class of vessel also proved to be satisfactory.[2.12.4] [2.12.5] The TRICAT waterjet structure, which pumps 23t of water every second for 18 hours a day every day for the past 12 years, has not to the authors' knowledge reported a single crack. The principals behind this approach are to reduce or completely eliminate all cut-outs in the transverse frames and, where possible, to be transversely framed—that is, no longitudinals.

Longitudinal intercostal Tee bars were introduced to the strut region of the SLICE® to increase the structural redundancy and the shear paths available—the greater the shear paths available the greater the load shedding. This means that should a crack occur, it will propagate into a region of low stress and become more of an inconvenience than a catastrophic failure.[2.12.6]

Reducing the cut-outs in the transverse frame has two positive effects from a fatigue perspective.[2.12.7] First, a cut-out creates a notch in the frame, which is a localized stress raiser. Second, it allows the welder to perform a much longer weld run that significantly reduces the number of stop-starts. A stop-start is notorious for being the site of crack initiation. In regions where welding has limited access, it is near impossible for the welder to guarantee that a stop-start has been sufficiently ground down to remove any voids left behind the two runs. Reducing the number of stop-starts also produces a better quality weld since the welder is concentrating on the weld being deposited and not the correct length of the stitch weld.[2.12.8]

Access for fabrication is near impossible in some regions of a SWATH; this aspect becomes a very important variable for the structural design. Access for the welder and ease of welding in the SLICE was made difficult owing to its unique geometry. The struts are barely wide enough for a person to traverse the haunch down into the pod. In order to reduce the likelihood of a stop-start fatigue crack, this region was fully welded. The areas of high stress we also welded using an Argon-Helium mixture to ensure good penetration and fusion. The use of 5356 filler wire on a 5083 plate also produces a better quality weld than would 5183 filler wire. Therefore to minimize internal weld defects, 5356 filler wire was chosen despite its lower "as-welded" strength compared to the ubiquitous 5183.[2.12.3]

The strut connection region is not just subjected to excitation by the sea loads. The internal machinery and appendages are sources of hull excitation. The internal sources of vibration needed investigating to establish whether this would promote a global hull excitation. An FE study for the modes of natural frequencies of the hull girder was conducted. The principal sources of forcing frequencies are:

Engine	31.7 Hz
Upper g/box	15.9 Hz
Lower g/box	6.3 Hz
Propellers blade passing	24.8 Hz

Clearly the lower propeller shaft would be the most critical. Global hull girder modes of vibration are invariably in the region 4–8Hz. Therefore, the vibration study focused upon the modes of vibration at and around this frequency range.

The first modes of hull girder vibration were calculated to be below 6.3Hz. A vertical mode—this is where the entire pod and strut is displaced in a vertical manner, with an effective axis in the raft structure—was shown to be less than 1Hz (see Fig. 2.12.14). A "yaw" mode, where the hull effectively twists about the vertical center of the pod (see Fig. 2.12.15), and a "rolling" vibration, where the pod is rolling about its own axis, were all below 2Hz (see Fig. 2.12.16).

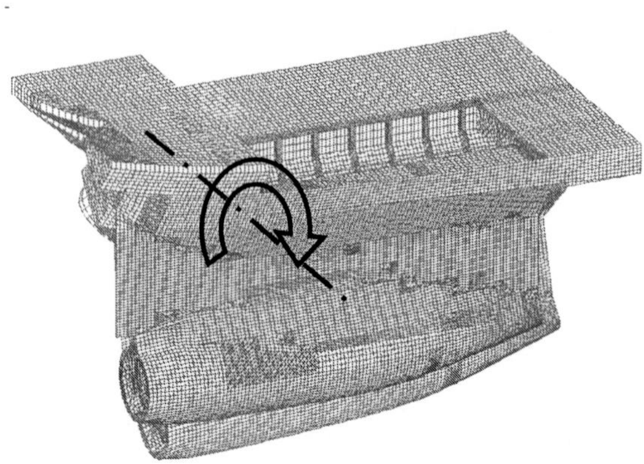

Fig. 2.12.14. Vertical mode of vibration

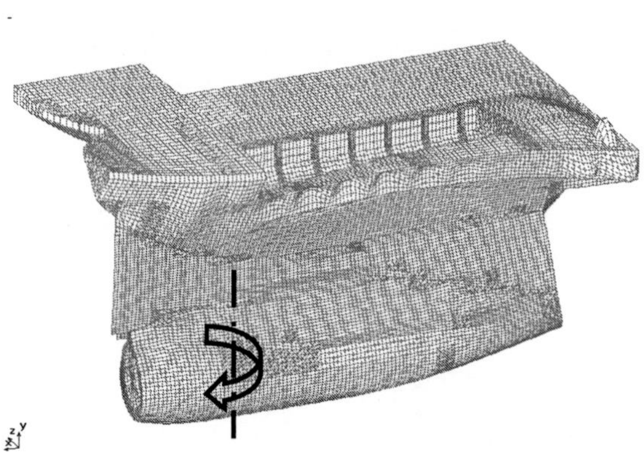

Fig. 2.12.15. “Yaw” mode of vibration

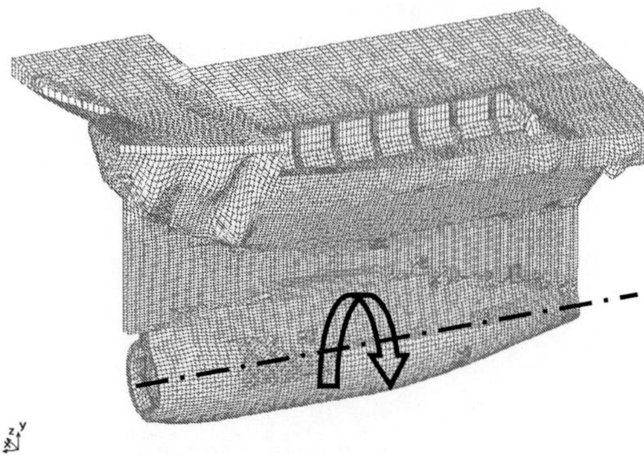

Fig. 2.12.16. “Rolling” mode of vibration

At higher frequencies, the hull modes became less dominant and the local effects such as frame warping and hull plating between stringers “dishing” increased. These effects were still below the 6.3Hz, albeit at very low displacements. At and very close to 6.3Hz, small localized vibrations and displacements were also observed. However, such localized effects, if problematic on sea trials, are easily remedied. No such effects occurred.

2.12.5. Propulsion system

The geometry of a SWATH poses the fundamental problem of where to put the main engines. Other SWATHs have used a diesel electric drive, an inclined shaft,[2.12.2] engines in the lower hulls, and, as chosen on the SLICE, a Z drive. Standard marine diesel engines are each rated at 1,343 kW (1,800 BHP) at 1900 ROM drive via a flexible coupling to a right angle gearbox. This gearbox with a reduction ratio of 2:1 has a steel casing, semi-elastic mounts and PTO’s for the thruster hydraulic pumps and gearbox oil pumps. It also has a clutch.

A vertical intermediate shaft takes the drive to the lower gearbox, which has an aluminum casing and is also semi-flexibly mounted. This gearbox has a ratio of 2.5:1 and incorporates a thrust bearing.

The propeller shaft has a disc brake while the controllable pitch propeller has four blades with a diameter of 1830mm. For a craft of this size, this is a very large slowing running propeller. Together with a reasonably symmetrically input wake pattern, the propeller efficiency is very high.

Such a complex high powered arrangement set into a very confined lightweight aluminum structure was cause for concern at the early design stages. Accessibility, produceability, and possible local structural vibration were examined closely as noted in section 3. So too was the torsional vibration characteristics of the drive train. Trials validated the detailed design.

2.12.6. Ship systems

The SLICE® Crew Transport Vessel systems reflect the typical systems found on this class and size vessel with the addition of the sophisticated ride control system. The unique SLICE® four-hull configuration does require some system components to be split among the hulls. One advantage of this is a level of redundancy that might not normally be implemented into a similar monohull vessel. Viz:

1. The propulsion system consists of independent propulsion engines each driving a controllable pitch propeller through two 90° gearboxes.
2. Separate bow and stern thrusters augment the propellers for station keeping and docking.
3. The fuel oil system consists of four fuel oil tanks (two aft, port and starboard; and two forward, port and starboard) and two day tanks. Two fuel oil purifiers assure the day tanks contain clean fuel. The purifiers are cross connected to permit filling both day tanks if one of the purifiers is inoperable. A pair of three-way valves and pumps permits moving fuel fore or aft to aid trim.
4. The bilge system uses independent bilge pumps in all four pods to keep the bilges dry.
5. Two ship's service generator sets feeding power through independent switchboards provide for all electrical needs. Each generator set is capable of supplying 100% of the vessel's electrical power. A shore power connection provides power at the dock.

Other ship systems included:

6. Five battery systems provide 24 VDC for engine start-up, generator start-up, and back-up power for essential navigation, propulsion/steering, machinery control systems, and switchboards.
7. The integrated bridge provides command, control, monitoring, and communication functions.

8. The heating, ventilation, and air conditioning system provides for crew and passenger comfort as well as engine combustion air and equipment cooling and ventilation.
9. The seawater system provides water for machinery cooling, drinking water via a reverse osmosis unit, and sanitary flushing.
10. Dual, cross-connected fire pumps supply seawater for the fire main system.
11. Four independent ballast systems permit filling and empting the four ballast tanks to keep the vessel trim. The aft and forward ballast system are cross connected port and starboard to permit ballasting operations if there are pump failures.
12. A freshwater system supplies potable water from two independent storage tanks. The tanks can be filled at the dock or from a reverse-osmosis-type water maker at sea. Ultraviolet units assure the water is potable.
13. Two independent, cross-connected air compressor systems supply ship service air for sea chest blowouts, shaft breaks, and utility air.
14. The two engine/generator rooms are protected with an FM200 fire extinguishing system.
15. A complete complement of life saving equipment is aboard, including five inflatable life rafts, two Jason's Cradles, and life rafts and life buoys.
16. The sewage system consists of two independent marine sanitation devices and black and gray water tanks.
17. Separate tanks are used for lube oil, waste oil, and waste oily water.
18. A public address and telephone system is augmented with an entertainment system that pipes live television, recorded video, or music throughout the ship.

2.12.7. Noise

Limits for HSC craft under the IMO Code are nominally 75dB in the passenger areas and 65dB in the wheelhouse. This was a challenge for a craft with an arrangement whereby the engines are adjacent to the accommodation and wheelhouse rather than low down within the hulls. Care was taken in a number of details of the final design. Extra divisions were incorporated so that there was a double leaf effect between the engine room space and the wheelhouse and after accommodation. All rotating machinery was flexibly mounted. The exhausts were taken outboard rather than discharged under the raft. Care was taken to avoid short circuit situations that could carry structure borne noise into the wheelhouse. The fire insulation covering all the engine room internal surfaces down to 300mm below the light waterline creates good damping of the aluminum structure and of course some absorption attenuation.

The results on trials at full power were very satisfactory as the following figures show:

Wheelhouse	65 dB
Aft saloon	68–72 dB

The level in the engine rooms is particularly low—normally with a high-speed diesel installation one would expect 105–10 dB. The noise level in the Engine rooms measured 103 dB. The level in the saloon is also better than one generally expects to find in a small HSC passenger craft. The crew cabins come outside the HSC regulations since, while underway, the crew are attending to their duties. The cabins are for occasional overnight use while at anchor.

2.12.8. Ride control/steering

The customer's desire for improved sea-keeping required that the maximum significant vertical acceleration would be less than or equal to 0.5 meters per second squared, root mean square (RMS) on average for a period of 24 hours. This acceleration would be measured at the passenger seat furthest from the center of gravity of the vessel. To achieve this sea-keeping requirement a unique ride control system augmented the already superior performance of the SWATH-type hull form.

The patented integrated steering and motion stabilization system includes one rudder/stabilizer assembly mounted on the tail cone of each aft hull and one canard assembly mounted inboard near the bow of each forward hull. The rudder stabilizer assembly consists of two tapered control fins that are mounted on opposite sides of the tail cone and connected by a common shaft that runs transversely through the tail cone (see Fig. 2.12.17). This whole assembly is canted down from the vertical to permit the dual functions of steering and motion stabilization.

Fig. 2.12.17 Rudder/stabilization

The canards and rudder/stabilizer control surfaces provide forces and moments that can stabilize the vessel motions and steer the vessel as shown in Table. 2.12.1.

The integrated steering and stabilization system incorporates these control surfaces in a feedback control system to perform the following functions when the vessel is underway at forward speed:

- Operator helm steering using follow-up proportional control of the rudder/stabilizers
- Automatic heading (i.e., autopilot) control
- Automatic trim, and list stabilization
- Automatic pitch, roll, heave, and yaw motion damping
- Operator trim or list control using a two-axis joy stick

In addition, the integrated steering and stabilization system includes the following manual and local control functions:

- Helm steering using non follow-up control of the rudder/stabilizers
- Local steering using non follow-up control of the rudder/stabilizers
- Aft steering stations in each aft pod
- Manual control of the canards
- Local control of the canards

Operation of the integrated steering and stabilization system is provided in the automatic and manual modes from the forward control station in the wheelhouse. Local control of the rudder/stabilizers is provided from the thruster/steering gear rooms. Local control of the canards is provided from the canard rooms.

Sea-trial data confirmed the operation and performance of the integrated steering and stabilization system. During one such trial in the Philippines, the vessel operated around 19kn in sea state 5, while experiencing accelerations less than 0.4m per second squared RMS, fig. 2.12.18.

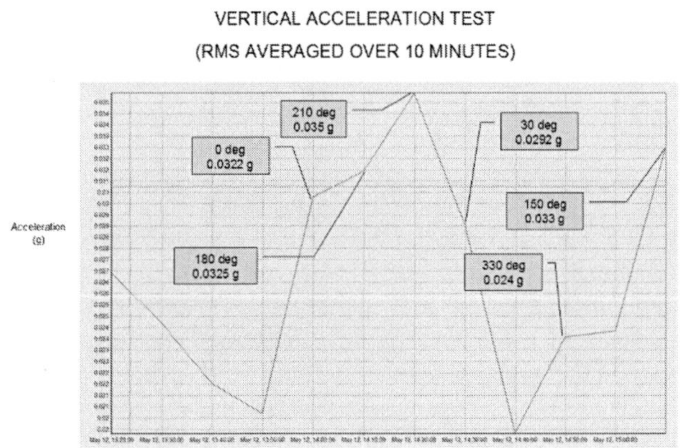

Fig. 2.12.18. RMS values

2.12.9. Conclusions

In conclusion, the SLICE® hull design has combined the superior sea-keeping capability of a SWATH with high-speed operability. This makes it ideal for many government, military, and commercial applications.

Despite such complex hull geometry, it is possible to satisfy the requirements of the HSC code. Careful attention to detail in the early design stages mitigated the technical risks, while maintaining the design intent of SLICE®.

Designing a small fast vessel with sea-keeping that is superior to that of vessels that are significantly larger is possible. The sea-keeping of SLICE® exceeded all expectations without the need to reduce speed in higher sea states. The extremely low waterplane area coupled with lifting surfaces demonstrates that it is possible to control the motions with ease.

SLICE® is a technology that allows SWATH vessels to obtain high speeds without sacrificing efficiency. The inherently small size of SLICE® identifies it as an ideal candidate, meeting the high-speed and low-motion requirement of commercial, government, and military markets.

Table 2.12.1. Matrix of rudder and stabilizer controls

	Trim stabilization	Pitch motion damping	Heave motion damping	List stabilization	Roll motion damping	Steering	AP heading stabilization	Yaw motion damping
Combined canard deflection	x	x						
Differential canard deflection				x	x			
Combined rudder/stabilizer deflection	x	x						
Differential rudder/stabilizer deflection				x	x	x	x	x
Combined canard and rudd/stabilizer deflection			x					

References

[2.12.1] T. Schmidt, "Technology for the 21st Century, SLICE®: A Revolutionary New Ship," United States Japan Natural Resources/Marine Facilities Panel, 17 May 2000.

[2.12.2] N. Warren, J. Kecsmar, & A. Crawford, "Revisiting a SWATH 7 Years On," RINA. SURV 6, Conference, 17–18 March 2004.

[2.12.3] DNV HSLC Rules, 2005.

[2.12.4] N. Warren, J. Kecsmar, & N. Sims, "Waterjet Propulsion a Shipbuilders View," RINA. Conference, 1–2 December 1994.

[2.12.5] J. Kecsmar & R. A. Shenoi, "Some Notes on the Influence of Manufacturing on the Fatigue Life of Welded Aluminium Structures," *Journal of Ship Production* 20 (3), August 2004, SNAME publication.

[2.12.6] J. Kecsmar, "Fatigue of Welded Aluminium Structures in High Speed Marine Vessels," thesis for master of philosophy, Southampton University, England.

[2.12.7] A. Fyfe, G. L. Hawkins, R. A. Shenoi, W. G. Price, P. Temarel, P. J. C. L. Read, & J. Kecsmar, "Fatigue Performance of Welded Aluminium Tee Connections," Fast '97.

[2.12.8] J. Kecsmar, "Guidance on How Weld Quality Influences and the Fatigue Life of Aluminium Structures," Autumn Meeting of JASNAOE, Kobe, Japan, November 2006.

CHAPTER 3

HYDROFOIL CRAFT

3.1

HYDROFOIL CRAFT

QIJUN NI

China Ship Scientific Research Center, China

Introduction

As a form of high-performance ship, thousands of hydrofoil crafts have been built and operated all around the world. The hydrofoil is a ship with an airfoil-like lift device mounted on its bottom. During navigation, the hydrofoil can lift the hull above the water and reduce the resistance, so the hydrofoil craft can achieve a high speed.

The hydrofoil is designed as a wing with a camber. When the flow past the wing is at a certain angle of attack, the pressure on the above surface of the wing drops and the pressure on the down surface of the wing rises; this causes the difference of pressure on both sides to provide the lift for the hydrofoil. Because the density of water is 800 times that of air, the hydrofoil with a small area can provide greater lift in water than in air. When the hydrofoil is close to the free surface, the lift on the hydrofoil decreases dramatically, this is called the "shallow dipping effect." And cavitation will have emerged on the suction of the hydrofoil when the craft sails at a very high speed, which will severely influence the lift and resistance of hydrofoil crafts. Thus, hydrofoils should be designed carefully to enable a high speed craft.

The different types of hydrofoil craft can be distinguished by the type of hydrofoil and its layout on the bottom of the ship. The single hydrofoil craft has only one hydrofoil mounted below the bow and the two-hydrofoil can lift the craft's hull above the water. Two-hydrofoil craft can be adopted for a plane layout, a canard layout, or a tandem layout. There is also a triple- hydrofoil craft.

According to mechanisms of keeping stability, the hydrofoil consists of a self-stabilized hydrofoil, a self-control hydrofoil, and a self-stabilized with self-control hydrofoil. The self-stabilized hydrofoil is a kind of surface-piercing hydrofoil that uses the change in wetted area of the side wing to keep stability. The shallow dipping self-stabilized hydrofoil uses the "shallow dipping effect" to obtain stability. And the hybrid self-stabilized hydrofoil has both a wetted surface effect and a shallow dipping effect. Inland ships usually adopt the self-stabilized hydrofoil, and sea ships often use the self-control hydrofoil to improve sea-keeping performance. The sea craft usually set some lift control devices on the hydrofoil to adjust lift continually in order to keep the hull above water, which can lessen the disturbance of waves and maintain the stability of hydrofoil craft at sea.

The most famous international self-control hydrofoil passenger ship is the JetFoil, designed in America, whose patents have been sold to a Japanese shipyard that produces two crafts per year. CSSRC developed its own technique to design hydrofoil crafts. By using the same engines and waterjets as a JetFoil, the engineers of CSSRC designed and manufactured the PS-30 series of self-control hydrofoil passenger ships. The ships *Bei-Xing* and *Nan-Xing* sail on the route between Hong Kong and Macao. The displacement is 3 tons larger than that of a JetFoil; furthermore, passenger seating is increased by 12%, the sailing speed is 1 knot faster than a JetFoil, and its performance at take-off has also been improved. For details of the PS-30 refer to *Research on Developing a Fully Submerged, Automatic Controlled and Waterjet Propulsion Hydrofoil Craft, Model PS30*.

The triple hydrofoil self-stabilized passenger ship is a new kind of ship that developed quickly in recent years; in particular, the fully-transparent cruise ship is of note. It allows passengers to sightsee conveniently as the roof of the cruise ship is fully transparent. Considering the ship's longitudinal strength, the layout of the triple hydrofoil is applied, which is beneficial for the strength of the ship's structure. The layout and interactions of the triple hydrofoils result in the lowest resistance and the greatest lift. Some small triple hydrofoil ships are currently operating in the upper reaches of the Yangtze River in China, on which see the paper "Research of New Type Hydrofoil Craft with Triple Hydrofoils."

Because the hydrofoil's hull is above the water when it sails at an economical speed, resulting in less speed loss in waves, good sea-keeping performance, and good economics, hydrofoil crafts are operated in many countries. The water in the upper reaches of the Yangtze runs very quickly and there are too many ships sailing. Thus, a hydrofoil passenger ship is highly suited to being applied in the river's upper reaches. The *Yuan Zhou 2000* inland hydrofoil passenger ship is widely recognized because of its excellent performance. Due to the completion of the improvements to the highways and the realization of the importance of energy saving, hydrofoils can play a very important role in river transportation on the Yangtze. This may transform the concept and mode of water tourism, and may be helpful to developing the overall industry structure. More details can be found in the paper "Development of *Qinzhou II*, All-Transparent Tri-hydrofoil Sightseeing Boat."

With varied applications in rivers and oceans, the research and design of hydrofoil craft have been fully developed. The cavitation of the hydrofoil should be taken into account when designing a high-speed hydrofoil. On a folded hydrofoil, see the papers "A 2-D Anti-Cavitating Hydrofoil Design by Section Reshaping" and "The Introduction of a New High-Speed Passenger Ship—Hydrofoil Catamaran *Super Hydrofoil-40*." *Super Hydrofoil-40* is a successful example of a hydrofoil passenger catamaran that combines a catamaran with a hydrofoil.

At the same time, many countries are accelerating their research on composite hydrofoil craft, including channel hydrofoil craft and multihull hydrofoil craft. See the papers "Development of Modern Hydrofoil-Assisted Multihulls" and "Motion Prediction of a High-Speed Hybrid Hydrofoil Catamaran."

Although catamarans have so many advantages, their natural frequency of rolling is close to natural pitching frequency, which easily results in the coupling motion (synchronism) of rolling, pitching, and heaving. The coupling motion is very harmful to the ship and one of the leading causes of seasickness. For catamarans, it is hard to combat coupling motion through conventional anti-roll methods (such as bilge keels and anti-rolling fins). The hydrofoil stabilizer is a new measure to effectively improve the motion performance of high-speed catamaran on the sea.

The present author presents the bottom hydrofoil stabilizer system. Through analysis, we know the features of the bottom hydrofoil:

1. The hull bottom split hydrofoil is installed on the bottom of the ship, which satisfies the requirement of installation depth and effectively isolates wave interference. Thus, the stabilizer effect of the hydrofoil is ensured and the range of sea conditions to which the hydrofoil can adapt is expanded.
2. The hull bottom split hydrofoil installed on the two sides of the bottom of ship will not only take full advantage of the transverse space, but also increase the rolling arm to the hilt, which will redouble the roll damping (and may reach more than five times that of the linked hydrofoil system). A hydrofoil with the simplex anti-roll or anti-pitch function will be promoted as a hydrofoil containing anti-roll and anti-pitch functions, and it will play an

important role in restraining the synchronism that is most harmful to the catamaran.

According to classification, a single hydrofoil craft may only use a hydrofoil as an accelerating method for a high-speed planing boat by adding a set of hydrofoils on the bow.

When the hydrofoil craft slides at high speed, the hydrofoil will raise the bow up over the water while the stem is planing, which will decrease resistance, increase navigational speed, and improve sea-keeping performance. This kind of single hydrofoil craft is simple and practical. Please refer to the paper "Two Kinds of High Speed Passenger Ship with Bow Hydrofoil." When the hydrofoil is arranged as a multihull-ship, it greatly improves the ship's performance. However, the hydrofoil system and its arrangement will take years to be researched and developed to improve its rationality. Please refer to the paper "Development of Modern Hydrofoil-Assisted Multihulls."

3.2

SINGLE HYDROFOIL AND DOUBLE HYDROFOIL CRAFT

3.2.1. Two Kinds of High-Speed Passenger Ship with Bow Hydrofoil (2005)— Zhang Daxiong (China Ship Scientific Research Center, China)

Abstract
Indonesia is a country of one thousand islands, which has prompted it to develop high-speed passenger ships to ensure its economic development. Their passenger ships need to take into account various special needs: These ships can't install waterjets due to the large quantities of floating garbage in the water. The ships must be able to sail in shallow water and be able to "bow dock" at quays. And many passengers carry bicycles or motorcycles, and so on. To satisfy all those difficult requirements, we designed two types of high-speed passenger craft, which can carry about 250 and 200 passengers, respectively. With bow hydrofoils, resistance at high speed is greatly reduced, service speed is improved, and comfort in waves is also increased. Customers are very satisfied with our designs.

Indonesia is a country of one thousand islands, where ships are one of the most important transportation tools. Alongside Indonesia's economic development, the country demands more high-speed passenger crafts. Because Indonesia is a developing country, there are many special requirements for high-speed passenger crafts that will satisfy local people.

We designed two types of high-speed passenger craft, each with different requirements, for an Indonesian company.

The first is the VS-250 type (see Fig. 3.2.1.). The principal dimensions and characteristics of the VS-250 are as follows:

L_{oa}	36.5m	Displacement	100t
B_{oa}	7.2m	Passengers	244
Mean draught	1.1m	Crew	5–6
Max. draught	1.8m	Main Engine power	3 × 1,050kW
Max. speed	37.5kn		

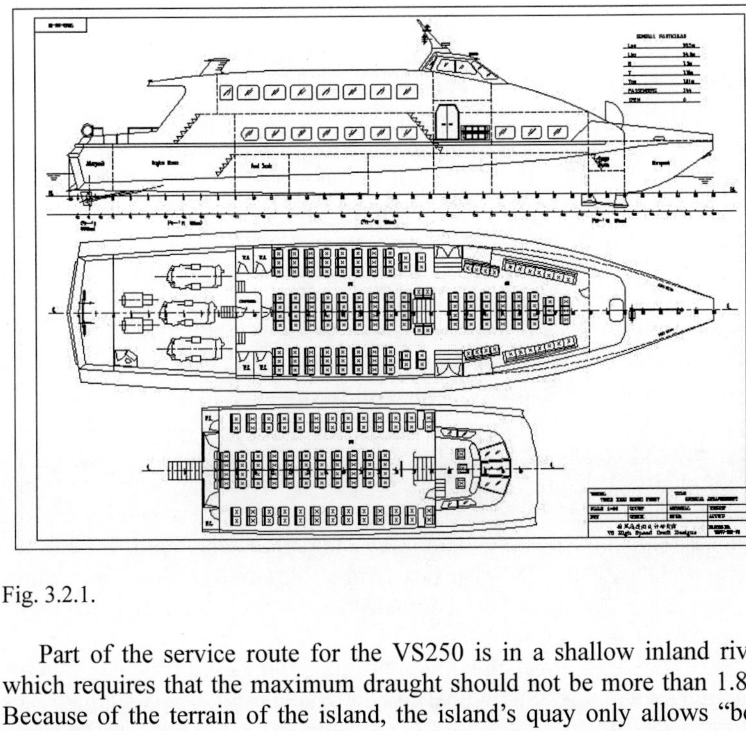

Fig. 3.2.1.

Part of the service route for the VS250 is in a shallow inland river, which requires that the maximum draught should not be more than 1.8m. Because of the terrain of the island, the island's quay only allows "bow docking"; thus, a square bow deck was adopted. Because some passengers usually ride motorcycles or bicycles, and will carry their cycles when they are on board, a wide stern deck with fixed rails for cycles was adopted. And as passengers needed to push their bicycles forward from stern to bow to get off the craft, we designed a 0.75m-wide walking lane and handrails at both sides of the craft. The endurance journey of the ship was about 300 nautical miles. The ship could run at an overload condition for service on the long journey, and an auxiliary fuel tank was ready for a further 6t of diesel.

Another difficulty in design is how to attain a high speed. According to the resistance data of regular high-speed crafts, it is impossible to reach maximum speed above 32kn with the power of existing main engines. In order to improve the speed, we particularly focused on carried resistance experiments on ship models in the towing tank. After stages of improvement and optimization, the resistance was reduced about 12%. After the data of the model test was correlated to the full ship and new propellers were designed, the maximum service speed could reached 32kn. To improve the speed of the ship, we developed a new layout by adding a hydrofoil mounted at the bow, which improved the speed-power performance dramatically. When the bow hydrofoil undertakes 20% displacement, the bow was raised by a small amount, the trim angle was increased to 2.2°–2.5° and the length of the wetted stern was shortened. The ship would be in a planing state on the water and the resistance was reduced dramatically. If the data of the model test correlated to the full ship, and a new propeller was designed, the forecast maximum service speed could reach 37.5kn. The owner of the ship was very satisfied by these renovations. Furthermore, in the longitudinal regular waving test (the testing tank can produce longitudinal waves only), the vertical acceleration at the middle of the ship was reduced by adding a bow hydrofoil, which means that the bow hydrofoil could improve comfort in waves. Finally, the layout with the bow hydrofoil was recommended to the owner of the ship, which could improve not only the speed of ship but also the ship's sea-keeping performance. But this kind of hydrofoil had to be designed carefully: its structure strength needed to be evaluated using a special method. The hydrofoil also needed to be built and tested precisely.

This craft was equipped with three MTU/DDC 16V 2000 M70 diesel engines and a ZF BW255 gear box, propelled with a straight line oblique axis propeller. All the whole equipment was robust with high reliability and designed with advanced technology.

The craft had an aluminum alloy hull with a longitudinal framing structure system; the longitudinal and horizontal trusses were supported by the good rigidity and stability of the aluminum alloy pipes. The locations of stress concentration were strengthened partially and were fully in compliance with the Rules and Regulations for Construction and Classification of Sea-Going High Speed Craft.

The craft below the freeboard was divided into eight watertight tanks to satisfy the requirement for two cabins to be light and not damage stability. The engine cabin was equipped with fire separation, a smoke-detecting alarm system, and stationary type CO_2 fire extinguishers. This boat satisfied the restricted high-speed passenger ship requirements for fire prevention, fire extinguishing, safeguarding, and safety. To reduce the vibration and noise of the ship, a flexible support system was installed in the main engine and sub-engine, and noise absorbing and separating system was equipped in the engine cabin. The equipment and material used in all the cabins was very light, durable, and caused less noise, which guaranteed the comfort of the passengers.

After the VS250 was built and was in operation, the tips of the propeller (the lowest part of the boat) sometimes still touched the bottom of the river and were damaged. The owner of the ship then required a new type of high-speed passenger craft with a lesser draught—the requirements were for the lowest point of the craft under the water to be within 1.6m yet for the craft to have the same high speed, sea-keeping, and economy. This was a new challenge for the designer.

Obviously, the oblique axes propeller could not be used anymore. However a waterjet system could not be used either because there was so much floating garbage on the water. And an air propeller would be low efficiency and very noisy. Finally, we designed the VS200 shallow-draught high-speed passenger craft by using a surface-piercing propeller (partially submerged propeller).

The principal dimensions and characteristics of the VS200 are as follows and are shown in Fig. 3.2.2:

L_{oa}	32.0m	Max. speed	35kn
L_{wL}	27.4m	Main engine	4 × 790kW
B_{oa}	7.09m	Passengers	VIP 17
T	1.55m	First cabin	38
Displacement	74.5t	Economic cabin	132

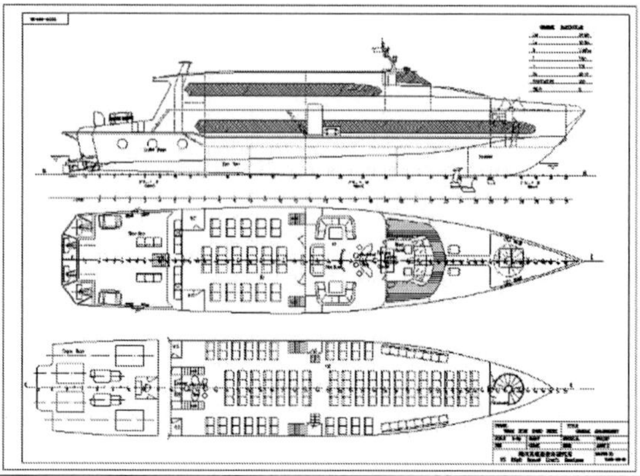

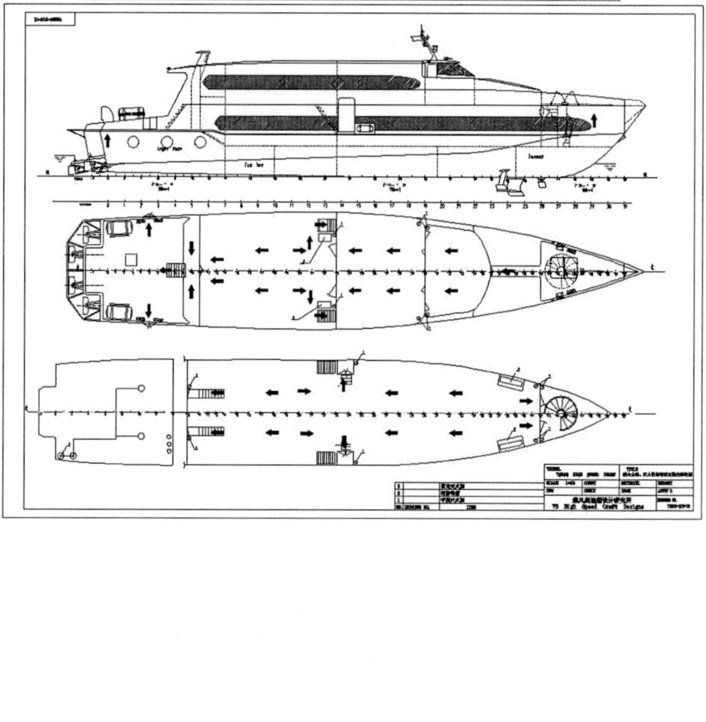

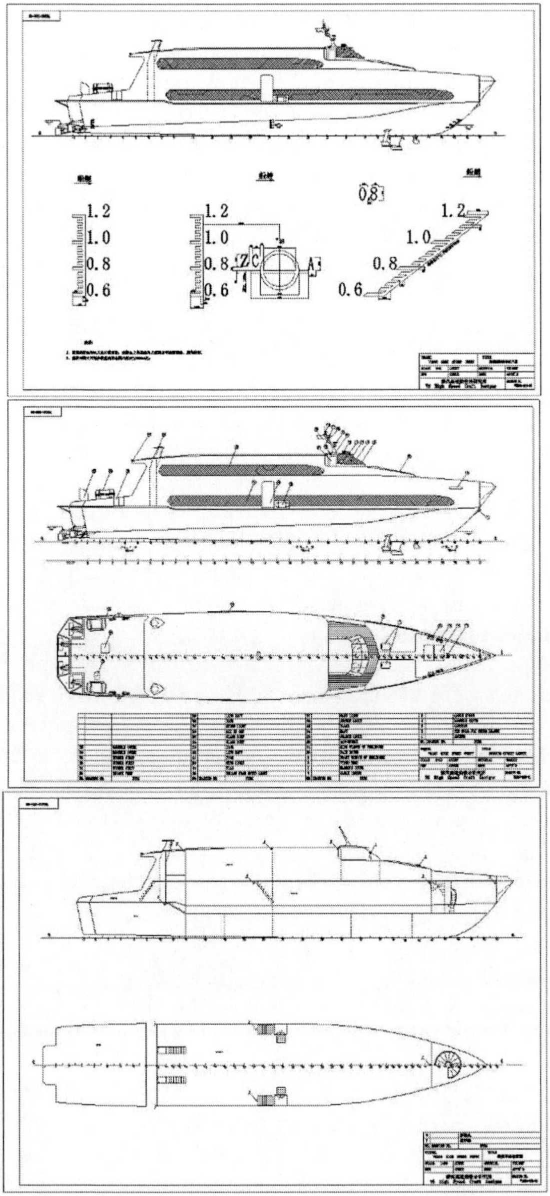

Fig. 3.2.2.

The hull lines referred to the VS-250. Most of the craft bottom adopted a straight V shape, which was easy to build and had high hydrodynamic efficiency in planing conditions. The bow line was increased; the bow deck became as high as 4.1m, and the behavior in the seaway was also improved. Because some quays could only allow “bow docking,” the bow was still square shaped. This non-streamline shape was not as beautiful and was disadvantageous in reducing wind resistance. Then, the elevating bow was designed and equipped according to the needs of the ship owners.

To reduce resistance and improve speed when it was sailing, the auxiliary bow hydrofoil was mounted in the bow of the ship. The curve of resistance versus speed for the ship were provided, which could be used for the design of the surface-piercing propeller.

The selection of a power system for the high-speed passenger boat was very important, which should include economic efficiency, reliability, and practicability. A very careful and serious investigation and comparison was carried out.

Many high-speed passenger boats use two large and powerful diesel engines, such as the MTU 12V4000 M70 with total power of 3,480kW, which is high performance and can be used for a very long time. Two main engines could easily be arranged in the engine room. However, the price would be more than twice as much as that of four MTU 12V2000 M70 with a total power of 3,150W. And the price was even more expensive if the gas turbines were used. If three MTU 16V2000 M70 engines with a total power of 3,150kW were used, it was also very expensive. As the middle surface-piercing propeller would produce side force, the course of the ship would be kept by rudder control, and the resistance of the ship was increased. The design using three engines was abandoned. Finally, four MTU 12V 2000 M70 engines were adopted by the designers, which needed four Sea Rider Ti30 surface-piercing propellers. These details in the design satisfied the requirement for sailing in shallow water. Four main engines with surface-piercing propellers were very economical with high hydrodynamic coefficiency and high reliability. The only difficulty was how to arrange the four main engines in the engine room.

Although it was a difficult task, the main engines and propellers were installed at the stern of the ship, which made the arrangement of the shafts easy. In the middle of the stern a special hull was adopted that extruded to the aft of the ship. Thus the outside two engines and propulsion systems were 1m in front of the inside two engines and propulsion systems. This allowed enough space for rudder manipulation by turning the propeller systems and the arrangement of the cabins.

The two gear boxes inside connected with the main engines. These were fixed to the boat rigidly. Thus the gear boxes could receive the thrust of propulsions. The connection between the gear boxes and the main engines was elastic, and the middle main engines could be supported elastically with a vibration-reducing structure. The outside propulsion systems were connected using thrust bearing, and then were connected to the long shaft with CENTA DISC carbon fiber. Finally, the outside propulsion systems were connected to the gear boxes and the main engines. The designs could satisfy the requirement of a 1m displacement between the outside and inside main engines, which lost less power in the propulsion system with and weighed less compared with the universal joint shaft system. The arrangement of propulsions avoided the resonant vibration from transverse vibration and torsional vibration, which could satisfy and adjust the vibration characteristics of the shaft system.

The bow hydrofoil was also equipped to the VS200 high-speed passenger boat. The higher sailing speed could be reached by supporting the bow hydrofoil and stern planing surface. The maximum draught was less than 1.6m. The bow hydrofoil was a hollow structure welded with stainless steel. The durable structure of the hydrofoil was lightweight. The hydrofoil was mounted to the bottom of the bow, which could be adjusted to fixed location during the sea trial. Steady hydrodynamics could be achieved by the adjusted hydrofoil at a particular depth in the water, which could automatically adjust the boat's navigation status. The bow hydrofoil not only could reduce the resistance of the ship but also provided extra damping force in waves, which could retard acceleration and swing. The design of the hydrofoil for the VS-200 was simple, reliable, and effective.

3.2.2. Research on Developing a Fully Submerged, Automatically Controlled, and Waterjet-Driven Hydrofoil Craft, Model PS30 (2000)— Zhang Daxiong (China Ship Scientific Research Center, China)

3.2.2.1. Introduction

In 1973, the first passenger hydrofoil craft, the JetFoil, with a fully submerged and automatically controlled hydrofoil system and a waterjet, was completed in the United States. Its sea-keeping performance at a high speed of 42kn was extremely good, providing a high level of comfort to passengers. It was a milestone in high-level technology in the development of passenger hydrofoils. The Far East Hydrofoil Co. Ltd. (FEHC) ordered

14 hydrofoil crafts of that kind, which were operated on the route between Hong Kong and Macao.

In 1990, the China State Shipbuilding Corp. (CSSC) was authorized by FEHC to build two passenger hydrofoil crafts similar to a JetFoil. However, the owner hoped that the displacement, the number of passengers, and the craft's speed could be increased and that take-off performance could be improved by using the same engines as a JetFoil. Obviously, it was a serious challenge. CSSC attached great importance to the manufacture of high-tech crafts and the institutes and factories in the CSSC group organized together and cooperated with one another to complete this difficult challenge.

To satisfy the owner's requirements, the research projects were launched in the China Ship Scientific Research Center (CSSRC); these included optimization of the hull form, design of the hydrofoil and struts, and optimization of the intake waterjet. Numerical simulation of the strength of the hull and hydrofoil and the fatigue life of hydrofoil structures had been carried and compared with the rules. The CSSRC's engineers accomplished the projects and satisfied the owner's requirements.

Between 1994 and 1995, two new craft, model PS-30, were built and successfully completed sea trails. They were named *Balsa* and *Praia* (Fig. 3.2.3). The new craft had a 3t displacement and had 23 more seats and could go 1kn faster than a JetFoil. These two new crafts together with 14 JetFoils satisfactorily operated on the route between Hong Kong and Macao.

Fig. 3.2.3. PS-30 hydrofoil craft

Principal dimensions and characteristics:

L_{oa}	29.10m
B_{oa}	9.20m
Beam	8.60m
T	4.51m
Δ	118t
Service speed	43kn
Passengers	294
Crew	9
Main engines	Allison-501 KF, 2 × 3184kW

3.2.2.2. Take-off performance

Sometimes JetFoils had difficulties taking off in rough sea during the summer, because the high-temperature air made the main engines produce lower power. In such cases, the craft had to return to the harbor and to begin sailing again after reducing passenger numbers. The engineers of CSSRC investigated these phenomena and devised the following improvements:

1. Lengthening the ship length to increase L/B
2. Fitting a fixed aft hydrofoil instead of the retractable aft hydrofoil, and removing the slot at the stern
3. Designing a new form of stern hull
4. Strengthening the effect of the auxiliary intake flow of the waterjet propulsion system
5. Modifying the program that automatically controls the flaps

Methods 1–3 were designed to reduce the peak resistance of the hull while taking off. During take-off, the majority of the resistance was due to the ship hull, which was much greater than that of the hydrofoils, the struts with the pods, and the air resistance of the superstructure. Seventy-three model tests were carried out to investigate and analyze the effects on the resistance with different combinations of flap angles, changes in LCG position, and displacement.[3-2-1] Finally, the perfect design was achieved: the peak resistance at take-off was 5.3% lower than that of a JetFoil with 3t more displacement. The peak resistance would be 7.9% lower (Fig. 3.2.4 and Fig. 3.2.5) than that of a *JetFoil* with the same displacement.

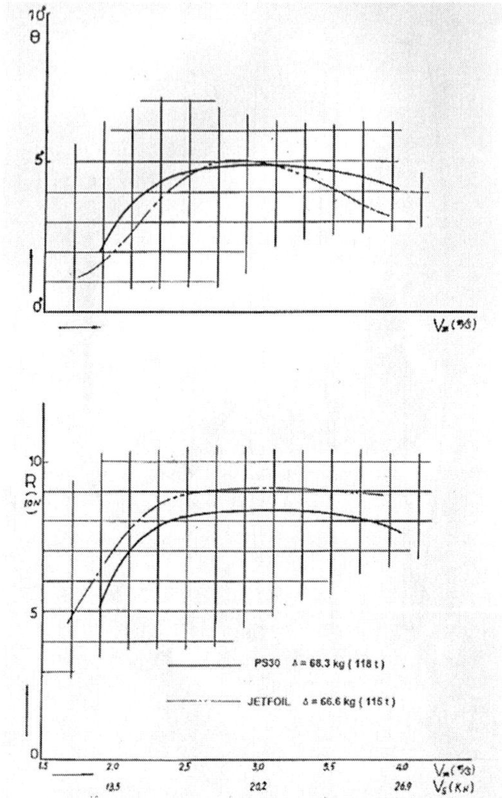

Fig. 3.2.4. Resistance and trim at take-off

Fig. 3.2.5. The test model

3.2.2.3. Foil system

Improving the hydrofoils' performance was the key to increasing the ship's speed and improving the other dynamic characteristics. The aft hydrofoil system was redesigned and was different from that used for the JetFoil, and two struts were adopted for the aft hydrofoil instead of the three used for the JetFoil.

(1) The section of the aft foil

The hydrofoil section was designed using Eppler's method.[3-2-2] A favorable pressure distribution on the surface of the hydrofoil could be achieved, which would enlarge the cavitation bucket and delay the cavitation inception at the design angle of attack (Figs 3.2.5 and 3.2.6). The separation in the boundary layer was also controlled at design conditions, which showed that the cavitation inception at off-design conditions could also be delayed. It is suitable to pitch in wave conditions and to operate the flaps. The section form is shown in figure 3.2.5. The thickness ratio was 0.08, which was thicker for two struts rather than three.

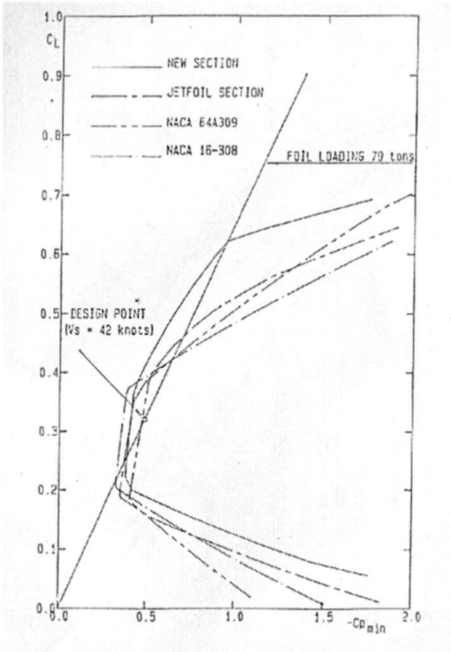

Fig. 3.2.6. Cavitation buckets at design angle of attack (C_L=0.32, Vs=42 kn)

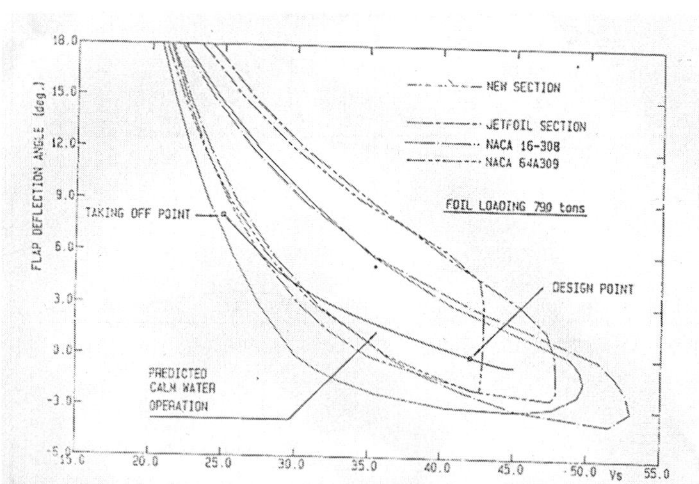

Fig. 3.2.7. Comparison of cavitation limits of the four section

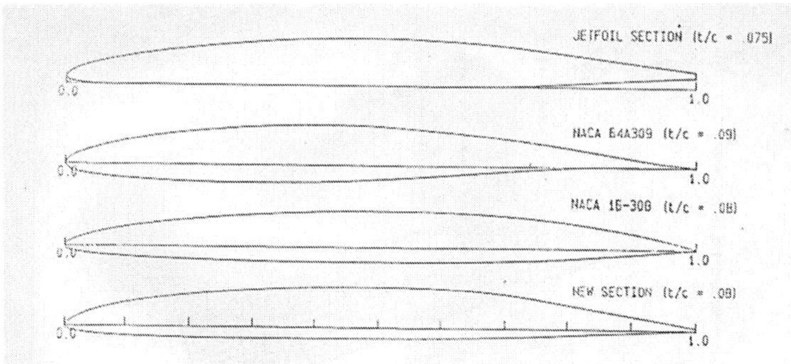

Fig. 3.2.8. The shape of the sections

To check results of the theoretical design, the cavitation inception tests of 2-D foils with these sections were conducted in a cavitation tunnel, and the lift and the drag of these 2-D foils were measured and recorded.[3-2-3] The sea trial of the PS30 at light load conditions attained a speed of 50kn without any abnormal phenomena due to cavitation.

Because there was not a sufficiently large water tunnel with a free surface in China, it was impossible to conduct the performance tests on the isolated three-dimensional hydrofoil system with the combination of the foil, the struts, the flaps, and the pods. Numerical simulations of the pressure distribution on the aft and fore hydrofoils system were carried out to analyze their performance. The simulations were done by using the program DSD, based on the surface panel method,[3-2-4] which could satisfy our needs in designing hydrofoils. The numerical results could provide the pressure distribution on the surfaces of the systems (Figs 3.2.7 and 3.2.8). It shows that there was a suction peak on the designed fore foil and multiple suction peaks on the lateral side of the pod due to the interactions on the low pressure induced by the foil, struts, and pod itself. However, the numerical results showed that no cavitation would occur within the designed speed range, which was later proven in the sea trial.

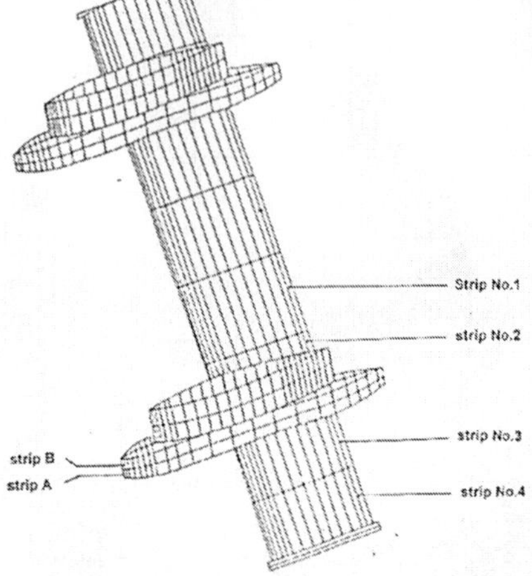

Fig. 3.2.9. The checkerboard of the aft foil system

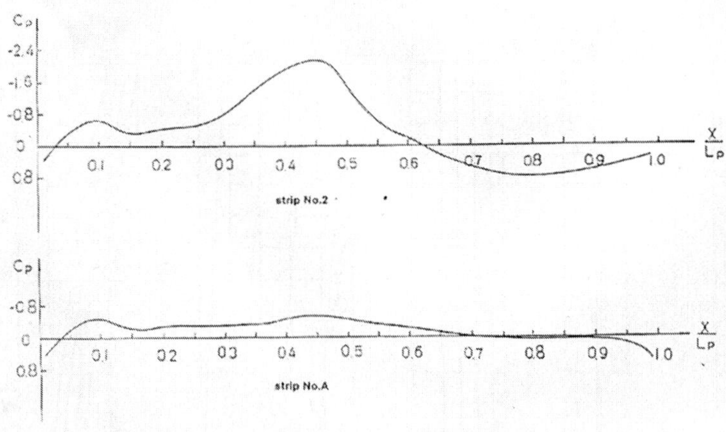

Fig. 3.2.10. Pressure distribution on the aft foil

(3) Model test on assembly of the hydrofoil, struts, and pod
The assembly model tests with the combinations of the foils with the flaps, the struts, and the pods were conducted to analyze the hydrodynamic characteristics of the designed foil system in the deep towing tank of CSSRC.[3-2-5] A dynamometer was specially designed and prepared for these tests. Fig. 3.2.11 shows the models in the deep water towing tank. The results are shown in Figs 3.2.12–14.

Fig. 3.2.11. The integrated model tested in the towing tank

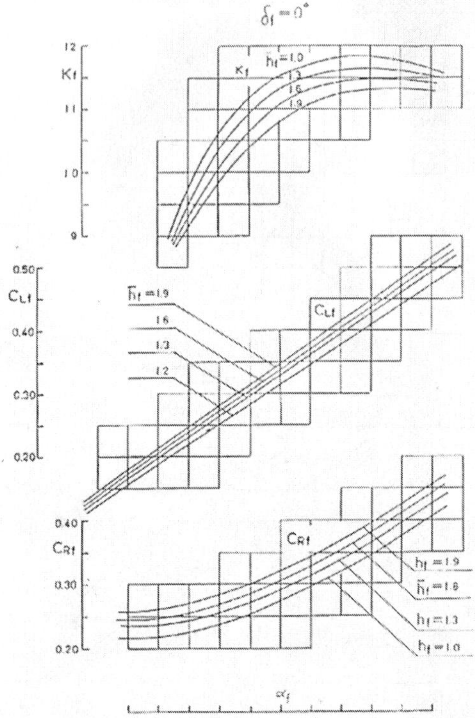

Fig. 3.2.12. The lift and drag coefficient of the fore foil

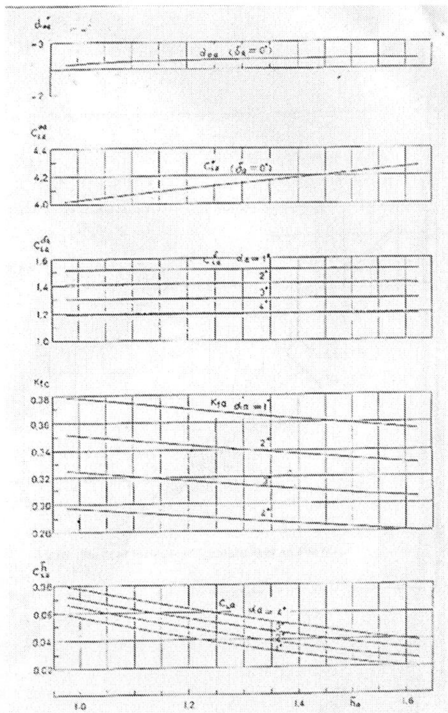

Fig. 3.2.13. The lift and drag coefficient of the aft foil

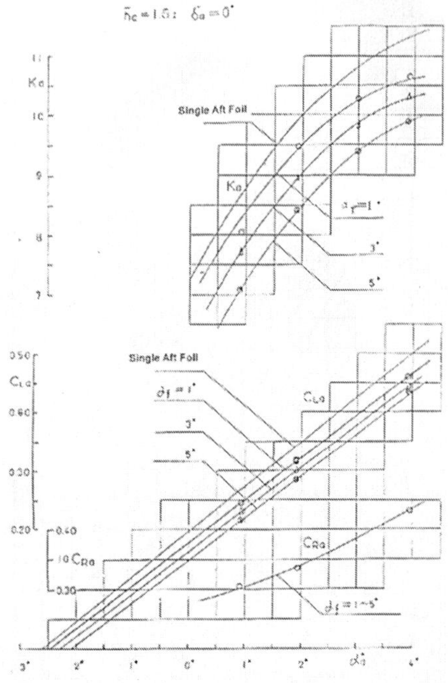

Fig. 3.2.14. The lift and drag coefficient of the aft foil running behind the fore foil

According to the data from model tests, the installation angle and height of the foils could be initiated, the mathematical model and the related parameters could be set up for control automation, and the program for controlling the flaps for take-off and for foil-born navigation were worked out. All these were checked and adjusted, then finally confirmed after the tests on the automatically controlled model in the towing tank.

The flaps of the foils were controlled by the hydraulic system. The hinge moment of a flap must be provided by the design of the hydraulic system and the selection of its mechanical components. Therefore, the tests for the hinge moment were conducted.[3-2-6] The results are shown in Fig. 3.2.15.

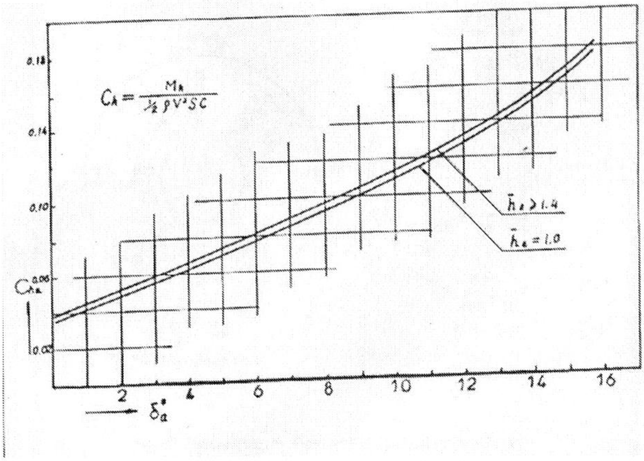

Fig. 3.2.15. The coefficients of the hinge moment of the aft foil flap

3.2.2.4. Waterjet propulsion system

The waterjet system of the JetFoil was equipped with a single inlet. The duct following the inlet was connected with two branches, each of which guided the water into the impeller of the waterjet. The singe duct and inlet system had a disadvantage that brought the inflow into a turbulent flow. The PS30 was equipped with two inlets and duct systems, with each connected to a pump independently, which could avoid making the inflow too turbulent. These smart designs could reduce the energy loss inside the ducts and improve the propulsion efficiency. The PS30's aft foil had two struts instead of three in the JetFoil, which could reduce resistance. A series of experiments were carried out to analyze the performance of the waterjet propulsion system.

(1) Intake

The tests in the large cavitation tunnel of CSSRC showed that no cavitation occurs on the designed intake at the condition that is the same as the craft running or taking off with a straight course. But when the speed of the craft exceeded 45kn and the pitching angle was greater than 4°, or the yawing angle was greater than 5°, cavitation would occur on the intake of the inlets.[3-2-7] It was advised that such an operation should be avoided as far as possible.

(2) Internal duct

There was no branch in the PS30's internal duct system, so a larger bend angle was designed that could reduce flow resistance and was beneficial to efficiency. Model tests in the wind tunnel showed that the total head loss was 10% lower at the designed speed, and 15% lower at take-off compared with that of the JetFoil (Fig. 3.2.16). The experimental results indicated that even if there was no guide vane installed at the bend, the increase of head loss was also neglected. To reduce the difficulties in fabrication and installation of the vanes, it was decided no guide vane would be installed in the duct of the PS30's waterjet.[3-2-8] The inflow to the pump of the PS30's waterjet was more uniform than was the JetFoil's (Fig. 3.2.17), which was beneficial to vibration and noise.

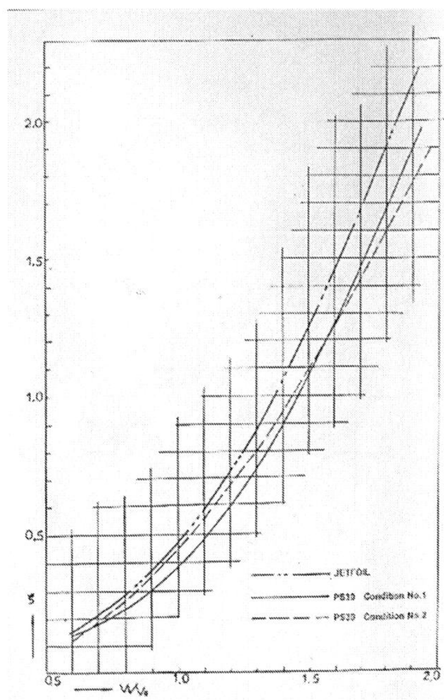

Fig. 3.2.16. Total head loss of the inlet duct

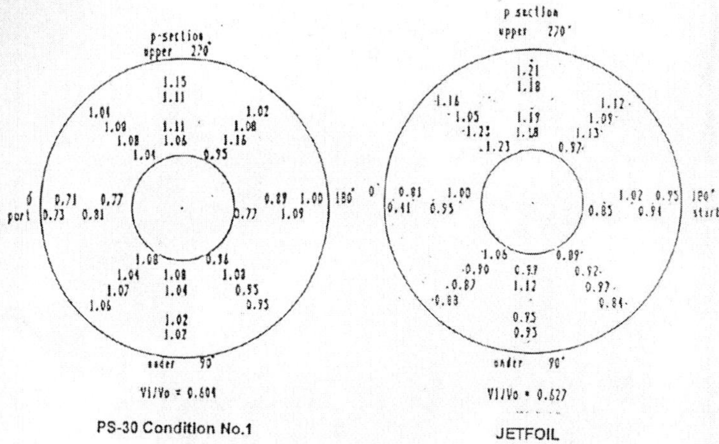

Fig. 3.2.17. Inflow field at the suction side of the pump impeller

(3) Pump for the waterjet

Because the operating condition of the 7PS30 was very close to that of the JetFoil, the PS30 was equipped with the same type of waterjet as the earlier craft. However, a cavitation test of the waterjet was carried out on CSSRC and the results showed that its specific speed could reach 3000, which could provide good cavitation performance.[3-2-9]

(4) Integration of the aft foil and the waterjet propulsion system

The performance of only one aft foil and one waterjet propulsion system was analyzed and given above. Next, an integration model of the aft foil, the struts, the pods, and the inlets was manufactured, and a special dynamometer was designed to measure the necessary data. In the model tests, a pump system was installed to simulate the flow rate running through the inlets in order to investigate their interactions. The integration model was as one unit to bear lift, drag, and thrust. And the lift and net thrust was measured (Figs 3.2.18 and 3.2.19). The net thrust was the result and included the interactions on all components, which had been compensated with the drag from the inlets. Beside the interference on the aft foil and the front foil, it was predicted that the speed of PS30 would reach 42kn.[3-2-10] The sea trial results showed that the prediction was only a little underestimated.

Fig. 3.2.18. The integration model

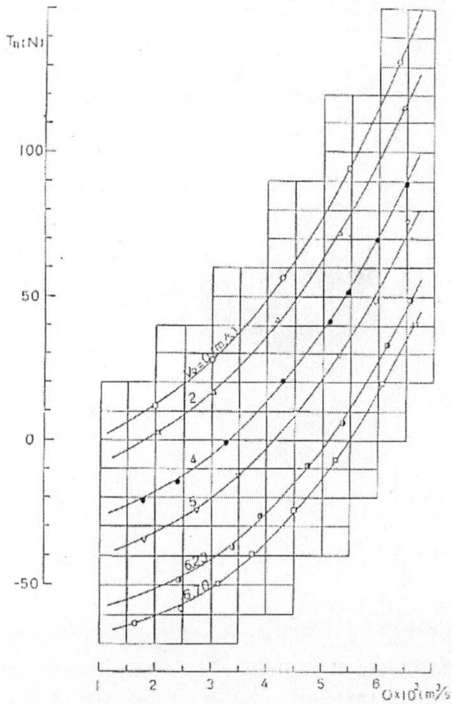

Fig. 3.2.19. The net thrust at different speeds (α=1.68°)

The experimental results indicated that the lift of the foil would be reduced by about 3.4% as the flow entered the inlets. Therefore, the attack angle of the foil would be increased to compensate for the loss of lift.

3.2.2.5. Test of the automatic controlled model

The take-off mode tests on PS30 were carried out under the following conditions: the flap angles of the fore and aft foil were set at certain fixed angles, which were different from the real status of operation. The automatic controlled model fully simulated the processes of real operation of taking off and foil-born running. The trim, hull clearance, flap angles, the submergence of the fore and aft foil, and the resistance were measured. And the optimized parameters of the automatic control system could be gotten from model tests, the data from which was necessary for adjusting and testing the automatic control system[3-2-11] (Fig. 3.2.20).

Fig. 3.2.20. The automatic control model test

3.2.2.6. Strength of the foil structure

By using the SAP5 code, the integration of the fore foil and the strut and that of the aft foil with the struts were analyzed by the finite element method under the given loads. The extreme cases had been considered which were maximum lift, maximum side force, and large inclination when one end of the foil was broaching. The analysis showed that the safety margin of each integration was larger than zero.[3-2-12] The case that assumed the fore or the aft foil integration was affected by some floating bodies with an assumed impact load, and was calculated by using the analytical method. It showed the safety margin was also greater than zero.

3.2.2.7. The fatigue life of the aft foil integration

Since PS30's aft foil system was different from a JetFoil's, its strength must satisfy requirements and its fatigue life must be checked. First, the dynamic analysis was done to get the free frequency (Fig. 3.2.21). Since the first harmonic of the free frequency when submerged was 5.22Hz, which was much lower than the wave's encounter frequency, the fluctuations of the lift could be treated as a quasi-steady case to determine the stress level in fatigue analysis. According to the sea condition on the route between Hong Kong and Macao, the N–S curve of the material, and Miner's theory of accumulated damage, the finite element method used to determine the stress of hot spots concluded that the fatigue life of the aft foil system was more than 10,000 hours in sea conditions with a wave height 1.2–1.5m and the factor of safety was 1.5. Therefore, the fatigue life was acceptable.[3-2-13]

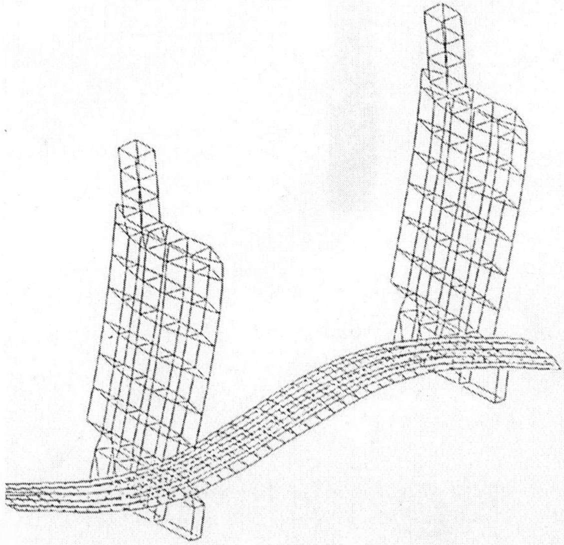

Fig. 3.2.21. The first mode of the aft foil system vibration

3.2.2.8. Conclusion

On the basis of a great deal of arduous works, the technical difficulty in the design of the PS-30 was solved. Two PS-30 craft named *Balsa* and *Praia* were successfully navigating on the route between Hong Kong and Macao, which proved that all the innovations in the research and the design of the PS-30 were successful. CSSRC engineers and researchers developed new test equipment and test technology that had supported the research and design of the PS30.

References

[3-2-1] Zhang D. X., "The Take-Off Performance of PS-30," *CSSRC Technical Report* no. 92174, 1992.

[3-2-2] Dang J. & Tang D. X., "Research and Design on the Hydrofoil Section," *CSSRC Technical Report* no. 91178, 1991.

[3-2-3] Dang J. & Tang D. X., "Design on the Hydrofoil Section and Hydrodynamic Characteristics Experiment of 2-D Model," *CSSRC Technical Report* no. 91179, 1991.

[3-2-4] Zhu D. X. & Li B. C., "Numerical Prediction of the Pressure Distribution and Hydrodynamic Characteristics of Hydrofoil-Strut-Pod System," *CSSRC Technical Report* no. 91612, 1991.

[3-2-5] Lan S. X. & Wang S. C., "Hydrodynamic Characteristic Experiment of Fore and After Hydrofoils of PS-30 Separate and Combinative," *CSSRC Technical Report* no. 91293, 1991.

[3-2-6] Wang S. C., "Test of the Flap Hinge Torque of Hydrofoils," *CSSRC Technical Report* no. 91316, 1991.

[3-2-7] Zheng Z., "Cavitation Test of Water Inlet of Ps30," *CSSRC Technical Report* no. 91322, 1991.

[3-2-8] Feng P. F. & Xa B. X., "Test of the Water-jet Inlet Channel of Ps30," *CSSRC Technical Report* no. 92082, 1992.

[3-2-9] Feng P. F., "Test of Water-jet Pump of Ps30," *CSSRC Technical Report* no. 91190, 1991.

[3-2-10] Li B. C., "Test of Hydrofoil Water-jet Inlet-Pod System of Ps30," *CSSRC Technical Report* no. 92030, 1992.

[3-2-11] Wang Z., "Test of PS30 Ship Model with Autocontrol System," *CSSRC Technical Report* no. 92043, 1992.

[3-2-12] Gan X. L., "Numerical Prediction of Hydrofoil Strength," *CSSRC Technical Report* no. 92595, 1992.

[3-2-13] Gan X. L. "Numerical Prediction of Hydrofoil Fatigue Strength for Ps30," *CSSRC Technical Report* no. 92594, 1992.

3.3

MULTIHULL HYDROFOIL CRAFT

3.3.1. Development of Modern Hydrofoil-Assisted Multihulls—

Dr Günther Migeotte
(CAE-Marine, Mechanical Engineering Laboratories, Stellenbosch University, South Africa)
and Prof. Dr Nikolai Kornev
(University of Rostock, Germany)

Abstract
Hydrofoil assistance has found application over the last two decades in a variety of high-speed craft including workboats, military craft, pleasure craft, and high-speed ferries. The application of these hybrid high-speed ships is discussed with reference to recent milestones in their development and application. Prospects for their future are discussed with reference to current design capabilities and challenges for future development.

3.3.1.1. Introduction

The development of hydrofoil-assisted catamarans has a long history spanning back to the early 70s and the first displacement catamarans to be built, such as the *USNS Hayes*,[3-3-1] which used hydrofoil-assistance to rectify its poor sea-keeping and wet deck slamming problems. The Soviets also conducted research at this time (by Yermolayev[3-3-2]) on hydrofoil systems for high-speed planing vessels and also developed a series of hydrofoil-assisted monohulls. Since these early developments, significant progress has been made in the application of hydrofoil systems for both sea-keeping and ride control and also for reductions in resistance. This paper focuses on the application of hydrofoils for reductions in resistance rather than for ride control.

The most common hydrofoil-assisted multihulls are:

- So-called *hydrofoil catamarans* that make use of hydrofoils and propulsion systems that allow the hulls to be lifted completely clear of the water. These vessels are not inherently stable and need expensive ride-control systems to keep them stable. These vessels are in essence similarly to conventional hydrofoil craft such as the Boeing JetFoil.

- *Hydrofoil-assisted catamarans* that make use of hydrofoils to lift the hulls only partially out of the water. The fact that the hulls remain in contact with water means the vessels are inherently stable and can make use of conventional catamaran propulsion systems such as flush inlet waterjets.
- More recently, a number of *hydrofoil-assisted trimarans* have been built. These vessels typically make use of a central hull and two small amahs. Hydrofoils are typically installed in each of tunnels.
- *SWATH type hydrofoil-assisted catamarans* have recently been introduced that make use of very large voluminous hydrofoils that provide significant buoyancy in addition to dynamic lift.

Notable milestones in the development of hydrofoil-assisted multihulls are as follows:

- 1980: launch of the first planing hydrofoil-assisted catamaran in Cape Town, South Africa, to the newly patented Hysucat design.[3-3-3] To date over 300 have been produced. Vessels ranging from 5m to 36m in size have been built to this design making it the most successful hydrofoil-assisted catamaran design so far.
- 1989: development of the first fast passenger ferries making use of hydrofoil-assistance by Hitachi Zosen in collaboration with the University of Tokyo.[3-3-4] This design was groundbreaking in the sense that it combined SES-type slender demi-hulls with hydrofoils and an active ride control system for improving passenger comfort. To date, eight 35m vessels of this type have been built as well as one 40m vessel.
- 1989: development of the first planing hydrofoil-assisted catamarans in China. A development similar to the Hysucat development for planing boats was undertaken from 1989 to 1997.[3-3-21]
- 1991: development of the Norwegian Foilcat series of hydrofoil catamarans[3-3-5] by Kvaerner Fjellstrand and Westamarin. A 35m and a 40m version were developed, neither of which are still in production. Four vessels in total have been built, three by Kvaerner and one by Westamarin.
- 1993: development of hydrofoil-assisted catamarans in Korea. KRISO in collaboration with Hyundai Heavy Industries,[3-3-7] and

Daewoo Shipbuilding and Heavy Machinery[3-3-8] both developed different types of hydrofoil-assisted catamarans. Hyundai and Daewoo both built one vessel each (40m and 45 m L_{OA} respectively) for use as fast passenger ferries.

- 1993: development of the Mitsubishi Super shuttle hydrofoil catamaran in Japan.[3-3-9] Two of these vessels have to date been built, the first in 1993 and the second in 1998 to the same operator.
- 1994: design and construction of Hydrofoil-supported ferries in Germany. A development was undertaken by various shipyards including Lürrsens and Henze Shipyard for the development of hydrofoil-assisted catamarans. Two ferries (20m and 30m L_{OA}) were built by Henze shipyard[3-3-10] making use of the Hysucat foil system.
- 1999: first HYSUWAC-type hydrofoil-assisted catamaran launched. The 45m, 170t vessel built by Halter Marine in the USA achieved 42kn with 2 × 2000kW.
- 2001: first hydrofoil-assisted trimaran launched. The vessel is a 55m ferry operating between Korea and Japan[3-3-11] and was built by North West Bay Ships.
- 2002: launch of the world's fastest passenger ferry, a 40m, 50kn hydrofoil-assisted catamaran developed in Russia and built by Almaz in St. Petersburg.[3-3-12]
- 2003: first lifting-body-type hydrofoil-assisted catamaran technology demonstrator launched making use of SWATH lifting body technology.[3-3-20]

The milestones above, spanning over two decades, represent only the most significant vessels. To date it is estimated that over 600 hydrofoil-assisted vessels have been built worldwide ranging from pleasure craft, workboats, and military craft to fast ferries. Reference 3-3-13 provides some additional information about recently built hydrofoil-assisted catamarans.

3.3.1.2. Types of hydrofoil systems for multihulls

In general, a number of different foil systems have found application. These range from simple passive monofoils to complex tandem foil systems that integrate ride control into the foils design through adjustable foils or flaps. Figs 3.3.1 and 3.3.2 show examples of fixed and adjustable foil systems. The configurations of the foil systems also vary considerably from vessel to vessel. Fig. 3.3.3 summarizes the different foil systems currently in use.

The types of hydrofoil configurations shown in Fig. 3.3.3 consist of two fundamental types: (1) Those with hydrofoils located *between the semi-hulls* of a catamaran or trimaran, such as those illustrated in the left column of Fig. 3.3.3. (2) Those that have hydrofoils that *extend below the keel line* of the hull, such as those illustrated in the right column of Fig. 3.3.3.

Fig. 3.3.1. A hydrofoil with adjustable incidence angle

Main foil for a 118-foot (36m) power catamaran

Fig. 3.3.2. A fixed Hysucat hydrofoil mounted between the semi-hulls of a 36m catamaran

Foil configurations with the foils located between the semi-hulls have the advantage that the foils are well protected by the hull and there is low risk of damage during operation or docking of the vessel. Locating the foils in the tunnel does however limit the amount of load the foils can carry and in general the lift is usually limited to a maximum of about 70% of the displacement weight for small planing catamarans, down to about 25% for larger semi-displacement catamarans. To carry more load in an efficient manner on hydrofoils, it is necessary to make use of hydrofoils that are more deeply submerged. This means locating the foils below the keel line on struts that allows the hull to be lifted farther out of the water, reducing the wetted area and increasing the load fraction on the foils up to 100% if necessary.

In general, planing type hydrofoil-assisted catamarans make use of hydrofoils that are located in the tunnel between the semi-hulls and do not protrude below keel level. For planing speeds, this type of foil system produces sufficient lift to provide very good improvement in planing performance without introducing the complication of having foils protrude below the keel. Such designs are very practical for pleasure and military craft where they can be used in a variety of applications and can for example be beached or tailored without difficulty.

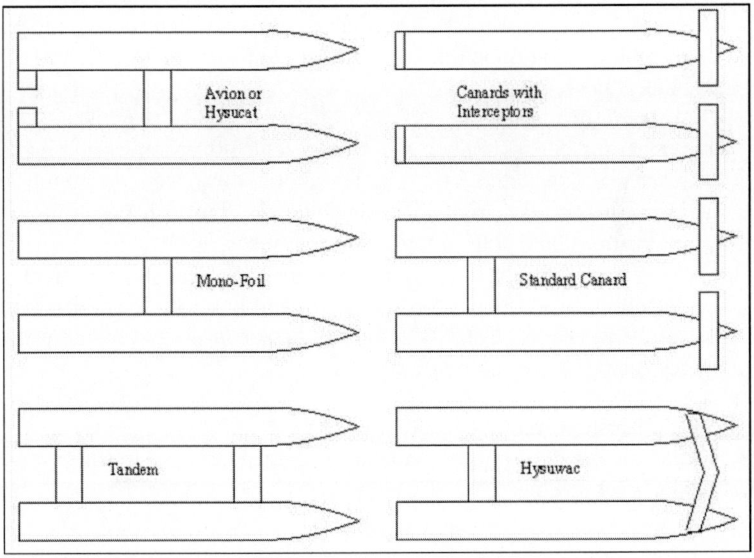

Fig. 3.3.3. The most common configurations of hydrofoil-assisted craft currently in operation

Semi-displacement and semi-planing multihulls usually make use of hydrofoils located below keel level. Hydrofoils located at keel level are only capable of bringing very limited reductions in resistance that often do not justify the extra cost of the foil system. To gain more substantial benefit from a foil system, a higher lift system is required, such as those shown in the right hand column of Fig. 3.3.3. By positioning the foils below the keel, they are more deeply submerged and suffer less from the "free surface effect" and thus are capable of generating more lift and are better able to reduce resistance at non-planing speeds.

3.3.1.3. Resistance tendencies

All hydrofoil-assisted vessels have the same general resistance tendencies:

- At speeds below the displacement hump, there is no improvement for resistance compared to the same hull without foils. In many cases the foils are a source of additional appendage drag and the total resistance of the vessel is increased by 5%–10%. Only the most efficient foil systems cause no increase in resistance at low speeds. The foils do nevertheless provide significant motion damping and sufficient lift is generated for them to be efficient ride control devices.
- At semi-displacement and semi-planing speeds (e.g., speeds of typical fast ferries) only the most efficient foil systems bring useful resistance reduction of up to about 40%. The amount of improvement is highly dependent on hull shape, foil–hull interactions, and configuration of the foil system.
- At planing speeds it is relatively easy to obtain improvement in performance. At these speeds the lift produced by the foil system can be in excess of 60% of the boat displacement and reductions in resistance can be up to 50%.

3.3.1.4. Foil systems mounted at keel level

Fig. 3.3.4 shows the typical resistance tendencies for a foil system mounted at keel level.

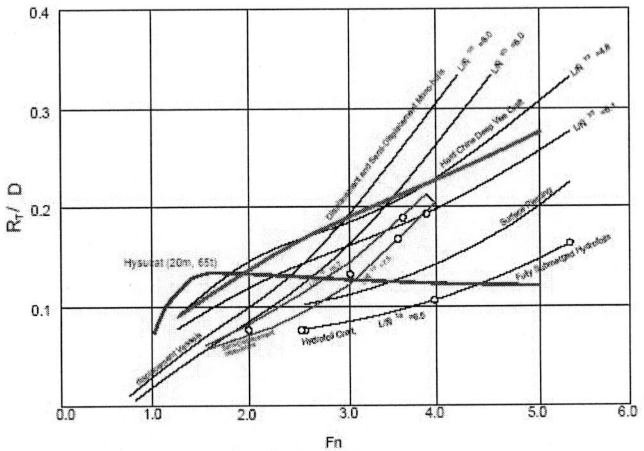

Fig. 3.3.4. Resistance displacement ratio of a Hysucat for different volumetric Froude numbers in comparison with different types of craft

The figure clearly shows the increase in resistance due to the hydrofoil system around hump speed and the strong decrease in resistance above $F_{n\nabla}$= 2.02. Hydrofoil systems such as the Hysucat have a relatively flat and sometimes downward sloping resistance curve, as shown in the figure.

3.3.1.5. Foil systems with foils mounted below the keel

Fig. 3.3.3 illustrates the wide variety of foil systems that are in use making use of foils under the keel. The resistance tendencies of these vessels are quite varied and are highly dependent on the configuration of the foil system. Fig. 3.3.5 illustrates these resistance tendencies. Three distinctive phases of operation are present, displacement, transition, and planing. These are shown in Fig. 3.3.6. Of particular interest is the *transition speed range* as this is the range in which the majority of fast ferries operate. In the transition phase, the resistance is highly variable and a transition resistance hump is common. The extent of the hump is largely dependent on the hull shape and the foil system layout. Vessels with canard foils forward (see Fig. 3.3.3) typically have the highest hump as the small foils do not have the lifting capacity to lift the hull sufficiently out of the water at transition speeds. The Hysuwac foil system offers the best transition hump-resistance characteristics. Due to the high aspect ratio front foil, it can provide good lift at transition speeds. The optimum resistance curve in Fig. 3.3.5 is representative of that achievable with the Hysuwac foil system.

Fig. 3.3.5 also shows the general relationship between waterjet thrust and resistance. If the transition hump is under predicted, the vessel may not have any thrust margin to overcome the hump even though, if this hump could be miraculously overcome, a much higher speed could be maintained in the planing phase. A number of vessels have had difficulty overcoming the transition hump due to poor design and this has resulted in some hydrofoil-assisted catamarans being complete failures.

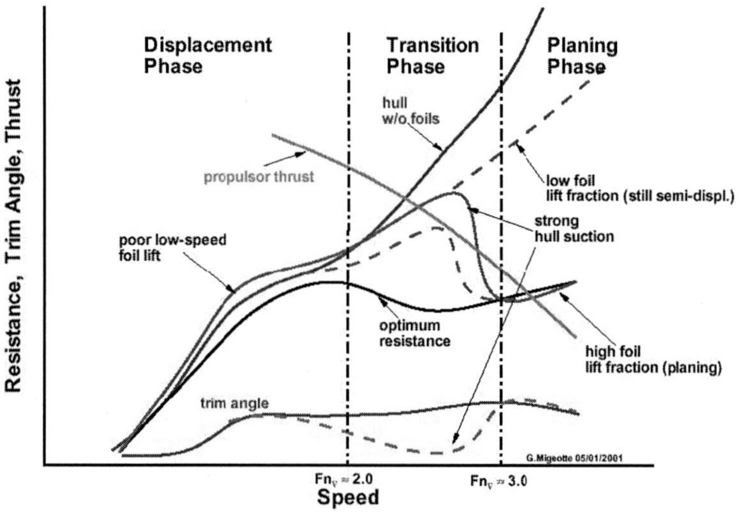

Fig. 3.3.5. Resistance tendencies for hydrofoil-assisted catamarans, which have the foil located below the keel level

3.3.1.6. Propulsion systems for hydrofoil-assisted catamarans

In general, carrying 70% of the displacement weight on foils implies that the hull has been lifted almost all the way out of the water and in such cases the propulsion systems often need careful consideration to ensure that no ventilation occurs. Flush inlet waterjets are the most susceptible to this problem. Fig. 3.3.7 shows Kvaerner's solution to this problem for the FoilCat: integrating the water inlet with the strut of the rear hydrofoil. This is not dissimilar to the intakes used by conventional waterjet propelled hydrofoil craft such as the Boeing JetFoil. While this solution does ensure that no waterjet ventilation takes place, the overall propulsion efficiency is significantly reduced.

Many of the hydrofoil-assisted catamarans currently in operation today are vessels that have been retrofitted with hydrofoils. For such vessels it is not possible to easily modify the waterjet intakes and the hydrofoils need to be carefully designed so that the best reduction in resistance is obtained without lifting the hull out of the water to the extent that the waterjets ventilate.

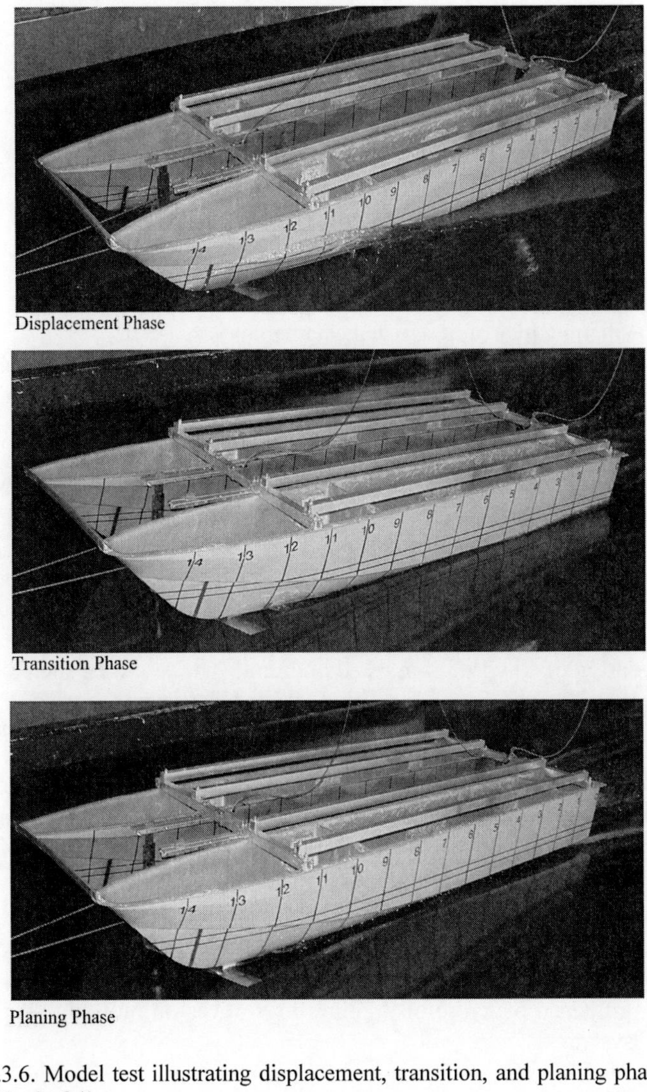

Fig. 3.3.6. Model test illustrating displacement, transition, and planing phases for hybrid hydrofoils

Fig. 3.3.7. Waterjet intake of the Kvaerner FoilCat

3.3.1.7. Planing hydrofoil-assisted catamarans

At planing speeds it is well known that multihull interactions are generally small and if present are usually beneficial,[3-3-14] as each planing hull can be modeled by the lifting analogy of Wagner.[3-3-15] Similarly, the presence of hydrofoils also creates positive interference effects. Investigations by Hoppe[3-3-16] have shown that the positive interference amounts to about 20% of the drag reduction achieved with the foil system. During the design of a planing hydrofoil-assisted multihull it is therefore not necessary to pay detailed attention to the complex hull–foil interactions. Conventional methods for designing planing hulls, such as Savitsky's method,[3-3-17] combined with classical hydrofoil theory, usually gives satisfactory results to the design of a hydrofoil-assisted multihull. The results of such methods in experienced hands provide reliable results eliminating the need to model test the vessel.[3-3-18]

3.3.1.8. Non-planing hydrofoil-assisted catamarans

Designing foils for non-planing speeds is hydrodynamically much more complex as at these speeds there are a number of detrimental interactions between the multiple hulls and also between the hulls and the foils.
These interactions include:

- Suction or lift on hull due to the hydrofoils
- Influence of hydrofoil trailing tip vortices on hull and propulsion system
- Wave interference between hull and hydrofoils

- Interaction between fore and aft hydrofoils: a thorough understanding of these interactions is required to design a foil system that can provide useful resistance reduction and importantly be free of any kind of dynamic instability.

Designing a suitable foil system for non-planing hydrofoil-assisted multihulls requires a computational method that can capture the complex interactions between the hull and foils. A computational code that has the capability to model free surface effects, lifting surfaces, and calculate the exact path of the hydrofoil trailing tip vortices is required. Methods such as classical hydrofoil theory have no way of accounting for such complexities and one is forced to make use of CFD. Our experience has shown that methods solving potential flow equations provide the correct level of complexity for design purposes. The authors make use of the non-linear vortex lattice method (NVLM) as implemented in the AUTOWING software by Kornev.[3-3-19] A potential flow method such as AUTOWING allows for reasonable computational times needed for design.

AUTOWING has the capability to model wave interactions, the vortex roll-up process, and the interactions between multiple hydrofoils satisfactorily for design purposes. Figure 8 shows a typical AUTOWING result showing the free surface of a HYSUWAC type hydrofoil-assisted catamaran. Note the downwash in the tunnel between the semi-hulls created by the hydrofoils, which is not present on conventional catamarans.

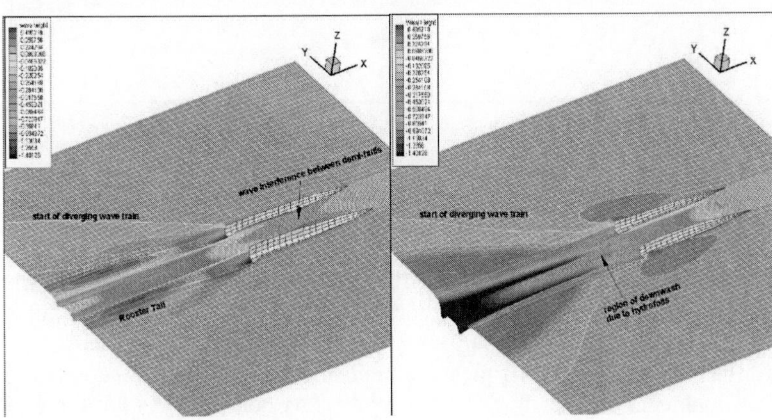

Fig. 3.3.8. Comparison of free surface of a semi-displacement catamaran with and without foils using AUTOWING

The results of the NVLM as implemented in AUTOWING have been found to agree well with experiments. Fig. 3.3.9 shows a comparison between measured and calculated resistance for a hydrofoil-assisted catamaran. As can be seen, even sudden changes in the resistance, which due to complex non-linear hull–foil interactions are captured by the method.

As with all significant ships that have been built, model testing remains the most valuable design tool for any new development. The accuracy of model testing hydrofoil-assisted catamarans is usually somewhat less accurate than that achieved with conventional ships due to laminar flow effects over the hydrofoils. Often laminar separation of the flow over the foils is a problem. The hydrofoils are usually small in relation to the waterline length of the hull and it is therefore impossible to secure turbulent flow over the foils. Special corrections need to be made for this.

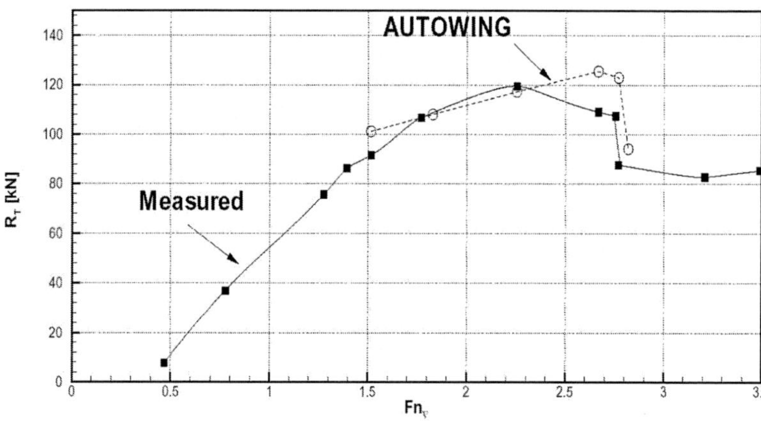

Fig. 3.3.9. Resistance comparison of AUTOWING in relation to experiments

3.3.1.9. Future prospects and conclusions

Hydrofoil-assisted multihulls have been in use for over two decades. Early developments in the 80s were mainly on small planing type and initial developments took place worldwide including South Africa, Russia, and China. Vessels for use as pleasure craft and military craft predominated developments at this time. In the late 1980s and early 1990s the first catamaran fast ferries making use of hydrofoil-assistance were developed.

The sizes of hydrofoil-assisted craft have also steadily increased and the largest hydrofoil-assisted multihull built to date is the 50m trimaran built by North West Bay ships. It is likely that hydrofoil assistance will incrementally be applied to larger and larger vessels as these vessels are designed to attain speeds in the order of greater than 60kn.

To date the fastest hydrofoil-assisted vessel, built by Almaz, has achieved speeds of over 50kn and it is expected that future hydrofoil-assisted craft will exceed these speeds as ways and means are found to overcome cavitation limitations at high speeds.

New developments are likely to take place—and are already taking place—in the design of trimarans and monohulls that make use of hydrofoil-assistance. Also, hydrofoils are being integrated with SWATH technology, which will introduce a new range of SWATH/hydrofoil hybrids. Navatek Ships is spearheading this development.[3-3-20]

Hydrofoil-assisted vessels over the past two decades have been shown to possess various favorable characteristics compared with conventional monohulls and multihulls, such as lower resistance, better sea-keeping, better maneuverability, and lower wake wash. All these are contributing to renewed interest in these vessels in the new millennium and will likely result in increasing numbers of hydrofoil-assisted vessels being utilized in all applications of high-speed ships.

References

[3-3-1] J. B. Hadler et al., "Ocean Going Seakeeping Design, Based on the Experiences of USNS Hayes," *SNAME Proceedings*, pp. 126–61, 1974.

[3-3-2] S. G. Yermolayev et al., "New Prospective Types of High-Speed Vessels: Gliding, Foiled, Catamarans, Catamarans with Hydrodynamic and Aerodynamic Lifting," *Transactions NTOSP* 186, pp. 24–32, 1972.

[3-3-3] K. G. Hoppe, "The Hydrofoil Supported Catamaran, Progress Research Report," Dept. of Mechanical Engineering, Stellenbosch University, 1980.

[3-3-4] H. Miyata, "Development of a New Type Hydrofoil Catamaran," *Journal of Ship Research* 33 (2), 1989.

[3-3-5] K. J. Minsaas, "Design of Hydrofoil Catamarans in Norway," *FAST '93*, pp. 83–89.

[3-3-6] N. Gee, "Passive Foilcats—Design, Practice and the Future," 9th International Surface Craft Conference, 1993.

[3-3-7] K.-S. Min, "Long Range High Speed Catamaran Ship Design," *FAST '91*, pp. 591–606, 1991.

[3-3-8] B.-S. Kim et al., "Improvement of Hydrodynamic Characteristics of Catamaran with Hydrofoil," *FAST '93*, pp. 631–42, 1993.

[3-3-9] K. Kihara, “Diesel Driven Fully Submerged Hydrofoil Catamaran: Mitsubishi Supershuttle 400, ‘the Rainbow,’” *FAST ’93*, pp. 139–150, 1993.

[3-3-10] H. G. Gerdsen, “Design and Construction of Fast Hydrofoil-Supported Ferries,” *Maritime Transport International*, pp. 180–82, 1994.

[3-3-11] J. D. Adams and G. Migeotte, “Hydrofoil-Assistance for Multihulls,” Workboat Show and Conference, New Orleans, 1991.

[3-3-12] Speed at Sea Superfoil 40.

[3-3-13] K. G. Hoppe, “Recent Applications in Hydrofoil-Assisted Catamarans,” *Fast Ferry International*, pp. 30–36, September 2001.

[3-3-14] D. Savitsky and A. Dingee, “Some Interference Effects between Two Flat Surfaces Planing Parallel to Each Other at High Speed,” *Journal of the Aeronautical Sciences* 21, pp. 419–20, 1954.

[3-3-15] H. Wagner, “Ueber Stoss- und Gleitvorgaenge an der Wasseroberflaeche von Fluessigkeiten,” *Zeitschrift fuer Angewandte Mathematik und Mechanik* 12 (4), 1932, pp. 193–215.

[3-3-16] K. G. Hoppe, “The Hydrofoil Supported Catamaran,” Progress Research Report 1980–82, University of Stellenbosch, Mechanical Engineering Department, 1980.

[3-3-17] D. Savitsky, “Hydrodynamic Design of Planing Hulls,” *Marine Technology* 1, 1964.

[3-3-18] K. G. Hoppe, “Hydrofoil Supported Catamaran Developments in South Africa,” HIPER ’99, Cape Town, 1999.

[3-3-19] http://www.argo-group.de/autowing.

[3-3-20] T. J. Peltzer, “Lifting Body Technologies for Transformational Ship Designs,” 9th Naval Platform Technology Seminar, 2003.

[3-3-21] L. Zhao, “Experimental Study on the Performance of Channel Hydrofoil Type Planing Boats,” International Conference on New Ship Technology into the 21st Century, Shanghai, 1994.

3.3.2. A Brief Introduction to a High-Speed Passenger Craft—Hydrofoil Assisted Catamaran *Superfoil 40* (2005)—

Gui Mo
(Director of Port Division, Headquarters of Shanghai Deepwater Port Construction),
Stanislav P. Pavlov
(Director, MTD Ltd., Russia, Chief Designer of *Superfoil 40*)
and Yun Liang
(Engain Science & Technology Ltd, Hong Kong; FRINA)

Abstract
The performance, design, and sea trial results of high-speed CAT are described in this paper. *Superfoil 40* was imported in August 2004 and operated between Small Yang San Island of Zhe Jiang Province and Lu Cao Port of Shanghai City. The craft was designed by the Marine Technology Design Company (MTD), Russia, and completed by Almaz Shipbuilding Corporation, Russia, in 2002. The automatic riding control system type FICS-3 was arranged to control the bow hydrofoils and stern interceptors during running take off and it is foil borne both on calm water and in waves, which ensured the craft achieved the maximium speed of 55kn (making it perhaps one of the fastest passenger craft in the world). The craft could even keep to a speed of 45kn in waves with a significant wave height 2m (SS4). And the vertical acceleration in such bad weather conditions could satisfy the requirements of the standard ISO2631/3-85; no seasickness of riders appeared in two hours of operation.

3.3.2.1. Introduction[3-3-2-25]

In 2003, the headquarter of the Shanghai Yang San Deep Water Port proposed a project for constructing or importing a 100–150-passenger craft with a high speed of 45–50kn, which could be operated in open seas or the coast line near Shanghai with a significant wave height equal to or below 2.0–2.5m. Passengers would not be seasick on the two-hour high-speed journey. That is to say, the owner wanted a passenger craft with high speed, good sea-keeping, and economical efficiency, and the craft needed to be delivered before the middle of 2004.

It was well known that there were various high performance marine vessels (HPMV) with specific purposes and functions. However, it was very difficult to build a 100t craft with high speed, good sea-keeping, and economical efficiency. The advantages and disadvantages of different HPMVs are listed in Table 3.3.1.

Table 3.3.1. Characteristics of different HPMVs

ACV/SES	Hugh speed (50–60 kn), poor sea-keeping quality, serious speed loss in waves
High-speed planing hulls	High speed (40–50 kn), poor sea-keeping quality, high slamming load, large speed loss in waves
HSCAT	Medium speed (30–40 kn), high ratio of seasickness for passengers in waves
SWATH	Good sea-keeping quality, low speed (20–25 kn)
Wing in ground effect craft	Under-developed technology for 100t craft, with high risk,
Wave-piercing craft	Medium speed (30–40 kn), medium sea-keeping quality
Deep submerged automatic control hydrofoil craft	High speed (40–50kn), fine sea-keeping quality, expensive

Therefore, it was very difficult to find existing high-performance vessels from domestic providers to satisfy the needs of the customer and the ship owner. The best way was to order a high-performance craft under construction or a second-hand high-performance craft.

A high-speed hydrofoil-assisted CAT named *Linda Express* operated between Tallinn in Estonia and Helsinki in Finland; it was designed by the Marine Technology Development Corporation (MTD) and manufactured by the Almaz Shipbuilding Corporation. The two corporations were from Russia. The craft was completed in July 2002 and operated on the planned route between 2002 and 2003. However, there were too few passengers for so large a craft to carry and the craft could not always keep to its maximum speed of 55kn along the partially planned route in coastal waters. The Estonian owner wanted to sell the craft as the advantages of the craft could not be put to good use. This hydrofoil-assisted catamaran was ordered and purchased by the Chinese customer in August 2004, and was renamed *Superfoil 40.*

3.3.2.2. Brief introduction to the *Superfoil 40* automatic control hydrofoil-assisted catamaran

(1) General

The hydrofoil-assisted catamaran had a structure welded with aluminum alloy which was equipped with a bow hydrofoil and stern interceptor and was propelled by four waterjet propulsions driven by high-speed diesel engines. The semi-displacement of the catamaran adopted a semi-planing demihull profile.

The passenger cabin was arranged on the main deck, which was divided into economical and business classes. The navigation room was arranged on the superstructure with a wide field of vision, in which there were three armchairs for the captain, deputy captain, and chief engineer.

With the help of the automatically controlled bow hydrofoil and the stern interceptor, the craft could run as the bow and a major part of the demi-hulls' stern were planing above the water surface, and only a very short part of the aft was submerged. So the craft could achieve a high speed of 55kn, making it one of the fastest passenger ships around the world. The automatic riding control system FICS-3, provided by Aurora Science and Technology Corporation was used to control the hydrofoils and stern interceptors during the operation of taking off, when foil borne, and during turning and emergency stopping. The automatic ride control in waves with various directions was the most important part of the design of this high performance catamaran, which provided a perfect riding experience and comfort for passengers while operating in the Gulf of Finland in 2002–3.

(2) Principal dimensions and characteristics (Figs 3.3.10 and 3.3.16)

L_{oa}	41.00m
L_{wL}	40.28m
L_{pp}	6.63m
B_{oa}	12.40m
B	11.0m
T_m (at midship section)	3.55mm
T foil borne, full displacement	1.20m
When hull borne, full displacement, foil in operation	3.40m
Hull borne, full displacement, foil retracted	2.10m
The height above the waterline (not including radio antenna)	11.40m
Passengers	
Economic class, seat spacing 900mm	224
Business class, seat spacing 1m	62
Total passengers	286

(3) Loads

Payload:

Crew	14, 75kg for each 900kg
Passengers	286, 75kg for each 1,450kg
Luggage,	2,860kg
Fuel	4,400 liters 3,740 kg
Fresh water	1,000 liters 1,000 kg

Duty free goods 3,050 kg
Total 33,000 kg

(4) Performance

(a) Speed

With four MCR 1,740kw high-speed diesel engines, the craft could achieve a maximum speed of 55kn in calm water, the total load of which was 33t. The relationship between resistance and Froude number is shown in Fig. 3.3.12. The result shows that when the craft reached the maximum speed, R/W = 0.108, Kn = W Vs/102 N = 6.45, and Kp = P V/N = 4.16.

Where W was the weight of craft (kg), Vs was speed in m/s, V was the speed in km/h, N was the power of engines (kw), and Kn and Kp were hydrodynamic coefficiency and transport efficiency, respectively. It was found that the efficiencies were higher compared with other HPMVs.

(b) Cruising range

140 miles with full displacement at a maximum speed of 55kn.

(c) Sea-keeping quality

In sea trials, the craft could be operation at V = 45–47 kn (following sea), 45kn (beam sea), and 44kn (head sea) in SS4 (significant wave height 2.3m), and the mean was 45kn (Fig. 3.3.15). The speed loss in waves of SS4 was about 18%, which was rather smaller compared with ACV/SES (30–50%), with a high-speed planing and mixture type of MH (30–40%).

Meanwhile, the vertical acceleration (RMS) in 1/3 octave frequency with the craft operating at MCR in waves of a significant wave height of 2.0m was below the limitation and thus complied with ISO2631/3-85. The passengers would not be sick after two hours high-speed riding.

The estimation of sea-keeping quality and seasickness rate are shown in Fig. 3.3.13, where the abscissa were 1/3 oct encounter frequency of the craft and RMS of vertical acceleration, curve 1 was the two-hour seasickness limitation from the ISO2631/3 standard, curve 2–6 were the vertical acceleration of the craft at a significant wave height of 2m with the speed of 40–45kn, and the craft was quartering, heading, beaming, quartering following, and following waves, respectively. The results in Fig. 3.3.13 showed that passengers would be seasick on a two hour journey when the craft was operating at the speed of 40–45 kn.

Fig. 3.3.14 shows the limited operation hours for passenger without seasickness in significant waves with a height of 2.0m: it found the limit was two hours for all wave directions and 5.1 hours for following waves.

When the wave height was larger than 25m (significant wave height), the craft could be foil borne and the outlet of the engines dropped down.

When the significant wave height was larger than 3.0m, the craft could be hull borne with retractable foils at the speed of 18kn. Figs 3.3.13 and 3.3.14 show estimations according to the model test results, automatic efficiency, and the sea-trial data of the craft in the Gulf of Finland (Fig. 3.3.15), where (a) and (b) respectively denoted the measured vertical and lateral acceleration of the craft in waves with significant wave heights of 2.3m in directions of following, beaming, and heading waves. It was obvious that all the data satisfied the limitation for seasickness in two hours of operation as outlined in ISO2631/3.

(5) Engines and propulsions
The craft was equipped with four MTU 12V4000M70 high speed diesel engines, each of which could develop 1,740kW power at the speed of 1,900–2,000rpm, and the specific fuel consumption was 206g/kWh. The craft was propelled by two MJP J550RDD and two MJP J550DBD waterjet propulsions with rudder and reverse system driven by four main engines, respectively.

(6) Rudder
The main rudders were integrated with waterjet propulsions. There were two auxiliary spade-shaped balance rudders installed at each stern of the demi-hull, which would improve the maneuverability and course stability of the craft, particularly in following waves. The rudders were made of titanium alloy to improve their anti-erosion capability.

(7) Bow hydrofoils and stern interceptors
The two bow hydrofoils were made of high strength titanium alloy and were installed at the bow of each demi-hull, which were the deep submerged type especially designed for this craft. The submerged part of the hydrofoils were solid and without movable flaps. The foils would be retracted when the foils collided with floating objects, which could protect the hulls and save the cost of repair.

The attachments of the bearing, foundation, hydraulic parts, and so on were arranged in the hulls and above the water line at still condition.

When the craft was operated in shallow water, the foils could be retracted wholly till above the waterline, which was convenient for the inspection and repair of the hydrofoils. The retractable system was driven by the hydraulic system.

The attack angle of the foils could be controlled by an automatic riding control system, FICS-3, for smooth operation at take-off and when foil borne in calm water and in waves. The retractable range of the foil was +90° (forward) to -13°, and the operation range of foil was +7° (forward) to -11°, the max rotation rate of the foils was 11°/s.

(8) Stern interceptors
There was one specially designed interceptor mounted at the stern of each demi-hull (Fig. 3.3.17), which was high efficient for the hydrodynamics of the craft. The interceptors could control the stern draft, which could reduce the drag of the craft and carry out the riding control of the craft.

The interceptors were made of high-strength titanium alloy and connected with transoms of each demi-hull by bolts, which were controlled by the automatic riding control system FICS-3 (Fig. 3.3.17). The interceptors became the high-pressure planing surface with the surface of the stern and skegs at the rear of each demi-hull, which could increase the lift and reduce the drag by the wetted surface, and increase the lift–drag ratio.

When the craft was operated in waves, the submerged depth of interceptors and the incident angle of the bow foils were controlled to ensure a low response of the craft, which would improve the sea-keeping quality of craft and reduce the seasickness rate of passengers.

(9) Automatic control system[3-3-29]
The automatic riding control system FICS-3 was designed to control the movement of the foil and interceptor and was manufactured by Aurora Science and Product Corporation, Russia, which was one of the special and most important parts in the design of *Superfoil 40*.

The diagram of the system is shown in Fig. 3.3.18, where: (□) FCI basic control panel; (□) PC2 reserve control panel; (□) FC3 central control panel; (□) FC4 linear acceleration probe equipment for the directions of X, Y, Z; (□) FC5.1 mechanical control equipment for bow hydrofoils; (□) FC5.2 mechanical control equipment for stern interceptors; (□) PP-UBC2 emission and receiver of altitude; (□) A052-2 antenna of emission and receiver for altitude; (□) FC6 upper limit of power equipment; (□) sensor for motion parameters of mini-navigator; (□) MMP12-1 plane screen of TV (system basic monitors); (□) electric hydraulic cylinder; (□□) gyro-compass; (□□) hydraulic cylinder; (□□) interceptors; (□□) bow hydrofoils; (□□) GPS.

The principle programs of the system were as follows: At first, the captain should preset the submerged depth of the bow hydrofoil and the trim angle of the craft to keep the expected riding attitude on basic control panel FC1 (□) (or reserved control panel FC2 [□], if the former was out of order). Then the data would be inputted to the central-control computer FC3 (□). Meanwhile, the linear acceleration of X, Y, Z, then directions from FC4 (□), and the altitude of the craft from PP-UBC2 (□), and antenna of emission and receiver could give the altitude of the craft and the submerged depth of the hydrofoils. Two sets of such equipment were available, one of which was auxiliary. The data from the gyrocompass (□□), GPS (□□), and mini-navigator (□), which also had ancillary equipment, would be sent to the central control system. All the data should be processed to the control information, which would be inputted to the bow hydrofoil mechanical control equipment FC5.1 and the stern interceptor mechanic control equipment FC5.2. Thus, the hydraulic actuators would drive the bow hydrofoils and the stern interceptors automatically.

Since the attack angle of the bow hydrofoils and the submerged depth of the interceptors were changed continuously, the given riding attitude of the craft could be kept approximately constant and would need little acceleration in calm water and waves. And the motion parameters would be shown to the captain on the main monitor of the system.

There were six cameras in the craft to monitor the submerged depth of the hydrofoils on the port and starboard sides, the wetted length of the stern demi-hulls, and the working status in the main engine rooms. If the wetted length of the semi-hull was too short, the air would be entrained into the intake of the waterjets. The captain could adjust and shorten the submerged depth of the interceptors to increase the wetted length to prevent entrained air. At the same time, the submerged depth of the foils also could be adjusted according to the different loading of the craft and hydrological conditions.

The main functions of the riding control system were: (a) providing the craft with steady motion using a preset average riding attitude in calm water and waves (significant wave height up to 2.5m) which could maintain the trim angle of the craft and the submerged depth of the foils to maintain the space position of the craft; (b) improving pitching stability; (c) improving rolling stability; (d) improving heaving stability and vertical acceleration, which should be lower than the standard of ISO2631/3-85 for two hours riding; (e) improving the maneuverability of the craft by reducing the turning radius up to 1.5 to 2.5 times by tumbling automatically: if the turning angle velocity is greater than 0.5°/s, the system tumbles automatically; (f) allowing the take-off program to make a quick take-off and the submerged depth of the interceptors to increase while the craft is accelerating; (g) automatic monitoring for the failure of the system and control loop; (h) providing the algorithm for emergency stops.

The characteristic of the FICS-3 system can satisfy the stability of both longitudinal and transverse motion and the comfort standard of ISO2631/3-85. As these two requirements were contradictory in some cases, the selection of the correct algorithm was very important to the whole control system!

The altitude signals were from altitude equipment located at the bow of the craft. And the linear and angular signals were from the mini-navigator. The optimization and combination of the signals in the control program ensured the good stability of the craft and the comfort of passengers.

(3) Sea trial results[3-3-27]

The sea trial of the *Superfoil 40* was carried out in the service zone at Tallinn, Estonia, near the Gulf of Finland. The loading conditions in the sea trial were as follows:

Crew	7
Passengers	174
Fuel/liter	8,800
Displacement/t	-150
Fresh water/liter	2,200
Lubricated oil/liter	185
Weather & waves:	
Wind velocity/m/s	10–13
Wind direction	west-southwest
Significant wave height/m	1.5–1.8–2.0 (eye measurement)
Wave length/m	- 7 (eye measurement)
Engines	four engines with 2,000 rpm (MCR) four waterjets

The test results were as follows:

Route no/	1	2	3	4	5
Wave direction	Beaming, from port	Following	Beaming, from starboard	Heading	Following-quartering, from port 45°
Rolling/deg.), minus to port down	-1.0—+1.8	-0.3—+0.3	-1.9—+0.9	-0.3—+-.3	-0.2—+0.2
Pitching/deg., Minus to bow up	-2.7	-2.7	2.7	3.2	2.7
Speed/kn)	45	47	46	42	44
Remark				1.5 minutes for take off	1.2 minutes for take off

Route No.	6	7	8	9	10
Test conditions	All passengers gathering at port	All passengers gathering at starboard	Full speed turning left	Full speed turning right	Emergency stop
Heeling/deg.	-0.7	+0.7	Turn angular velocity, 100–150°/s	Turn angular velocity, 100–150 °/s	From full speed ahead to full speed astern
Take off time /min	1.2	1.2			
Speed up/kn	0–46	0–46			
Foil borne speed/kn	46	46	46	46	
Trim/deg.	-2.7	-2.7	-2.7	-2.7	Inertia distance, 100 m

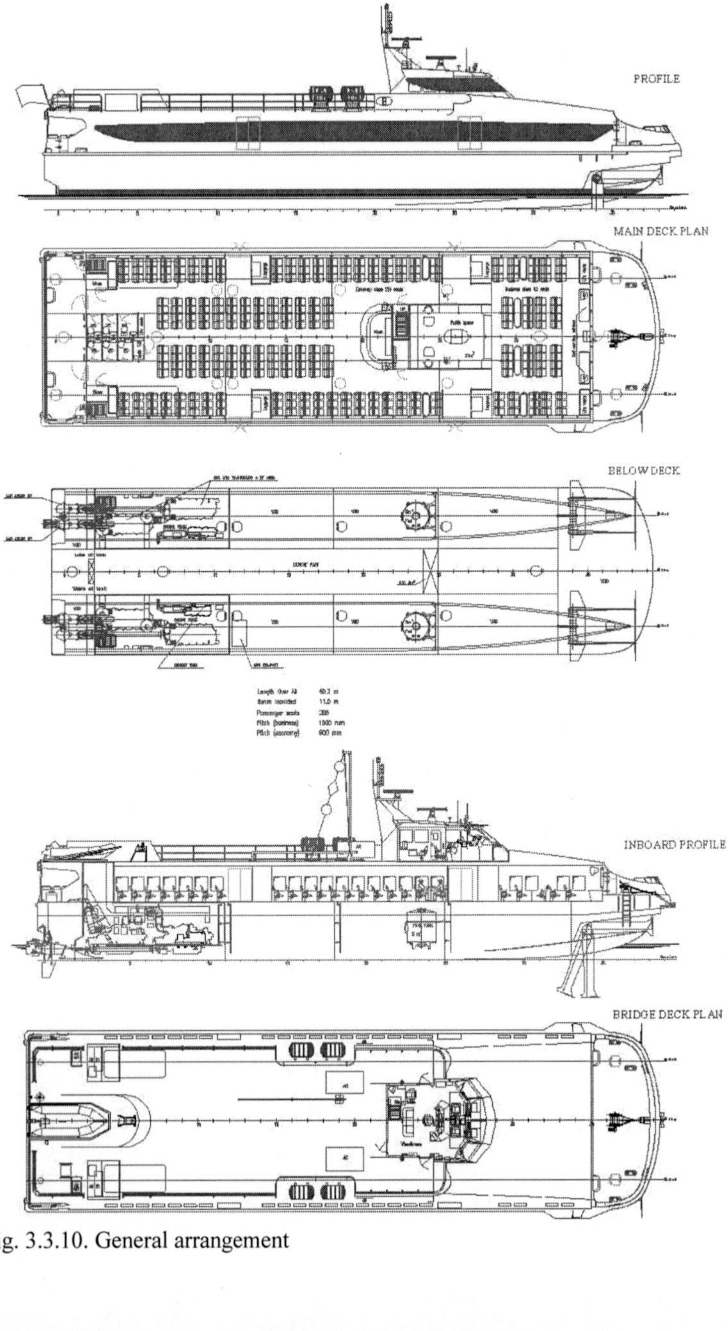

Fig. 3.3.10. General arrangement

Fig. 3.3.11

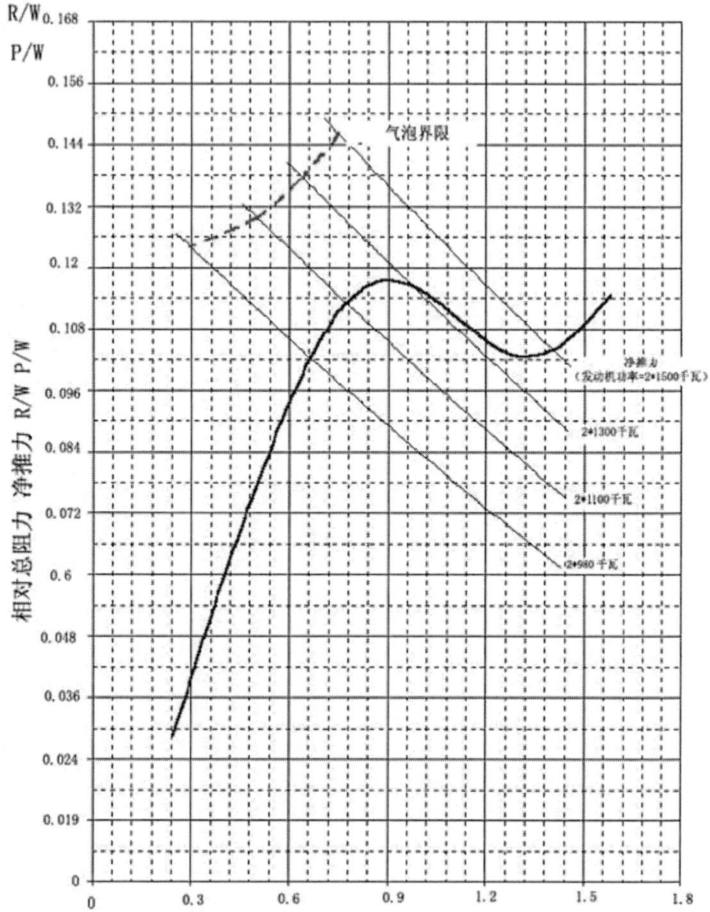

Fig. 3.3.12(a). Relative resistance P/W versus Froude number Fr

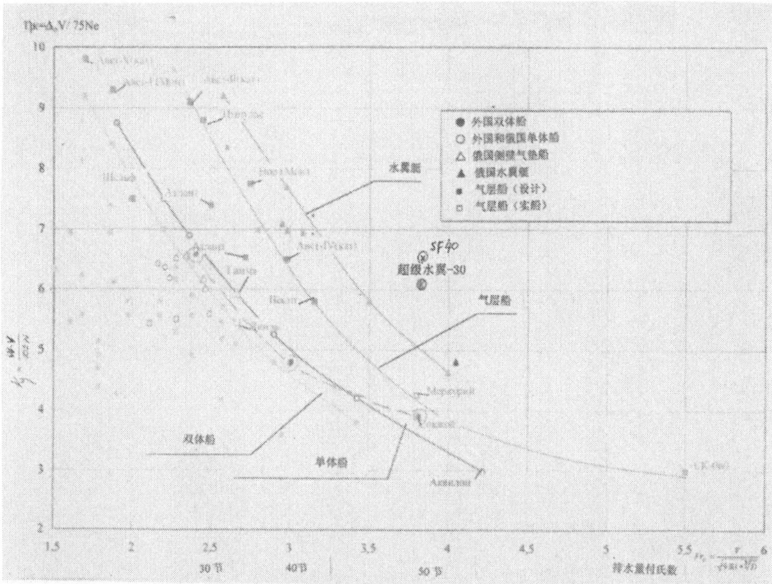

Fig. 3.3.12(b). Comparison of hydrodynamic coefficiency between *Superfoil 40* and other high-speed crafts

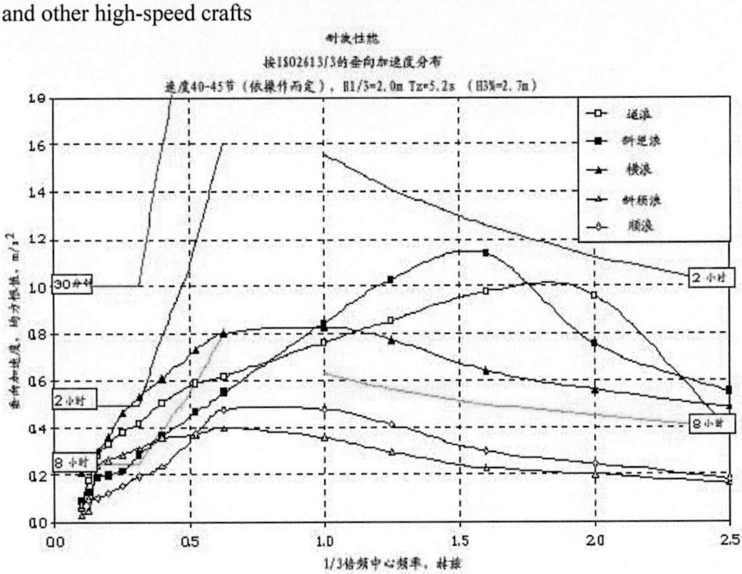

Fig. 3.3.13

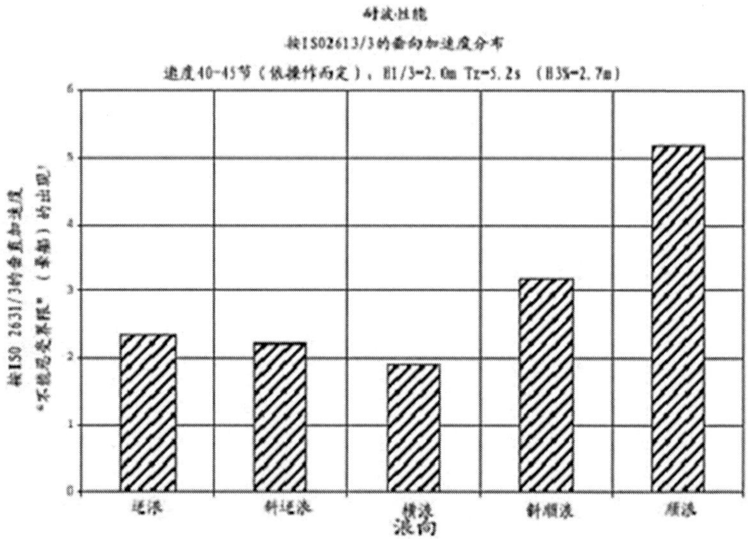

Fig. 3.3.14.

Vertical acceleration

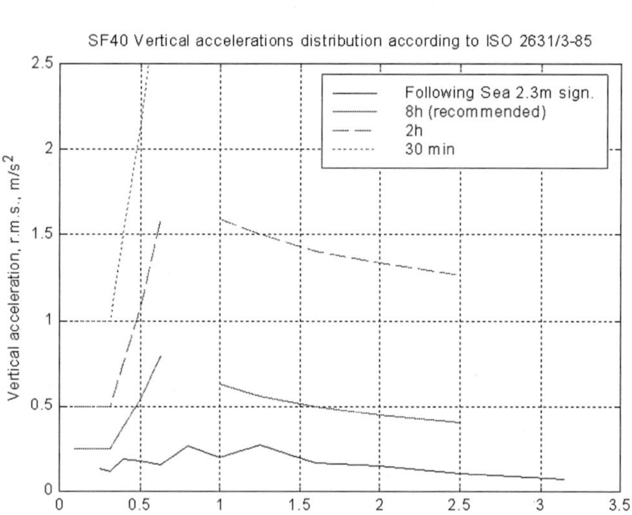

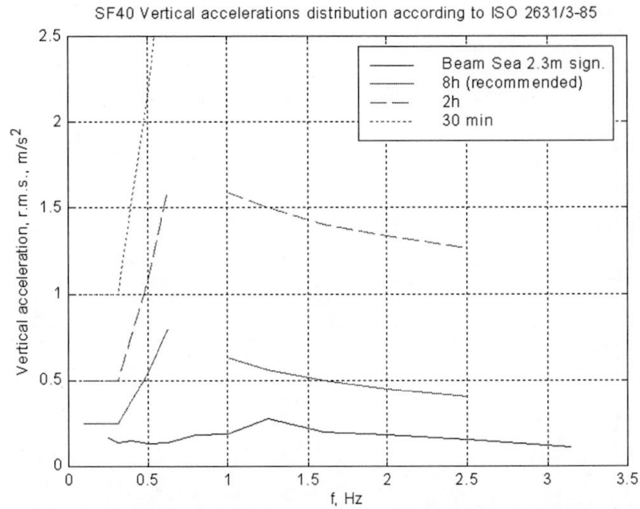
SF40 Vertical accelerations distribution according to ISO 2631/3-85
Beam Sea 2.3m sign.
8h (recommended)
2h
30 min
Vertical acceleration, r.m.s., m/s^2
f, Hz
0
0.5
1
1.5
2
2.5
3
3.5

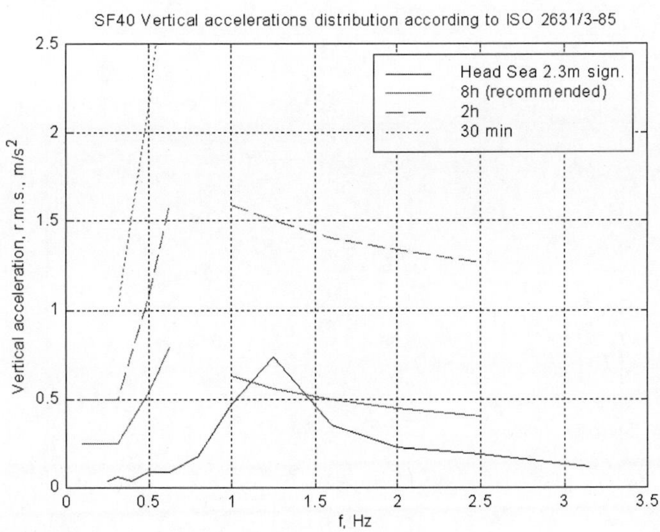
SF40 Vertical accelerations distribution according to ISO 2631/3-85
Head Sea 2.3m sign.
8h (recommended)
2h
30 min
Vertical acceleration, r.m.s., m/s^2
f, Hz
0
0.5
1
1.5
2
2.5
3
3.5

Fig. 3.3.15 (a)

Transverse acceleration

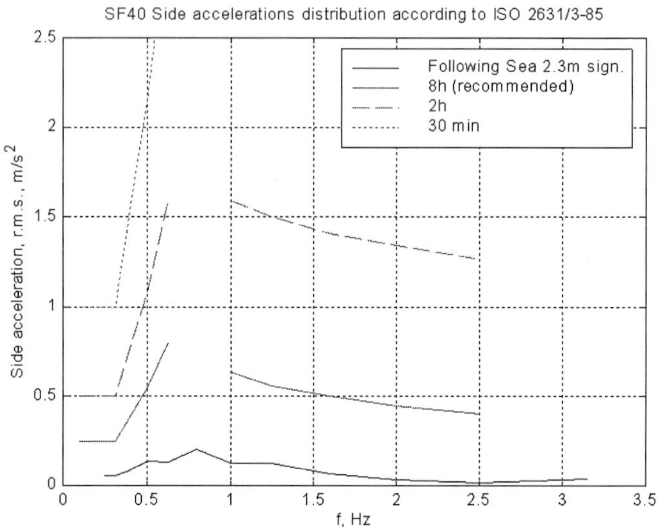

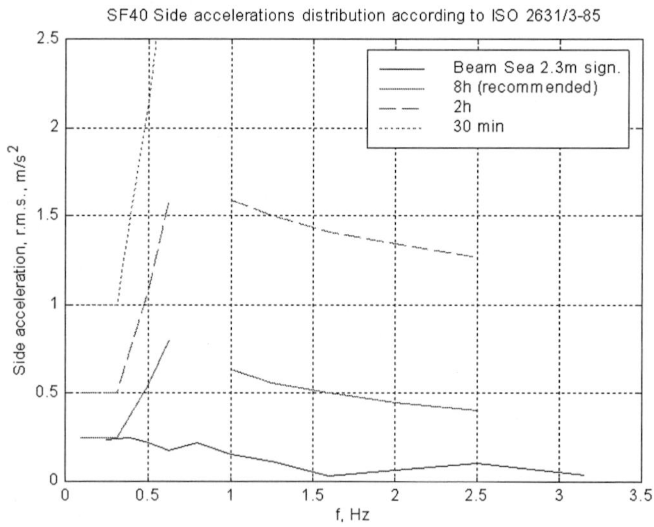

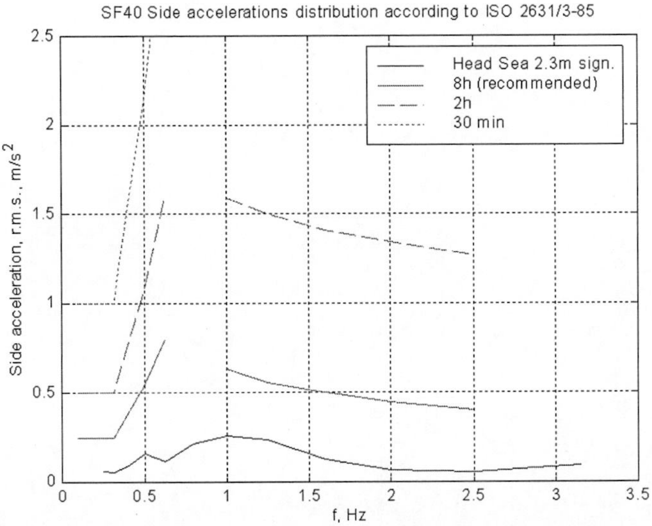

Fig. 3.3.15 (b)

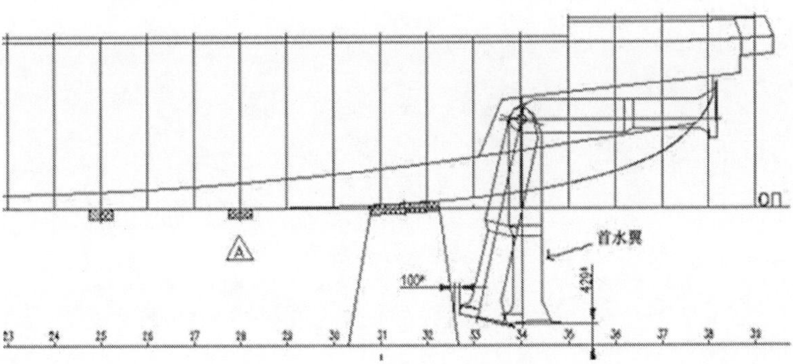

Fig. 3.3.16

Fig. 3.3.17

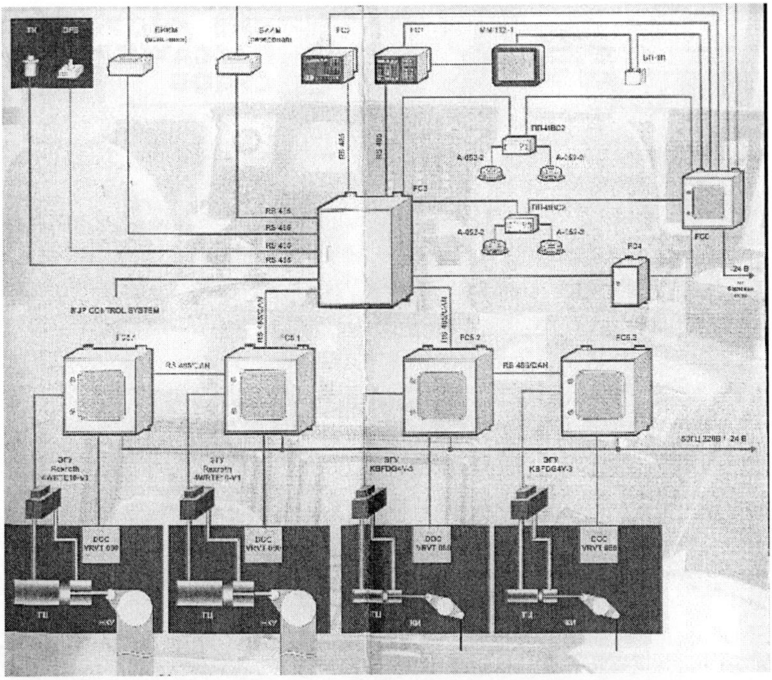

Fig. 3.3.18. Riding control system FICS-3 for the hydrofoil-stern interceptor

References

[3-3-22] Engineering & Construction Headquarter of Shanghai: Wave and Weather Condition at Yang San Area, 2003.

[3-3-23] Yun L., "Proposal for Importing the Hydrofoil Assisted CAT from Russia or Ukraine," Engain Science & Technology Company (HK), May 2003.

[3-3-24] Yun L., "Brief Introduction to the Hydrofoil & Hydrofoil Assisted CAT," Engain Science and Technology Company (HK), May 2004.

[3-3-25] "Linda Line Introduces Almaz Marine Superfoil 40," *Fast Ferry International*, July–August 2002.

[3-3-26] Specification of "Superfoil 40, 40 m Length High Speed Passenger CAT with Bow Hydrofoil and Stern Interceptor," Engain Science & Technology Company (HK), April 2004.

[3-3-27] Allan Koitmae, "Report of Sea Trial, Carried Out in Finland Gulf Near Tallinn," Engain Science & Technology Company (HK), Tallinn, Estonia, June 15, 2004.
[3-3-28] Marine Technology Design (MTD) Reports, May 2004.
[3-3-29] Motion Control System for High-Speed Passenger Vessels, by Aurora Corporation Science & Production.

3.3.3. Design and Motion Prediction of a Hybrid Hydrofoil-Assisted Catamaran (2015)— Gao Chengjun, Chen Haigang, Pan Guoping and Zhao Fa-ming (China Ship Scientific Research Center)

Abstract
In this paper, 2.5D theory with viscous effect is used to predict the motion responses of a high-speed hybrid hydrofoil-assisted catamaran in regular waves, and the numerical results agree well with the experimental data. Various ratios of lift and buoyancy are calculated, and the complicated hydrofoils are replaced by simple foil models. It is found that the viscous effect cannot be ignored in calculating longitudinal motion characteristics of high-speed catamarans and hybrid hydrofoil-assisted catamarans. The results show that the method can be used to predict the characteristics of motion response, which can improve motion prediction of high-speed hybrid hydrofoil-assisted catamaran.

3.3.3.1. Introduction

In the operation and design of high-speed crafts, better sea-keeping is a very important challenge for better maneuverability and safety because of the large motions and accelerations of craft. It was known that the motion characteristics of a craft are critical to its successful utilization on a special route because of the effect of wave length (C. C. Fang, 2004). On the other hand, catamarans have better sea-keeping performance in ground swells, which cause coupling motions. The craft may pitch, roll, and heave in short waves (K. G. Hoppe, 2001; Sim Yong Lee, Key Pyo Rhee, 2002). The engineers of CSSRC carried out research to improve the sea-keeping performance of a high-speed catamaran with hydrofoil and begin the design of a high-speed hybrid hydrofoil-assisted catamaran. There are three hydrofoils longitudinally arranged under each slice of the catamaran, which are used to counteract the waves. Fig 3.3.19 shows the arrangement of the foils on the hybrid hydrofoil catamaran slice. The measured data of the model test shows that the hybrid hydrofoil catamaran has better sea-keeping performance than a catamaran. Model tests on sea-keeping in a wave tank are the best option; nevertheless, such tests are very expensive and are limited by the dimensions of the tank and the speed of the carriage. The quality of the tests is determined by the model size and the number of oscillations in the tests (F. Pe'rez Arribas, 2006). On the other hand, open water tests in irregular waves are limited by the need to obtain an accurate measurement of the sea state, which needs to obtain the transfer function of the motions. Therefore, new methods should be proposed to predict the sea-keeping of high-speed hybrid hydrofoil catamarans.

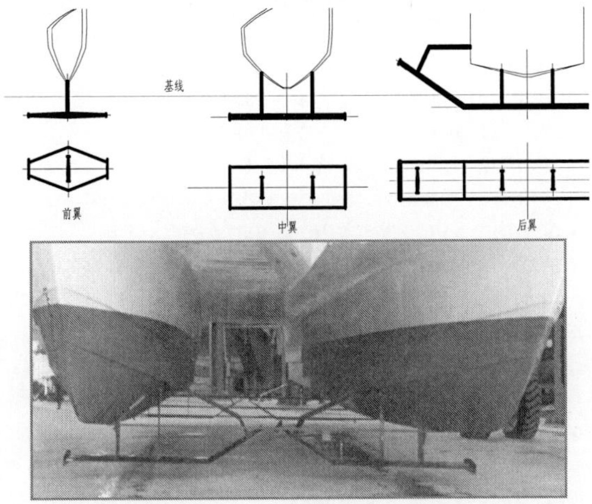

Fig. 3.3.19. The foil configuration of the hybrid hydrofoil catamaran

There are three essential theoretical methods used to predict the motion characteristic of a ship: strip theory (Henzhe Jin, 2005), the 3D method (M. Edward, 2008), and the 2.5D method (Wenyang Duan, 2001). Strip theory can predict the linear motion response of a conventional monohull at low and moderate Froude numbers. But for high Froude numbers, the three-dimensional effect of a free surface becomes dominant and strip theory generally fails to predict the hydrodynamic performance of high-speed ships. Additionally, it usually gives spurious oscillations in the motion response for multihull vessels. Although the three-dimensional method can simulate the spatial effects of the flow and the forward speed effect of a free surface. But the 3D method is very complex, time consuming, and numerically unstable in some cases. The 2.5D method is a newly developed numerical approach, where the boundary condition of velocity potential is based on two-dimensional slender theory and the free surface condition is defined in a three-dimensional system.

The motion of a surface craft is highly sensitive to its speed and hydrodynamic appendages, especially for high-speed ships at high Froude numbers. While the speed increases and the hydrodynamic forces increase quickly, the wave-damping no longer predominates and the viscous contribution has to be taken into account in the research of ship motion (Jide Li, 2008; Yuefeng Wei, 2007; Xueliang Wang, 2007; R. Centeno, 2000). Considering hydrodynamic appendages (Yadong Liu, 2005), the hull motion becomes more complicated and difficult to predict. When the catamaran is in longitudinal motion, the hydrodynamic damping produced by waves is smaller than for conventional ships in the same state. Therefore, the viscous damping produced by the catamaran is larger than for lots of conventional ships in the same state. If the damping force produced by the viscosity of the water is ignored, the numerical results would be larger than the true value. Hereby, when we calculate the motion characteristics of a catamaran, the viscous effect must be included. The above-mentioned three theory methods can't predict this effect. Therefore, classical ship motion theory and most existing sea-keeping algorithms do not consider these effects, which mainly influence the damping of vertical motion. It is necessary to present a modified method to simulate the motion of high-speed crafts, especially for catamarans with hydrodynamic appendages.

In this research, the motion response of a high-speed hybrid hydrofoil catamaran is calculated with a modified method based on the 2.5D method. The numerical results are compared with experimental data from the CSSRC tank.

3.3.3.2. 2.5D theories: hydrofoil and viscous effects

Consider a hybrid catamaran traveling in regular waves of small amplitude in deep water with the forward speed *U*. The right-hand coordinate system O-XYZ is defined to move with the ship. The O-XY plane coincides with the calm water surface, the OX axis points to the direction of ship travel, and the OZ axis points upwards to the center of gravity. In this paper, attention is focused on the vertical motions such as heave and pitch in heading waves. It is assumed that there is no coupling effect between the longitudinal motion and the transverse motion. If that is the case, the motion equations of heave and pitch can be written as:

$$\begin{cases} ([M]+A_{33})\ddot{x}_3 + B_{33}\dot{x}_3 + C_{33}x_3 + A_{35}\ddot{x}_5 + B_{35}\dot{x}_5 + C_{35}x_5 = f_3 e^{i\omega t} \\ (I_5 + A_{55})\ddot{x}_5 + B_{55}\dot{x}_5 + C_{55}x_5 + A_{53}\ddot{x}_3 + B_{53}\dot{x}_3 + C_{53}x_3 = f_5 e^{i\omega t} \end{cases}$$

(3.3.1)

where $[M]$ is the mass of the ship, A_{ij} is the added mass coefficients, B_{ij} is the damping coefficients, C_{ij} is the hydrostatic restoring coefficients, f_j is the swing of the incident wave force or force moment, I_5 is the inertia moment of the ship about the y axis, and ω is the encounter frequency of the waves. Indices 3 and 5 refer to heave and pitch motions. C_{ij}, I_5, A_{ij}, B_{ij} can be expressed as:

$$\begin{cases} C_{33} = \rho g A \\ C_{35} = C_{53} = -\rho g S_y + C_{35}^h + C_{35}^v \\ C_{55} = \rho g \forall h_y + C_{55}^h + C_{55}^v = \rho g (J_y + (z_B - z_G)\forall) + C_{55}^h + C_{55}^v \end{cases}$$

(3.3.2)

$$I_{55} = \iiint_{v_b} \rho_b (x^2 + z^2) dv = J_y \qquad (3.3.3)$$

$$\begin{cases} A_{ij} = \dfrac{1}{\omega^2} \mathrm{Re}(H_{ij}) + A_{ij}^h \\ B_{ij} = -\dfrac{1}{\omega} \mathrm{Im}(H_{ij}) + B_{ij}^h + B_{ij}^v \end{cases} \quad i = 3,5; j = 3,5 \qquad (3.3.4)$$

where A, S_y are area and static moment of water plane of the ship respectively, $\forall$ is the displacement of the ship, z_B, z_G are center of buoyancy and gravity respectively, h_y is the longitudinal metacentric height of the ship. The calculation of H_{ij} can be referred to Yuefeng Wei (2007).

(1) Hydrofoil effect

When a hybrid hydrofoil catamaran oscillates in waves, the angle of attack is always changing. When this happens, the vertical lift and trim moment are changed. To evaluate the effects of the hydrofoil on the catamaran, we suppose that the sailing state remains unchanged and the complicated hydrofoils are replaced by rectangular 3D simple foil models in calculation. By calculating the vertical lift and trim moment of hydrofoils under each slice, we can obtain the added mass, damping, and hydrostatic restoring coefficients as follows:

$$A_{33}^h = 2\sum_{j=1}^{3}(m_j^h + a_{33j}^h), \qquad A_{35}^h = -2\sum_{j=1}^{3} l_j(m_j^h + a_{33j}^h) = A_{53}^h$$

$$A_{55}^h = 2\sum_{j=1}^{3} l_j(m_j^h + a_{33j}^h), \qquad B_{33}^h = \rho\sum_{j=1}^{3} A_j^h[UC_{hj} + \frac{8}{3\pi}C_{Dj}^h\left|\dot{Z}_{hj}\right|]$$

$$B_{35}^h = -\rho\sum_{j=1}^{3} l_j A_j^h[UC_{hj} + \frac{8}{3\pi}C_{Dj}^h\left|\dot{Z}_{hj}\right|] = B_{53}^h, B_{55}^h = \rho\sum_{j=1}^{3} l_j^2 A_j^h[UC_{hj} + \frac{8}{3\pi}C_{Dj}^h\left|\dot{Z}_{hj}\right|]$$

$$C_{35}^h = \rho U^2\sum_{j=1}^{3} A_j^h C_{hj} = C_{53}^h, \qquad C_{55}^h = -\rho U^2\sum_{j=1}^{3} l_j A_j^h C_{hj}$$

(3.3.5)

where A_{ij}^h, B_{ij}^h, C_{ij}^h are the added mass coefficient, the damping coefficient and hydrostatic restoring coefficient of the foil, m_j^h, C_{hj}, A_j^h are the mass, lift coefficient, the plane area of the j foil, $\dot{Z}_{hj}$ is relative speed, C_{Dj}^h is the transverse viscous damping coefficient, and l_j is the distance from axes of the j foil to the center of gravity.

(2) Viscous effect

The hydrodynamic coefficients can be gotten by using potential theory. For most conventional ships, the wave-making damping plays a dominant role in the overall damping, but for twin-hull configurations the wave-making damping is not so important and the viscous damping is relatively larger because the area of the waterline plane of the twin-hull is smaller. To include this damping component, the viscous effects should be coupled into the mathematical model by using a cross-flow approach developed from simplified aerodynamic theory.

The total force due to viscous lift and cross flow drag is expressed as

$$F_{v3} = \frac{1}{2}\rho U^2 A_p \sin\alpha\left|\sin\alpha\right|(a_0\left|\cot\alpha\right| + C_D) \tag{3.3.6}$$

Where α is the angle of attack, A_p is the projected area, and a_0 and C_D are the lift and drag coefficient.

Equation 3.3.7 can be developed and expressed as follows:

$$\begin{cases} F_{v3} = B_{33}^v\dot{\xi}_3 + B_{35}^v\dot{\xi}_5 + C_{35}^v\xi_5 + F_{3v}^{(e)} \\ F_{v5} = B_{53}^v\dot{\xi}_3 + B_{55}^v\dot{\xi}_5 + C_{55}^v\xi_5 + F_{5v}^{(e)} \end{cases} \tag{3.3.7}$$

where B_{ij}^v is the viscous damping coefficient, C_{ij}^v is the viscous restoring coefficient, and ξ_i is the vertical displacement of water particle. The coefficients B_{ij}^v and C_{ij}^v can be calculated with the following formulas:

$$B_{33}^{v} = \rho a_0 U \int_L B_m(x) + \frac{4}{3\pi}\rho C_D \int_L B_m(x)(|\dot{z}_{1s}| + |\dot{z}_{1p}|)dx \tag{3.3.8}$$

$$B_{35}^{v} = -\rho a_0 U \int_L xB_m(x) - \frac{4}{3\pi}\rho C_D \int_L xB_m(x)(|\dot{z}_{1s}| + |\dot{z}_{1p}|)dx = B_{53}^{v}$$

(3.3.9)

$$B_{55}^{v} = \rho a_0 U \int_L x^2 B_m(x) + \frac{4}{3\pi}\rho C_D \int_L x^2 B_m(x)(|\dot{z}_{1s}| + |\dot{z}_{1p}|)dx \tag{3.3.10}$$

$$C_{35}^{v} = \rho U^2 a_0 \int_L B_m(x)dx \quad , \quad C_{55}^{v} = -\rho U^2 a_0 \int_L xB_m(x)dx \tag{3.3.11}$$

Where $B_m(x)$ is the projection of the transverse section in the horizontal, and $\dot{z}_{1s}$, $\dot{z}_{1p}$ are the vertical speed of the transverse section of the left and right slice in the x position.

3.3.3.3. Numerical results and analysis

In this paper, we calculate the longitudinal motion characteristics of high-speed catamaran and hybrid hydrofoil-assisted catamaran. Then the heave and pitch motion response of high-speed hybrid hydrofoil catamaran are analyzed with different ratios of lift and buoyancy. The main dimensions of the hybrid hydrofoil catamaran model are described in Table 3.3.2.

Table 3.3.2. Main dimensions of the hybrid hydrofoil catamaran model

L_{pp} /m	B /m	T /m	Δ /kg	B_S /m	B_{SS} /m
4	1.2	0.163	256	0.335	0.865

where B_{SS} is the distance of the mid longitudinal sections of two slices

(1) Motion response of high-speed catamaran

To analyze the viscous effect, we first compare the result of the heave and pitch motion response of catamarans with or without the viscous effect, based on the 2.5D method with experimental data at different F_n (Figs 3.3.17–20).

The peaks of the heave motions are reduced as speed is reduced. When the viscous effects are not included, the peaks of the heave motions are larger than the experimental data at high F_n, and the value of λ / L is larger near the peak. In addition, the numerical result shows the spurious oscillations phenomenal in low F_n. When the viscous effects are considered, all the results are changed. The modified 2.5D method can predict characteristics of longitudinal motion and pitch motion of high-speed catamaran. Unfortunately the error between the numerical and experimental data of heave-motion response becomes bigger when the F_n is lower.

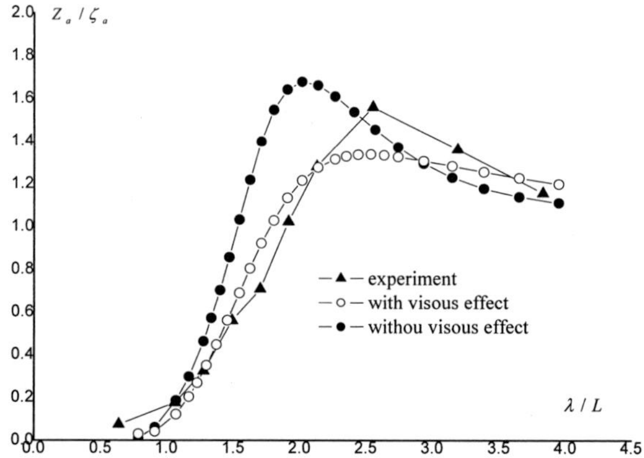

Fig. 3.3.20. Heave response of catamaran (F_n=0.9349)

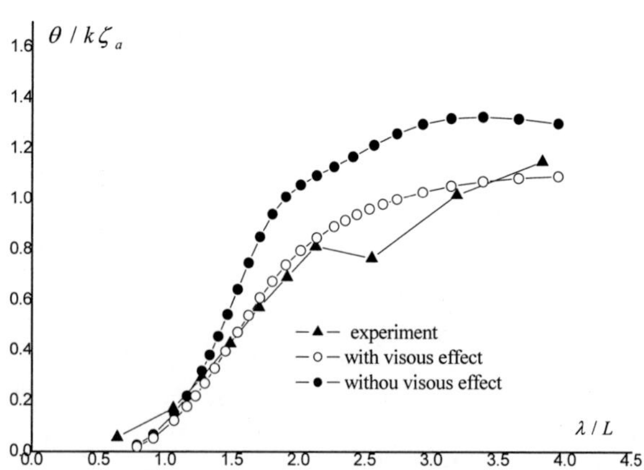

Fig. 3.3.21. Pitch response of catamaran (F_n=0.9349)

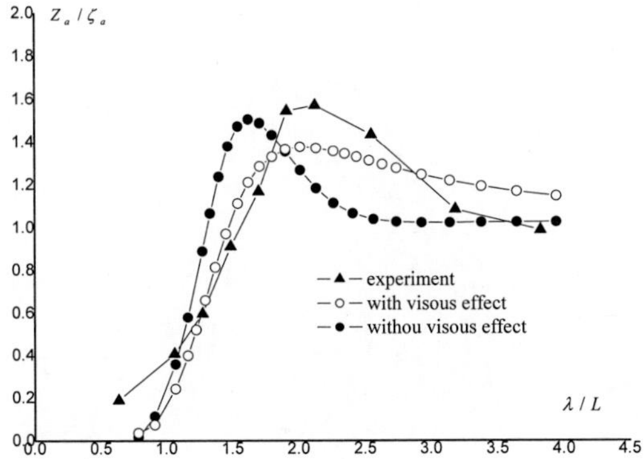

Fig. 3.3.22. Heave response of catamaran (F_n=0.7272)

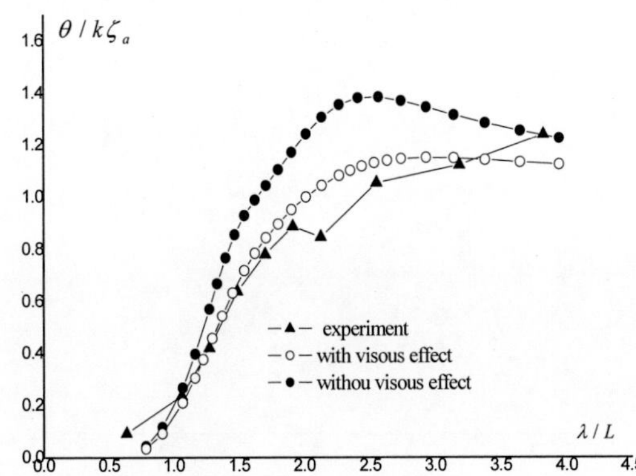

Fig. 3.3.23. Pitch response of catamaran (F_n=0.7272)

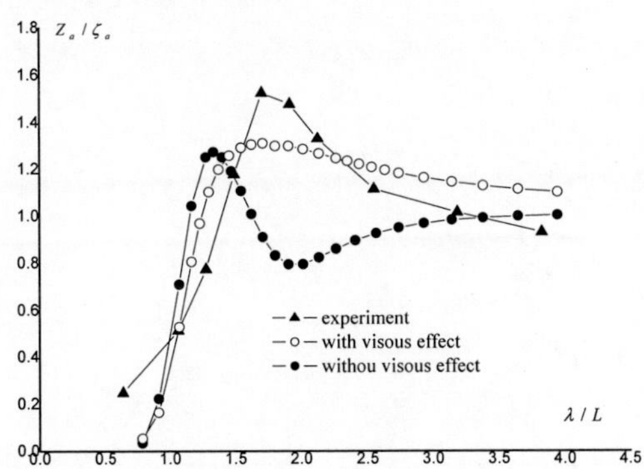

Fig. 3.3.24. Heave response of catamaran (F_n=0.5713)

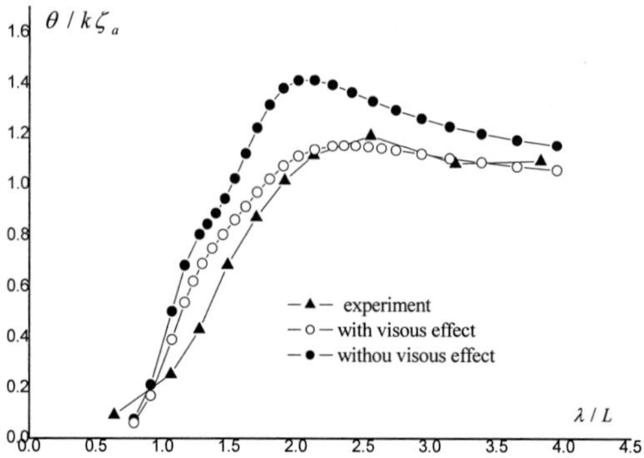

Fig. 3.3.25. Pitch response of catamaran (F_n=0.5713)

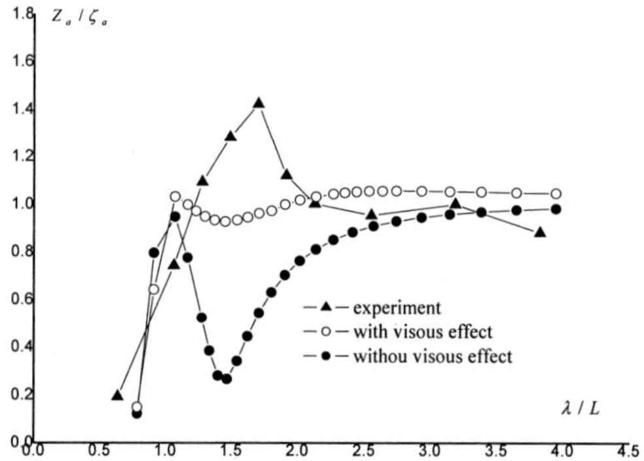

Fig. 3.3.26. Heave response of catamaran (F_n=0.3896)

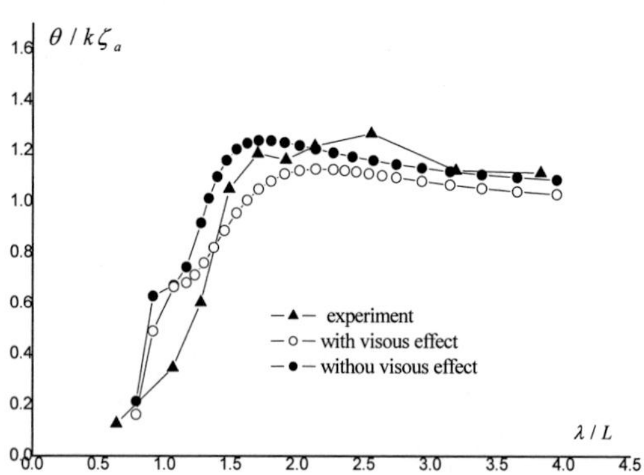

Fig. 3.3.27. Pitch response of catamaran (F_n=0.3896)

(2) Motion response of high-speed hybrid hydrofoil catamaran
In order to analyze the motion characteristics of a hybrid hydrofoil catamaran and the damping impact of hydrofoils, the heave and pitch motion responses will be simulated. To reveal the advantage of considering the viscous effect, we calculate these responses by considering the viscous effect. Only two F_n (0.9427, 0.6752) in head waves are calculated. We first show the hydrodynamic coefficients.

Figs 3.3.28 and 3.3.29 illustrate the change in hydrodynamic coefficients for different F_n. The added mass increase is small after the hydrofoils are mounted because the wetted surface and the mass of the hydrofoils are very small relative to the hybrid hydrofoil catamaran. When the F_n becomes less, the curve of the pitch added-mass coefficient becomes constant. But the damping coefficient of the hybrid catamaran is larger than the naked catamaran, which shows the preferable damping function of the hydrofoil. Thus, the hydrofoil is deemed to be a familiar instrument for reducing motion and resistance.

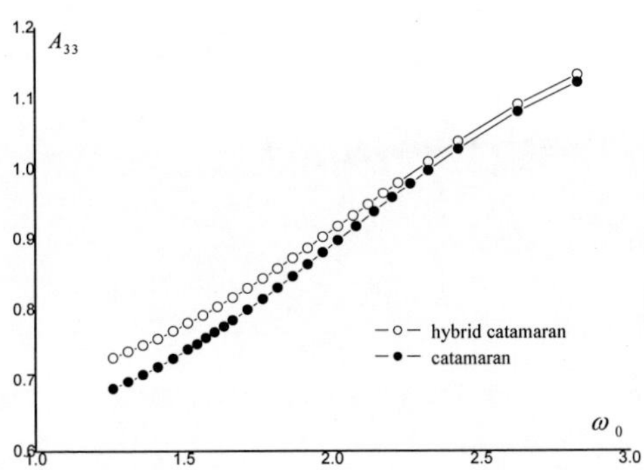

Fig. 3.3.28. Hybrid catamaran hydrodynamics coefficient (F_n=0.9427)

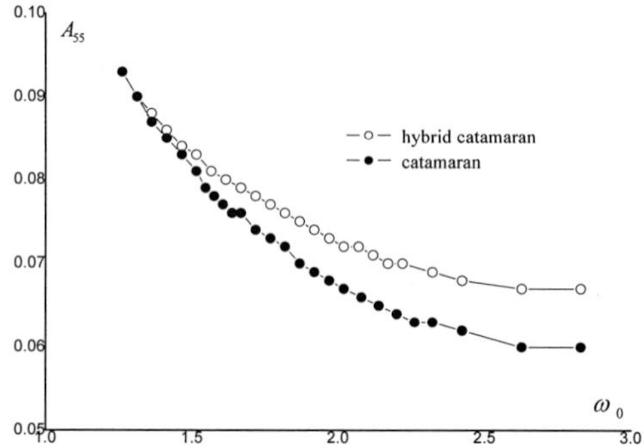

Fig. 3.3.29. Hybrid catamaran hydrodynamics coefficient (F_n=0.9427)

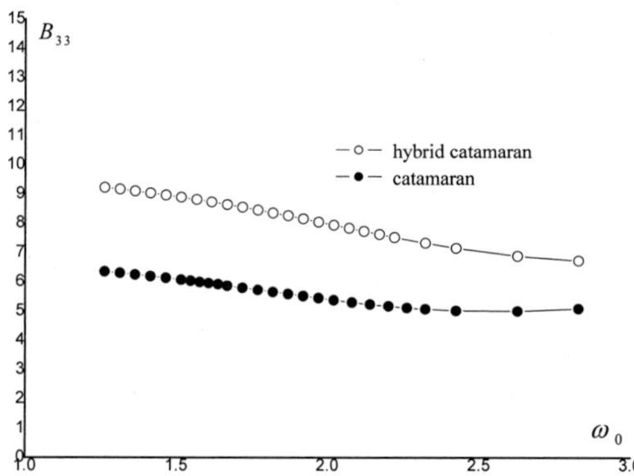

Fig. 3.3.30. Hybrid catamaran hydrodynamics coefficient (F_n=0.9427)

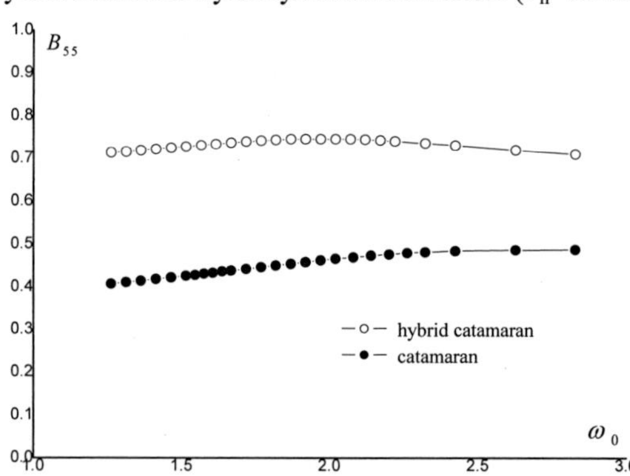

Fig. 3.3.31. Hybrid catamaran hydrodynamics coefficient (F_n=0.9427)

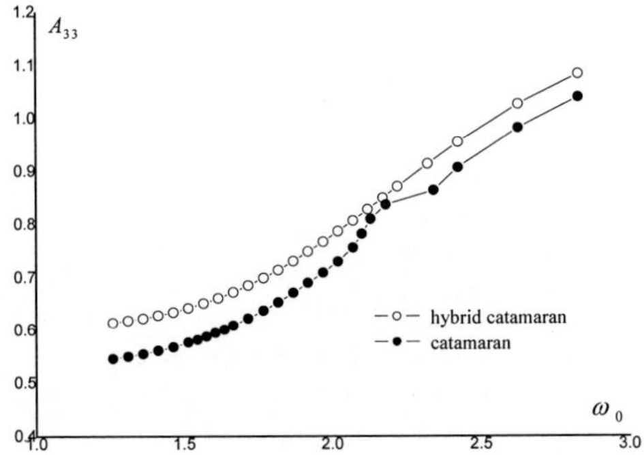

Fig. 3.3.32. Hybrid catamaran hydrodynamics coefficient (F_n=0.6752)

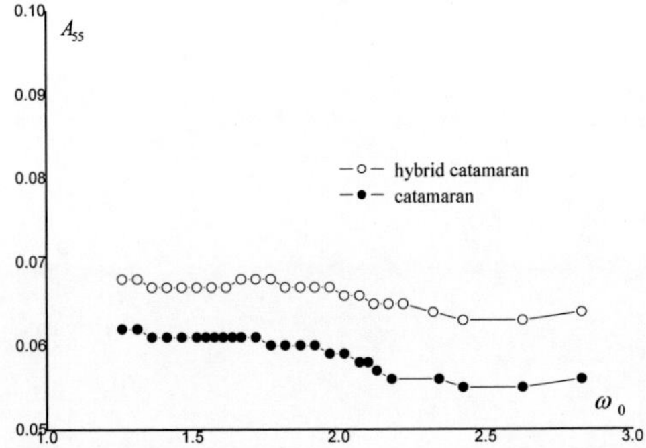

Fig. 3.3.33. Hybrid catamaran hydrodynamics coefficient (F_n=0.6752)

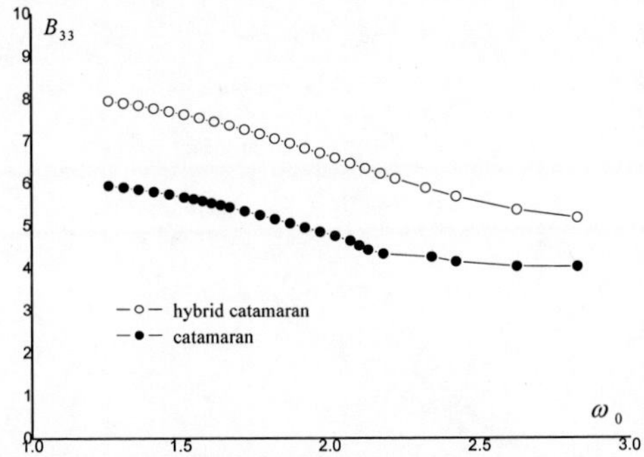

Fig. 3.3.34. Hybrid catamaran hydrodynamics coefficient (F_n=0.6752)

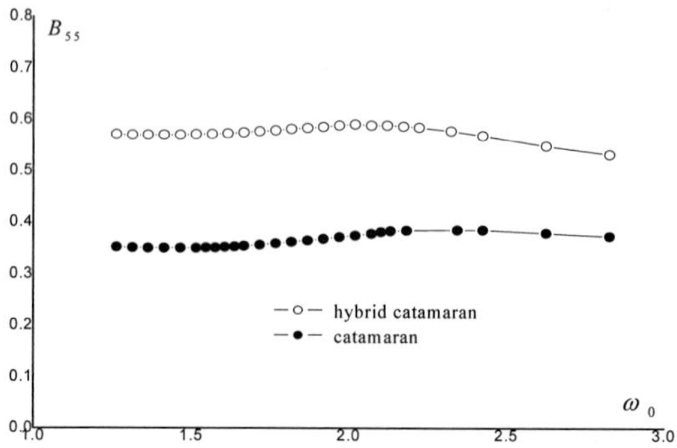

Fig. 3.3.35. Hybrid catamaran hydrodynamics coefficient (F_n=0.6752)

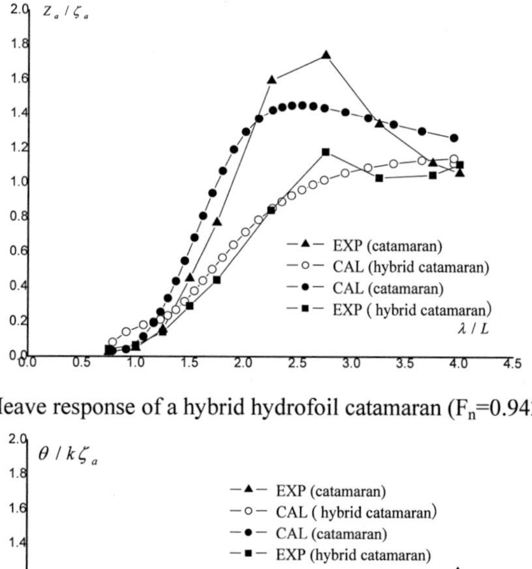

Fig. 3.3.36. Heave response of a hybrid hydrofoil catamaran (F_n=0.9427)

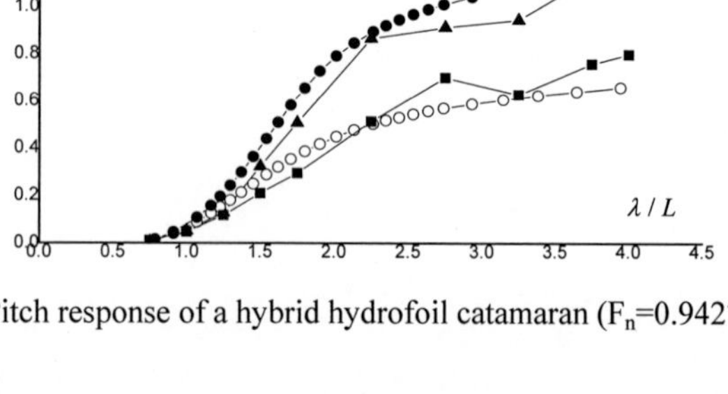

Fig. 3.3.37. Pitch response of a hybrid hydrofoil catamaran (F_n=0.9427)

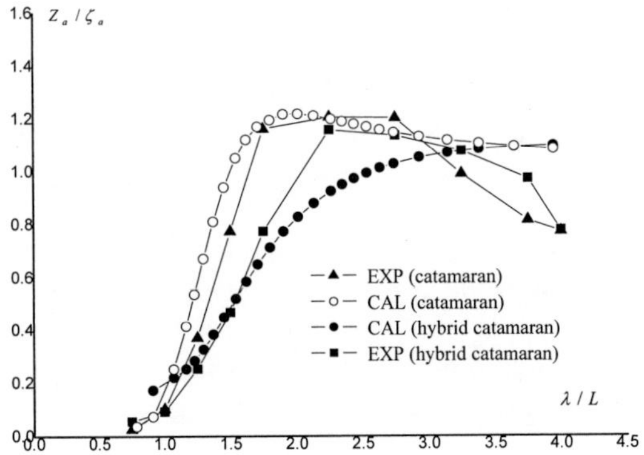

Fig. 3.3.38. Heave response of a hybrid hydrofoil catamaran (F_n=0.6752)

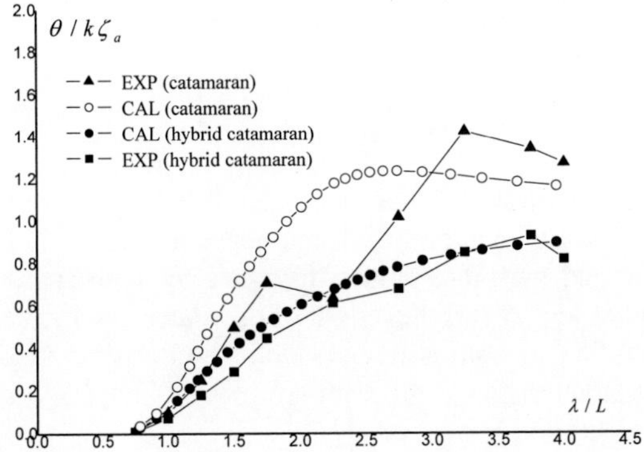

Fig. 3.3.39. Pitch response of a hybrid hydrofoil catamaran (F_n=0.6752)

Figs 3.3.36–39 show a comparison between the numerical and experimental results of heave and pitch motion responses of a hybrid hydrofoil catamaran in head waves. From the results one finds that the peaks of heave motion or pitch motion in different F_n reduce rapidly. And the numerical results and the trend of the curve agree closely with the experimental data, except for more than one peak in the experimental data.

For F_n = 0.6752, the motion response of the pitch reduces quickly with the hydrofoil, and the peak of the heave motion response falls finitely. There are small differences between the calculation and the experiment. In particular, the peak wave length of the heave motion response in calculation is different from the experimental data.

There are various reasons for the difference in numerical and experimental data. First, there are various hypotheses concerning the simplified hydrofoil model in calculation. Then the ratio of wave amplitude and the length of the waterline is larger than 1/50. But the trend of the curve is almost the same and the numerical method of the longitudinal motion characteristics of the high-speed hybrid hydrofoil catamaran can improve the precision of motion prediction.

(3) Comparison on different ratio of lift and buoyancy

To analyze the effect of the size of the hydrofoils and the ratio of lift and buoyancy on motion response, two ratios of lift and buoyancy and the relevant size of the hydrofoils are chosen. The lift coefficients of the foil are described in Table 3.3.3.

Table 3.3.3. Lift coefficients of the foil

	C_{h1}	C_{h2}	C_{h3}
Lift/Buoyancy=1/4	0.062	0.1342	0.129
Lift/Buoyancy=1/3	0.061	0.185	0.132

Figs 3.3.40–43 show the numerical results of motion response. When the ratio of lift and buoyancy is 1/3, the heave and pitch motion response reduce rapidly and the pitch motion response tends to smoothness while the ratio of wave length and the length of the water line increases. And the decrease in motion responses reduces with the decrease in F_n. Thus, the sea-keeping characteristics for the ratio of lift and buoyancy is equal to 1/3 and the ratio of lift and buoyancy is 1/4.

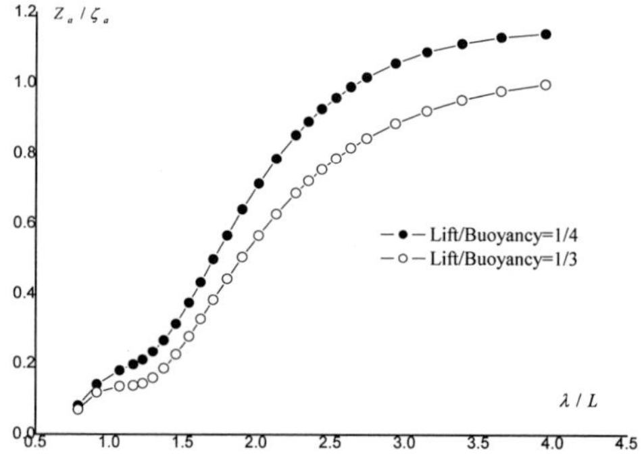

Fig. 3.3.40. Heave response of a hybrid hydrofoil catamaran (F_n=0.9427)

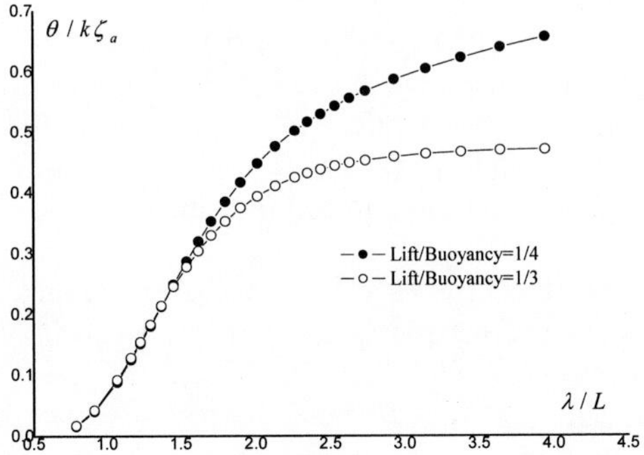

Fig. 3.3.41. Pitch response of a hybrid hydrofoil catamaran (F_n=0.9427)

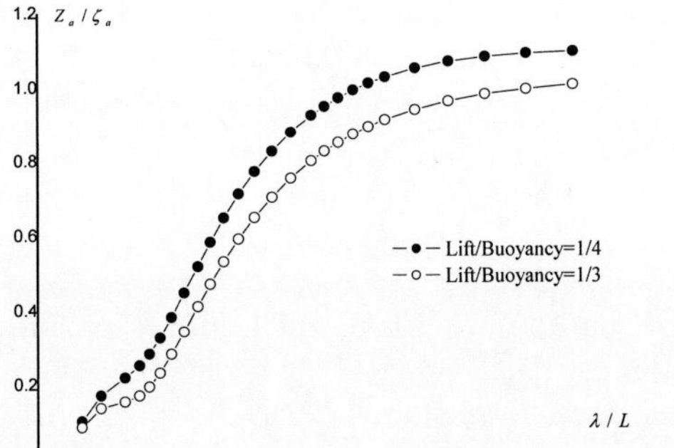

Fig. 3.3.42. Heave response of a hybrid hydrofoil catamaran (F_n=0.6752)

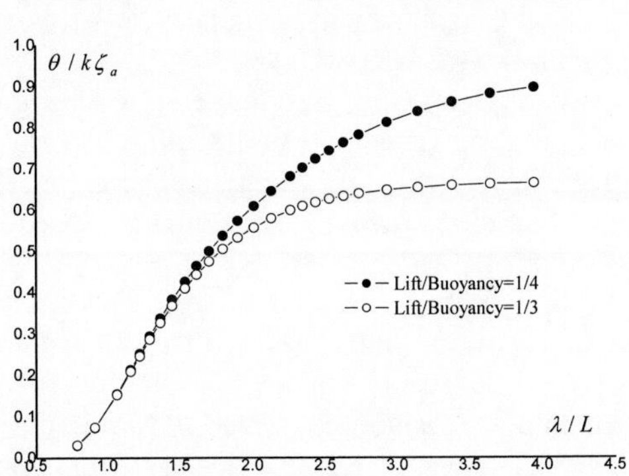

Fig. 3.3.43. Pitch response of a hybrid hydrofoil catamaran (F_n=0.6752)

3.3.3.4. Conclusions

In this paper, the 2.5D theoretical method including hydrofoils and the viscous effect is used to calculate the longitudinal motion characteristics of high-speed catamarans and hybrid hydrofoil catamarans. The effects of different ratios of lift and buoyancy on motion responses are compared. Some conclusions can be drawn from the comparison between numerical and experimental data.

It is found that the viscous effect cannot be ignored in calculating longitudinal motion characteristics of high-speed catamarans and hybrid hydrofoil catamarans.

This method can be used to predict motion characteristics of high-speed catamarans and hybrid hydrofoil catamarans. But for conditions at low Froude numbers, it will produce a larger difference between numerical and experimental data.

The hybrid hydrofoil catamaran has better sea-keeping characteristics than the high-speed catamaran for installing hydrofoils.

The sea-keeping characteristics will be better when the ratio of lift and buoyancy is larger.

References

[3-3-30] C. C. Fang & H. S. Chan, 2004, "Investigation of Seakeeping Characteristics of High-Speed Catamarans in Waves," *Journal of Marine Science and Technology* 12 (1).

[3-3-31] K. G. Hoppe, 2001, "Recent Applications of Hydrofoil Supported Catamarans," *Fast Ferry International*, pp. 30–36.

[3-3-32] Sim Yong Lee & Key Pyo Rhee, 2002, "Design of Ship-Motion Regulators for Foil Catamarans in Irregular Sea Waves," *IEEE Journal of Oceanic Engineering* 27 (3).

[3-3-33] Faming Zhao, Dong Guo, Guoping Pan, & Chengjun Gao, 2008, "Research of New Type Triple-hydrofoil Craft," Superfast 2008.

[3-3-34] F. Pe'rez Arribas & J. A. Clememte Ferna'ndez, 2006, "Strip Theories Applied to the Vertical Motions of High Speed Crafts," *Ocean Engineering* 33, pp. 1214–29.

[3-3-35] Henzhe Jin, 2005, "Research on the Conceptual Design and Performance of Hyswath," PhD thesis, Shanghai Jiaotong University, China.

[3-3-36] Edward M. Lewandowski, 2008. "Multi-vessel Seakeeping Computations with Linear Potential Theory," *Ocean Engineering* 35, pp. 1121–31.

[3-3-37] Wenyang Duan & Wuzhou He, 2001. "Motion Characteristics of Fast Displacement Vessels," *Tsinghua Univ (Sci & Tech)* 41 (12), pp. 82–85.

[3-3-38] Jide Li & Heng Zhang, 2008. "Prediction Motion of Hybrid Monohull Considering Viscous Effect," *Journal of Ship Mechanics* 12 (2), pp. 180–87.

[3-3-39] Yuefeng Wei, Wenyang Duan, & Shan Ma, 2007, "Trimaran Motions and Hydrodynamic Interaction of Side Hulls," Fast 2007, pp. 401–9.

[3-3-40] Xueliang Wang, Xuekang Gu, & Quanming Miao, 2007, "Prediction of Hydrodynamics Performance of Catamarans Accounting for Viscous Effects," Fast 2007, pp. 410–16.

[3-3-41] R. Centeno, N. Fonseca, & C. Guedes Soares, 2000, "Prediction of Motions of Catamarans Accounting for Viscous Effects, *Int. Shipbuild. Progr* 47 (451), pp. 303–32.

[3-3-42] Yadong Liu, 2005, "Research on the Ship Form of Hyswath," PhD thesis, Shanghai Jiaotong University, China.

3.3.4. Research on a New Type of Tri-hydrofoil Craft (2015) —Zhao Faming, Guo Dong, Huang Quanshui and Pan Guopin (China Ship Scientific Research Center)

Abstract

A hydrofoil craft with triple longitudinal hydrofoils is a new type of high-speed craft; the characteristics and mechanics of this tri-hydrofoil craft are analyzed in this paper. Compared with a traditional hydrofoil craft, a tri-hydrofoil craft has the same resistance performance and better sea-keeping performance without a control system. And the stress and deformation of the structure are reduced. Two fully-transparent glass-loaf tourism crafts with a triple foil have been designed and built. Therefore, the traditional hydrofoil craft can be replaced by the new type of craft with triple longitudinal hydrofoils.

3.3.4.1. Introduction

Two tandem hydrofoils were used to support the weight of the traditional hydrofoil craft, which was called the "tandem hydrofoil craft." Traditional hydrofoil craft have been used in military and commercial journeys for almost 50 years. The new type of triple-hydrofoil craft is mounted by three hydrofoils longitudinally located on bow, middle, and stern to support the craft's weight.

Theoretical research and initial model tests indicated that a hydrofoil craft with triple longitudinal hydrofoils would have better sea-keeping performance, which can achieve good motion performance without a control system. A reasonable arrangement of the three hydrofoils could improve the performance in terms of speed and power, and a hull supported by three longitudinal hydrofoils could bring benefits to the vessel structure. On the other hand, an adaptive triple-longitudinal-hydrofoil system made it possible for the design of larger hydrofoil crafts.

3.3.4.2. Sea-keeping analysis

The triple-longitudinal-hydrofoil craft is little different from tandem crafts that can bring achieve a better performance. In Fig. 3.3.44, the result of the first sea-keeping model test is shown and compared with the traditional craft. On the basis of the results of the theoretical analysis, the characteristics of the hydrofoil craft with triple longitudinal hydrofoils are as follows:

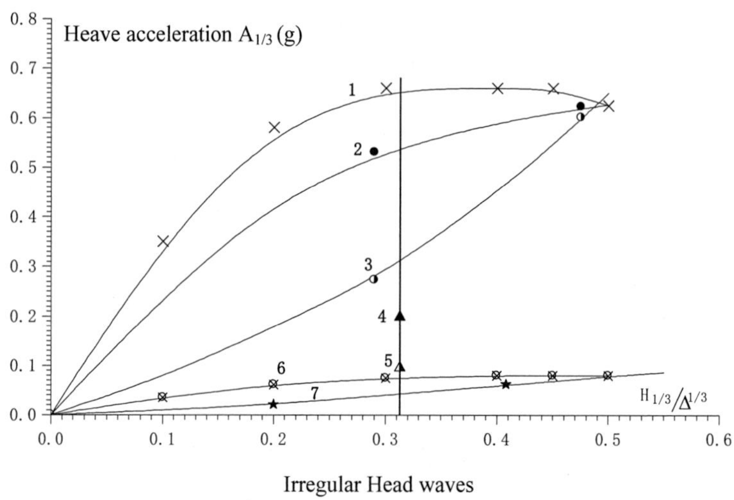

1. Hovercraft (25-40kn)
2. Self-stability twin-hydrofoil craft at wave(33-40kn)
3. Self-stability self-control twin-hydrofoil craft at wave(39-40kn)
4. Conventional wave-piercing catamaran(26-36kn)
5. Stabilization wave-piercing catamaran(26-36kn)
6. Submerged self-stability twin hydrofoil craft(45kn)
7. The test tri-hydrofoil craft

Fig. 3.3.44. Comparison on significant value of heave responses of different crafts in irregular waves ($H_{1/3}/\Delta^{1/3} = 0.315$)

(1) Reduction of longitudinal interference moment

First, the surge movement of the hydrofoil craft is caused by the interference and heave of the front hydrofoil. The motion characteristics could be improved if traditional craft adopted a canard configuration, which means that the load of the front hydrofoil will decrease and the interference moment will be reduced. The load of the front hydrofoil could be designed to be small enough in a tri-hydrofoil configuration, but its longitudinal effect on the craft's Cg will not be increased with the reduction of the load ratio (this is difference from the traditional supporting method). Therefore, the longitudinal interference moment of a triple longitudinal hydrofoil craft will be smaller in the same conditions.

(2) Adaptability to wave disturbance

When the load F_f of the front hydrofoil in a twin-foil configuration has been given, the load ratio will remain unchangeable with the same Cg, and the needed lift force will always be equal to its load F_f. When the lift force might increase to ΔL_f as the front hydrofoil been interfered, the relatively interference force $\frac{\Delta L_f}{L_f} = \frac{\Delta L_f}{F_f}$. In a short time, the relative interference force will control the heave movement of the front hydrofoil.

When the craft adopts the configuration of a triple longitudinal hydrofoil, the hull of the craft can be taken as a static unbalanced simple beam. The load F_f of the front hydrofoil is assigned a given value even in a dynamic condition; if the Cg remains unchanged, the load will be redistributed according to the attitudes of the hull and the movement of the water surface. Specifically, if one of the three hydrofoils is interfered with and its lift force is changed, the force that supports the structure will be changed and the load that supports the structure by the hydrofoils would be changed, which will decrease the interference force caused by the hydrofoil. For example, if the lift force $\Delta L_f > 0$ is the front hydrofoil interference, it makes the supporting force (L_f) increase to ($L_f + \Delta L_f$). And the load of the front hydrofoil assigned by the structure ($F_f = L_f$) will increase to ($F_f + \Delta F_f$)> F_f, the relatively interfered force is not

$\frac{\Delta L_f}{F_f}$, which causes the front hydrofoil's heave movement, but the relatively smaller one $\frac{\Delta L_f}{F_f + \Delta F_f} < \frac{\Delta L_f}{F_f}$. In other words, the interfered force borne by the front hydrofoil has decreased (refer to Fig. 3.3.44). The remaining two hydrofoils have the same results.

The hydrofoil craft with triple longitudinal hydrofoils has the characteristic of decreasing the relative wave interference force, which can be adapted to wave disturbance.

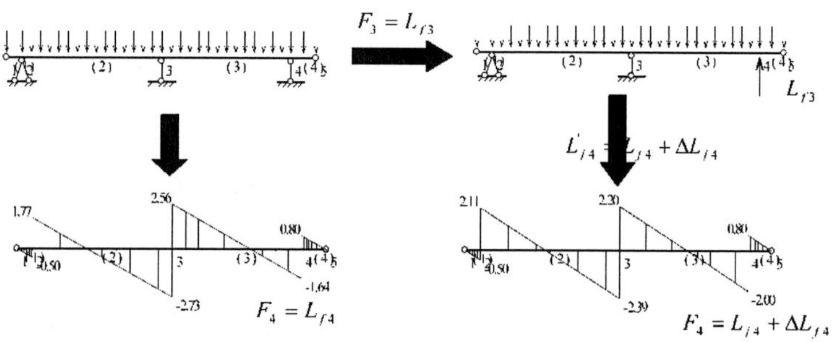

Fig. 3.3.44

(3) Longer inherent movement period

Triple hydrofoil-supported craft have changed the design concept of traditional hydrofoils with the surface-piercing system. The mechanics of the tri-hydrofoil craft joins the traditional surface-piercing system and fully-submerged hydrofoil system. So the longitudinal and transverse inherent movement period could be longer than that of the traditional hydrofoil system, where the peak of movement acceleration decreases and enhances the gentleness of the movement.

3.3.4.3. Resistance

The resistance from the model test (Fig. 3.3.45) showed that the triple longitudinal hydrofoil craft has the same resistance performance compared with traditional hydrofoil craft, which can be explained by the qualitative analysis as follows:

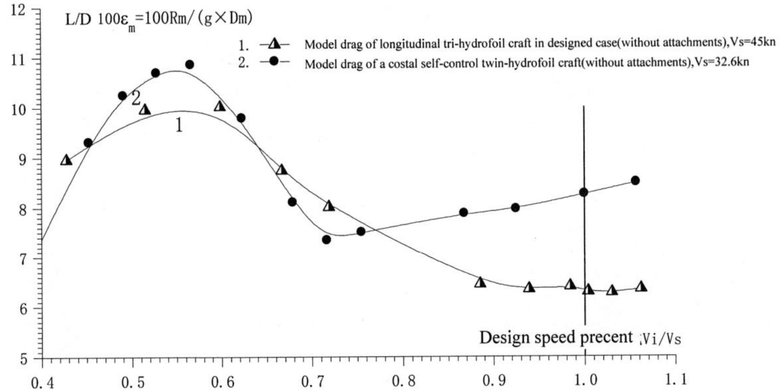

Fig. 3.3.45

(1) Disadvantages

Because the space between hydrofoils obviously is reduced in triple-hydrofoil systems, the middle foil and stern foil may work in a deteriorated wake, inevitably causing additional interference resistance.

(2) Advantages

(a) The propellers of hydrofoil craft are usually installed below the stern hydrofoil in the triple hydrofoils supporting system, the wake after the middle hydrofoil brings a stronger downwash, which improves the propeller inflow and its efficiency.

(b) Having no flaps set on hydrofoils to control movement will bring advantages by reducing the resistance of the hydrofoil system.

3.3.4.4. Stress and deflection of the hull

Traditional hydrofoil craft can be taken as a two-point-supported beam (Fig. 3.3.46). But hydrofoil craft with triple longitudinal hydrofoils can be taken as a triple-point-supported static volatile unbalanced simple beam (Fig. 3.3.44), which can sharply reduce the longitudinal stress and deflection and can also reduce deflection stress and deformation of the hull in roll motion. The configuration of triple hydrofoil craft can reduce the weight of the structure, increase capacity and extend the fatigue life of the hull. The main engine and propellers will work reliably.

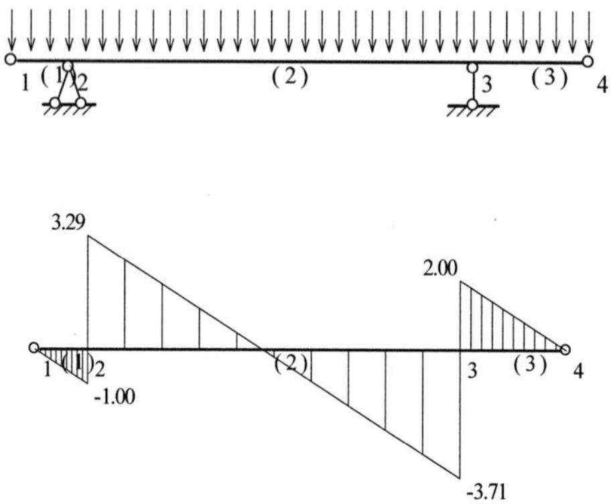

Fig. 3.3.46

Theoretical calculation indicated that the longitudinal deflection could be reduced about 70% compared with traditional tandem hydrofoil craft, which was validated by measuring the deformation measurement in the real boat.

3.3.4.5. Ship design

The first fully-open sightseeing tourism craft with triple hydrofoils in the world was designed with a fully-open sightseeing cabin. Conventional hydrofoil craft supported by tandem hydrofoils make a larger sagging deformation when foil borne at high speed; tandem-hydrofoil crafts have to use a fully-enclosed steel-welded upper structure to bear this deformation, which will restrict the application of hydrofoil craft in tourism. But the design of configuration by using triple hydrofoil can sharply reduce structural deformation: the upper structure can use tempered glass to meet the requirements of sightseeing and tourism, which was the first application of a triple-hydrofoil craft in this new field.

With the breakthrough of the key technology in the design of triple hydrofoil craft by researchers at CSSRC, the first two high-speed fully-open sightseeing and tourism crafts, named *Qinzhou II*, with triple hydrofoil were designed and constructed by CSSRC Beijing Yuanzhou High-Performance Vehicle Co. Ltd. These craft will be used to meet the requirements of tourism in the Three Gorges of the Yangtze River. The result of the first trial ride and operation showed that general performance was steady and reliable and that it can satisfy the standard of design. The craft were built in 2008–9; the operation of the craft in the Three Gorges brought about economic returns and social befits.

The primary parameters of *Qinzhou II* high-speed fully-open sightseeing tourism craft with triple hydrofoils are as follows:

$L_{o.a}$ 32.85m
Length 31.50m
Waterline 28.2m
$B_{o.a}$ 8.10m
Beam design 5.00m
Depth design 2.00m
Draft (floating state) 0.75m
Capacity 10.0t
Velocity 72 km/h
Engine power 2 × 890kW/2,300rpm

Fig. 3.3.48. Photos of *Qinzhou II*

3.3.4.6. Conclusion

The sea-keeping and resistance performance of the triple-hydrofoil craft was better than a traditional craft. The stress and deformation of the hull was reduced by avoiding a control system, which was more simple, economical, and reliable. Triple-hydrofoil high-speed craft will be more attractive to potential users and will be applied widely in new fields.

References

[3-3-43] Lan Shaoxiang, Wang Xiaochuang, Zhang Wenzhong, & Zhao Faming, "Model Drag and Seakeeping Test Study on the High-Speed Hovercraft with Longitudinal Tri-Hydrofoils," High Performance Ship Key Technology Study, Hovercraft special research report, 2003.

[3-3-44] Lan Shaoxiang, Wang Xiaochuan, & Zhang Wenzhong, "Transverse Stability Design Technology Study on the Piercing Hydrofoil System," Hovercraft Special Research Report, 2001.

[3-3-45] Li Baiqi, "Dimensions Choice and Foilborne Power Calculation," *Ship Performance Research* 4, 1978.

[3-3-46] Su Hengyu & Tan Linsheng, *Structural Mechanics of Ships*, Wuhan: Press of Huazhong University of Science.

3.4

THE RESEARCH AND DESIGN OF HYDROFOIL CRAFT

3.4.1. Design of 2D Anti-Cavitation Hydrofoil by Section Reshaping (2012)

—Wu Wei and Xiong Ying
(Naval University of Engineering, China)

Abstract
To improve the anti-cavitation performance of the hydrofoil, the appropriate transformation was carried out on the suction side of the two-dimensional foil. With integration of the geometric parameterization, the panel method and the optimization algorithm, the optimization method of the anti-cavitation hydrofoil was reconstructed. The modification was achieved by setting up a camber on the suction side of the foil section and by using a cubic B-spline to parameterize and express the shape. The basis data for the hydrodynamic performance of the foil for the object function of the genetic algorithm was gained by using the panel method, which was used to optimize the camber distribution of the NACA0010 airfoil. The simulation of an unsteady flow-field with/without cavitation around the original and modified foil was carried out by using RANS in Fluent. The size and shape of the cavitation on the foils can be given and compared by the volume fraction of gas, which showed that the proposed optimization method could be used for the design of the anti-cavitation hydrofoil. The method was basic to the design of the three-dimensional anti-cavitation hydrofoils.

3.4.1.1. Introduction

(1) Design and cavitation of hydrofoil

Hydrofoils are widely used on ships, including hydrofoils, rudders, oars, and stabilizing fins. The design of the hydrofoil profile aimed for a hydrofoil that was in the given flow-field and would provide the needed hydrodynamic performance.[3-4-1] The design methods for foils include the direct numerical optimization method, the inverse design method, and the panel method. Integrating the optimization strategies with a proper parameterization in the expression of the foil makes the design more and more flexible. Li Shengzhong et al.[3-4-2] took the uniformity of the wake and ratio of the drag and lift as the object function in the design of the foil. They used Bezier curves to express and reconstruct the geometry of the foil. And they adopted DOE and NSGA II algorithms to the multi-objective optimization of the hydrofoils. Xu Weibao, et al.[3-4-3] used the particle swarm optimization (PSO) integrated with the boundary element method to design an airfoil section for a high ratio of lift and drag. Zeng Zhibo[3-4-4] gave a comparative analysis of two kinds of parameterization methods (B-spline and Eppler methods) in the optimization design of foils.

Different from aerofoils, hydrofoils operate under the water. Cavitation occurs when the partial pressure is lower than the saturation pressure of water. The occurrence of cavitation changes the characteristics of hydrofoils and brings about some serious problems. If the cavitation occurred on a high-speed operation propeller, it would produce a large area of sheet cavitation, cloud cavitation, and then cavitation break-up. The cavitation seriously affects the efficiency of the propulsion. In addition, the deterioration of cavitation caused by the rotating blades is harmful to the vibration of the hull and the stealth of the ships. So it is important to understand the mechanism and characteristics of unsteady cavitation, which were carried out by a lot of researchers. Potential theory had been used to predict the cavitation on the propeller. Xiong Ying[3-4-5] predicted the developing procedure and volume pulse of cavitation on the propeller by using the boundary element method. Cheng Xiaojun et al.[3-4-6] applied re-entrant jet and the pressure closure model to simulate partial-cavitation on two-dimensional hydrofoils. Kinnas and Fine[3-4-7] used the nonlinear boundary integral method, based on velocity potential, to study partial- or super-cavitating around the two-dimensional hydrofoil. And they gave meaningful conclusions that could be applied to the research on three-dimensional hydrofoils. Mishima,[3-4-8] Lee,[3-4-9] and Krishnaswamy[3-4-10], [3-4-11] used a similar approach to the research on partial- and super-cavitation of a hydrofoil. On the other hand, many researchers applied the viscous solver to the problem of cavitation in detailed discussions.[3-4-12–14]

(2) Set of block flow on airfoil

From the experimental observation,[3-4-15] the re-entrant jet occurs at the end of the partial cavitation, which is similar to the separation in the boundary layer. In the experiment, the speed of the re-entrant jet is approximately 10% of the speed of the external mainstream. The frequency of occurrence of the re-entrant jet is almost the same as the frequency of the cavitation volume. The re-entrant jet made the cavitation bubbles merge and collapse in the wake of the hydrofoil, which leads to the formation of cloud cavitation behind the sheet cavitation. By using a high-speed video camera, it was found that the re-entrant jet is closely related to the generation of cloud cavitation.[3-4-16] On the basis of this understanding, the control of the re-entrant jet could lead to the reduction of cloud cavitation. Usually, there are two methods to control the re-entrant jet. One is setting up barriers to the surface of the foil, the other is setting a reverse air jet. Gu Wei et al.[3-4-17] found that the unsteady cavitating flow had the characteristics of the eddy flow with an open wake. Especially under the large angle of attack, the vortex in a cavitation bubble spread from the leading edge of the hydrofoil to the end of the cavitation and gathered there. They would trigger the burst of reverse flow that results in the shedding of cloud cavitation and the collapse of cavitation attached to the leading edge of the foil. In their experiments, a bar was set on the hydrofoil surface, which led to some improvement in the cavitation flow within a certain range of cavitation numbers. Zhou Li et al[3-4-18] used CFD software to simulate the suppressed cavitation flow around the airfoil set with a small bump.

(3) The main work

According to the literature and experiments,[3-4-15–18] the foil is reshaped by setting a vaulting barrier on the suction side, which not only blocks the re-entrant jet, but also avoids the singularity on the suction surface that changes the stability of the flow field. It is very important to set a proper barrier on the foil; thus, the main content of this paper aims to establish the optimal method to find the parameter of the vaulting barrier for the NACA0010 under a given condition. The optimization flow chart is shown in Fig. 3.4.1.

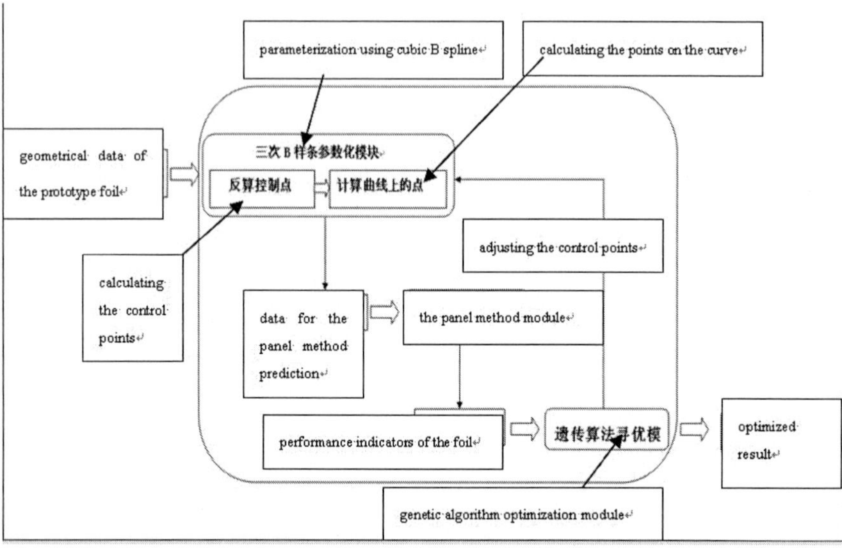

Fig. 3.4.1. Optimization procedure

3.4.1.2. Parameterization of hydrofoil geometry by B-spline

The numerical optimization design of the hydrofoil is based on the parameterization of foil geometry. In the method, the original foil should be discretized and expressed in parametric functions. By selecting the appropriate parameters of the airfoil as variables for the design and by using the optimization algorithm, the optimum design of the foil could be gotten and the flow-field around the foil could be calculated. The optimization scope is limited by the parametric expressions of foils, which will determine the availability of optimal hydrofoil under a given target. Airfoil or hydrofoil parameterization methods include polynomial fitting and linear superposition of analytic functions, such as the Hick-Henne shape function method,[3-4-3] the Bezier curve,[3-4-2] and the B-spline curve.[3-4-4] Hick-Henne made the foil curve superimposed on a large number of local perturbation functions; it is difficult to make the curve smooth everywhere. And the establishment of the parameters and the geometric properties in the optimization process is particularly important to avoid the blindness in the optimization. Therefore, Bezier curves and B-spline curves are used to deal with the movement of the control points that can ensure the smoothness of the hydrofoil in optimization and satisfy the need of the given target. The Bezier curve has some shortcomings compared with the B-spline curve. First, it cannot accurately describe the conics. Also, the parameters of the Bezier curve are global. In other words, when one control point changes, the entire curve will be changed.[3-4-19] At the same time, the degree of the basis function of a Bezier curve with N control points is N-1, which cannot be derived to its expression in some geometry. It is possible for certain processing to describe the curve in segments and to make connection points to meet the C_2 conditions that will make the parameterization inconvenient. On the other hand, the degree of the B-spline curve basis function k has nothing to do with control point N. The B-spline curve can provide a good description of the control parameters and geometry. So the authors used a cubic B-spline in the hydrofoil parametric expression. The B-spline curve can be written as follows:

$$p(u) = \sum_{i=0}^{n} d_i N_{i,k}(u) \qquad (3.4.1)$$

where, d_i (I = 0,1,..., n) are the control points and $N_{i,k}(u)$ (I = 0,1, ..., n) are the B-spline basis functions. It is called the node vector of non-descending order of parameters U : $u_0 \leq u_1 \leq \cdots \leq u_{n+k+1}$, which are determined by the sub-polynomial spline. According to the above equation, the deformation of the curve could be changed by adjusting the control points.

The vaulting barrier is set up on the suction side of NACA0010. The points of the pressure surface and the leading edge and trailing edge of suction (around 15% each) and the maximum thickness (from the lead edge of about 30%) are unchanged. First, the value of control points d_i (i=0, 1, …, n) are calculated. Then a spline curve is constructed through the control points by using cubic B-spline basis functions. Thus, the data can be used to input the prediction for the boundary element method. Then by giving a displacement vector $\vec{P}(\Delta x, \Delta y)$ to the control point, which corresponds to the change in the maximum thickness of the suction side, so the modification of the foil by the barrier is achieved to block the flow. The expression of the B-spline and displacement of control points on the hydrofoil are shown in Fig. 3.4.2. The red curve of the suction side reflects the changes. It is easy to find that the B-spline works well in the parameterization of the design of the hydrofoils.

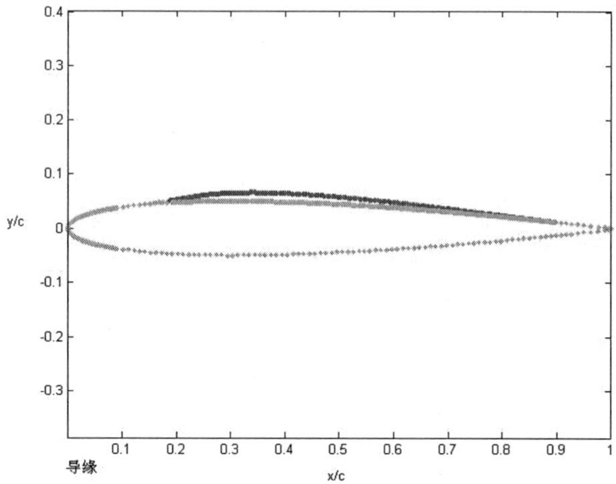

Fig. 3.4.2. Parameterization of airfoil by B-spline

3.4.1.3. Panel method

The panel method is an effective tool for solving the flow-field around hydrofoils. The panel-method-based source, sink, and vortex is used in this paper. The surface of the body can be divided into *m* pieces from the down side to the upper side of the foil; the first point is at the trailing edge. Each segment can be approximately taken as a short line. Source strength in each line is σ_j (i= 0, 1… m) and the vortex strength is γ_j. So there are m+1 unknowns to be solved. To satisfy the boundary conditions of the body on the control points (which often lie in the mid-point of the element), it needs to create *m* control equations. The Kutta condition is an additional boundary condition and can be written as another control equation. Then m+1 equations simultaneously would be solved to obtain all unknowns. The source/sink and vortex laid in the control point of short line j will interact with those of any short line i (x_i, y_i). Thus, the velocity potential can be written respectively as:

$$\frac{1}{2\pi}\int_{panel_j} \ln r_{ij} \cdot \sigma_j dS_j \quad \text{and} \quad \frac{1}{2\pi}\int_{panel_j} \gamma \cdot \theta_{ij} dS_j \tag{3.4.2}$$

Where, $r_{ij} = \left\{ \left(x_i - x_j\right)^2 + \left(y_i - y_j\right)^2 \right\}^{\frac{1}{2}}$, $\theta_{ij} = \arctan \frac{y_i - y_j}{x_i - x_j}$

When the angle of attack is α and the speed of flow is V_∞, the velocity potential can be generated on any control point $\phi(x_i, y_i)$. Because the normal velocity at control points are zero, the equations are:

$$\frac{\sigma_i}{2}+\sum_{\substack{j=1\\i\neq j}}^{m}\frac{\sigma_j}{2\pi}\int_{\Delta s_j}\frac{\partial}{\partial n_i}\left(\ln r_{ij}\right)dS+\sum_{\substack{j=1\\i\neq j}}^{m}\frac{\gamma}{2\pi}\int_{\Delta s_j}\frac{\partial}{\partial n_i}\left(\arctan\frac{y_i-y_j}{x_i-x_j}\right)dS=-V_\infty\left[\cos\alpha\cos(n_i,x)+\sin\alpha\sin(n_i,x)\right]$$

(i=0, 1, …, m) (3.4.3)

The Kutta conditions at the trailing edge give another equation:

$$\sum_{j=1}^{m}\frac{\sigma_j}{2\pi}\int_{\Delta s_j}\frac{\partial}{\partial n_0}\left(\ln r_{0j}\right)dS+\sum_{j=1}^{m}\frac{\gamma}{2\pi}\int_{\Delta s_j}\frac{\partial}{\partial n_0}\left(\arctan\frac{y_0-y_j}{x_0-x_j}\right)dS=-V_\infty\left[\cos\alpha\cos(n_0,x)+\sin\alpha\sin(n_0,x)\right]$$

(3.4.4)

From equations 3.4.3 and 3.4.4, the values of σ_j (i = 0, 1… m) and γ can be solved. And then the distribution of velocity and pressure on each side of the foil can be obtained by the following formula:

$$V_i=V_\infty\cos\alpha\cdot\sin(n_i,x)+V_\infty\sin\alpha\cdot\cos(n_i,x)+\sum_{\substack{j=1\\j\neq i}}^{m}\frac{\sigma_j}{2\pi}\int_{\Delta s_j}\frac{\partial}{\partial S_i}\left(\ln r_{ij}\right)dS+\frac{1}{2}\gamma+\sum_{\substack{j=1\\j\neq i}}^{m}\frac{\gamma}{2\pi}\int_{\Delta s_j}\frac{\partial}{\partial S_i}\left(\arctan\frac{y_i-y_j}{x_i-x_j}\right)dS$$

(3.4.5)

$$C_{pi}=1-(V_i/V_\infty)^2$$ (3.4.6)

The force f_x, f_y on the entire airfoil was obtained by superimposition of the force on each segment. The lift (or pressure drag) is a perpendicular (or parallel) component to the flow direction. Fig. 3.4.3 shows the distribution of pressure along the surface of NACA0010 at the angle of attack of 6°, which was obtained by using the panel method. The result showed that the distribution pressure was very close to the data from the X-foil (V6.96) and the data obtained by using Fluent. Because of the singularity at the leading edge of the airfoil, the value of C_p obtained by the panel method is different from that of RANS in Fluent. Strictly speaking, the results of RANS coincided more closely with the real value. The trailing edge was not absolutely closed by the interpolation of the B-spline, which affected the accuracy at the trailing edge. Comparison of the numerical results of the lift coefficient and the pressure drag coefficient are shown in Table 3.4.1. The numerical results of using the panel method agree well with those from using other software. In Fig. 3.4.4 the different pressure distribution acquired by moving the control points is shown, which is due to the deformation of the suction side. Although the theory of potential flow cannot explain the relationship of the barrier and re-entrant jet, the numerical results showed that the barrier made the pressure coefficient before the barrier higher than the original, thereby delaying the inception of the cavitation or reducing the area of cavitation on the hydrofoil. But the pressure coefficient after the barrier was lower than the original, which is harmful to the cavitation performance. And the reduction of the "surplus space" can change distribution of the pressure coefficient in the front of barrier, which may improve cavitation performance of the hydrofoil. In this way, the proper distribution of pressure can be given to the optimization and design of the hydrofoils.

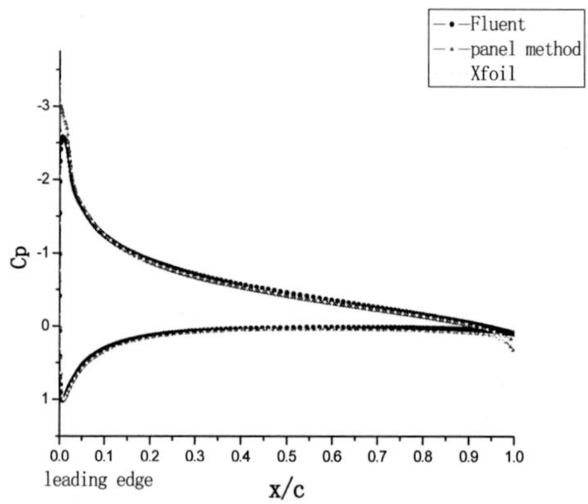

Fig. 3.4.3. Pressure distribution obtained by using different software

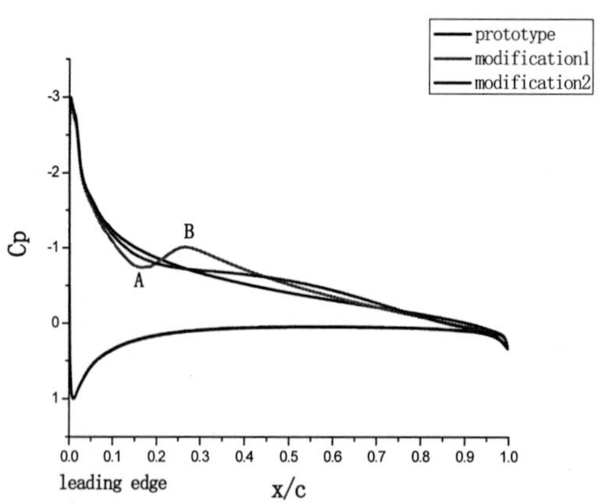

Fig. 3.4.4. Comparison of pressure distribution on different hydrofoils

Table 3.4.1. Lift coefficient and pressure drag coefficient by using different software

	Lift coefficient C_L	Pressure drag coefficient C_{Dp}
Boundary element method	0.69433	-0.00099
X-foil	0.7124	-0.00094
Panel method	0.71963	———
Fluent	0.690156	———

3.4.1.4. Genetic algorithm

There are several algorithms to be used in various optimizations. A genetic algorithm is a good way to solve the optimization of complicated systems. If the genetic algorithm is used to find the maximum value of the function, the mathematical model can be described as the following:

$$\begin{cases} \max\ f(X) \\ s.t.\quad X \in R \\ \qquad\ R \subseteq U \end{cases} \qquad (3.4.7)$$

where $X = [x_1, x_2 \ldots, x_n]^T$ are decision variables, $f(X)$ is the objective function, U is the basic space, R is a subset of U or a set of constraints. The search for the optimal solution will imitate the process of biological evolution. The genetic operators act on the group $P(t)$, which is made up by a series of individuals. The new generation $P(t+1)$ with higher fitness will be given by applying a selection operator, a crossover operator, and a mutation operator in the operations of genetic evolution. After a number of genetic manipulations, the optimal individual will be found in the final group, which will reach or be close to the optimal solution.

(1) Objective function

The control points in the scope of $\Delta x \in [-0.25C, 0.25C]\ \Delta y \in [0, 0.04C]$ are moved to reshape the suction side of the hydrofoil. The barrier will increase the coefficient of pressure in front of it. Because it is singular at the leading edge of the hydrofoil in potential flow theory, the average of Cp is initialized as the target function of optimization from a 3% chord length after the leading edge. Function 3.4.8 gives the increment of the pressure coefficient on the fore part of the hydrofoil.

$$f(x) = \max(\Delta \bar{h}) - (B) \quad (3.4.8)$$

On the other hand, the lowest point B on the body after the barrier will lead to a serious risk of cavitation (Fig. 3.4.4). Thus the value of point B should be controlled and given an appropriate penalty into function. The value of point B was limited to the same pressure coefficient at the 15% chord length of the suction side of the original hydrofoil.

(2) Optimal results

The optimal vector $\vec{P}(\Delta x, \Delta y)$ is (-0.067, 0.04) by using the genetic algorithm. The final shape and the distribution of the pressure coefficient on an optimized foil are given in Figs 3.4.5 and 3.4.6. The lift coefficient is 0.741, the pressure drag coefficient is -1.01×10^{-03}, which are larger than those on the original one. As NACA0010 is a symmetrical wing, the barrier change is the camber of the foil. When the reshaped foil is operating at the same angle of attack, the lift coefficient will be increased. The modification of the camber has less impact on the lift characteristics of the foil. And the final hydrodynamic performance of the optimal hydrofoil will be almost unchanged for little change in the thickness.

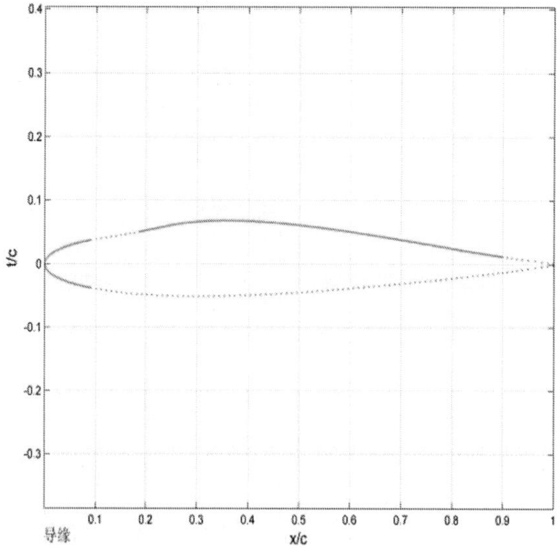

Fig. 3.4.5. Optimized shape of the hydrofoil

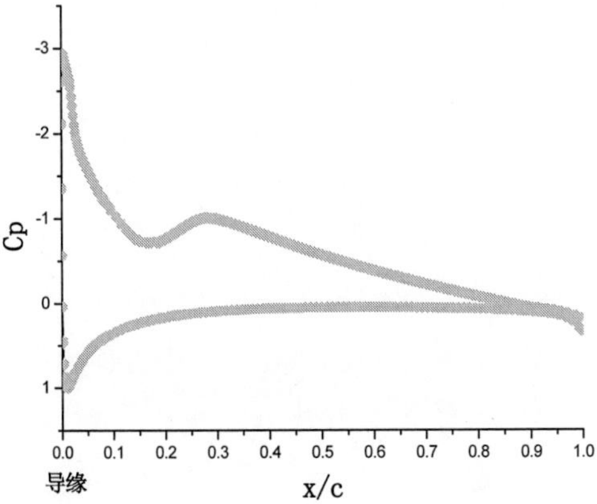

Fig. 3.4.6. Pressure distribution on the optimized hydrofoil surface

3.4.1.5. Result of RANS

The RANS equations are used to simulate the cavitation flow around the original and optimized hydrofoils by using CFD software. The chord length is 0.1m in numerical simulation. The "C" type mesh is applied in the computational domain and the grid number is about 80,000. The computational domain and the grid are shown in Fig. 3.4.7. The inlet and upper and lower boundaries of the domain are set to the velocity inlet boundary conditions. The outlet is set to pressure outlet boundary conditions. The surface of the hydrofoil is set to non-slip wall boundary conditions. The velocity V of the inflow is 10m/s. and the angle of attack is 6° and $Re(Re = \frac{V_\infty C}{\upsilon})$ is 1 × 10^6. The cavitation number $\sigma(\sigma = \frac{p_\infty - p_v}{\frac{1}{2}\rho V^2})$ is controlled by changing the outlet static pressure p to get the desired cavitation number σ. Saturated vapor pressure in the calculation is $P_v = 3540\ pa$. Using Reference 3-4-13, the RNG $k-\varepsilon$ turbulence model for closure of RANS was used in the simulations. When the cavitation number is 1.3, the numerical results of the unsteady flow-field with cavitation are shown in Figure 3.4.8. The volume fraction of vapor shows that the modified hydrofoil can suppress the development of cavitation especially in the initial stage because of the control of the peak at point B in the optimization.

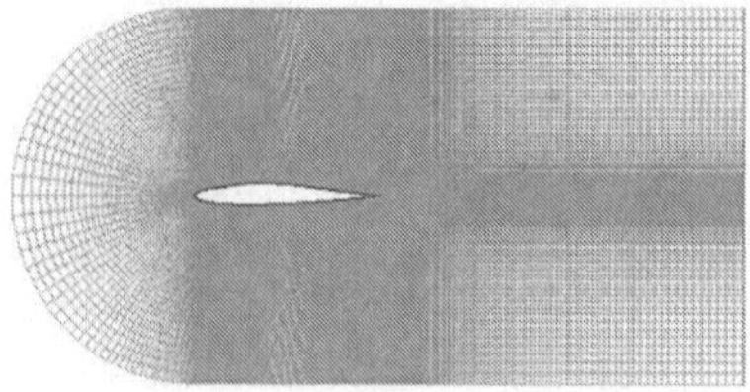

Fig. 3.4.7. Grid around the modified hydrofoil

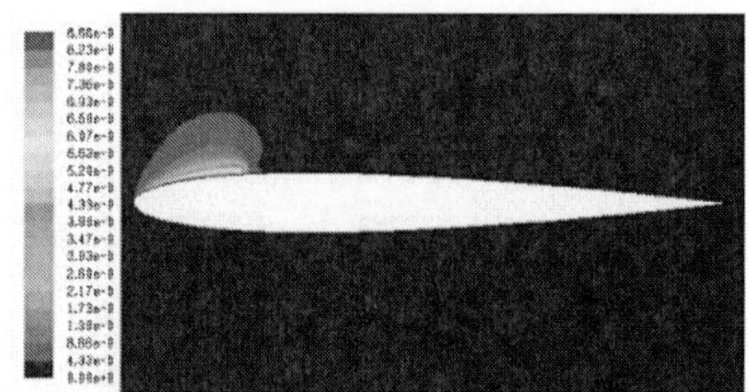

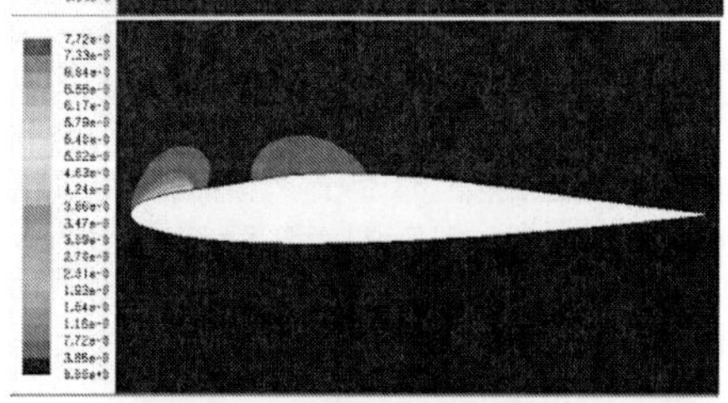

t1

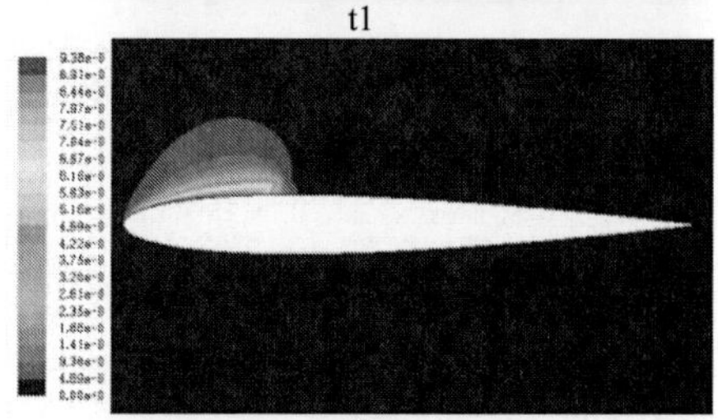

t2

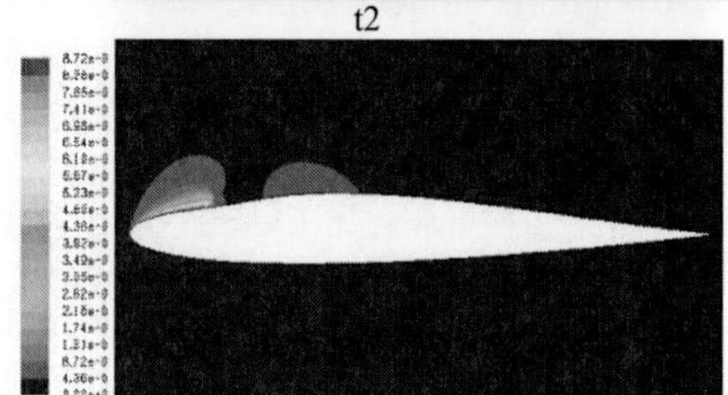

t3

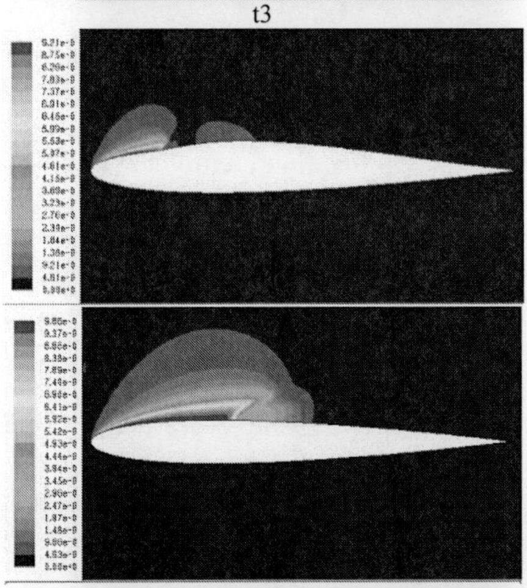

t4

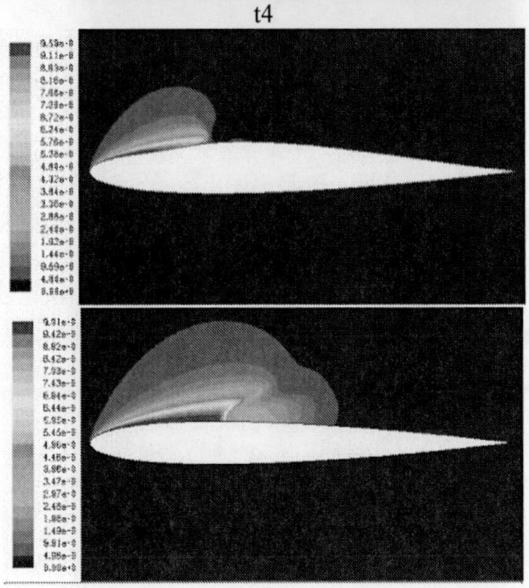

t5

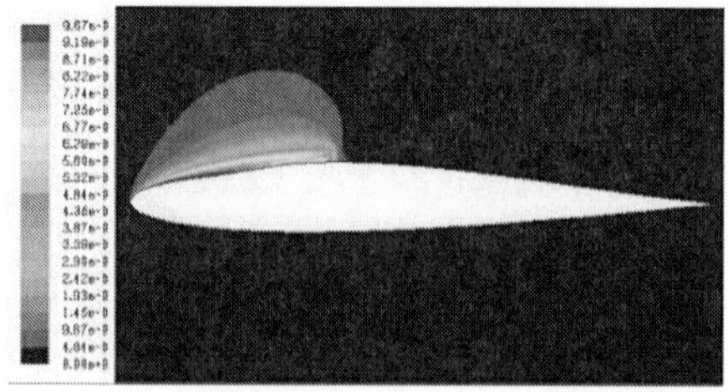

Fig. 3.4.8. Contours of vapor fraction around the hydrofoils

3.4.1.6. Conclusion

1. In this paper, the optimization method based on a genetic algorithm is proposed to design an anti-cavitation hydrofoil. With integration of geometric parameterization, the panel method, and the genetic algorithm, the optimized shape of the NACA0010 airfoil was achieved, which could be a new method for the design of hydrofoil.
2. Setting a barrier on the suction side can suppress cavitation of the hydrofoils. The change in the shape can change the hydrodynamic performance of the hydrofoil. Nevertheless, the optimization is only used to suppress the cavitation on a two-dimensional hydrofoil, which can provide a reference for the design and optimization of three-dimensional hydrofoils.

References

[3-4-1] Wang Xian-fu, *Marine Air-Hydrofoils Theory*, National Defence Industry Press, p. 10, 1998.

[3-4-2] Li Sheng-zhong, Zhao Feng, & Yang Lei, "Multi-Objective Optimization for Airfoil Hydrodynamic Performance Design based on CFD Techniques," *Journal of Ship Mechanics* 14 (11), pp. 1241–48, 2010.

[3-4-3] Xu Weibao, Wang Chao, & and He Bao, et al., "Application of Particle Swarm Optimization Theory in the Hydrofoil Design," *Journal of Ship Mechanics* 15 (6), pp. 598–604, 2011.

[3-4-4] Zeng Zhibo, "A Comparative Investigation on Optimization of Propeller Blade Section Design," *Journal of Ship Mechanics* 15 (12), pp. 1344–52, 2011.

[3-4-5] Xiong Ying, Ye Jinming, & Wang Dexun, "Prediction of Unsteady Cavitation on a Hydrofoil Section," *Journal of Fluid Mechanics* 240: pp. 59–96, 1992.

[3-4-6] Cheng Xiaojun & Lu Chuanjing, "On the Partially Cavitating Flow around Two-Dimensional Hydrofoils," *Applied Mathematics and Mechanics*, 21 (12): pp. 1310–18, 2000.

[3-4-7] S. A. Kinnas & N. E. Fine, "Nonlinear Analysis of the Flow around Partially- or Super-Cavitating Hydrofoil by a Potential Base Panel Method," Proceedings of the IABEM-90 Symposium, Rome, Italy, October 15–19, 1990.

[3-4-8] S. A. Kinnas & S. Mishima, "Systematic Design of Optimum Cavitating Sections," in *Second International Symposium on Cavitation*, pp. 107–112, original conference held in Tokyo, Japan, April 5–7, 1994.

[3-4-9] C. S. Lee, Y. G. Kim, & J. T. Lee, "A Potential-based Panel Method for the Analysis of a Two-Dimensional Super- or Partially-Cavitating Hydrofoil," *Journal of Ship Research*, 36 (2): pp. 168–81, 1992.

[3-4-10] P. Krishnaswamy, "Flow Modeling for Partially Cavitating Hydrofoils," PhD thesis, Department of Naval Architecture and Offshore Engineering, Technical University of Denmark, 2000.

[3-4-11] P. Krishnaswany, P. Andersen, & S. A. Kinnas, "Re-entrant Jet Modeling for Partially Cavitating Two-Dimensional Hydrofoil," CAV2001 session B1.005, Fourth International Symposium on Cavitaion, June 20–23, California, USA.

[3-4-12] Han Bao-yu, Xiong Ying, & Chen Shuangqiao. "Numerical Simulation of Cavitation around 2-Dimensional Hydrofoil," *Chinese Journal of Hydrodynamics* 24 (6): pp. 740–46, 2009.

[3-4-13] Sheng Huang, Miao He, Chao Wang, & Xin Chan, "Simulation of Cavitation Flow around a 2-D Hydrofoil," *J. Marine. Sci. Appl.* 9: pp. 63–68, 2010.

[3-4-14] Liu Chengjiang, Wang Yongsheng, & Liu Jubin, "Numerical Simulation of Cavitating Flow around Two-Dimensional Hydrofoil," *Journal of Naval University of Engineering* 20 (5): pp. 95–100, 2008.

[3-4-15] Wang Xiangfu, *Cavitating and Supercavitating Flow Theory and Applications*, National Defense Industry Press, 2: pp. 11–14, 2009.

[3-4-16] G. Kuiper, "New Developments and Propeller Design," *Journal of Hydrodynamics* 22 (5): pp. 7–16, 2010.

[3-4-17] Gu Wei & He Yousheng, "Flow Control on Unstable Cavitation Phenomena," *Acta Mechnica Sinica*, 33 (1): pp. 19–27, 2001.

[3-4-18] Zhou Li & Zong Zhi, "A Numerical Study of Propeller Noise Prediction and Control," *Shiport China*, pp. 187–92, 2004.

[3-4-19] J. Lepine & J.-Y. Trepanier, “Wing Aerodynamic Design Using an Optimized Nurbs Geometrical Representation,” 38th Aerospace Science Meeting and Exhibit, Reno, NV, January 10–13, 2000.

3.5

APPLICATION OF HYDROFOIL CRAFT

3.5.1. Design of *Qinzhou II* Fully-Transparent Tri-Hydrofoil Sightseeing Tourist Boat (2015)

—Wang Xiaochuan and Guo Dong
(China Ship Scientific Research Center)

Abstract
The *Qinzhou II* fully-transparent tri-hydrofoil sightseeing tourist boat is the first ship in the world of this new type. It is the first time a fashionable and fully-transparent cabin has been designed for a hovercraft, and it is specially tailored to satisfy the new requirements of tourism in the Three Gorges. The new tourist boat's special advantages make it suitable for fully-transparent sightseeing, fast arrival and slow tours, and trunk- and branch-line direct transportation. This boat was designed utilizing a new concept and high-end technology; it stands as a successful and innovative model in the planning of a new ship type in a regional zone in China. In this paper, the details of the *Qinzhou II* are introduced. It is necessary to build and operate in batches, which will develop a new recreation industry in the Three Gorges.

3.5.1.1. Introduction

The *Qinzhou II* fully-transparent tri-hydrofoil sightseeing tourist boat is the first ship of this new type to be developed in the world. Traditional hovercrafts are supported by twin hydrofoils at fore and stern. When the crafts are in a foil-borne condition at high speed (twin-pivot dynamic support), there is larger sagging than when in a floating condition (uniform buoyancy support). Then the superstructure of the cabins of the twin-hydrofoil hovercrafts should be fully closed to bear the deformation of the steel (Fig. 3.5.1), which does not satisfy the requirements of a tourist boat and limits its application in the field of tourist sightseeing.

Fig. 3.5.1. Second-generation high-speed hovercraft *Yuanzhou-2000* in China

Yuanzhou-2000 was the second-generation of high-speed hovercraft in China (Fig 3.5.1); it operated in inland waters but could not satisfy the requirements for a tourist boat with a fully-closed metal superstructure.

According to the development of tri-hydrofoil crafts, the third hydrofoil was mounted under the midship to enhance the rigidity of the hull, which would reduce the sagging deformations of hybrid hydrofoils. The cabin of the superstructure can adopt the glass frame. The fully-transparent cabin satisfies the requirements of a sightseeing tourist boat (see Fig. 3.5.2).

Fig. 3.5.2. Third-generation trunk-line fully-transparent tri-hydrofoil sightseeing boat

The boat is designed by China Ship Scientific Research Center (CCSRC) to fulfill the new needs of tourists in the Three Gorges Dam area. CSSRC developed and completed the research project "Analysis and Study of a New-Type Longitudinal Tri-Hydrofoil Hovercraft by Model Tests," which was issued in 2005–10. The hydrodynamic arrangement of the tri-hydrofoil hovercraft was optimized to improve the performance of power and speed and reduce unfavorable interference that was the result of the shorter distance between the hydrofoils and the additional struts. The new tri-hydrofoil hovercraft was self-adaptive to the wave interference. The superstructure of the tri-hydrofoil hovercraft is suitable to be designed with a fully-transparent frame of glass. The breakthrough in the key technologies needed for the design of the high-speed tri-hydrofoil hovercraft had been achieved in load distribution of hydrofoils, the selection of hydrofoil type, and in the hydrodynamic parameters, all of which were evaluated and appraised as the achievements of the scientific and technological department of the Ministry on June 29, 2004. The experts of the appraisal committee appraised that the innovations in design of the first hovercraft were independent intellectual property rights, which was the highest level in China. The innovations have been awarded a national patent on September 22, 2014 (Patent No. ZL03 2 78250.0).

3.5.1.2. Design concept of the fully-transparent tri-hydrofoil sightseeing tourist boat *Qinzhou II*

To fulfill the needs of public tourist boats in the domestic market, the international popular concept and technology of fully-transparent superstructures were used in the research project and design of the series of tourist boats named "Qianzhou."

The main design concept of the series of "Qinzhou" tourist boats focused on affability and a passenger-friendly design.

1. The passengers could feel intimate with nature through the fully-transparent sightseeing cabin.
2. The passengers would be provided with hospitality and safe services in the people-oriented cabin.
3. The analysis of economic efficiency and the market showed that the ship owners would receive a good return on their investment in tourist boats.

By adopting the above design concept, the first *Qinzhou I* fully-transparent low-speed tourist boats, named *Guangdong* and *Small Three Gorges* were designed and operated in the area of the Small Three Gorges in 2005 (Fig. 3.5.3). The experience in designing *Qinzhou I*, with its fully-transparent cabin and large-cold-capacity central air conditioner, was used and adopted in the design and research of the fully-transparent tri-hydrofoils hovercraft *Qinzhou II*.

Fig. 3.5.3. *Qinzhou I* fully-transparent low-speed tourist boats

After the completion of the Three Gorges Reservoir, a new craft was proposed to fulfill the new requirements of touring in the area. The conceptual design of the fully-transparent tri-hydrofoil sightseeing tourist boat *Qinzhou II* was then completed (see Fig. 3.5.4). Its characteristics are as follows:

Fig. 3.5.4. The third-generation of *Qinzhou II* series of fully-transparent tri-hydrofoil sightseeing yachts. The poster reproduced in the figure features the yacht *Longzhou*

(1) Very large deck and fully-transparent cabin with central air conditioning and upholstered seats
Triple hydrofoils were adopted to avoid the sagging deformation of twin-hydrofoil hovercraft when in foil-borne condition. This allowed the configuration of the fully-transparent frame to be used in the design, which featured the large-cold-capacity central air conditioner and aviation-style upholstered seats. The fully-transparent cabin combined comfortable sightseeing conditions with panoramic views for the tourists. All passengers could walk on the very large deck outside the cabin when the boat was at low speed.

(2) Fast traveling and slow touring
The journey was about 286km between Wanzhou and Yibing (Maoping), and the journey without visits to various special locations was about 186km. There were two operational conditions for the boat: one was foil-borne for fast travel at the maximum navigation speed of 70km/h; the other was for sightseeing and touring at the low service speed of 12km/h. This was the concept of fast traveling and slow touring, which greatly increased traveling efficiency.

(3) Trunk-line and branch-line non-stop journey
At present, none of the tourist boats could be operated in the branch rivers in the area of the Small Three Gorges and Shennong Stream; the visitors would therefore have to transfer to other boats that could be operated in the branch rivers. The new tourist boats could be operated in both the trunk and branch lines by controlling the principal dimensions with better maneuverability; this would avoid the need for visitors to transfer between boats and made the journey more interesting to the visitors.

The fully-transparent superstructure and tri-hydrofoil were the first in the world to use this technology, and ensured that the boats operated with riding comfort and economical efficiency in the market. The normal operation of the boats promotes new and continuous developments of the technologies used in the design of future tourist boats.

3.5.1.3. Importance to developing the emerging industry of fast tours in the area of the Three Gorges
The storage water level of the Three Gorges Reservoir reached 135m on June 15, 2003, and the spectacular scene of the dam lake occurred in the high gorges. Then the waterway broadened naturally and the flow in the reservoir became slow. If the dangerous shoals in the Yijijiang River

disappeared, the river would become the golden waterway and the express waterway would become a reality.

The old scenery in the Three Gorges remained and the new scenery in the Three Gorges was more charming. When the Three Gorges reservoir was filled, the artificial and natural scenery of the highest dam and the largest reservoir became a trademark world-class scenic spot in China.

The tourist boats were irreplaceable transportation tools for touring the Three Gorges. It would be impossible to replace the tourist boats with the buses running on the new expressways along both sides of the reservoir.

(1) Quick in and out by high-speed tourist boats in the Three Gorges

The present tourist opportunities in Three Gorges were important components of tourism in western China. But the tourist journeys then offered were completed as simple tours on low-speed boats running between Chongqin and Yichang in the Three Gorges. Given the time tourists would spend, the old tourist style could be suited to the pattern of fast traveling in the new Three Gorges, where improvements to tourist opportunities in western China has been limited by the amount of time needed. Too great a cost and too long a time hindered potential visitors traveling in the new Three Gorges.

In the past, the beginning and end point for tours of the Three Gorges were Chongqin and Yichang. After the multi-level transportation by air to Wanzhou and the railway and expressway had been formed, Wanzhou was intended to be the new beginning of tourism to Chongqin Municipality by April 2006. This was beneficial to shorten tour times and reduce the cost of tours and would lay good foundations for fast-in and fast-out tourism in the Three Gorges.

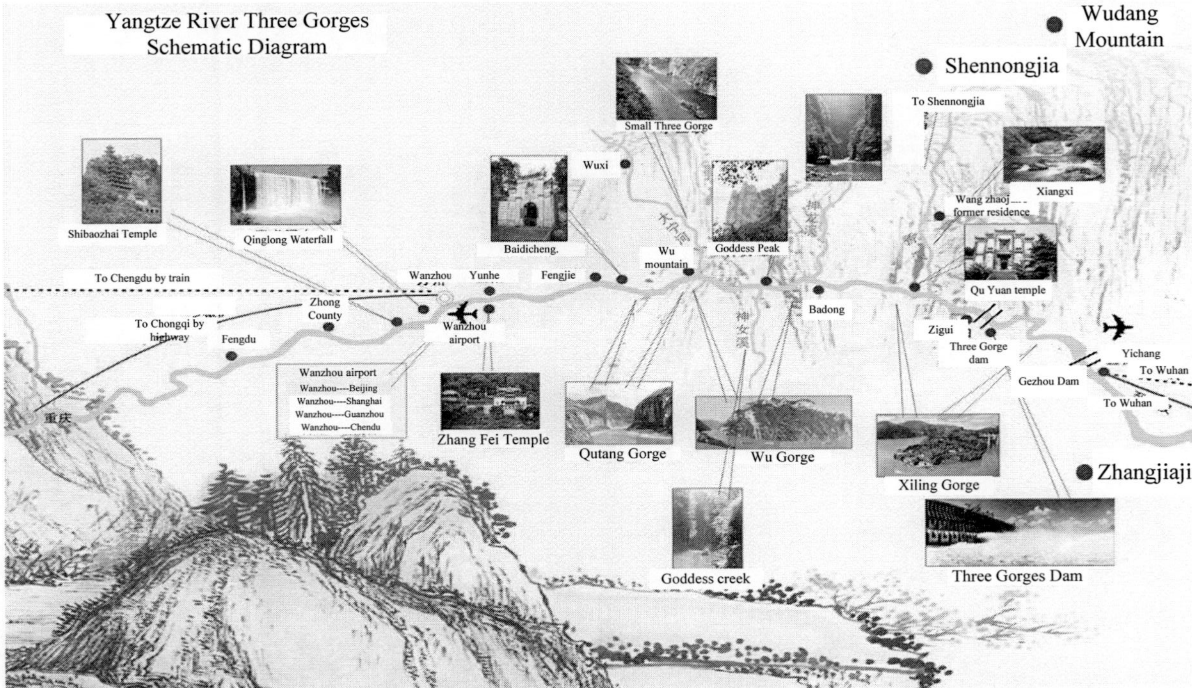

Fig. 3.5.5. The range of tourist opportunities in the Three Gorges area

When the fully-transparent high-speed hydrofoil sightseeing tourist boats had been built and were operational, the new pattern of transportation to connect the Wanyi express waterway with the Chengyu expressway could be formed, which would make a reality of fast travel and slow touring in the whole area around Chendu and Yichang. This allowed more visitors to reach more tourist spots in the same or shorter time at a lower cost. More personalized and plentiful tourist opportunities could be designed. As the quality of existing tour products was greatly improved, this brought a positive influence to all tourism products related to the Three Gorges in western China, which was the best solution to satisfying the requirements of various choices for tourists in the area of the new Three Gorges. It was very important to stimulate the consumption of tourists in the Three Gorges, especially increasing tourists among workers on paid leave.

(2) Harmonious development of Three Gorges tourism and the local economy

While the boats enjoyed more than 80% capacity of tourists on traditional tours to the Three Gorges, the tourists could only enjoy the natural landscape and could not experience the local history and culture in the area of the Three Gorges. The only possibility was selling tickets to traveling spots for the harbors along the way, in the old tourist style. However, this greatly restricted the harmonious development of the Three Gorge tours and the local economy.

The central government attached great importance to the problem of immigrants and brought forward the immigrant policy of "affirmative removing, steady settlement, getting rich." The present focus was to solve the problem of getting rich. And the ultimate solution was to integrate the industry with the advantages in the area around the reservoir.

By integrating fully-transparent high-speed hydrofoil tourist boats into abundant tourism resources in the areas around the reservoir, the emerging industry of express tourism was initiated in the area around the reservoir near Wanyi, which made a reality of the idea to develop new daily "one-day panorama tours by boat." The new idea of sightseeing on the boat by day and relaxing on the shore by night would be reached. The tourists could experience at their leisure the local history and culture in the area of the Three Gorges. On the other hand, the harmonious development of Three Gorges tours and the local economy could be promoted by attracting tourists to consume on the shore and by bringing forward job opportunities for immigrants.

(3) A good chance for owners of tourist boats with high-quality yacht companies and an increase in the economy in the area around the Three Gorges

The significant differences between *Qinnzhou II* and low-speed tourist boats are as follows: the benefit would be significantly improved; the whole turnover of tourist boats each day would be higher than that of normal low-speed tourist boat over 2–3 days. The details are as follows:

Ship type	Number of ships for same investment of 56 million yuan	Price of full navigation	No. of days for full navigation	Max. turnover of a full navigation
Yu-Yi Star	200 passengers per boat	1000 yuan	3 days	200,000 yuan
Qinzhou II	516 passengers in 6 boats	385 yuan	1 day	596,000 yuan
Comparison compared with low-speed boat for the same investment	passenger number increased by 158%	Price reduced 61.5%	Reduced by 2 days	Increase of 198.0%

1. The operating time of the whole journey was abundant, and the berth scenery was more than for the slow boats. The time for sightseeing was increased by about 60%, and there was a large enough time margin to deal with the suddenness.
2. More tourists could be attracted since the tour length was changed from one day to three days and the schedule of the tour could be better arranged. The tourist boats could be used to provide more long-distance and value-added tours to more visitors all around the world.
3. The boats could be provided for special tours to better fit the schedules of group tours.
4. The multi-level transportation on the Wanyi route could balance the bi-directional passenger flow, which would avoid the problem of few visitors traveling slowly in normal boats.
5. Express tourism in the area around the Three Gorges could be formed by integrating cheap tour resources on land, including travel agencies, scenic spots, buses, restaurants, hotels, and so on.

6. The tri-hydrofoil fully-transparent high-speed hovercraft was a unique new ship type in the world, which would be a new transport tool aimed at tours in the area around the Three Gorges.

The significant difference between *Qinzhou II* and hovercraft

Item	*Qinzhou II*	*Meteor* hovercraft
Attribute	tourist boat	Passenger ship
Speed/km/h	12–70	55–60
Operating characteristics	Fast traveling, slow touring	High-speed foil-borne operation, with long duration floating and navigation limited by the main engines
Engine	Interval heavy loads with extend overhaul period	Continual heavy loads with short overhaul period
Sightseeing	Fully-transparent cabin, all passengers can walk on the deck	Closed cabin, no passengers can walk on the deck
Passenger type	Group tourists with few individuals	Individuals with few group tourists
Ticket mode	Sold to travel agencies (remote, reserved), seldom sold locally	Sold at local ticket offices and the habors
Berth mode	Berths near the scenery not to delay the departure times	At passenger terminals, the departure time would influence the benefits
Difference of Fare	Two-way tickets charging full fare for all passengers	Majority were interval passengers paying partial fares
Departure mode	High-attendance departures depending on groups of tourists	Departure by schedule, attendance ratio could not be proved
Influenced by buses running in expressway	Convenient and linked to buses running on the expressway	Soon to be defeated by buses running on the expressway

The design and construction of *Qinzhou II* was an innovation in hovercrafts in China, which had some advantages over normal tourist boats and existing passenger hovercrafts. The different running modes and various opportunities for tourists would satisfy the new needs of tourists in the area around the Three Gorges, which would attract more tourists. If the business opportunities could be kept, the ship owners could get see new points for economic increase to become excellent companies.

3.5.1.4. Introduction of the fully-transparent tri-hydrofoil hovercraft *Qinzhou II*

(1) Ship-type, navigation zone, and usage

(1) The ship's type is a tri-hydrofoil high-speed hovercraft with an aluminum hull. The craft's design is that of a monolayer deck with fore fully-transparent cabin and aft sightseeing and walking deck, which adopts the arrangement of an aft layout with twin engines integrated with twin oars and twin rudders. The type of the ship is named *Qinzhou II*.

(2) The tourist boats have three obvious advantages, including a fully-transparent and air-conditioned cabin, the ability for trunk-line and branch-line navigation and for fast traveling and slow touring. It is a new-generation transport tool for tours specially aimed at the requirements of tours in the area from Wanzhou to Yichang (Maoping) around the reservoir. The boats are safe, economical, and environmentally friendly.

(3) The tourist boats were designed to meet the requirements of navigating the class B zone: its navigable water was freshwater in class B, C, and J navigation zones. The boats were suitable to be used as tourist boats and passenger ships in rivers more than 2.7m deep, including the trunk lines and branch lines of the Yangtze River, such as the Daning River, the Shennong Stream, the Xiangxi Steam, and so on. It can satisfy the requirement of safe daytime navigation at any time of the year, except for special climate conditions such as heavy fog.

(2) Parameters of design

(1) Principal dimensions

Overall length	32.85m
Overall length of the hull	31.50m
Waterline length	28.2m
Overall breadth (inc. fender of hydrofoil)	8.10m
Molded breadth	5.00m
Overall height (propeller tip to mast top)	8.50 m
Molded height (baseline to driving top)	5.35m
Molded depth	2.00m
Draught on floating condition	0.75m
Max. draught on floating condition	2.34m
Depth of waterway on floating condition	2.70m
Draught at foil-borne condition	1.48 m
Designed navigation zone	freshwater class B, C, J

(2) Payloads

1. Passengers

The supposed average passenger weight was 70kg and personal luggage weight was 10kg; the average weight of each passenger was thus 80kg. This meant that 86 passengers could be loaded, for a total load of approximately 6,880kg.

2. Crew

The supposed average weight of each shipman was 70kg and there were five shipmen:

Drivers	2
Marine engineers	2
Waiters	1
Total	5,350kg

3. Food and drink for passengers

The food and drink for each voyage totaled 180kg, allowing 2kg per passenger.

4. Tourist souvenirs

50kg per trip

5. Daily supplies and reserved goods

Cab	10kg
Engine room	50kg
Duty room	40kg
Total	100kg

6. Freshwater

200kg for lavatories per trip.

7. Standby lubrication oil in engine room

90kg for main engine, gear case, and steering engine

8. Main and assistant engine oil

Standard load for main and assistant engine for endurance of 300km and 10% repertory was 1,750kg

9. Total payload 9,600kg

(3) Performance

1. Speed

a. High-speed foil-borne condition with double propeller running:

The rotating speed of the double engines was 2,300rpm and the temperature in the engine room was not greater than 55°C. The continuous speed when foil-borne with the designed payloads in calm water was up to 70km/h, which was for high-speed riding on a trunk line.

b. Slow-speed floating conditions with a single propeller running:
The continual speed of floating conditions with the designed payloads in clam water was 12–20km/h, which was for slow-speed sightseeing on a branch line. The angle of the rudder was about 5° in the navigation in a floating condition. Left and right engines should be used alternately to keep each engine running in the same interval.

2. Endurance

The maximum endurance of the tourist boats was up to 300km by continuous foil-borne speed at a rotating speed of 2,300rpm with double main engines.

3. Navigability

The following should be satisfied by the designed payloads. When the yacht travels in a foil-borne condition with full power in waves of $H_{1/3} \leq 0.8$m, the maximum angle of roll is $\theta 1/3$□4°, the pitch is $\alpha 1/3$□3°, and there was no drastic heave or roll motion.

4. Turning performance

The rotating speed was 1,000 rpm, in a floating condition with double engines, and the turning diameter was less than seven times that of the ship length (rudder angle was ±3.5°). The rotating speed was 2,300rpm in a floating condition with double engines, and the turning diameter was less than 32 times that of the ship length (rudder angle was ±8.0°).

5. Stability

The stability in a floating condition satisfied the "Rules of Legal Inspection Technology Inland River Ships" (2004). The stability in a foil-borne condition with the designed loads satisfied the "2000 HSC Code" (IMO).

6. Unsinkability

There were 10 watertight cabins under the main deck. The unsinkability satisfied requirements that the hull would not sink when one tank was destroyed.

7. Cabin noise

a. Foil-borne condition with twin-propeller running
With the main engine rotating at a speed of 2,300rpm and the doors and windows closed, the average noise level was 75–78dB(A).

b. Floating condition with single-propeller running
With the main engine at a rotating speed of about 1,000rpm and the doors and windows closed. The average noise level was 65–68dB(A).

8. Refrigeration effect of air conditioning
    a. Refrigeration cabins: passengers' cabin, shop, toilet, and cab.
    b. Refrigeration effect: When the outer temperature was ≤40°C, the main engine rating speed was 2,300rpm and the number of passengers was 86, the average temperature in the cabin should be ≤28°C after the air conditioning system had been working for 30 minutes. The air conditioning was not active when the boat was at berth. The air conditioning should be started 30 minutes prior to departure using shore electric power.
    c. The air conditioning system had electric heating equipment, the power was □30kW, the average temperature inside the cabin was 5°C–8°C higher than that outside the cabin.
9. Environmental index
    a. The discharge of engines satisfied the requirements of the "Marine Diesel Engine Nox Emission Control Technology Rules" (IMO).
    b. To satisfy the requirements of the "Rules of Legal Inspection Technology Inland River ships" (2004), the oily water, domestic sewage, and garbage were gathered and discharged to special dock or collection ships.

(4). General layout (Fig. 3.5.6)

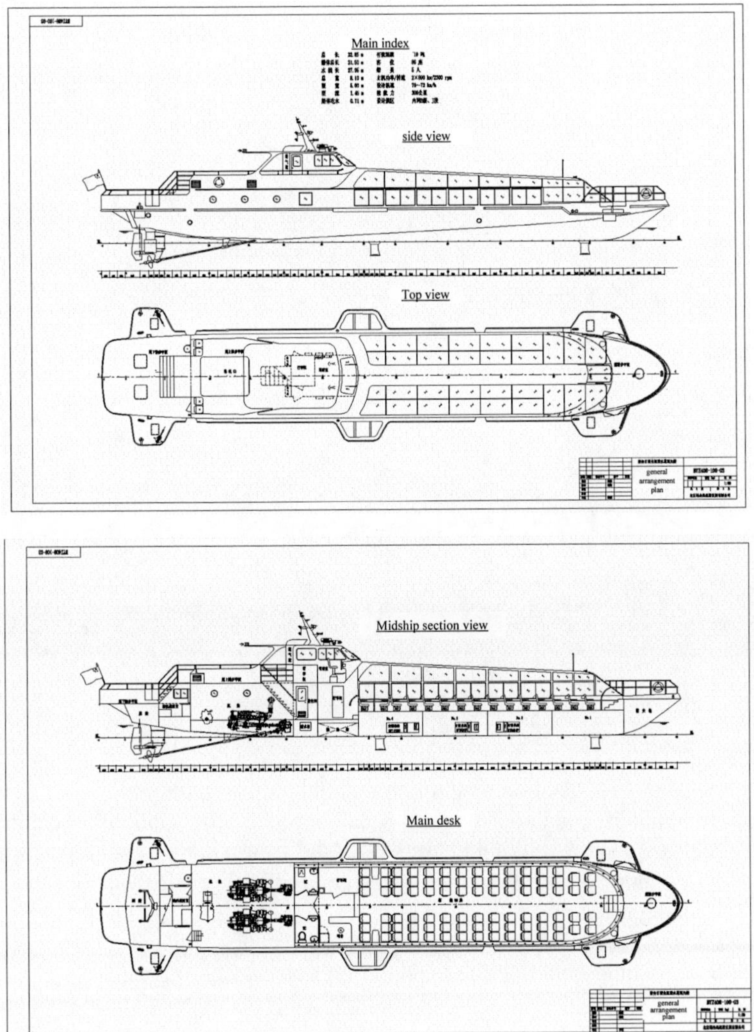

Fig. 3.5.6. General layout of *Qinzhou II*

3.5.1.5. Characteristics of the design of *Qinzhou II*

(1) Tri-hydrofoil (Fig. 3.5.7)

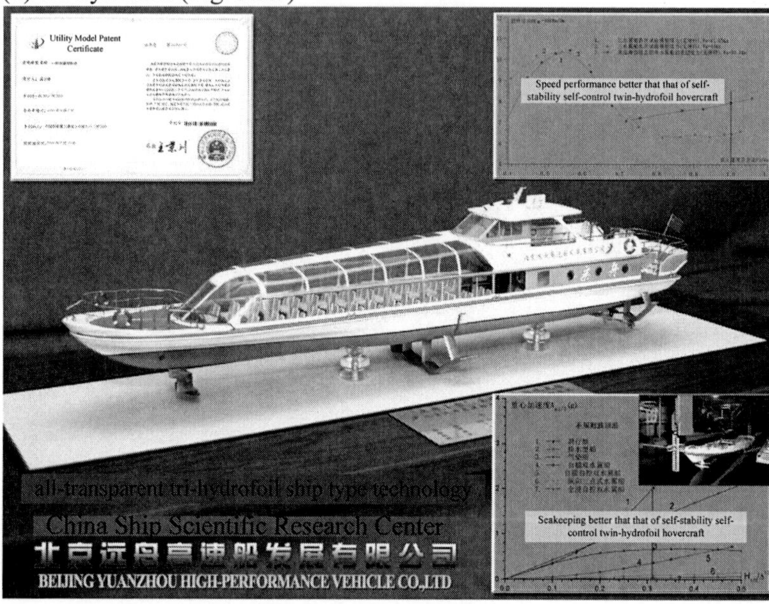

Fig. 3.5.7. Patent for tri-hydrofoil ship type

(2) Sightseeing (Fig. 3.5.8)

1. The tri-hydrofoil system was adopted to strengthen the rigid structure and to make true the design of the fully-transparent cabin.
2. Fore and aft walking decks satisfy the requirement for all tourists to be able to exit the cabin for sightseeing. The aft deck is shaded from the sun.
3. Bow bulwark and wave suppression plates prevent green water splashing onto the fore deck during floating conditions.

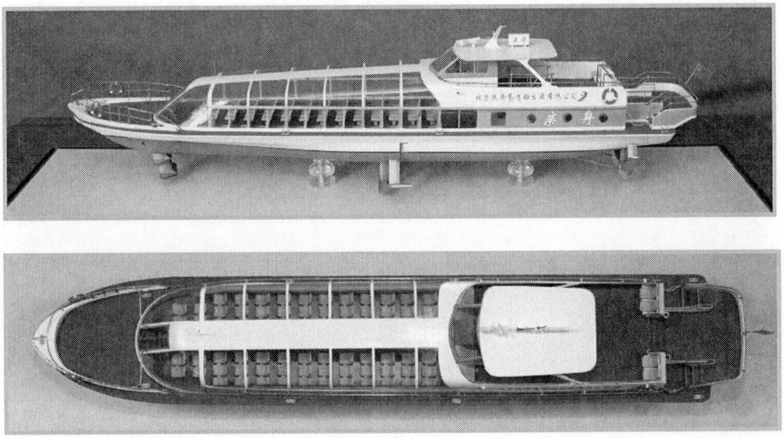

Fig. 3.5.8. Fully-transparent cabin with fore and aft walking decks

(3) Design for fast traveling and slow touring

1. There are two operating conditions: one was at the speed of 72km/h with twin-propellers running, the other was at the speed of 12km/h with a single-propeller running. The high-speed ride could satisfy the requirements of fast traveling on a trunk line, and the slow-speed ride could be used to tour and sightsee in the big Three Gorges slow tour and in the small Three Gorges.
2. The navigation room is arranged at a high level to satisfy the requirement of high-speed riding.

(4) Design for through a trunk line and through a branch line

1. There are three chairs on either side of the aisle, which reduces the breadth of the boat.
2. Three hydrofoils are mounted on the bottom of the hull, which reduces the breadth of the hydrofoils.
3. The design of the ship's hull is integrated with a hovercraft and wave-damping ship, which could damp the waves after the stern and reduce the harmful influence on other ships.

(5) Environmentally friendly design

1. The diesel engine satisfies the discharge requirements of the IMO.
2. The design for the discharge of oily water and domestic sewage satisfies existing national standards.
3. Water lubrication is used to cool the shaft bearings, which removes the pollution from leaking oil.

(6) Design of service facility

There should be plentiful supplies of all necessary service facilities for a tourist boat, such as transparent windows for sightseeing, a comfortable environment, and relevant living service facilities, and so on. More details of the humanization design had been included in the general design of the boat to satisfy tourist requirements. The comprehensive service facilities to satisfy the needs of tourists in the boat include a baggage room, a photoelectric button calling system for each seat, and wireless broadcast throughout the boat.

(7) Reduction of vibration and noise

1. The main-engines were arranged at the stern of the hull.
2. Elastic vibration damping was used in the design of systems of power and propeller shafting. (Three 5-blade low-vibration and low-noise propellers were designed.)
3. A noise insulation chamber was placed between the passenger cabin and the engine room.
4. Advanced damping structural material was used to reduce the structural noise.

(8) Safety

1. Transfer alarms and a camera monitoring system guarantee the safety of the boat during high-speed and slow-speed navigation.
2. The hull will not sink even if one tank is destroyed.
3. The boat is equipped with enough devices for fire extinguishing and rescue to meet the rules.
4. Handrails are arranged outside the sightseeing deck according to the rules.
5. The height of the freeboard aisle was well designed to avoid the glass frame outside the passenger cabin colliding with other ships in branch-line rivers. The hydrofoils are protected by a fender bracket with a winding steel cable.
6. Triplex glass guarantees the safety of the top of the passenger cabin.
7. The boat was designed to be operated for two hours at night, which is convenient for emergency navigation if the tour is delayed.

CHAPTER 4

AIR LUBRICATION CRAFT (ALC) AND AIR CAVITY CRAFT (ACC)

4.1

INTRODUCTION TO AIR LUBRICATION CRAFT (ALC) AND AIR CAVITY CRAFT (ACC)

LIANG YUN

Marine Design and Research Institute of China (MARIC)

Introduction

To reduce the drag of crafts, researchers used to take steps so that the crafts would take off from the water surface; thus, over the past half century, various novel crafts appeared, such as ACV, SES, HYC, WIG, and so on. However, the challenge for such craft was that various special pieces of equipment had to be mounted, such as an air cushion system and a skirt on ACV/SES, hydrofoils on HYC, air wing on WIG, and so on, which served to complicate the structure and equipment and increase the cost.

Therefore, ship researchers considered whether it was possible to take steps to reduce the drag of the craft not by the craft taking off from the water surface but by making a thin layer under the bottom of the craft, and also between the water surface and the craft bottom, to generate air friction drag instead of water friction. Because air density is only 1/800 of water density, the craft's friction can be reduced significantly. This is the basic principle of air lubrication craft and air cavity craft.

4.1.1. The basic principle of air lubrication craft and air cavity craft and their classifications

It is well known that while a solid body moves in the water with a definite high speed, the flow after the moving body might be separated, thus forming an air cavity after the body and making the body separate from the water, so as to reduce the friction.

This is the basic principle of air lubrication craft, which may be divided into two groups: air lubrication craft and air cavity craft.

1. Air lubrication craft

Craft of this type used to be thought of as large ships with low speeds, as shown in Figs 4.1.1 and 4.1.2.

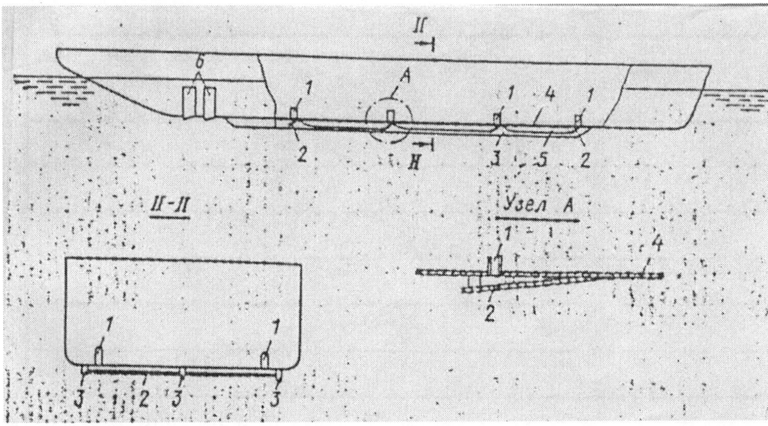

Fig. 4.1.1

Fig 4.1.1 shows protruding plates arranged under the ship's bottom to reduce the friction of the ship through lower speed.[4-1-2] The numerals in the figure mean the following: 1. hole for feeding pressurized air; 2. inclined plate under the ship bottom for forming the air cavity; 3. vertical keel; 4. ship bottom; 5. plate for protecting the water propeller from air suction.

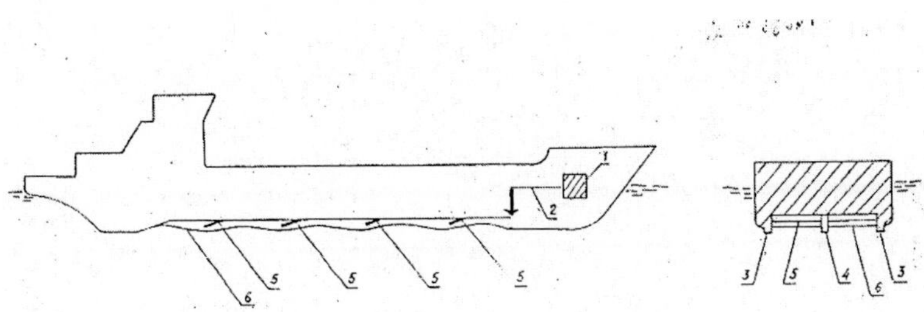

Fig. 4.1.2.

Fig 4.1.2 shows the schematic design of the device for forming an integrated air layer under the bottom of a ship.[4-1-2] The numerals in the figure mean the following: 1. compressor; 2. system of air supply under the bottom; 3. side keel; 4. longitudinal keel; 5. inclined plates for the formation of the air layer; 6. the air layer.

From the figures, one can see that the air layer can be formed under the ship's bottom to reduce the friction drag of the ship. The air cavity under the bottom is continuously formed, and the water surrounds the air cavity. Since the air layer was fed by a compressor after the plates, it can be called an artificial air cavity.

The distance between the two plates—that is, the length of the artificial air cavity—can be calculated in terms of the incoming ship speed and the inclination angle of the plate.

Such air lubrication theory used to be applied to low-speed ships, particularly on ships with flat bottoms and large block coefficients, such as cargo barges.

2. Air cavity monohulls

Air cavity monohulls can be applied to high-speed craft with transverse steps, the schematic configuration of which is described in Fig. 4.1.3. As the pressure of the water flow after the step is low or even negative, the pressurized air can be fed into this area from the holes in the bottom, so as to make an air cavity under the bottom and reduce the drag by up to 10–30%.

Since the air layer is so thin, and the air flow rate is small, the power for the air compressor is also small, only 1–3 % of the total power output of the craft, compared with ACV (25–40 %) and SES (10–25%), respectively.

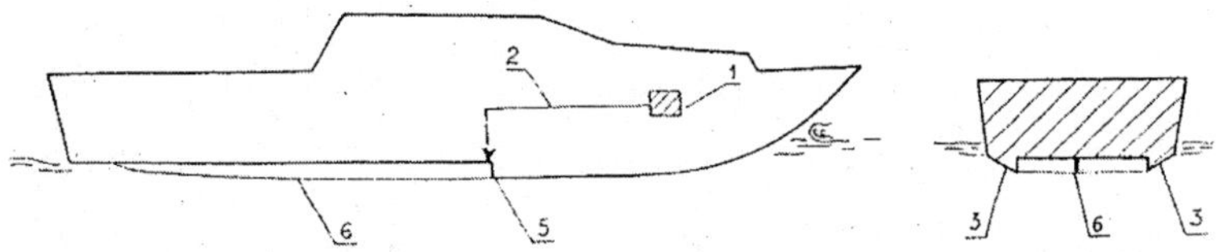

Fig. 4.1.3

Fig. 4.1.3[4-1-2] shows the schematic figure of an air-cavity planing hull. The numerals in the figure mean the following: 1. air compressor; 2. air-feeding system; 3. side keel; 4. central longitudinal keel; 5. longitudinal step; 6. air layer.

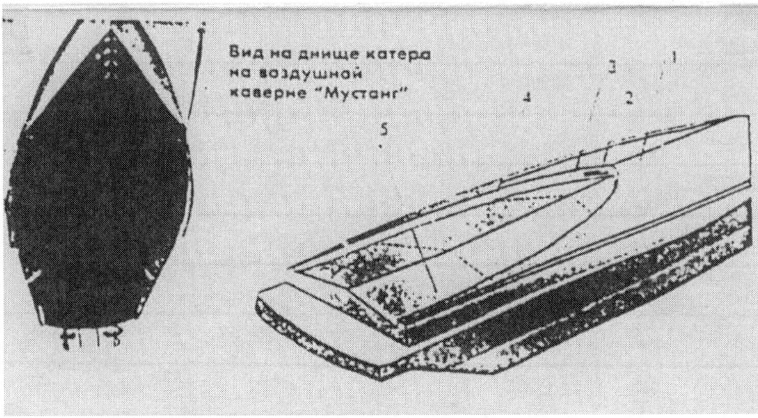

Fig. 4.1.4

Fig. 4.1.4.[4-1-5] shows a view of the bottom of a Russian ACC-type *Mustang*. The numerals in the figure mean the following: 1. craft bottom behind step; 2. hole for feeding pressurized air to the bottom after the step; 3. arrow-like steps; 4. bottom; 5. side skeg.

3. Air cavity catamaran
The basic principle of the air cavity catamaran (ACCAT) is the same as for the air cavity monohull: to reduce the friction drag of the CAT. It is of particular importance for the CAT since the CAT's form divides the ship beam into two parts along the longitudinal direction, to form two side hulls and a high length–beam ratio for the side hull, thus decreasing the wave-making drag. However, this increases the wetted surface of the CAT, and also the friction drag. The air-cavity configuration can improve the friction drag, so as to improve the power performance significantly.

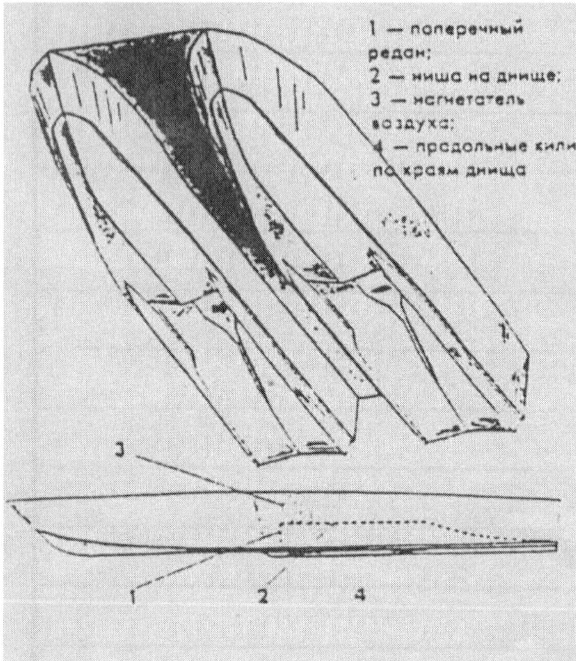

Fig. 4.1.5.

Fig. 4.1.5. shows the bottom profile of a Russian Air Cavity CAT.[4-1-5] The numerals in the figure mean the following: 1. transverse steps; 2. craft bottom; 3. air compressor; 4. side wall.

4.1.2. Worldwide development of air cavity craft

1. Russia and Ukraine[4-1-1]

In Russia, air cavity technology was firstly successfully investigated in Krylov Central Ship Scientific Research Institute. In 1961, research on establishing an artificial air cavity on a high-speed planing hull was carried out at the institute; since 1993, the institute has conducted research on using air-cavity technology on transit-speed type crafts and CAT, and the design and construction of such type crafts was begun at the Alexeev Hydrofoil Design Institute and the high-speed craft factory, respectively, in the former USSR city of Gorky (now Nizhny Novgorod).

At present, Russia has constructed about 70 air cavity craft, and completed 30 design projects for such craft. The craft have a displacement of 14–360t and a speed of 30–52kn, and have both civil and military applications.

2. USA and Norway
In the 1990s, Europe Surface Effect Ltd. received the support of the US navy to develop an SES without a skirt—that is, an air cavity craft—with a length of 40m and a speed of 70kn, a large cargo ship with a length of 150m and a speed of 50kn, and, of course, naval ships.

3. China
Model experimental and theoretical investigations of air-cavity-craft technology has been carried out for years in various institutes and universities, such as the Maine Design and Research Institute of China (MARIC), the China Ship Scientific Research Center (CSSRC), the China Naval Engineering University (CNEU), and the Shanghai Ship and Shipping Research Institute (SSSRI).

Now, we would like to introduce several papers concerned with the performance, technology, and application of ACC, which were presented at the International Annual Shanghai HPMV conference in past years. These chapters concern the following subjects:

1. The research and development of high-speed ACC in Russia and Ukraine, in sections 4.2 and 4.3
2. The research and development of high-speed air supported vessels (ASV or ACC) in Norway and the United States in section 4.4
3. The practical design of high-speed ACC in China in section 4.5
4. The research and design of low-speed ACC in China in section 4.6
5. Fundamental hydrodynamic research on ACC in section 4.7
6. Experimental investigations on the drag of ACC in various Chinese research institutes and universities in sections 4.8 and 4.9
7. Theoretical investigations on ACC in section 4.10.

References

[4-1-1] L. Yun et al., "Development of Russian WIG and Air Cavity Craft in 2nd Generation," Proceedings of HPMV Conference, April 2002, Shanghai, China.
[4-1-2] "Principle and Performance of Air Cavity Craft," Science and Technology Report, Flying Dragon Science and Technology (Hong Kong), 2003.

[4-1-3] L. Yun et al., "Practical Technology for Air Cavity Crafts," Science and Technology Report, FDST (Hong Kong), 2003.
[4-1-4] Brochure by FDST for HPMV, 2003, Shanghai.
[4-1-5] A. Sverikov, "Ship on Air Cavity," *Boat and Yacht* 1 (179), 2002 (in Russian).

4.2

THE CONTRIBUTION OF RUSSIAN INVENTORS TO THE DEVELOPMENT OF ARTIFICIAL CAVITY SHIPS (2002)

GENNADY A. PAVLOV

Technical Director of the industrial complex comprising Hong Kong Engain Science & Technology Ltd., the joint stock company Research and Production Complex "Trek," Russia, and the Ukrainian Feodosia Shipbuilding Company "Morye"

Abstract

The idea of using "air lubrication" to reduce resistance in the movement of ships was offered for the first time in 1882 by Goustav de Laval. However, significant results in this area were only achieved much later due to the occurrence and development of a fruitful idea about the creation of the special flow under the bottom of a vessel now known as "artificial cavitation flow."

Due to intensive research begun in the 1950s at the Krylov Shipbuilding Research Institute (KSRI), Russian scientists and designers were the first in the world to have achieved impressive results in the use of artificial gas cavities to reduce the resistance of ships in a wide range of Froude numbers. Today Russia is the country in thc world with the longest experience of successfully reducing the resistance of dozens of high-speed ships with the aid of artificial gas cavities—up to 20–30 %. But although the achievements of Russian researchers in this area are reflected in numerous publications, the contribution of Russian inventors to this area still awaits elucidation.

This paper is devoted to the story of the most important inventions of the Russian scientists and designers who have ensured effective practical use of an artificial gas cavity to reduce resistance to the movement of high-speed ships. The paper contains numerous bibliographic references and is well illustrated.

The idea of using air lubrication to reduce the resistance of ships was offered first in 1882 by Goustav de Laval. Since then, theoretical and experimental research on this area has been widely developed. At present, Russian inventors offer many well-constructed decisions for the practical realization of this principle. Let's consider only the most typical one of them, which appeared at the end of twentieth century.

According to this invention,[4-2-1] published in 1971, the special nozzles are placed behind the transverse step of a high-speed boat with a profiling bottom. With the aid of air channels and air entry holes at the bow, the nozzles are connected to the atmosphere (see Fig. 4.2.1.).

It was the idea of the inventor that atmospheric air under the action of the air velocity head and ejection effect would create a flow if it could get under the bottom of a boat, which would form air film and isolate the bottom from the water.

Another invention,[4-2-2] published in 1986, made a special cavity with lateral and stern fences under the bottom of a high-speed vessel behind the cross step. The holes are made in the bottom before the longitudinal step (Russian-Redan), which communicate with the atmosphere with the aid of channels in the hull. The fairing has been located before each hole as a bevel-cut cylinder (see Fig. 4.2.2).

According to the inventor's plan, negative pressure will be created by a moving boat with fairings of holes on its bottom. Therefore, thanks to the ejection effect, under the bottom of the boat, air will enter the holes and the air cavity after the step as well as isolate part of the bottom from the water surface after the step.

A third invention,[4-2-3] published in 1981, uses the energy of waves to force air under the bottom of the boat. For this purpose, a special box is mounted on the fore part of the hull, which connects with the fore edge of the basic hull with the aid of special hinges and with the stern edge via spring shock-absorbers. On all perimeters, except for the stern side, the cavity between the basic hull of the boat and the mentioned box is linked with atmosphere via the air channel, on which a non-return valve ensures that only air passes into the specified air cavity (see Fig. 4.2.3).

Inventor's certificate USSR № 318 501

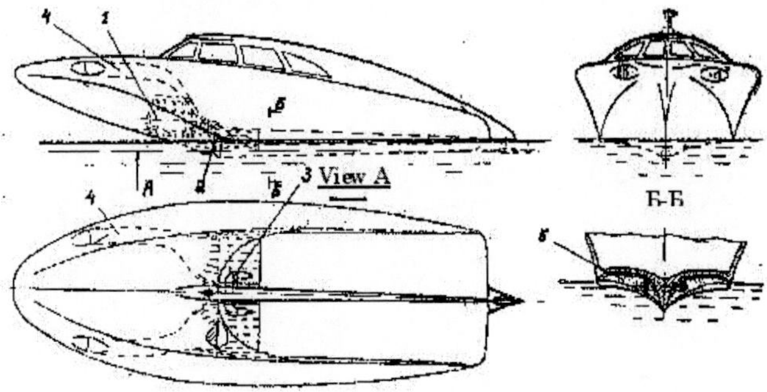

1-engine; 2-hydraulic pump; 3-jet nozzle; 4-air channel; 5-air lubrication boat

Drawing 1

Inventor's certificate USSR № 1 273 292

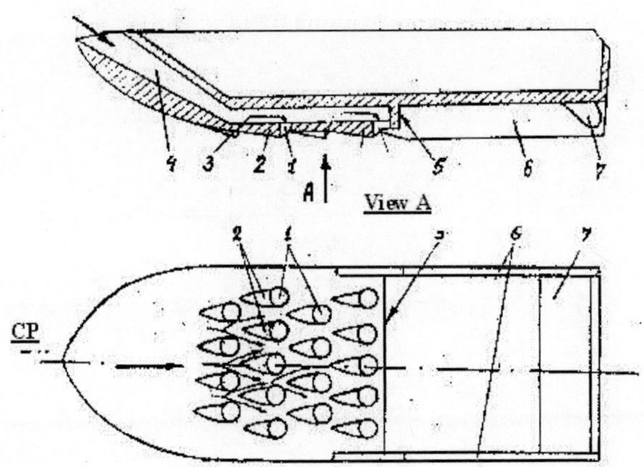

1-hole; 2-fairing; 3-air cavity; 4-air channel; 5-cross redan; 6-side restrictor of cavity; 7-stern restrictor of cavity.

Fig. 4.2.1 (top) and Fig. 4.2.2 (bottom)

Inventor's certificate USSR № 796 049

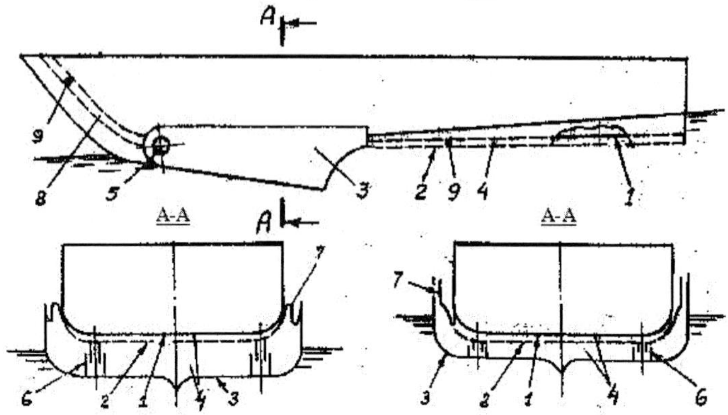

1-hermetical bottom; 2-perforated bottom; 3-box; 4-air cavity; 5-hinge; 6-absorber; 7-flexible hermetical plate; 8-air channel; 9-non-return valve.

Drawing 3

Inventor's certificate USSR № 85 781

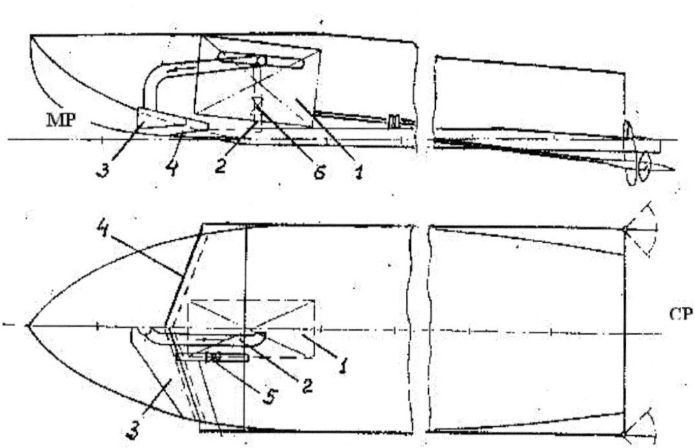

1-main engine; 2-gas exhaust line; 3-gas box; 4-slit; 5,6-valves.

Fig. 4.2.3 (top) and Fig. 4.2.4 (bottom)

Under the action of the impact of an incoming wave, the box compresses the shock-absorbers close to the bottom of the basic hull and pushes out air located in a cavity between it and the basic hull, at the stern part of the bottom. Due to the absorption effect of this device on a high-speed boat, the air lubrication effect of the bottom is successfully realized.

Another invention,[4-2-4] published in 1951, uses gas aeration under the bottom of the glider, with a cross redan (step) both before and after the redan space. Because the exhaust gases of the main engine of the boat are used, the gas travels via the pipeline to enter into a special gas box; from this box the gas is directed partially to the space after the redan (step) and partially through a slit in the specified box to the bottom near the fore edge of the bottom surface before the redan (step) (see Fig. 4.2.4).

However, the many inventions, including those described above, that aim to realize the useful effect of air lubrication have not managed to achieve essential results in the reduction of the resistance of vessels. As was found out later, the reason for the inefficiency of these projects, despite their inventive cunning, was that they did not use the special flow, investigated for 40 years by the well-known German engineer Reichard, which subsequently received the name "artificial cavitation flow." Reichard used the submission of air to the surface of rotating bodies to improve the form of engines at high-speed planes.

At the beginning of the 1960s, in the A. N. Krylov Central Research Institute (Russia) under the initiative of Professor A. N. Ivanov, a group of scientists developed research on use of an artificial cavitation flow to reduce the resistance of surface vessels. The basic work on realization of this idea was developed by A. A. Butuzov.

In Fig. 4.2.5 the scheme for the artificial cavitation flow with reference to the hull of a high-speed vessel with cross redan (step) is given. According to this theory, depending on the position of parts BC and DE of the vessel bottom with respect to each other and the speed of the incoming flow, it is possible to get the static air laminar between points C and D, where the boundary line is formed, which separates an external water flow above line CD. In the specified area, the liquid flow is different to the basic flow. In particular, it can be at rest. This case is of great interest.

Scheme of stream near gliging surface with cavity on lower side.

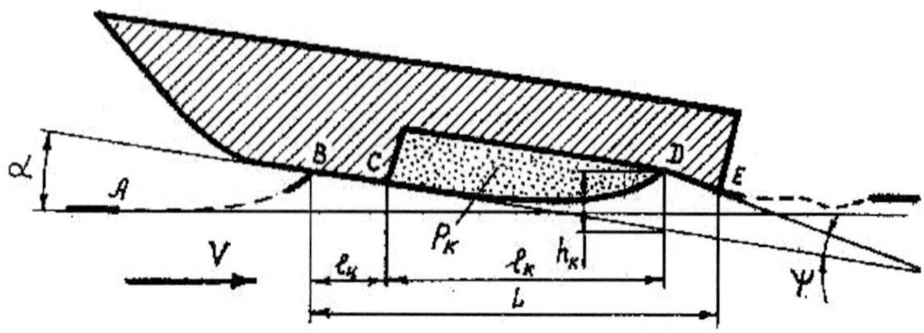

Fig. 4.2.5

In order to create a rested air laminar between the water flow CD and bottom DE, the parameters h_k and Ψ are extremely important, as the names of the basic parameters of longitudinal profiling. At these values of the parameters the residual resistance will be at a minimum and the pressure in a stagnant zone will be at maximum and thus greater than the atmospheric pressure.

It is obvious that in the real liquid the speed in the mentioned zone will never be equal to zero because of viscosity effects on its boundary. To reduce these effects to a minimum, in a stagnant zone the air or other gas is aerated, which only supersedes the liquid from the zone, isolating the hull of the vessel from contact with water. It is clear that the necessary air consumption will therefore be small, as will the power expenses for the creation of the flow, as the capacity required for it is proportional to the product of back pressure (i.e., pressure in a cavity P_k) on the air consumption Q.[4-2-5]

On the basis of Russian research results, inventions that determine these or other parameters that profile the bottom of a ship with an artificial cavity (SAC) have been developed. Most typical is another invention,[4-2-6] published on November 10, 1995. The invention concerns the SAC hull, in which on the bottom behind the cross redan (step) there is a cavity limited on each side by lateral skegs; when the ship moves, the cavity is aerated by the compressed air (see Fig. 4.2.6).

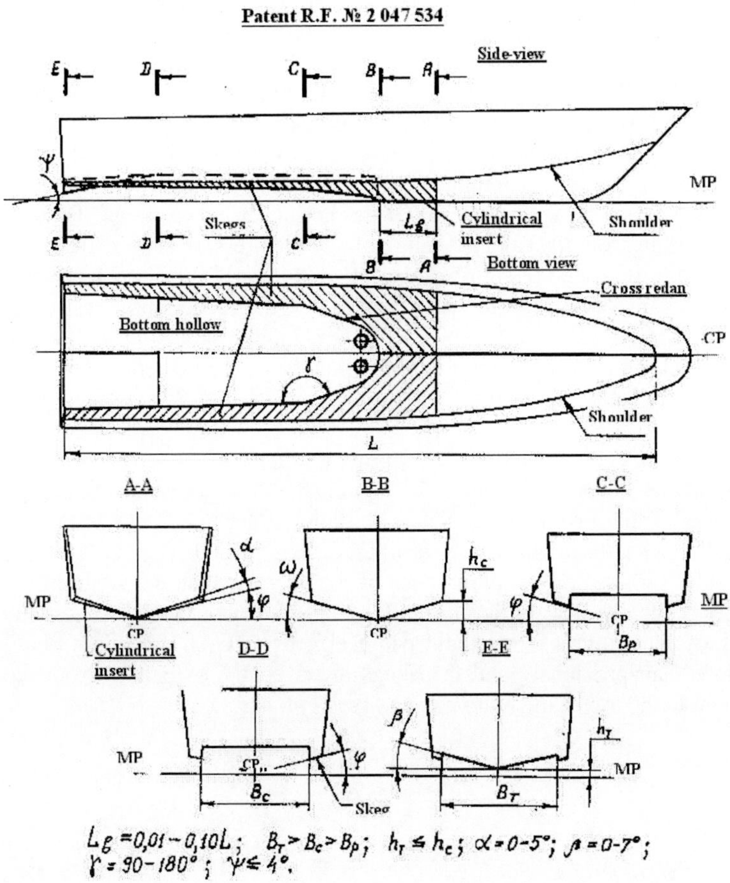

Fig. 4.2.6.

The essence of the invention depends upon a part of the bottom in front of the cross redan (step), conditionally an insert with 0.01–0.1 of the hull length, which is cylindrical in form or with a divergence angle of no more than $\alpha = 5$. The intersection point of the bottom edge of the transom with the centerline plane (CP) of the vessel hull moves from its base plane (BP) a distance that does not exceed the value of ½ the height of a bilge above BP in the area of the beginning of the cross step. A part of the bottom between the side skegs in front of the transom is fulfilled with deadrise $\beta =$ 0–7° and with obliquity to a BP of no more 4. Thereby the width of the off-step hollow (cavity) in any cross section does not exceed its size in a plane of transom, and the angle in the plane between stall forming edges of the cross step and side skegs is in a range $\delta = 90–180°$.

When a vessel with such a hull is in motion, the aerating of air in its off-step hollow develops and steadily supports the artificial cavity. The free surface of the cavity smoothly continues to moisten the surface of the bottom in front of the redan and does so smoothly without harmful spray becoming isolated on a stern part of the bottom between skegs. Thereby the loss of air from the cavity is minimal and the pressure is maximal.

For this reason, the significant part of the surface of the bottom is isolated from contacting the water surface and the reduction of resistance of such vessels achieves 25–30%. The gas bubble under the bottom of SAC also promotes an essential (approximately 1.5 times) reduction in the value of vertical overloads during the movements of a vessel on a rough sea.

The even greater increase of seaworthiness of SAC is provided according to a further invention[4-2-7] published on August 9, 1995. The essence of the invention is that a hollow under the bottom for the formation of an artificial cavity is limited in front and from the sides in particular by the stall forming edges of the skegs, which converge in the fore part of the SAC hull. The hull itself is wedge-shaped, so that the width of the bottom of the shoulder in each cross section, from fore to aft, is more than previously and the skegs are executed extending from the transom to the top of the bottom cavity (see Fig. 4.2.7).

Patent R.F. № 2 041 116

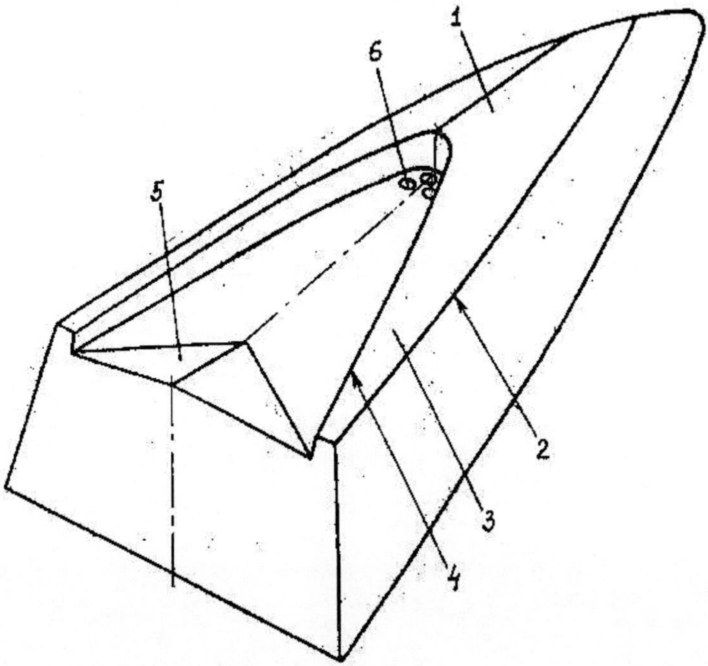

1-bottom; 2-shoulder; 3-skeg; 4-stallforming edge of skeg; 5-stern part of bottom; 6-hole for feed of compressed air to cavity.

Fig. 4.2.7.

As the description of the invention shows, for such profiling of the bottom the value of overloads tested by SAC in rough seas is 1.5–2.0 times lower than for other SAC craft. The further increase of seaworthiness of the SAC by an increase in the deadrise angle of the bottom in the front of the redan and by an increase in the waterline sharpening of the fore part of the hull collides with the usual form of a cylindrical insert in the front of bottom cavity.

This contradiction was solved in an invention[4-2-8] published on May 27, 1999. According to this invention, the cylindrical insert was carried out in the planned horseshoe-shape on the part of the fore and side in lines, which were equidistant in a longitudinal direction to the stall forming edges of the redan and side skegs. Thereby, the bottom branches of the frames that were taking place through a cylindrical insert, at the left and to the right of it, were carried out with a more intensive rise to the side shoulders than within the range of the cylindrical insert (see Fig. 4.2.8).

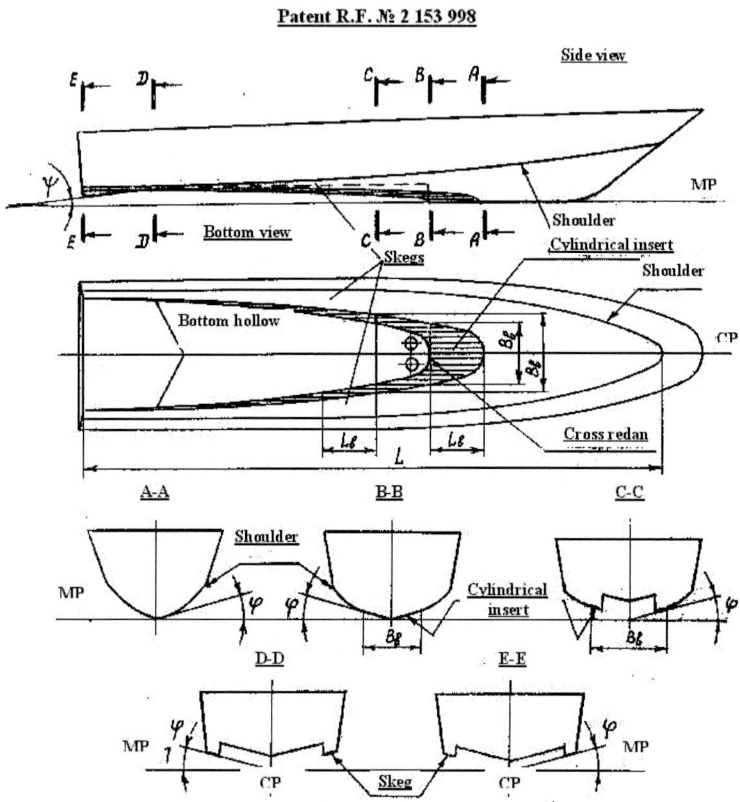

Fig. 4.2.8

A towing tank test of the SAC model with a horseshoe-shaped cylindrical insert showed that it formed at a flow on the bottom with such profiling that the artificial cavity has a more smooth free surface that is more smoothly isolated on the part of the bottom close transom.

The formation of a steady artificial cavity ensuring water is isolated from most of the SAC bottom surface, together with the lowered wave resistance of the hull, sharpened from a fore-body, on the real boat was achieved with a displacement of 25 tons and a speed of 35 knots, and the hydrodynamic property value is equal to 8.7.

The first experimental SAC were equipped with the fan, hung on the main engine, aerating the atmospheric air in the bottom cavity. It appeared not always to be optimal, as the fan required the additional area of a motor compartment and was a source of additional noise and vibration, and its mechanical driver carried with itself additional mechanical losses of power of the main engine of the SAC. It was decided to try to aerate the exhaust gases of the main engine in the bottom cavity to form an artificial cavity. Theoretical calculations and test results showed that the exhaust gases of the main engine of the SAC quite sufficed to ensure the vessel was steadily supported by a well-designed artificial cavity even at its maximum speed with maximum gas consumption. Thereby the value of the power of the main engine spent on aerating the exhaust gases in the cavity, as a rule, is no more than 2–3%. The power plant of the SAC thus becomes more compact and creates much less noise and vibration.

However, in the experimental operation of such an aerating system a shortcoming is also found. Not even the pressure of the exhaust gas of each main engine is capable of overcoming, during the start-up operation, the back pressure in a cavity. And in case of starting the main engine, the additional vibrations of the hull will happen due to the pulsation pressure of gas in the cavity.

To eliminate these shortcoming, according to the invention[4-2-9] published on April 15, 1990, the basic pipelines of the gas exhaust are supplied with side branch pipes located in the sides of the vessel and above the designed waterline, which are connected with the basic pipelines with two running valves. Meanwhile, the basic pipelines of the gas exhaust are mounted on the bottom linked to the gas cavity, and the pipeline axes are inclined to BP under an angle not exceeding 45° (see Fig. 4.2.9).

Inventor's certificate USSR № 683 148

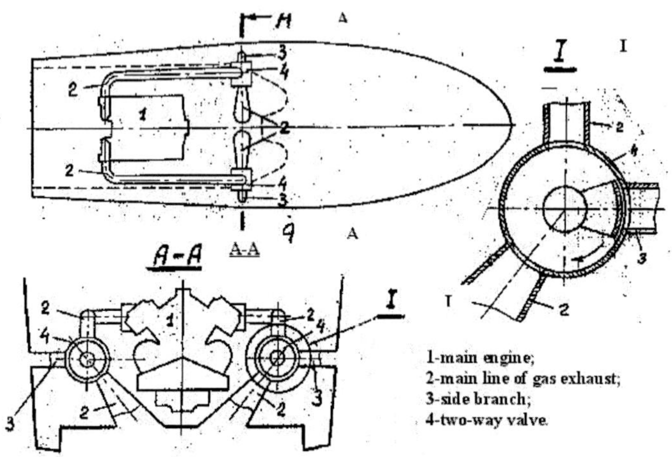

1-main engine;
2-main line of gas exhaust;
3-side branch;
4-two-way valve.

Drawing 9

Patent R.F. № 2 116 219

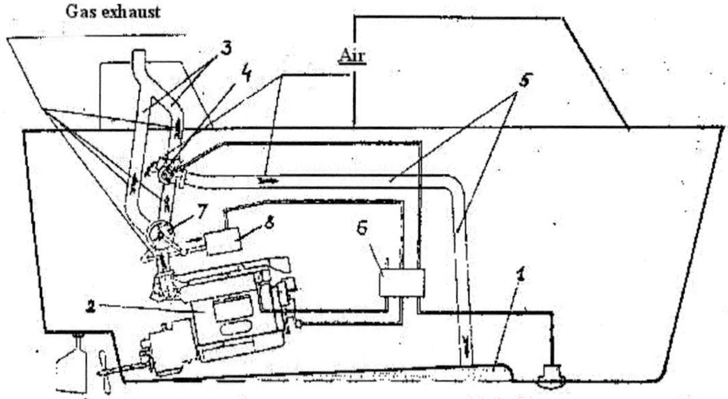

1-cavity; 2-main engine; 3-line of gas exhaust; 4-turbocharger; 5-air pipe line; 6-micropocessor; 7-controlled valve; 8-drive of valve.

Fig. 4.2.9 (top) and Fig. 4.2.10 (bottom)

In this case, when the main engine starts, the exhaust gases with the help of two running valves are directed through side branch pipes to the side of the vessel. This allows the back pressure in the gas exhaust system to be lowered and the raised vibration of the SAC hull to be lowered and to stop with the working main engine.

The carrying out of basic gas exhaust pipelines extending in the direction of the bottom cavity allows the speed of the exhaust gases and the value of the back pressure to be reduced. Together with an inclination of the pipelines to BP, it also prevents harmful deformation of the free surface of the artificial cavity under the bottom of the vessel.

However, despite the higher perfection of the aerating system of the gas cavity described above, there also are some disadvantages. The exhaust gas near the water surface damages the ecological cleanliness of the water space. There is no opportunity to control the gas quantity aerated to the cavity; this is in order to minimize loss of power to the main engine's exhaust.

These drawbacks were eliminated by a new invention[4-2-10] published on July 27, 1998. This invention installs a turbocharger on the pipeline that carries the exhaust gases from the main engine, which aerates compressed air to a cavity. It completely excludes any pollution of the environmental water space.

Optimizing the aeration of the compressed air to the cavity depends on the speed of the SAC; for this reason, the turbocharger can be controlled by a microprocessor. In this case, the turbocharger is equipped with a gauge of revolutions and the controlled choke with the actuator, the signals from the microprocessor depend on the speed of the vessel and the frequency of revolutions of the main engine, which are then used to adjust the aeration to supply the necessary quantity of exhaust gases to the turbocharger.

In each concrete case it allows one to determine the operation mode of the turbocharger, which supports optimum pressure in the artificial cavity with minimal power from the main engine (see Fig. 4.2.10).

To obtain high propulsion performance from the SAC it is necessary not only to ensure the artificial cavity under its bottom is effective and stable, but also to reliably exclude any opportunity for it to harmfully influence the SAC's propellers, especially if submerged propellers and waterjets are used, as is usual.

If no measures are taken, during movement of an SAC with submerged propellers, it is possible that a break of air along the inclined driving shafts may cross a free surface of a cavity. Besides, rotating driving shafts promote destruction of the cavity and premature washing by water of the stern part of the SAC's bottom.

To prevent these undesirable phenomena, according to an invention[4-2-11] published on October 20, 1997, inclined driving shafts can be used on the SAC in special fairings. The fairing in a section with a diametrical plane is designed to be quadrangular with the stern edge, inclined to BP under a larger angle than the fore edge. Both edges of the wedge-shaped waterlines are smoothly connected among themselves along an axis of the driving shaft. Besides, in the stern part of the fairing along an axis of the driving shaft, the cylindrical expansion with a diameter equal to the diameter of the propeller boss is executed (see Fig. 4.2.11).

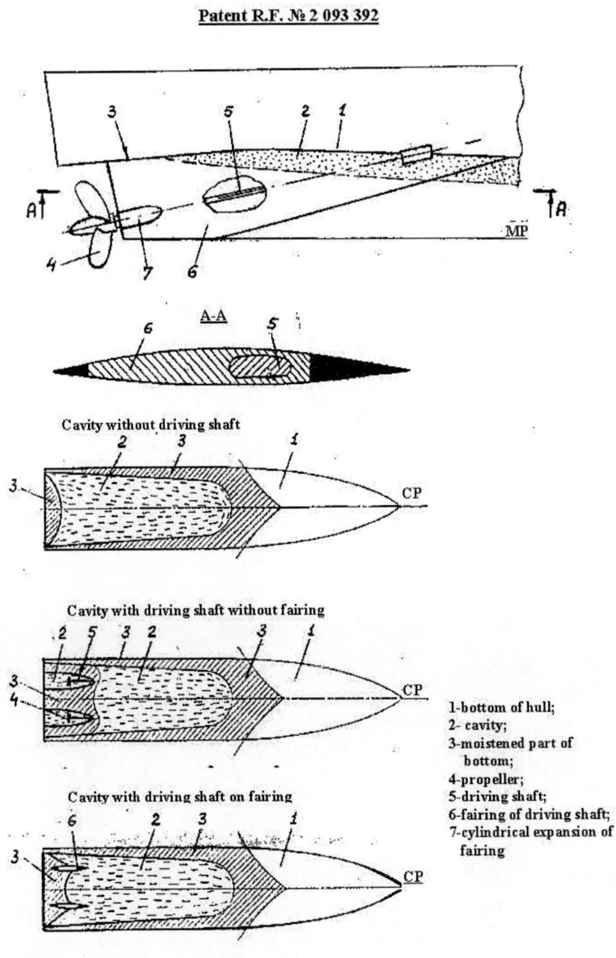

Fig. 4.2.11

However, as described above, the fairing of the driving shaft was problematic due to the reduction of its efficiency at lateral drift of an SAC; in this case, a part of the cavity located behind the fairing would be destroyed.

To eliminate this problem, an invention[4-2-12] published on November 20, 1999 offers the fairing in the place of an exit from it to the driving shaft with a width not less than 3.5 times the diameter of the shaft with a smooth reduction of it in the process of removal. Thereby, the bottom surface of the fairing in the area of the exit from it of the driving shaft should, if possible, repeat the shape of a free surface of an artificial cavity and be located a little bit below it (see Fig. 4.2.12).

Patent R.F. № 2 141 426

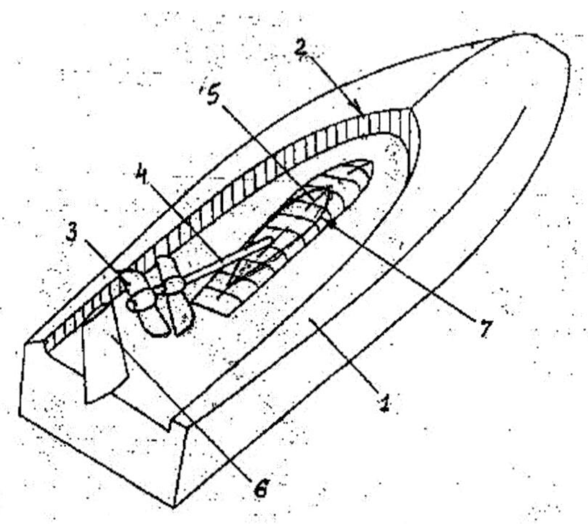

1-bottom; 2-bottom hollow for cavity formation; 3-propeller; 4-driving shaft; 5-fairing of driving shaft; 6-rudder; 7-line of intersection of free surface of cavity with fairing

Fig. 4.2.12

Thereby, the fairing of the inclined driving shaft of an SAC can be continued in the stern, forming itself simultaneously alongside the fairing of the bracket of the driving shaft and the fairing of the rudder stock, which allows it to prevent destruction of the cavity across the entire SAC bottom at all tracks of its movement.

In the case of SACs with waterjets, there are two basic problems. First, there is the probability of the ingress of air from the air cavity into the water scoop of the waterjet and, as a consequence of this, a great reduction in the propulsion property. Second, this incomplete use of the artificial cavity to reduce the resistance of the SAC—as a part of the bottom occupied by the fairing of the waterjet's water scoop and adjoined from both sides by the water moistened parts of the bottom—is excluded from the common surface of the bottom, isolated from contact with water at the expense of formation of the artificial cavity.

According to the invention[4-2-13] published on October 20, 1999, both these problems are solved as follows: the water scoop hole of the waterjet is supplied with a special fairing, the bottom surface of which in all modes of SAC movement is guaranteed by the free surface of an artificial cavity. In order not to moisten the bottom with water on both sides of the fairing of the water scoop, parts of the bottom on both sides from the fairing of the water scoop are close to the transom; the negative raise-of-floor is given in Fig. 4.2.13.

Due to the measure mentioned above, the probability of ingress of air from the cavity is prevented. Simultaneously, the prevention of moistening by water from the bottom next to the fairing parts of the SAC bottom essentially raises the efficiency of using an artificial cavity.

A more radical way of preventing the harmful influence of an artificial cavity on the propulsion performance of the waterjet was offered in the invention[4-2-14] published on June 27, 1995. This invention equips the SAC with a partially-submerged propeller (PSP) located behind the transom in a special cylindrical semi-tunnel that covers the propeller from above. The semi-tunnel is connected to the SAC hull with the aid of brackets (see Fig. 4.2.14).

Due to the design of a propulsor complex, it creates the ideal conditions for the complete use of an artificial cavity effect to ensure the maximal efficiency of the PSP; in addition, it also creates a significant reduction in the vibrating loadings created by the propeller on the hull of the vessel.

Today, vessels with the hydrodynamic scheme of an SAC in Russia are under construction in series. Their displacement ranges from 2 to 105 tons, and their speed ranges from 22 to 50 knots. On these vessels, various types of propulsors are applied: with water jets, ventilated water jets, PSPs, and also usual propellers on inclined shafts protected by fairings of special shape. In all cases, the effect of reducing the resistance of the real crafts proves to be true qualitatively and quantitatively.

Patent R.F. № 2 139 807

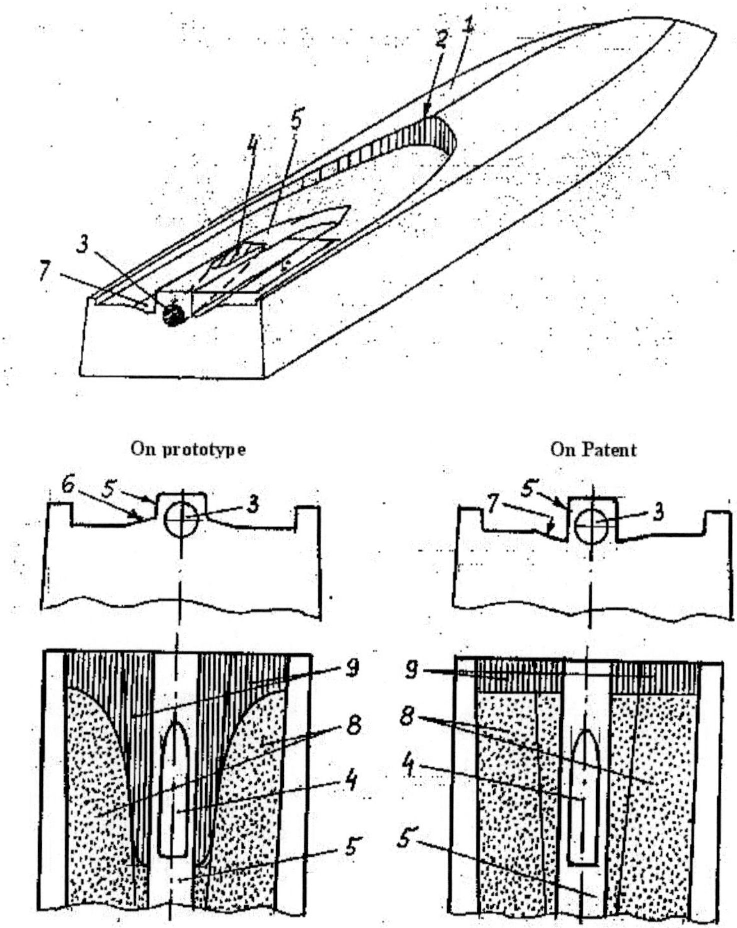

1-bottom of vessel; 2-bottom hollow for cavity formation; 3-water-jet; 4-water scoop h
5-faring of water scoop; 6-section with positive raise-of-floor; 7-section with negative r
of- floor; 8-cavity; 9-moistened section.

Fig. 4.2.13

References

[4-2-1] V. V. Ashinsky & V. I. Koniukov, "A Vessel with Air Lubrication of the Bottom," Inventor's Certificate USSR No. 318501; IPCB63B1/38; published October 28, 1971.

[4-2-2] A. E. Nedobezshkin, "A Vessel with System of Aerating of Air on Its Bottom," Inventor's Certificate USSR No. 1273292; IPC B63B1/38, B60V3/06; published November 30, 1986.

[4-2-3] O. M. Menchitz & V. I. Lisniak, "The Hull of a Vessel with the Double Bottom," Inventor's Certificate USSR No. 796049; IPC B63B1/18, B63B39/06; published January 15, 1981.

[4-2-4] B. A. Drugov, "Glider for Navigation with Application of a Gas Layer Under the Bottom," Inventor's Certificate USSR No. 85781; IPC B63B1/38, B63B1/18, B63H 21/34; published 1951.

[4-2-5] V. M. Pashin, A. N. Ivanov, C. G. Kaliuzhny, A. G. Lyakhovitsky, & G. A. Pavlov, "Hydrodynamic Design of Artificially-Ventilated Ships," SP2001, Lavrentiev Lectures, Proceedings of International Symposium on Ship Propulsion, June 19–21, 2001.

[4-2-6] A. N. Ivanov, V. G. Kaliuzshny, A. A. Butuzov, I. I. Matveev, A. N. Pavlenko, I. A. Zverev, & V. P. Kliukin, "The Hull of a High-Speed Vessel," patent of Russian Federation No. 2047534; IPC B63B1/38; published November 10, 1995.

[4-2-7] Y. V. Chaban, I. I. Matveev, S. I. Rogozshin, & M. V. Ezshov, "A High-Speed Vessel," patent of Russian Federation No. 2041116; IPC B63B1/38; published August 9, 1995.

[4-2-8] G. A. Pavlov, Y. P. Pridatko, & M. L. Apel, "The Hull of a High-Speed Vessel," patent of Russian Federation No. 2153998; IPC B63B 1/38; published May 27, 1999.

[4-2-9] A. I. Vilk, L. V. Kamenchuk, Y. A. Kudriavzev, B. S. Semenov, L. L. Heifetz, & K. A. Iahontov, "Glider," Inventor's Certificate USSR No. 683148; IPC 63 B1/38, B60V 3/06; published April 15, 1990.

[4-2-10] I. V. A. Tihomirov, "Vessel," patent of Russian Federation No. 2116219; IPC1/38; published July 27, 1998.

[4-2-11] K. V. Aleksandrov, L. K. Zagorulko, V. G. Kaliuzshny, Y. P. Mihailov, I. D. Novikov, A. N. Pavlenko, Y. V. Chaban, & M. D. Dmitrichenko, "A High-Speed Vessel," patent of Russian Federation No. 2093392; IPCB60V 3/06, B63B1/38; published October 20, 1997.

[4-2-12] A. A. Butuzov, "The Fairing of Inclined Driving Shaft of a High-Speed Vessel with Bottom Gas Cavities," patent of Russian Federation No. 2141426; IPCB63 B1/08, 1/38, 3/42; published November 20, 1999.

[4-2-13] A. A. Butuzov, A. V. Sverchkov, & S. A. Chalov, “A High-Speed Vessel with Gas Cavities and Waterjets,” patent of Russian Federation No. 2139807; IPC B 63B1/08, 1/10, 1/38, 3/42; published October 20, 1999.

[4-2-14] A. V. Romanov & A. N. Lopatin, “A Vessel,” patent of Russian Federation No. 2038255; IPC B63B 1/38, B60B 3/08; published June 27, 1995.

4.3

DEVELOPMENT AND PERFORMANCE OF AIR CAVITY CRAFT (2004)

GENNADY PAVLOV

Projects Chief Designer, Feodosia Shipbuilding Company "Morye," Ukraine

LIANG YUN

Technical Director, Engain Science & Technology Ltd., Hong Kong, China

Abstract

The development of a novel high-speed craft, the air cavity craft, and its performance in Russia and Ukraine are introduced in this paper. In various respects, such as resistance in calm water, sea-keeping quality, propulsion, engineering practice, and economy, the paper recommends that the air cavity craft will have a great bright prospect in the high-speed craft family. Actually, the craft has been put into practical operation for a while, both in Russia and Ukraine.

4.3.1. Introduction

People always seek higher transport speeds to shorten trip times and extend existing spaces. In the past half century, the operational speed of various transport tools has been enhanced to a great extent—that is, 20 times for airplanes, and 3 times for cars. However, for ships speed has only been enhanced two times because resistance increases drastically with speed—ship speeds have stayed at 30–35 kn.

It is well known that the density of air is smaller than water by 800 times; therefore, the best way to reduce water resistance is by taking ships out of the water.

Thus, in order to increase the speed of ships, since the middle of the twentieth century some novel ships have been created, such as ACV/SES, HYC, planing hull, and so on, which can reach 40–60kn in speed; WIG ships can reach even higher speeds of over 100kn. Nevertheless, these new technologies have their own particular problems, such as:

- More complicated technology and higher cost, as well as maintenance fees for lift and skirt systems for ACV/SES
- HYC not only have the same high costs and maintenance fees and complicated technology, but also have a cavitation problem on hydrofoils during high speed, which prevents the increase of craft speed
- WIG is a mixture of airplane and marine craft with a super high speed up to 100kn and more; however, they are also extremely high tech and cost accordingly
- Planing hulls used to operate on the water surface to reduce wave-making resistance to a great extent; in addition, this did not involve complicated equipment, such as the skirt-lift system on the ACV/SES and foil systems on the HYC and WIG. However, the great impact load and vertical acceleration acting on the bottom of the craft prevented its development and marketing

Despite that, the speed of ships has gone from 30–35kn to 40–60kn (ACV/SES/HYF), and even above 100kn for WIG.

Most novel crafts of these types operate off the water's surface by means of hydro-aerodynamic lift; thus, in the twentieth century they were only used on small-tonnage craft and large-tonnage ships still used conventional displacement types and therefore stayed at a speed of 30–35kn.

After the appearance of such novel crafts, ship designers and scientists began to consider that it might be possible to create a new craft with both high speed and simpler equipment and technology. For instance, a design without a skirt-lift system or foil system but that could be transferred to large craft. Such a novel type of craft has appeared—the air cavity craft (ACC).

4.3.2. The features of ACC and its advantages

1. The basic principles and advantages of ACC

The basic principle of high-speed ACC craft was introduced in 4.1 and 4.2. In order to improve sea-keeping quality, the transverse section of the hull can be designed with a definite deadrise angle; in addition, two sidewalls should be put at each side of the hull to keep the air layer under the bottom. Fig. 4.2.6 shows the typical profile of the ACC, and Fig. 4.1.4 shows the profile of a Russian ACC-type *Mustang* with both fine powering performance (up to 55kn) and fine sea-keeping quality.

Since the wave-making resistance of planing hulls is small, most of the drag of a craft is due to water friction and thus the total drag will be reduced due to the air layer. Fig. 4.3.1 denotes the relationship between engine power and the effectiveness of the air layer to the speed of a ACC weighing 25t and with 1,000 HP of total engine output.

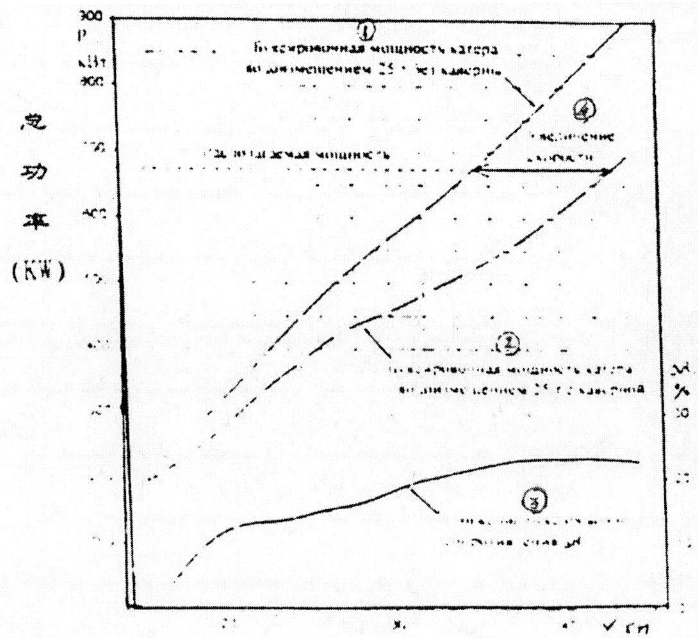

Fig. 4.3.1.

In the figure, the reduction of drag and increment of speed due to air lubrication have been denoted (ordinate denotes resistance, and abscissa speed):

1. Drag of craft without air lubrication
2. Drag of craft with air lubrication
3. Drag reduction due to air lubrication
4. Speed increment due to air lubrication on the same craft with the same total power

The figure shows that the increment of speed is up to 8kn (22.8%), from 35kn to 43kn. Meanwhile, since the air layer exists under the craft bottom, the impact of the load acting on the hull bottom will be decreased, which used to be extremely unfavorable to crew, passengers, and equipment. Perhaps, this is one of the main reasons for preventing the development of a planing hull, and why it used to be replaced by other high-speed crafts (HYF, CAT, SES).

2. To sum up, the advantages of ACC are as follows:

- Drag can be reduced down to 25–30%; that is, a speed increment of up to 15–20% at the same power. The suitable relative speed range is F_{nD}=1–4, and the lift–drag ratio can be as high as 10–12.5, drastically higher than a planing hull.
- The power for the air layer can be dropped down to 2–3% of the total engine power, compared with 10–15 % for SES and 30–40% for ACV.
- The hydrodynamic property of ACC is close to that on HYC and SES/ACV; however, it does not have complicated foil and lift/skirt systems, and attracts low cost and maintenance fees. In addition, because it has a shallow draught and no protruding objects (hydrofoils), it is easy for the craft to be alongside a pier and even to land without pier equipment.
- Since the air chamber is lower (10–15% H, where H is depth of ship board), the draught is also small, so the cabin can be accommodated more reasonably compared with SES.
- Owing to the large deadrise angle of the transverse section before the steps, and to the air layer after the steps, the sea-keeping quality is improved, being close to that of HYC but without an automatic control system.
- Various propulsion systems, such as conventional propellers, waterjet propulsion, semi-submerged propellers, air ventilation propellers, and so on can be used.
- In case of crafts in operation, the hull bottom used to be liable to corrosion, deformation, and erosion; the additional resistance caused used to increase and thereby decrease the speed. However, the additional resistance for ACC is reduced due to the insulation of the hull bottom from the water by the air layer. Transport efficiency may be improved up to 1.3 times or more due to the effectiveness mentioned above.

The proper operational range of ACC is at $\Delta V_S = 5.8kn$ or more. For the latter, as it is a planing hull, the whole length of the bottom after the step will be full of the air layer, so one step is enough, and the engine power savings may be as high as 30%. However, for the former, the craft is in semi-displacement operation mode, so the air layer may be more than one after the step, and the engine power saved may be up to 20–25%. Therefore, the ACC can be used on ships with small and medium tonnage.

To sum up, ACC is high-speed (50kn and more) simple equipment, it has a low cost, is easy to maintain, has a good sea-keeping quality, and is suitable to use on ships with small and medium displacements.

4.3.3. Development of ACC

Making an air layer on the hull bottom used to be an ideal for shipbuilding engineers and ship scientists, which had been offered by Russian and Swedish scientists.

Nevertheless, it is not easy to move from the idea to engineering practice. The reason is that the air layer is difficult to attach to the hull surface steadily, particularly with the interference of waves or other external circumstance. Besides, in cases where the air layer non-uniformly attaches on the hull surface, it will make the air rough, and even increase the resistance. In addition, the influence of the air layer on the work of the propellers is also a serious technical and engineering problem.

Therefore, only a few countries (e.g., Russia and Ukraine) have carried out such technology to the point of developing real ships; but research work on air cavities has been carried out in many countries (e.g. Russia, Ukraine, USA, Japan, Norway, China, etc.).

1. Russia and Ukraine

The development of air-cavity technology started with low-speed ships in the former USSR. In the 1930s, Russian scientist Lapshin welded an inclined plate onto the flat bottom of a low-speed ship to make a so-called artificial air cavity, as shown in Figs 4.1.1 and 4.1.2. The water flow separates after the inclined plate and generates an artificial air cavity where pressured air is fed into this layer. Since the length of a steadily generating artificial air cavity can be calculated, the inclined plates and the feeding air system can be arranged in every space on the hull bottom to form a regular air layer under the hull bottom, so as to reduce the resistance of the ship. Lapshin successfully carried out tests on 25m- and 75m-long barges in 1938 and 1957, respectively.

Tests on high-speed ACC were also successfully carried out by the Krylov Ship Central Research Institute in the former USSR.[4-3-8] [4-3-9] The artificial air cavity had been made steadily on a planing hull in 1961, and research on generating artificial air cavities steadily on high-speed displacement craft began in 1985. In 1993, they started research on an artificial air cavity on a planing hull and on mixed types of high-speed catamarans. The design and construction of ACC has been carried out at the Alexeev Central Hydrofoil Design Bureau and Shipyard in Gorky.

About 30 and more designs with 14–360t of displacement were carried out and more than 70 ACC with speeds of 30–52kn were constructed in Russia. Since Ukraine and Russia were formerly subordinate countries under the USSR, some ACC were also constructed in the "Morye" Shipyard in Ukraine.

Table 4.3.1 shows the ACC constructed in Russia, and Table 4.3.2 shows the ACC designed and constructed in Ukraine.

Table 4.3.1. ACC completed in Russia

Craft name	*Saigak*	*Serna*	*Linda*	*Muflon*	*Mercury*	*Mustang-99*
Completed year for first craft	1981	1992	1992	1992	1995	2000
Completed number*	>40	5	11	1	4	1
Displacement, D/ t	13	105	24.6	13.2	99.0	30.8
Max length/m	14.05	25.65	24.1	15.5	35.4	19.6
Max draught/m	0.65	1.52	0.95		2.0	0.9
Engine power/kW	735	2×2,430	660	1,100	2×3,670	2×1,220
Max. speed/kn	40	32	38	50	52	50
Propulsion	Water-jet	Air ventilated waterjet	Semi-submerg-ed propeller	Semi-submerg-ed propeller	Propeller	Semi-submerg-ed propeller

* still in continuous construction

Table 4.3.2. High-speed ACC completed and designed in Ukraine

Craft name	*Achill*	*Nautilaus 62*	*Wind*	*Storm*
Mission	Patrol boat	Luxury boat	Passenger craft	Passenger craft
Completed condition	Constructed	Constructed	Designed	Designed
Displacement/t	24	25	–	–
Max. length/m	18.6	19.0	36	39
Max. draught/m	0.85	0.8	1.3	1.4
Engine power/kW	2×1,007	2×588	2×1,399	2×1,399

Max speed/kn.	50	40–42	36	32
Propulsion	Arneson propeller	Propeller	Waterjet	Waterjet
Accommodation			200 passengers	250 passengers

To sum up, the ACC technology in both Russia and Ukraine is well proven and practical; it is used not only on monohulls but also on CAT. Various CATs have been designed that are 100m long, can accommodate 500 passengers and 130 cars, and can travel at 55kn.

Fig. 4.3.2(a), (b) show the operation mode and longitudinal section view of Russia air cavity passenger craft type *Linda* respectively.

Fig. 4.3.2(a).

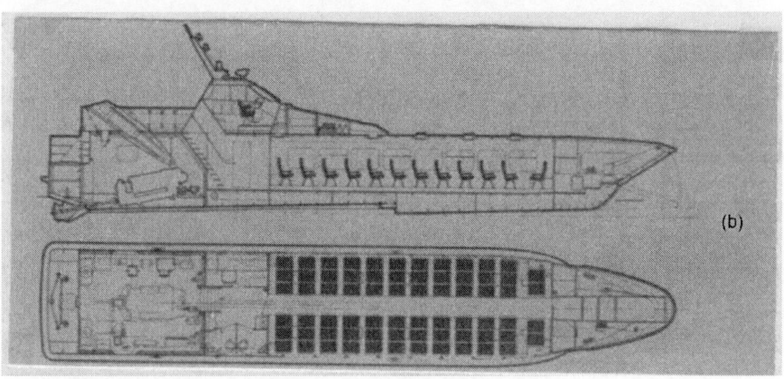

Fig. 4.3.2(b).

Fig. 4.3.3(a). Operation mode of Russia Landing craft

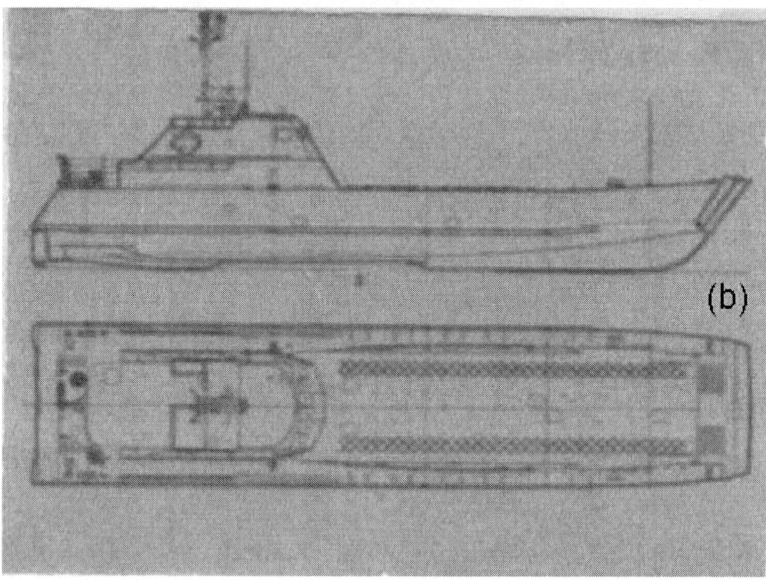

Fig. 4.3.3(b). Profile view of Russia Landing craft

2. USA, Norway, and others

In the 1990s, following the request and sponsorship of the United States Navy, the Norwegian SES Company carried out the R/D of a novel SES with no skirt; its development will be described in 4.4.

Japan and China also have carried out ACC technology investigations in research institutes and universities; however, many are still in the model experimental stage, and no real craft have appeared so far.

4.3.4. Comparison of performance between ACC and conventional ships

1. Power performance in calm water
Fig. 4.3.3 shows the running attitude and basic general arrangement of a Russian Air Cavity Landing Craft.

Fig 4.3.4 (a) denotes the resistance comparison between a 100t ACC and similar craft without AC technology. From the graphic, it is shown that:

1. The gain of air cavity technology resistance is reduced down to 40% (at 40 kn) and 26 % (at 30kn)
2. The speed increment is up by 5.8–6.4 kn (about 19%)

The resistance of the ACC, efficiency of the propeller, power, and hydrodynamic performance coefficient K_η can be obtained from the graphics. Where

$$K_\eta = Wv / 102N\left(\frac{kg \bullet m/s}{kw}\right) = \varepsilon \cdot \eta_p$$

W = craft weight, kg
V = craft speed, m/s
N = power, kw
ε = lift drag ratio
η_p = propulsion efficiency

Therefore, K_η is a non-dimensional coefficient, which properly denotes hydrodynamic property. From the figure, the K_η of the ACC *Mercury* is as high as 4.8–6.5.

2. Sea-keeping performance
It is well known that the criteria for sea-keeping performance of high-speed craft are greater, such as pitching and rolling angle, impact of load, vertical acceleration, speed loss, and so on. The factors influencing speed loss are:

- Resistance increment of craft operating in waves, which leads to speed loss automatically
- Speed drops down artificially, due to large impact of load, green water on deck, and unusual operation of propellers as well as uncomfortable crew and passengers. According to operational

experience, the latter is a very important reason causing speed loss.

In general, wave slamming or the impact of the load is a very important criterion for the sea-keeping quality of high-speed craft.

Thanks to the air layer attached to the bottom of the hull after the step of the ACC, which is a main lift surface, the deadrise angle of the hull bottom can be increased before the step; thus, the impact of the load of the ACC decreases drastically, probably dropping down to 1.5–2.0 times that of a conventional planing hull.

According to test results, the impact of the load coefficient at C.G $\eta_{3\%}$=1.4–1.5 (3% occurrence), where $\eta = A/g$, *A* impacts acceleration. It is obvious that the impact of the load is satisfactory. According to United States Navy requirements,[4-3-6] $\eta_{10\%}$ <2.5, equivalent to $\eta_{3\%}$ <2.61, for navy crews; and $\eta_{10\%} < 1.5$ equivalent to $\eta_{3\%}$ <1.57 for crews on passenger vessels operating for 1–2 hours. According to requirements for the Chinese Navy,[4-3-7] the η <4 at 1/3 craft length from bow.

Besides, according to some information, the impact of the load of a Russian ACC weighing 200t, with a 45kn speed, is about 1*g*, at wave height $h_{w3\%} = 2.0$ m, and pitching double amplitude less than 2°, the rolling angle is less than 5°. Fig 4.3.5[4-3-8] expresses the relation between the impact of the load of various high-speed crafts in waves and relative wave height. Where: n = a/g, $n_{3\%}$ denotes the maximum impact of the load with 3% occurrence, Δ volumetric displacement, and 1-Conventional planing crafts; 2-HYC without automatic control systems; 3-HYC with automatic control systems; 4-ACC.

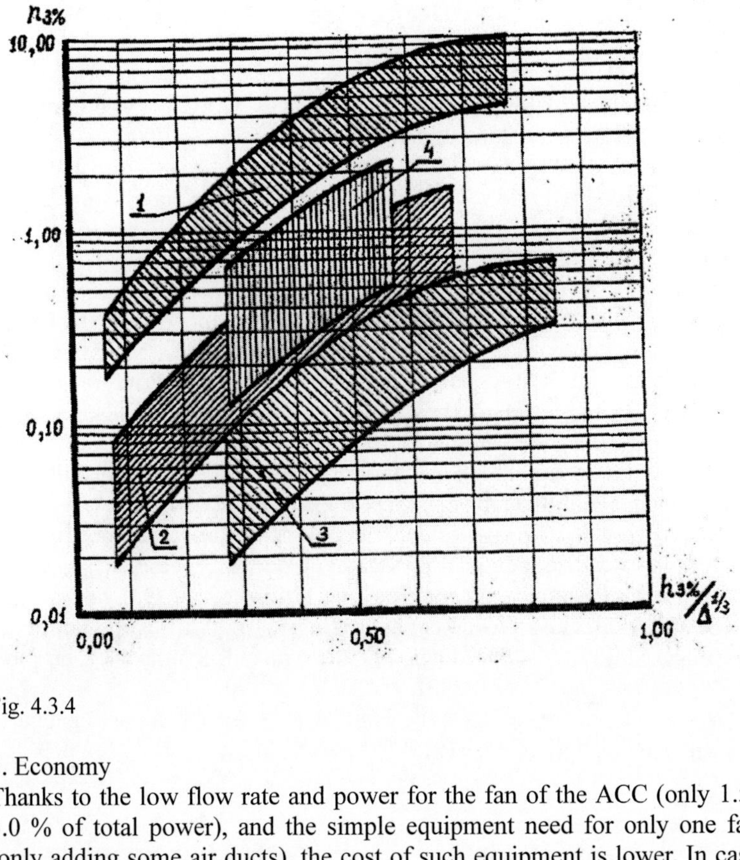

Fig. 4.3.4

3. Economy

Thanks to the low flow rate and power for the fan of the ACC (only 1.5-3.0 % of total power), and the simple equipment need for only one fan (only adding some air ducts), the cost of such equipment is lower. In case of using the exhaust gas of a propulsion engine as pressurised air, the cost will be even less.

Owing to less additional labor, speed will be increased and improved drastically, as will transport efficiency and sea-keeping quality, and the total engine power and cost will be reduced greatly compared with conventional crafts at the same speed. Therefore, the operational cost of ACC are much less compared with conventional planing hulls.

References

[4-3-1] G. Pavlov, "The Contribution of the Russian Inventors to Development of Ships on Artificial Cavity (SAC)," Proceedings of HPMV CHINA 2002, Shanghai, China, April 24–27, 2002.

[4-3-2] K. P. Wong, L. Yun, "An Introduction to the Novel WIG," Proceedings of HPMV CHINA 2002, Shanghai, China, April 24–27, 2002.

[4-3-3] A. Sverchkov, "Ship on Air Cavity," *Boat & Yacht* 1, 2002 (in Russian).

[4-3-4] L. Yun, "Development and Performance of Air Cavity Craft," Technical Report, Engain Science & Technology LTD, November 2002.

[4-3-5] Ulf Stephen Tudem, "New SES Technology: Without Flexible Skirts," Proceedings of HPMV 2000 China, Shanghai, China, April 19–23, 2000.

[4-3-6] E. P. Clement et al., "Optimized Design for Stepped Planing Monohulls & Catamarans," Proceedings of HPMV '92, USA.

[4-3-7] T. S. Gui, "Basic Design Principle of Naval High Speed Craft," Military Defense Industrial Press, July, 1992 (in Chinese).

[4-3-8] V. M. Pashin, A. N. Ivanov, V. G. Kaliuzhny, A. G. Lyakhovitsky, & G. A. Pavlov, "Hydrodynamic Design of Artificially Ventilated Ships SP2001," Proceedings of International Symposium on Ship Propulsion, June 19–21, 2001.

[4-3-9] G. A. Pavlov et al., "The Hull of High-Speed Vessel," patent of Russian Federation No. 2153998, May 27, 1999.

4.4

Air-Supported-Vessel (ASV) Technology with Improved Capabilities for a Wide Range of Fast Vessels Including Fast Offshore Crew boats and Support Crafts (2015)

U. S. Tudem

General Manager, Effect Ships International AS, Norway

T. K. Livgard

Technical Manager, Effect Ships International AS, Norway

Summary

Across the world, owners and operators of offshore service vessels of various kinds are searching for innovations and new technologies to improve the efficiency and capabilities of these vital vessels. In this paper, air supported vessel (ASV) technology will be presented as a challenger to conventional "wet" hull forms. Norwegian technology provider Effect Ships International AS (ESI) has carried out research and technology developments to prepare the new ASV monohull form for market introduction as a robust and capable platform for, among others, fast offshore crew boats and support vessels. To accomplish this technological change, the ESI has invested more than 12 million Euros over the last 10 years. The research is extensively backed by, among others, the European Union (EU), Innovation Norway, the Norwegian Research Council, leading maritime industrial companies, and academia. The current paper provides an introduction to ASV technology; in particular, it examines ASV soft-motion monohull forms, the present test results achieved, and the economic and other benefits from the use of the new patented hull forms for vessels operating several thousand hours per year.

4.4.1. Introduction

Effect Ships International (ESI) and its subsidiary, SES Europe AS (SE), have for the last decade developed and tested various air-assisted hull forms in catamaran and monohull configurations. The air-supported-vessel (ASV) technology, patented world-wide by ESI, is a high-efficiency skirt-less concept covering a variety of mono- and multihull variants.

A comprehensive test and evaluation process has concluded that ASVs exhibit several important and valuable benefits compared with conventional "wet" designs. Depending on the owner/operator's requirements, ASV hulls may be prepared to suit a wide range of applications and vessel sizes, covering all major market segments.

The current paper will focus on providing a general introduction to the new ASV monohulls designed primarily to deliver improved operational efficiency, high speeds, and pleasant onboard comfort in sea states; in short, it aims to be a robust and universally capable platform for, among others, offshore crew and support vessels. Less emphasis will be given to presenting topside designs and arrangements as, from a designer's perspective, the ASV monohulls will be similar to conventional monohulls.

4.4.2. ASV technology—cat versus mono

Both ASV catamaran hull forms and the new ASV monohulls may be successfully used for various types of offshore support vessels (OSV), fast crew boats, and other types of fast service crafts.

Although having developed, tested, and documented more ASV catamarans than just ASV monohulls, ESI has for various reasons now focused on ASV mono variants. The ASV monohull is virtually a mono and a cat combined. The general appearance above the water line is close to that of a conventional monohull, while the underwater parts of the hull have catamaran-like features.

Typically, an ASV mono has a more generous length-to-beam ratio (L/B) than does a conventional monohulled vessel. The vessel's sides are normally close to vertical all the way to the water surface and beyond. The resulting "water footprint" is larger than that of a conventional mono, allowing formation of a relatively large air cushion area underneath the vessel. Pressurized air, fed to an air cushion by means of a lift fan system, is contained within an integrated hull cavity. On either side of the cavity, a volume body-denoted propulsion gondola is adapted to and houses the selected propulsion system/engine/driveline. In the aft, a proprietary, solid and robust air cushion enclosure-flap arrangement seals off the aft section of the air cushion chamber. The lift fan system is located in the bow (smaller and medium size crafts) and feeds air into the cavity through a duct. Depending on operational conditions and other factors, the bow could have different configurations and shapes; these include but are not limited to a bulbous shape, an axe shape, or more conventional bow shapes.

Figs 4.4.1 and 4.4.2 are underwater pictures from the towing tank at SSPA Sweden; the first is a model of a 40m ASV cat (beam approx. 15 m), with a design speed of 55–70 kn and water jet propulsion. Its initial application was fast ferry use, but the hull could naturally also be used for various commercial, paramilitary, and navy applications.

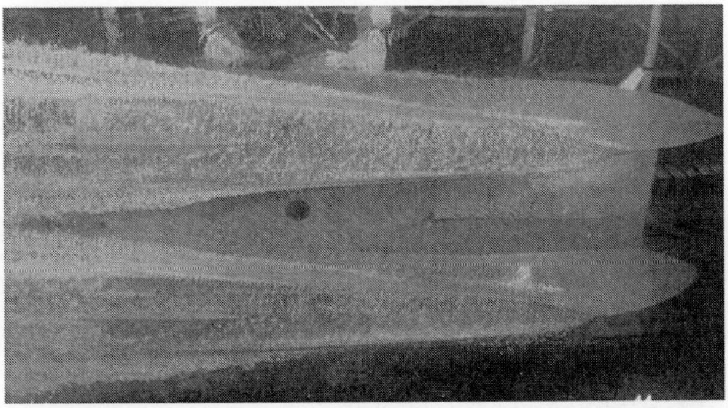

Fig. 4.4.1. Model testing of ASV catamaran at SSPA Sweden, showing an underwater picture at 70kn in calm water. The model is set up for WJ propulsion, with water pick-up from propulsion gondolas, one in each demihull. The length-to-beam ratio is approx. 2.7:1.

The second illustration is of a model of a 20m ASV commuter ferry (beam 6m); its design speed is 30kn, for fast commuter ferry applications (conventional diesel power or all electric [battery] power), configured for contra-rotating pod propulsion (Volvo Penta IPS or similar). It is suitable for several alternative applications, including work boats and pleasure crafts.

Fig. 4.4.2. Model testing of an ASV monhull (BB Green project) at SSPA Sweden, showing an underwater picture at 30kn in calm water. Length to beam ratio approx. 3.3:1

From a construction point of view, an ASV mono has a simpler construction than a twin demi-hull ASV cat. To build an equally long (or large) vessel, the ASV mono requires less material and fewer human hours. Weight and cost will be less and the payload fraction will be higher for the mono variant. In terms of usable volume and space, below the main deck the ASV mono is also a clear winner, offering even more usable space for accommodation, cargo, and so on than a conventional monohull with a similar beam.

4.4.3. ASV mono technology

1. Overall objective for the invention
Our objective was to deliver vessel technology to secure a robust and overall superior package of qualities and capabilities for the owner and operator. A paramount feature is decreasing hull water resistance through the use of air cushion(s) to minimize wetted surface areas and hull/water friction. This is particularly important for high-speed vessels. Another aim is to take advantage of "motion damping" effects from the air cushion and ASV hull elements, and to reduce pitching, slamming, and roll motions. All the above is without operational risk, major drawbacks, or a requirement for high maintenance or other costs. In short, the simpler the better.

2. Where does use of ASV technology make most sense?
With potential power and fuel cost reductions at or above 30–40%, it goes without saying that the more operational hours per year the vessel is in use, the greater the saving potential. Fuel cost is by far the most important

cost element for vessels operating several thousand hours a year at high MCR (maximum continuous rating). In several examples, the fuel cost represents 70% or more of the total operational cost. This is the simple reason why fast crew boats and others would benefit strongly, economically and operationally, from use of the new ASV technology.

3. Tailor making the ASV hull form to suit operation

ESI has developed and tested several ASV hulls, mono and catamaran. Fifteen different ASV hull forms have been extensively tank-tested at SSPA Sweden, a world renowned tank testing institution in Gothenburg, Sweden. More than 1,600 fully documented runs have been completed to this date, and additional tests are in preparation. What is interesting to notice is that all the tested designs have had significantly less hull resistance and additional benefits compared with comparable state of the art conventional "wet" hulls, confirming the design flexibility of the concept. The accumulated knowledge that ESI and SSPA today possess on optimization and tailored ASV manufacture will be matched with the owner's/operator's wish list to develop the most suitable ASV for each given application and/or vessel type. The optimization focus could be on, for example, speed capability (navy crafts), best possible efficiency at a defined speed (commercial projects), reduced onboard motions (comfort aspects), load carrying capability, onboard capacities (i.e., a larger deck/platform), to mention a few.

A typical start-up scenario will be for ESI to discuss possible projects with the owner/operator and yard in order to decide on a set of design and optimization features.

4. A typical ASV mono soft-motion hull

ESI has just completed phase 1 of a development process for a range of small- and medium-size coastal and dual-use ASV fast ferries (passengers and some cargo) for a leading Norwegian ferry operator—Torghatten Nord. The vessels will be in the 22–40m+ length range, to be constructed from carbon sandwich and with design speeds varying from 25–40kn or more. The vessels are primarily configured for the new range of commercially rated Volvo Penta IPS drives, a pod propulsion system with contra-rotating propellers that offers excellent efficiency in the targeted speed range as well as superb maneuverability and handling. The ASV hulls may also be adapted to other propulsors if preferred by the clients. The base design has a hull length of 21m with a beam of 6m.

During 2013, the design development of the hull and a few design and arrangement solutions were prepared with design assistance from leading Italian designer Studio Sculli. The hull has been extensively tank tested at SSPA in calm waters and in various waves. The results and findings will be discussed later in the paper.

Below is a rendering of the ASV soft motion model, here shown without the air cushion enclosure flap, propulsion, and other details.

Fig. 4.4.3. Rendering of tank-testing model ASV mono soft motion (21 × 6m), shown without air cushion enclosure flap, propulsion set-up, and some other details

5. The various ASV mono concept features and the reasons for them

(1) Lift fan system and function

An ASV mono has a lift fan system, typically located in the forward part of the bow section. The fan could either be hydraulically powered (from a smaller separate engine in the motor room or via PTO from the main engine[s]) or it could be electric. The fan is centrifugal and designed to deliver a defined volume flow (m^3/s) at a given pressure (kPa), depending on load, operational conditions, and so on.

Fig. 4.4.4. The hydraulically operated lift fan system in a 20m ASV mono test vessel. The dual inlet single inlet centrifugal fan is made from advanced composites

The main function of the fan is to secure and maintain a steady cushion of pressurized air underneath a major part of the underwater hull in calm water and in a sea state/waves. Establishing a stable layer of air between the water surface and the hull secures no direct contact between water and hull in said area. The pressure and volume of the air "pumped" into the air cushion/cavity will be balanced to achieve an optimum "filling" of the air cushion chamber. That is to fill up the chamber to the "edge" of the lower air cushion periphery, just before too high pressure results in massive ventilation over the edges. The vessel's air cushion support rate will be the fraction of the vessel's weight supported on air versus supported on buoyancy. For ASVs in the 20–40 m size range the support rate can be as high as 80% or more. The average air cushion pressure will be approx. 4–5 kPa for a 20m vessel and 7–8 kPa for a 40m craft. The air cushion will contribute to reducing the wetted surface area of the vessel, which is beneficial for minimizing hull–water resistance.

(2) Air cushion area and volume

A larger air cushion area allows use of a reduced air cushion pressure to obtain a defined air cushion support rate (say 80%). Reduced cushion pressure is also beneficial from a resistance point of view. Compared with most conventional monohulls an ASV mono will have a larger water footprint and more width. A conventional high-speed mono vessel will have reduced width at the waterline to achieve acceptable high-speed efficiency. More hull width for an ASV will secure a larger, safer, and more stable platform with more cargo capacity. The air cushion height will be decided on the basis of several factors. In a sea state, particularly at high speed, the cushion will need to have a height minimizing wave impacts through the air cushion and onto the cushion ceiling. At the same time, a larger air cushion height and volume of the air cavity will have a different effect in submersion of the vessel when on- versus off-cushion.

(3) Propulsion gondolas

The ASV IPR and patents cover a wide range of propulsion arrangements and locations. For an ASV mono the norm is a propulsion gondola on either side of the air cushion chamber for a twin driveline installation; alternatively, a third gondola at the vessel center line can be fitted for vessels with three propulsors. A single driveline with a propulsion gondola at the vessel center line and smaller-volume side bodies is also possible. A wide range of propulsors may be used; the propulsion gondola(s) will be adapted to the propulsor in question.

The propulsion gondolas have yet another function. The lower areas act as planing bodies, contributing dynamic support at speed, to further support the vessel. The relatively short length gives the planing areas a positive effect even at relatively low speeds.

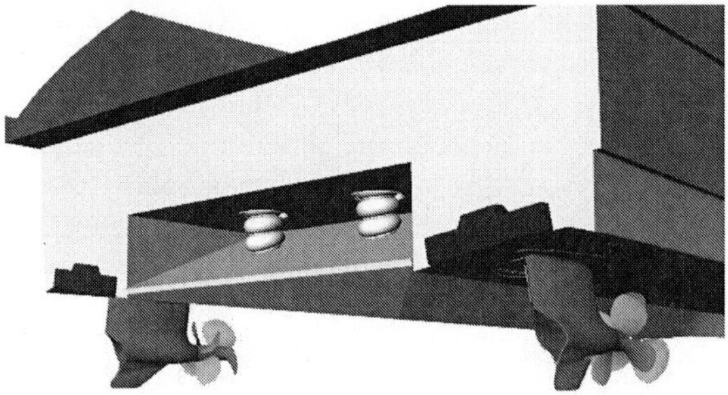

Fig. 4.4.5. Rendering of the aft section with twin propulsion gondolas. It is shown here with air cushion enclosure flap arrangement and optional Humphree Interceptors fitted on the propulsion gondolas to fine tune the trim. The illustration shows Volvo Penta Series 2 Pod propulsors

(4) Air cushion enclosure flap arrangement

A proprietary part of the ASV patents and IPR is the way that the concept entraps and maintains the air cushion intact without negatively influencing the vessel's motions, handling, and ride, particularly in a sea state. An ASV typically operates close to a neutral trim, and the air cushion periphery submersion relative to the water surface will be kept as even or horizontal as possible. In theory this will only function 100% in calm waters. In a sea state, ventilation from the air cushion will unavoidably be higher than in calm water. For the overall efficiency of the concept, it is important to keep the volume flow and low power.

Fig. 4.4.6. Picture from inside the air cushion chamber of one ASV monohull model, facing the air cushion enclosure flap

In the transverse aft section of the air cushion chamber an ASV will have an air cushion enclosure flap system, which, depending on the vessel's size and propulsion configuration, may consist of one or more flap sections. The flap closes the aft part of the air cushion without giving added resistance and without influencing the vessel's motions or ride. A wave passing through the air cushion chamber would, if the flap was fixed and solid, cause added resistance and induce pitch motions in the transom section of the vessel.

(5) ASV automation system

The flap and ASV system setting and function are controlled and monitored with a proprietary ASV automation system. An ASV should be as simple to operate as any conventional vessel. The ESI developed system handles all the ASV related functions, included lift fan system, flap angle, air cushion enclosure flap setting, flap spring rate, air cushion support rate, interceptor setting (optional), and so on. The system consists of a hardware and software package; it can be programmed individually to suit each vessel type, operation, and loading and operational conditions. Overriding the functions for all main settings allows the ship's operators to program their own special combination of settings or to quickly change the setting of one or more parameters if required.

4.4.4. Suitable propulsion systems for ASVs

For ASV mono fast crew boat or OSV application, the vessel size, speed, duty profile, and operation will be important to determine the suitability of each propulsion system as efficiency varies differently with speed for the respective systems. Other factors to be considered are availability/size, cost, maintenance, and capability beyond performance (i.e., ease of handling).

An ASV monohull and hull/propulsion integration will be tailored to the owner's operational requirements; in particular, the propulsion gondola shape and design will differ. Recommended propulsion systems include:

- Pod propulsion
- Waterjets (WJ)
- Fixed or variable-pitch propulsors

4.4.5. ASV monotank testing and results

During the last year, ESI has conducted tank testing with two different ASV monohulls at the SSPA Sweden towing tank in Gothenburg; this included tests on craft with a design length of 20–21m and with a beam of 6m. The ASV hull forms were designed for different operations and applications.

1. BB Green efficiency
The first was for the world's first 30kn-speed ASV battery powered fast commuter ferry, part of the 3.3 million Euro EU-funded project BB Green. Figure 4.4.7 is a rendering of the BB Green full-size demonstrator, to be completed in spring 2015. The vessel will be built from carbon sandwich. For more information on the project and technology, please visit: www.bbgreen.info, or contact the authors.

Fig. 4.4.7. Rendering of BB Green, the world's first 30kn fast battery-powered commuter ferry for sheltered waters. Design: Mauro Sculli, Italy

Tank-testing was carried out in calm water and in waves. Below, the hull-resistance results also present a comparison between ASV mono, ASV catamaran, and a state-of-the-art (SOA) conventional planing monohull.

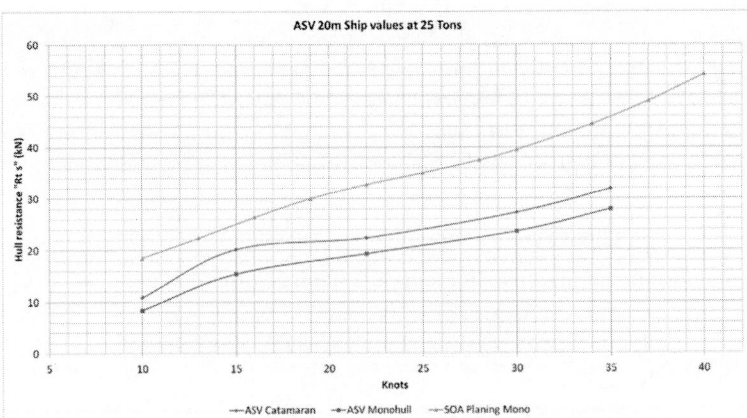

Fig. 4.4.8. BB Green ASV 20m @ 25t, hull-resistance comparisons: SOA planing monohull (green), ASV cat (blue), and ASV mono (red)—speeds 10–35kn

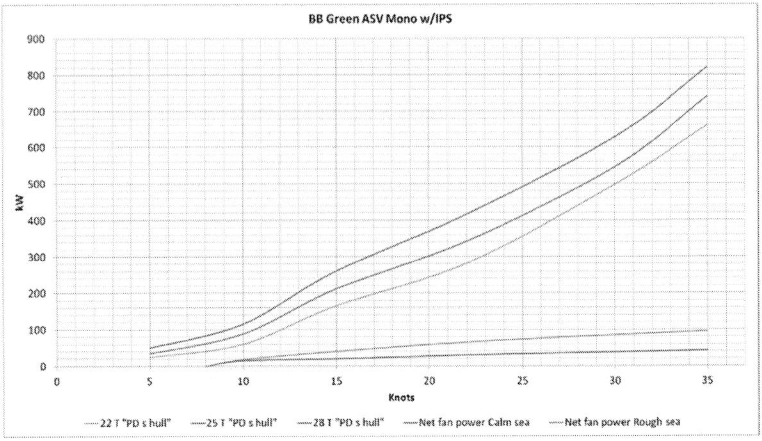

Fig. 4.4.9. BB Green ASV mono speed versus power with Volvo Penta IPS pod propulsion

Graph showing propulsion and lift fan (calm water and rough sea) powering requirements for three different operational weights (22.25t and 28t) and speeds from 5–35 kn in a 20m scale.

As will be seen from the above graphs, throughout the full speed range, the hull resistance and power requirement of an ASV mono is considerably below a comparable conventional SOA vessel. At design speed (30kn), the hull resistance is approximately 40% less. Adjusted for fan powering, a net saving of 35% is realistic.

2. ASV mono soft motion—Innovation Norway project

In cooperation with leading Norwegian ferry operator Torghatten, Diab (carbon sandwich engineering and materials), SSPA Sweden (tank testing and concept optimization), and Volvo Penta (propulsion/driveline system), ESI carried out design development and tank testing (phase 1 in the project) during late 2012 and 2013. Some of the results from the tank testing will be presented below. This new hull form has been developed for northern Norwegian Atlantic coast conditions, and will be suitable for a wide range of applications including fast crew boats, OSVs, fast patrol, and military crafts.

The illustrations below show two different superstructures and arrangements for a 21m ASV mono soft-motion coastal ferry.

Fig. 4.4.10. 21m ASV mono soft-motion Torghatten combination ferry (50 passengers and some cargo). Design: Mauro Sculli

Fig. 4.4.11. 21m ASV mono soft-motion Torghatten passenger-only ferry, with room for more than 100 passengers on two deck levels (main deck and lower deck). Design: Mauro Sculli

Some pictures and key results from the SSPA Sweden towing tank are presented below.

Fig. 4.4.12. ASV mono soft motion, towing at 35kn in calm water.

Fig. 4.4.13. ASV mono soft motion, towing 30kn in 1.8 m significant and irregular waves.

Fig. 4.4.14 shows the hull resistance test values for the ASV mono soft-motion concept, at three different displacements (design 25m^3, light 19 m^3, and heavy 31m^3) compared with a design SOA planing vessel at 25m^3.

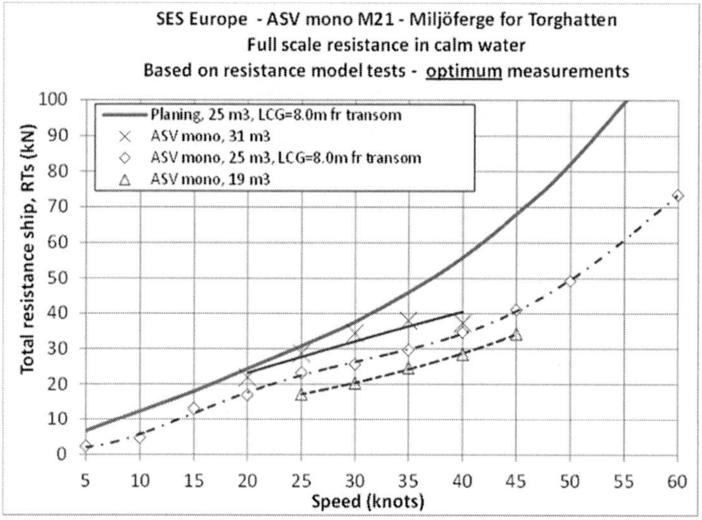

Fig. 4.4.14. Hull resistance comparisons for ASV mono soft motion

Please note that 45kn, 50kn, and 60kn with the ASV were not optimized. There is potential for further efficiency enhancements, which is particularly important for a navy and/or paramilitary application of the concept.

The hull resistance of the ASV mono SM at design speed is more than 30% less compared with SOA monohulls.

3. ASV mono testing in waves

The motivation for development of the ASV mono SM was the information received from leading US Navy and Israeli Navy experts on aspects that were unpleasant or dangerous for navy personnel and their experience with multiple G impacts when operating conventional high-speed planing monohulled vessels in waves, often over long distances and times.

Prior to this information, the US Navy had tested an early ASV catamaran test vessel in Norway and were impressed with the motion damping effect of the air cushion. They simply asked whether it would be possible to develop an ASV monohull with a similar motion-damping performance combined with the renowned benefits of a monohull (simple construction, more useable space/volume etc.).

The Torghatten 21m ASV mono SM project aims at operational speeds of 30–35kn. It was interesting to see how the new ASV hull form performed in tough waves under controlled conditions in the towing tank. The ASV SM hull shape was tested to the maximum of the wave maker's capacity, corresponding to irregular waves of up to 1.8m significance at a 21m size scale and at a speed of 30kn. The SSPA test team were very positive toward the findings, and the thorough slamming of a conventional vessel seems to be eliminated/reduced to a minimum.

4. A few additional comments

- Pitch, heave, and acceleration responses were lower for the ASV than statistics for conventional monohulls.
- The ASV mono SM was designed and tested at larger and more demanding seas than the BB Green counterpart. The advantages of the soft-motion (SM) hull form were more predominant in larger seas and higher speeds corresponding to high-speed navy/patrol vessel use.
- Evaluated against wave period, the SM has less response with increasing frequency/wavelength and typical higher sea states. At the least the pitch and heave motions and the level of accelerations at the vessel's forward perpendicular (F.P)/bow point in that direction.
- The accelerations at F.P are between two and four times higher than at AP (aft perpendicular).

- Since the SM has a bulbous bow protruding into the water and reduces the pitch motion in the forward section of the vessel, added resistance in calm water will be somewhat higher than with ASVs with a conventional bow shape.
- The true benefits with such a bulbous bow shape for an ASV will be particularly evident at a combination of high waves and sea states and at higher speeds.

4.4.6. ASV mono full-size demonstrator

With the first ever ASV mono hull full-scale test vessel (65 feet) tested and documented, ESI was awarded the prestigious European Power Boat of the Year Innovation Award in 2011. The performance and efficiency was proven to be far superior to that of market-leading conventional vessels. The hull form was configured to be suitable for a pleasure boat with large interior volumes and capacities. The design speed was approximately 30kn; its propulsion used a Volvo Penta IPS series 1 (2 × 320kW). A separate Volvo D3 and a Bosch Rexroth hydraulic system was used for the fan powering and fan power documentation.

Fig. 4.4.15. ASV 65' test vessel underway at 33kn. Observe the very low wake wash and neutral trim.

The graph below presents a direct comparison of fuel consumption between some of the leading brands, all based upon test results from leading European boating magazines and the UK-based *Motor Boat and Yachting*.

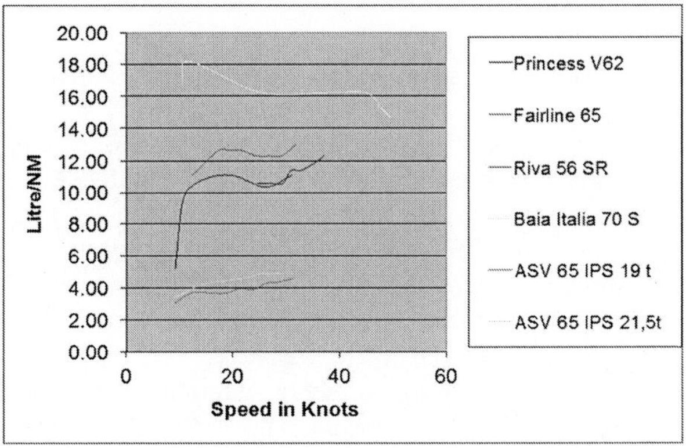

Fig. 4.4.16. Fuel consumption data on ASV Mono 65' test vessel compared with market leading European brands. Princess V62 and ASV 65 are directly comparable. Source: MBY test data 2010–12.

What can be learned from the above data and test results?

- There is a significant reduction in hull resistance—a very positive design spiral exercise (less installed power, less fuel capacity, less weight, further reduction in installed power, extended range, and more).
- At design speed, fuel consumption is reduced by 50%. Propulsion and lift fan powering consumption includes that for the ASV.
- There is virtually no hump speed/high resistance to overcome for the ASV. Lift up and go!
- It has a much lower fuel consumption at a large range of speeds—from slow to high speed.
- It has enhanced top speeds and enhanced high cruise speeds at very acceptable fuel consumption rates.
- There is no problem operating the vessel without the fan system engaged.
- The overall construction cost is close to or equal to that of a conventional vessel with the same performance/speed.
- As fuel cost is a predominant cost for workboats operating several thousand hours a year, "pay-back" time for the ASV investment is very short.

- Fuel-efficient vessels will have an increased economic life and a higher second-hand value than their "thirsty" and less efficient counterparts.

4.4.7. Construction materials—aluminum or composites?

ESI has campaigned for the introduction and use of a combination of efficiency improving solutions to boost overall capabilities and efficiency and reduce the operational cost of vessels. This philosophy will be relevant for most fast crew boats, OSVs, or other vessels aiming for higher speeds.

An ASV could be constructed from aluminum or from various kinds of composite materials; vacuum-infused carbon sandwich, reinforced for resilience, seems the most interesting alternative.

ESI is today involved in projects where aluminum construction is used, for example, a 22m ASV fast crew boat is under construction in Indonesia (see illustration of hull construction below). Another project is investigating use of stir-welded aluminum panels—a weight-saving solution—where panels/sections can be prefabricated to reduce man hours for welding at the yard (Marine Aluminum, Norway).

In addition there are advanced plans for several carbon sandwich ASV mono concepts: one is a 22m commuter ferry (BB Green), another is a fast coastal combination ferry (Torghatten), yet another is a fast crew transfer vessel (CTV). More details will follow as the various projects progress.

Fig. 4.4.17. 22m Aluminium ASV fast crew boat under construction in Indonesia.

For a vessel operated at high speed/load for many hours, the importance of weight saving for reducing fuel consumption will be easy to understand. Any vessel—conventional as well as ASV—will benefit from reduced ship weight. Efficiency and performance will be improved, fuel consumption will go down, and payload/transport efficiency will benefit.

The extra initial cost for more advanced materials will for most cases have a much shorter payback time than most anticipate.

A fast crew-boat/OSV should be thorough and robust. From a construction point of view, there will not be any significant difference between a conventional monohull and an ASV, either one built from aluminum or various kinds of composites, including carbon.

4.4.8. ASV designs and arrangements

ESI is a technology and concept provider, with a focus on developing and proposing the best possible hull forms and "platform" solutions for yards, owners, and operators, based upon a realistic "wish list."

Designing the topside and general arrangement of the vessel is normally left to the yard or the client to decide, naturally within some constraints (i.e., the main dimensions, propulsion system, and LCG location) set by ESI. Extensive research and concept documentation conclude that the ASV concept may be used on a wide range of vessel types and applications. In other words, the ASV monohull form is a pleasure to work with as the volumes and areas are generous and flexible in the design process.

In this paper, a few illustrative examples will be presented to highlight the "look" and layout of several ASV monohulled crafts. ESI has developed a close relation with, among others, the Italian design office Studio Sculli. Mauro Sculli and his team are used to working with ASV hull forms and would be a recommended designer for tailor-making ASVs for various markets and applications.

4.4.9. Fast crew vessels and fast offshore support vessels

In the opinion of the authors, the market is highly diversified, covering vessels with lengths from 15m up to more than 60m. As "fast" is a relative phrase, we consider "fast" for these types and sizes of vessels to be at least 25kn operational speed for a loaded vessel.

A wide range of vessel types are proposed and in use, and they come in a large number of sizes and shapes, with various highlighted key features adapted to general and local conditions and requirements. In general the hulls used are typically conventional "wet" hulls, either mono or cat configuration with limited "high-speed" efficiency. The yards and technology providers seem to market some capabilities to the owners and operators more frequently than others. In the opinion of the authors, the most common features are as follows:

1. Performance and capacities
   - Speed (the faster the better)
   - Endurance and range
   - Reliability and versatility (multi-purpose)
   - Capacities: deck area/space; personnel and cargo
   - Design and use flexibility; many roles for a given platform

2. Operation in sea states
   - Robust, rugged, and reliable vessel
   - Sea-keeping behavior and onboard motions
   - Comfort and passenger convenience
   - Stability and safety issues

3. Efficiency and cost issues
   - Construction (cost efficient and flexible design)
   - Economy in use/fuel savings
   - Efficiency (including reduced weight = greater payload)

4. Environmental issues
   - Reduced emissions in water and air
   - Greener operation

Prioritizing and ranking of the above will differ from operation to operation and case to case. Undoubtedly the economical and performance (including handling) issues will be ranked high, with reliability and motion aspects (comfort) following closely.

4.4.10. How does the ASV mono concept meet these requests?

1. Performance
With a significantly reduced hull resistance throughout the whole speed range for an ASV, the performance with a given installed power will be market leading. Look at Fig. 8 or Fig. 14 to see the difference in calm water speeds with the given power available.

Alternatively, top speed can be enhanced and/or the utilization/transport efficiency of the vessel could be extended. Regardless, an ASV will represent SOA.

2. Range
Reduced fuel consumption per NM enables an extended range.

3. Cargo platform size
Wider beams and catamaran like-pontoons (propulsion gondolas) give a larger usable platform with less roll-motion in a sea state. Volumes on or below deck will also be enhanced (passenger capacity/accommodation space). ASVs will be suitable for modular arrangement and multipurpose use.

4. Operation in a sea state
ASVs do not suffer from the severe drawbacks of the old types of air-assisted crafts. No flexible and high maintenance rubber skirts and the like are required. An ASV is a solid and reliable solution. An ASV may operate on a variety of air support rates depending on the actual conditions. The ASV air cushion dampens the ride and reduces onboard accelerations, motions, and impacts, as well as seasickness induced by pitch and roll. The claim is valid for on- as well as off-cushion use, which will be important for passenger and crew comfort as well as general operational safety.

5. Construction cost
Compared with a conventional monohull with a similar performance, an ASV will only be marginally more expensive.

6. Operational cost
Most installed main systems in an ASV will have a reduced size, and the fuel cost will be significantly reduced: 20–40% depending on operation and vessel type. Payback time for technology investment will be short.

7. Environmental issues
Less consumption equals reduced emissions into water and air. ESI is also engaged in full electric-battery-powered solutions (BB Green) and diesel/electric hybrid ASV designs.

8. Wake/wash
Results from the towing tank and full scale testing show the wake wash, particularly the wake wash energy, is less on an ASV compared with a conventional vessel.

4.4.11. ASV marteceranet project: 18m fast crew transfer

Effect Ships International AS, Norway, in collaboration with Tuco Marine, Denmark, and Diab AS, Norway, and with support from Norwegian and Danish Research Councils has started design development and construction of a new range of carbon sandwich offshore support/crew transfer vessels. The first vessel, an 18m fast crew transfer/offshore support craft is planned to go in the water during the summer of 2015. The range of ASV fast support craft will range from 15m to 60m hull length.

Figure 4.4.18 shows a profile image of a larger ASV Mono OSV, designed by Mauro Sculli. The vessel will have a length of approximately 60m with a beam of 15m–17m.

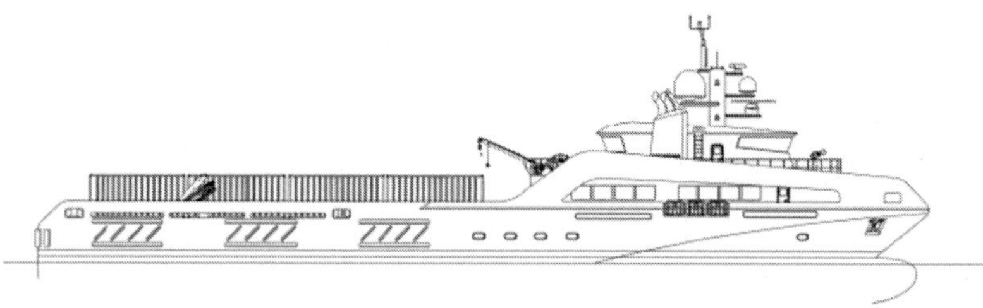

Fig. 4.4.18. Early design of a 45–60m ASV mono OSV. Design: Studio Sculli

4.4.12. Other ASV applications and projects

ESI may propose ASV mono designs from 12m upwards to more than 100m in L_{OA}. The company is also engaged in developments of ASV IPR and solutions for large commercial ASV-type ships.

The tank testing results from the ASV mono SM, depending on design particulars, may be scaled to suit vessels between 15m and more than 60m, with speeds up to 60kn at both a 20m scale and a 60m scale.

The high-speed capability in combination with a generous range makes ASV technology very suitable for navy and paramilitary use, including high-speed patrol vessels.

1. Fast ASV patrol crafts
Together with collaborating partner TBS Yard, Sweden, a range of fast and agile carbon patrol vessels are planned. Info on the first design, a 60m vessel with a 60kn speed, can be found at
http://www.tbsab.com/products/fpb-60-ehs.

2. Fast ASV landing crafts

Yet another navy application are ASV landing craft. In collaboration with Damen Shipyards, ESI is developing medium-load 30–40kn fast ASV landing craft (LC) for the naval sector. These vessels could be tailored to a wide range of sizes, from 20m up to 40m and above. The ASV LC can be configured for "drive through" or for single use. An ASV LC will have several benefits over a conventional LC, including speed capability, performance, low and variable draft, and improved beaching capabilities. Compared with current "Word War II" conventional LCs, the capabilities of the new ASV will be in another league.

Fig. 4.4.19. Design impression of ASV LC 26m and 40kn performance

3. Fast ASV mono ferries

For fast commercial ferries and passenger-only and car-passenger-cargo ferries, ASV monohull forms also have great potential. Figure 4.4.20 shows a 38m ASV fast ferry, with a capacity of 200 passengers on the main deck in addition to additional capacity on a lower deck.

Fig. 4.4.20. Design impression/rendering of ASV 38m fast passenger ferry. Profile view. Design: Mauro Sculli

4. Fast ASV pleasure crafts
As a platform for pleasure boats, ASV monohulls offer an excellent combination of space, operational behavior, performance, and operational economy.

Fig. 4.4.21. ASV Namaste 72-foot fast yacht. Design Mauro Sculli

4.4.13. Conclusions

As a result of significant R&D efforts, ESI has developed new state-of-the-art hull forms with its ASV mono concept. The technology is patented worldwide, is comprehensively documented, and the use benefits are considerable.

The exploitation phase has just started, and the company plans to enter into agreements and collaborations with leading yards, ship owners, and investors across the world.

Interested parties are encouraged to contact the authors to discuss exploitation of ASV technology, potential projects, and collaboration.

Acknowledgements

The authors would like particularly to thank SSPA Sweden AB, in particular project manager Hans Liljenberg for extremely valuable support in the concept design development and documentation phases.

References

[4-4-1] B. Allenstroem, H. Liljenberg, & U. Tudemu, "Concept Development and Model Testing—New Generation Air Assisted Vessels (AAV) with Water Jet Propulsion," FAST '03, Ischia, Oct 2003.

[4-4-2] A. Eilertsen, J. Griegeide, & U. Tudem, "Design Development of Advanced Air Supported Vessel (ASV) Designs," RINA WARSHIP, June 2004.

[4-4.3] A. Papanikolaou, "Review of Advanced Marine Concepts," (Transport Efficiency), HSMVO Naples, September 2005.

[4-4-4] U. Tudem & T. K. Livgard, "High Performance with Improved On-board Comfort: Introducing ASV 'Soft Motion' Mono with Water Jet Propulsion," RINA SURV 2009, Poole, UK, May 2009.

[4-4-5] U. Tudem, T. K. Livgard, & E. Taniputra, "Design Development and Construction of Air Supported Fast Crew and Offshore Support Vessels," RINA, Indonesia, November 2012.

4.5

Design and Exploitation of a New Air Bubble Navigation-Guide Ship (ACC) (2011)

Cheng Hongxia, Yang Dong, Ni Qijun and Xing Shengde

China Ship Scientific Research Center (CSSRC)

Abstract

This paper depicts the process of design for a new high-speed air bubble navigation-guide ship, which includes its exploration background, main purpose, performance characters, performance test, and power collocation. The test results validate the advanced performance of the bubble ship (ACC), and so it is good to promote this actively.

4.5.1. Exploitation background

The navigation-guide vessel is named a pilot vessel, which is usually used to taking charge of imports, exports, and shifting. The navigation-guide vessel is commonly abbreviated as a traffic boat.

Navigation-guide vessels have many tasks such as guiding, dragging, surveying shipping channels, and hydrological checks. But they are easily affected by weather and sea states; especially when under bold winds and boiling seas they need to cooperate with a towed ship to finish running.

Since 1998, Wuxi East high-speed craft-development company has manufactured 45 “East 780” for piloting in the Changjiang River estuary of Shanghai.

Aimed at the special needs of pilot vessels, we assume an exerting drag reduction with an air bubble on the original pilot ship. On the one hand, it can reduce resistance and improve speeds so as to shorten the arrival time of pilot vessel to the towed ship; on the other hand, it is expected to improve sea-keeping qualities using a gas layer on the ship bottom to build up its maneuverability under heavy winds and seas.

4.5.2. Main purpose

The ship will hopefully be used to replace existing pilot ships for piloting in the Changjiang River estuary, Shanghai.

4.5.3. Ship design

At present, the pilot vessel in the Changjiang River estuary, Shanghai, is a conventional hard chine craft with poor sea-keeping qualities, large speed loss in waves, and bad maneuverability in poor weather. The line plan is as follows:

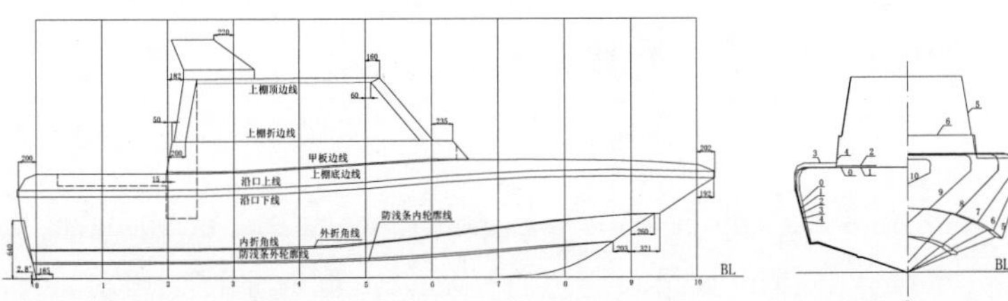

Fig. 4.5.1. Line of former pilot vessel

If we adopt an air bubble ship, the shortcomings of the chine ship can be worked out. The principle of the bubble ship is shown in Fig. 4.5.2, in which the ship bottom is a hollow cavity; when the ship slips at high speed, the ship bottom lies in a low-pressure fluid zone and the system of creating the air bubble will eject high pressure air into air compressor 1 from step 5 through pipe system 2, and then a thin air layer is formed on the ship bottom.

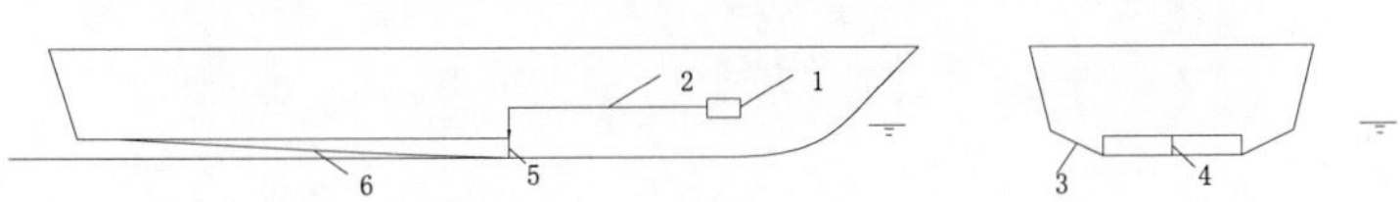

(1) air compressor, (2) system of creating air, (3) side keel, (4) longitudinal keel, (5) step, (6) air layer

Fig. 4.5.2. Schematic diagram of air bubble planing boat

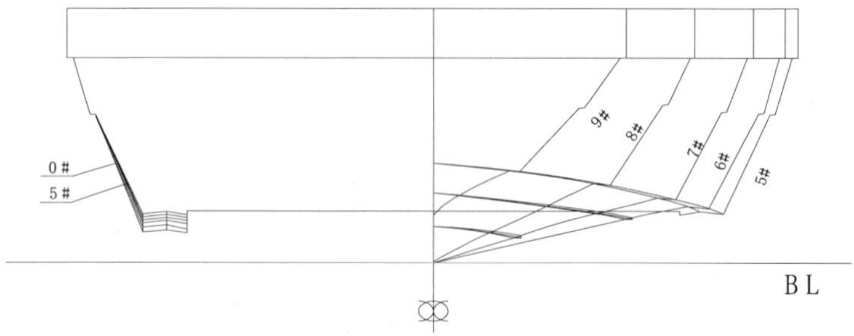

Fig. 4.5.3. Typical body plan of air bubble ship

Through comprehensive consideration, we adopted the bubble ship type to the pilot vessel for the first time, so as to acquire a better positive sea-going performance. To validate the bubble ship performance, we subsequently carried out correlative basin tests.

4.5.4. Tests

The tests were carried out in the ship model towing tank of Shanghai Jiao Tong University. The scan ratios (λ) of the bubble ship model and the chine ship model are all 5 (Lm = 1.56m, Bm = 0.494m, Δm = 32kg). The bubble ship is made with organic glass, while the chine ship is made with paulownia. The two model photos are as follows:

Fig. 4.5.4. Bubble ship model

Fig. 4.5.5. Conventional ship model

1. Resistance tests

The contrasts of the two ship models are shown in Fig. 4.5.6 and Table 4.5.1, where R is the ship total resistance and Δ is displacement.

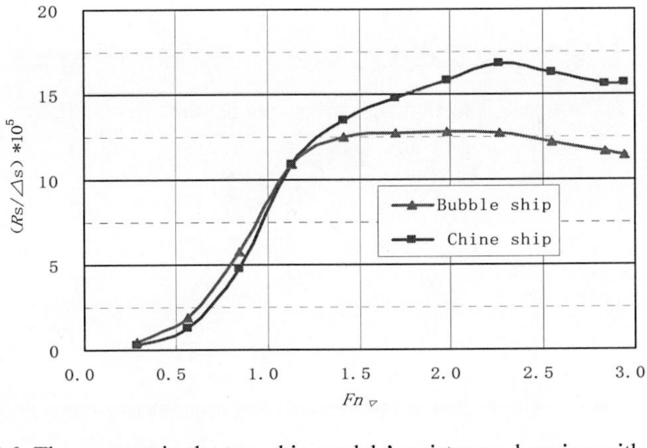

Fig. 4.5.6. The contrast in the two ship models' resistance changing with velocity in calm water

Fn∇	C_R (Bubble ship) R/Δ*105	C_R (Break angle Chine ship) R/Δ*105	Resistance reduction (deduction percent of bubble ship %)
0.3	0.42	0.30	-37.71
0.6	1.92	1.30	-47.45
0.9	5.75	4.77	-20.67
1.1	10.89	10.87	-0.12
1.4	12.52	13.52	7.41
1.7	12.69	14.78	14.14
2.0	12.77	15.87	19.55
2.3	12.71	16.79	24.29
2.6	12.23	16.27	24.83
2.8	11.65	15.63	25.46
2.9	11.47	15.70	26.97

Table 4.5.1. Resistance reduction effect of the bubble ship to chine ship

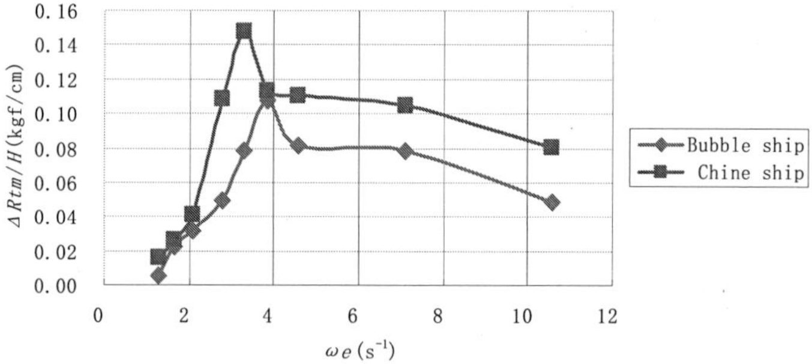

Fig. 4.5.7. Resistance increase response contrast in waves of the two ship models

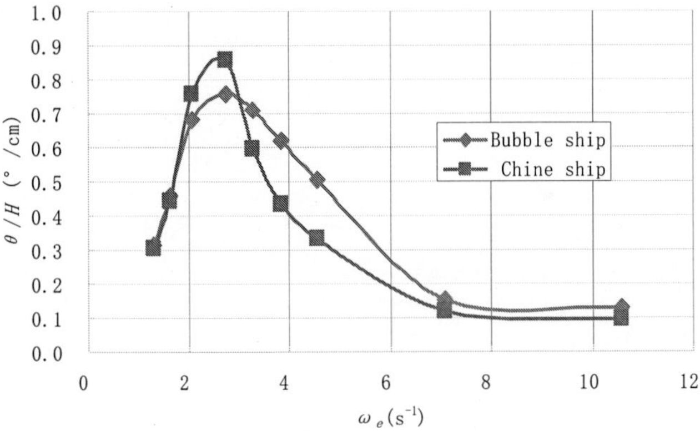

Fig. 4.5.8. Pitching response contrast in waves of the two ship models

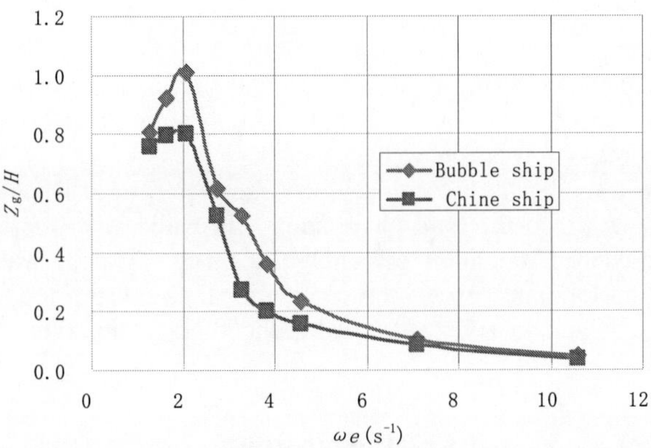

Fig. 4.5.9. Heaving response contrast at gravity center in waves of the two ship models

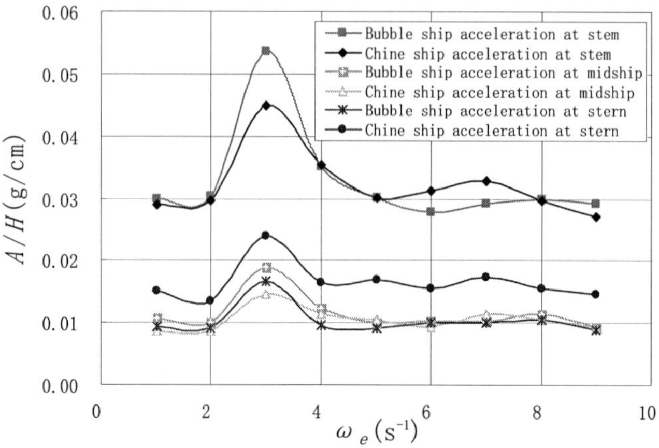

Fig. 4.5.10. Acceleration response contrast in waves of the two ship models

2. Experiments in regular waves
The results of the two ship models in regular waves are shown in Figs 4.5.7–10. The speed *Vm* is 4.0m/s, and the trim, heaving, acceleration, and resistance increase in waves have been tested.

From the tests results, we can see, in calm water, the air bubble ship resistance is less than the hard chine ship, especially when the speed is high, the resistance reduction percentage is up to 27%. In waves, the bubble ship performance excels the conventional chine ship too, and the reason is that steady air layer is formed at the bubble ship bottom both in calm water and in waves.

4.5.5. Power plant

Referring to the original pilot vessel configuration, and according to the resistance tests results, the bubble pilot ship is confirmed to adopt an outboard-inboard diesel engine, in which the mainframe is QSD2.0ES150 × 1, and the max power is 112kW; the rating rotate speed is 4,000r/min, and the stern engine suits an Alpha engine. This power configuration has the excellence that it can be handled simply and be used safely.

4.5.6. Conclusions

According to the air bubble ship model test results and the trial voyage results of the original pilot vessel, *East780*, the principal dimensions and hydrodynamic characteristics of the air bubble pilot ship need exploring, see Table 4.5.2.

Table 4.5.2. Performance comparison between the air bubble pilot ship and the original pilot ship

Item	East 780	Air bubble ship
Length overall/m	7.8	7.8
Molded breadth/m	2.48	2.47
Molded depth/m	1.28	1.55
Draft/m	0.51	0.525
Displacement/t	2.95	3.2
Mainframe	ME422STIP2	QSD2.0ES150
Mainframe power/kW	173	112
Fan power/kW	0	3.36
Ship speed/m/s	17.2	14.1
Passenger/p	6	6
$F_{r\Delta} = \frac{V}{\sqrt{g\Delta^{1/3}}}$	0.029	0.039
$K_\eta = \frac{WV}{102N}[\frac{KN\times m/s}{Kw}]$	1	1.345

In the Table, K_η is the hydrodynamic performance coefficient, W is the craft weight (kN,V is ship speed [m/s, N is power (kW)]). From the table we can see that the performance of the new air bubble pilot ship is predominant, so exploitation of the bubble ship should be extended vigorously, and the drag reduction technique with an air bubble is applicable to other ships to achieve drag reduction and seaworthiness amelioration, and then to improve sea-going performance.

References

[4-5-1] Cheng Lei & Lu Yueming, "Venture Analyses and Answer Measure of Run about Piloting Person at Changjiang River Estuary," *Art of Navigation*, 2009 (05), pp. 14–16.

[4-5-2] Sui Aiguo, Cheng Hongxia, & Ye Yonglin, "Design and Exploitation of a New Type High Speed Coastguard Vessel for Maritime Affairs," Vol. 1, HPMV 2010 (4).

[4-5-3] Ni Qijun, Cheng Hongxia, & Xing Shengde, “Key Technique Study Summarize of High Speed Air Bubble Ship,” inner information in CSSRC, 2006 (12).
[4-5-4] Cheng Hongxia & Xing Shengde, “Resistance Tests Report of Air Bubble Ship,” inner information in CSSRC, 2009 (06).

4.6

AIR FILM (LAYER) ENERGY-SAVING TECHNOLOGY AND APPLICATION TESTS ON SHIPS (2013)

CAI JINQI

Shanghai Qing Hang Air-filmed Resistance-reduced Ship Co., Ltd

Abstract

On the basis of the test results and analysis of air film (layer) resistance reduction technology on two real ships, this paper indicates the practicability of the technology in regard to energy saving. It is different from resistance reduction technology by air bubbles. When air is fed to the ship bottom and a thin layer of air is formed, the contact area between the ship hull and the water will be reduced by 25–50%, as a result 15–20% energy can be saved for both low- and high-speed ships. This energy saving technology can also be applied to old ships. The installation of the system on an old ship can be completed in around three days when the ship is at dock. During the design stage, the ship form and propulsion and the air-layer resistance reduction technology may be considered in a combined way to improve the overall energy saving effect of a ship. Only low investment is required when this energy-saving technology is applied to a ship and the ship bottom device resistance reduction is about 1.5%. Therefore, it is an economic and effective technical way to save ship energy.

In China, some valuable experiences of the study and application of air-film resistance reduction technology have been collected, which will be a good foundation for the wider application and further development of this technology.

As early as 1876, scientists found that water resistance of a solid body is 600 times greater than air resistance; thus, they suggested using a thin layer of air to separate the ship body from the water in order to reduce resistance. Recently, resistance reduction technology of ships using an air layer has achieved good results. Now there are three air-reducing resistance technology types on ships in use around the world.

The first method is air bubble technology for reducing resistance. It uses an installed tilt board and depends on ship speed to form an air bubble on the craft's bottom. The length of the air bubble is proportional to the square of the ship's speed.

The second method is micro air bubbles that use reducing resistance technology. It uses a ship bottom micro-hole device to form micro air bubbles on the ship bottom.

The third method is our energy-saving technology of a ship using a thin layer of air film. The author discovered the law concerning air film and invented an equipment system that could form a thin layer of air film on the ship bottom. This technology is patented and is one of the Shanghai government's high-tech transforming projects. The air pumped into the ship bottom can form a thin layer of air film when passing through the equipment installed. It can reduce the total area of the ship that is in contact with the water by 30–50%. The tests and applications show that it can help transportation ships and high-speed ships to save about 15–20% total energy. We have developed an economic way for ships to save energy.

4.6.1. The energy-saving device system of a ship using a thin layer of air film

The energy-saving device system of a ship using a thin layer of air film includes a blower, a transportation air pipe, and a device on the ship bottom. The air pumped into the ship bottom usually forms air bubbles instead of air films and the desired effects cannot be achieved. But the air pumped into the ship bottom passing through our installed equipment can form an air film and it can reduce resistance and save energy for the ship.

Our air film technology is used to install an air film device system for old ships to save energy, and to design and build new transportation ships and high-speed ships.

1. The air film device system has been installed on a 1,000t barge

The air film device system has been installed on a 1,000t barge after model tests. The towing resistance tests have been completed for one barge that has had the technology installed and one that has not. Table 4.6.1 shows the resistance towing test results and analysis of the barge in SS 2–3 waves.

Table 4.6.1. Towing test results and analysis of barge

Towing test results and analysis of barge						
	Speed/km/h	7	8	9	10	11
1	Resistance of inflated barge/t	1.481	2.09	2.815	3.72	4.82
2	Resistance of original barge/t	1.93	2.64	3.49	4.52	5.79
3	Resistance of uninflated barge/t	2.02	2.78	3.70	4.76	6.03
4	Reducing resistance for inflated and uninflated	26.7%	24.8%	23.9 %	21.8%	20.1 %
5	Air film energy-saving effect	15.9%	16.5%	16.3 %	15.6%	15.2 %

The tests show that resistance of the barge has been reduced by 20.1% at a speed of 11km/h because of the air film. It can save 15.2% energy when we deduct the power of the blower and the resistance of the bottom device.

If the friction resistance of this ship is 0.75 times the total resistance, it is

$R_f / R_t = C_f / C_t = 0.75$ (4.6.1)

Above, R_f, R_t, C_f, and C_t, respectively, are the friction resistance of the ship, the total resistance of the ship, the friction resistance coefficient of the ship, and the total resistance coefficient of the ship.

If reducing the resistance coefficient is

$k = (R_t - R'_t) / R_t$ (4.6.2)

where, R'_t is the resistance of the air film ship.

The friction resistance of the ship is

$R_f = C_f (1/2 \times \rho \Omega V^2)$ (4.6.3)

where, ρ, Ω, and V, respectively, are the mass density of the water, the ship wet surface area, and the speed.

The friction resistance of the air film ship is

$R'_f = C_f (1/2 \times \rho \Omega' V^2)$ (4.6.4)

where, Ω' is the air film ship wet surface area.

Because the total resistance of the ship is

$R_t = C_t (1/2 \times \rho \Omega V^2)$ (4.6.5)

If the total resistance of the air film ship is

$R'_t = C_t (1/2 \times \rho \Omega' V^2)$ (4.6.6)

because the residual resistance is the same for the inflatable ship and this is not the inflatable ship, then

$(R_t - R'_t) / R_t = (R_f - R'_f / R = (C_f / C_t) ((\Omega - \Omega') / \Omega)$ (4.6.7)

or

$(\Omega - \Omega') / \Omega = k / (C_f / C_t)$ (4.6.8)

The k is 20.1% for this air film barge, (C_f / C_t) is 0.75:
$(\Omega\text{-}\Omega') / \Omega = 0.201 / 0.75 = 26.8\%$ (4.6.9)
Our air film technology can reduce the contact water area of this barge hull by about 26.8%, reduce resistance by about 20.1%, and make energy savings of about 15.2%.

2. The air film device system has been installed on a 5,500t/424TEU container ship
The 5,500t/424TEU container ship is a coastal container ship, with an overall length of 122m and two main engines driving two propellers. We have designed an air film device system for the 5,500t container ship according to CCS class and its design has been improved and is better than a 1,000t barge. It can reduce installation time and it needs only about 2–3 days to install the air film device system on the old ship.

Fig. 4.6.1. Film device on a 424 TEU container ship bottom

The test on installing the air film device system on the 5,500t container ship was not perfect, because the valve of blower was nearly closed and the blower failed to output air normally—only a small amount of air overflowed along both sides of container ship—but when we opened the valve of the blower, a large amount of air flowed to the front and overflowed along both sides of the bow. It was about 10 times larger than the nearly closed valve. The test results were affected because the air could not flow to the front of the ship bottom. But we could analyze the energy-saving effect of the 424 TEU air film container ship according to this test result at main engine speeds of 430r/min and 450r/min when it pumped air and did not pump air.

The test result of the original 424 TEU container ship
Test date: February 23, 2011
Test area: East Sea coastal area of China
Wind: Grade 3–4

Table 4.6.2. Test result of 424TEU container

Name speed	Voyage	Left engine		Right engine		Speed for shore	Speed in calm power	Total power
		speed	shaft power	speed	shaft power			
r/min		r/min	kW	r/min	kW	kn	kn	kW
430	go	433.7	413.3	432.9	370.3	9.45	8.94	771.6
	return	427.8	380.2	436.0	379.4	8.43		
450	go	446.5	436.9	458.9	433.8	9.68	9.28	882.35
	return	453.0	462.5	455.3	431.5	8.87		

Table 4.6.1 shows the test results for the 424 TEU container when it pumped air and when it did not pump air.
Test date: July 7, 2011
Test area: East Sea coastal area of China
Wind: Grade 5–6, downwind grade 7 (waves showed white).

Table 4.6.3a. Test result when it did not pump air

Name speed	Voyage	Left engine		Right engine		Speed for shore	Speed in calm power	Total power
		speed	shaft power	speed	shaft power			
r/min		r/min	KW	r/min	KW	kn	kn	KW
430	Go	432.8	400.4	435.6	354.9	9.02	8.92	769.5
	Return	426.8	401.8	435.2	381.9	8.82		
450	Go	457.4	473.5	452.5	395.0	9.32	9.10	872.3
	Return	448.8	464.6	450.2	411.5	8.88		

Table 4.6.3b. Test result when it pumped air

Name speed	Voyage	Left engine		Right engine		Speed for shore	Speed in calm power	Total power
		Speed	Shaft power	Speed	Shaft power			
r/min		r/min	KW	r/min	KW	kn	kn	KW
430	go	431.7	401.3	435.6	330.7	9.07	8.95	744.85
	return	426.9	402.5	435.2	355.2	8.83		
450	go	457.2	477.4	452.5	361.8	9.18	9.09	843.65
	return	448.7	460.8	450.2	387.3	8.99		

The tests show that the shaft power has been reduced by about 3.2% when the blower inputs a small amount of air compared with when air is not pumped. It predicts that shaft power should be reduced by about 5–9 times more than when the normal air blower input quantity. It predicts that shaft power will be reduced by 19.2% when we calculate according to 6 times 3.2%. It can save shaft power 14.7% when we deduct the power of the blower and the resistance of the bottom device by about 4.5%. The fuel efficiency of the container ship at a speed of about 9kn is 0.65, therefore its energy-saving is about 10%.

This friction resistance of the container ship is about 0.7 of the total resistance. From equation 4.6.9, and when the shaft power is reduced by about 3.2%, we can get

$(\Omega\text{-}\Omega') / \Omega = 0.032 / 0.7 = 0.0457$ (4.6.10)

The wetted surface area of this container ship Ω is 2,750 m^2, so in this case the air film area at the ship bottom ($\Omega\text{-}\Omega'$) is about 126 m^2.

We designed the air film area at the container ship bottom to be 16.5m wide, 62m long, and have an area of 1,023m^2. The shaft power will be reduced 19.2% when we analyze on the basis of six times, and with this we can calculate formula 4.6.9, which shows that the air film area is 754$m^{2,}$, which is 74% of the total designing air film area. In fact, the energy-saving efficiency can also be increased if we manage reasonably, including the adjustment of floating state and in better sea conditions.

We can derive the following conclusions according to the tests results of the two air film ships:

1. The air pumped by the blower into the ship bottom can form an air film when passing through the device installed. It can save about 10–15% energy for the original ship installed with the air film device system.
2. The resistance of the air film device on the ship bottom is about 1.5% of total resistance.

3. Installing the air film device system on the original ship takes about 2–3 days.
4. It has some effect due to the fuel efficiency of the main engine for the energy-saving effect of the original ship using our air film technology.

Fig. 4.6.2. Air film research test

4.6.2. Micro-bubble reducing resistance technology

For micro-bubble reducing resistance technology, a micro-bubble device needs to be installed on the ship bottom. The air pumped by the blower into the ship bottom and installing device can form a micro-bubble air–water mixture to save energy.

According to formula 4.6.3, the frictional resistance of the micro-bubble area of the micro-bubble ship bottom is $R_f = C_f\ (1/2\times\rho'\Omega\ V^2)$. The ρ' is the mass density of forming the micro-bubble air–water mixture. The Ω is the micro-bubble ship bottom area. The frictional resistance of the micro-bubble area of the micro-bubble ship bottom can be reduced by about 50% when the ρ' is about 0.5 times the mass density ρ of water.

The mass density of air is about one eight-hundredth for water mass density ρ. We think that the frictional resistance of the air film ship bottom area is about zero.

In fact, the study also shows that the part of the micro-bubble on the ship bottom will be combined into forming air film and it can increase the energy-saving effect. The blower will pay more power due to the water surface tension of the micro-bubble.

The air film law and air film forming law are unique. We hope to develop this technology through cooperation with other units. Its benefits include saving energy for ships and reducing the emission of CO_2.

4.6.3. Design of the ship's air film energy-saving device

The design of the ship's air film energy-saving device asks for a reasonable combination of ship forms, propeller, and air film energy-saving technology. It can improve the overall energy-saving effect. A 35,000t air film cargo ship model comparison tests show that it can reduce resistance by 22% and it forecasts saving power of about 17%.

The design idea for the fast speed 4,200 TEU or 8,000 TEU air film energy-saving container ship is that the air film on the ship bottom can reduce the total wetted area of the ship hull by about 40%. It can save about 15–20% energy.

Our air film technology on high-speed ships includes reduction of resistance of high-speed ships and improvement of ship forms. The ship model comparative tests show that the resistance of our air film high-speed ship reduces by 22% more than the SES and the blower of the SES needs 15% of the main engine power; but the air film high-speed ship needs 1% only and the resistance of the bottom device on the ship bottom is about 1%.

Designing and building an air film energy-saving ship can save 15–20% energy. But the cost of the energy-saving devise system of our technology is about 0.5% of the shipbuilding cost. Our energy-saving technology in ships using a thin layer of air film will match developmental trends.

4.7

Mechanism Study of Use of an Artificial Bubble for the Drag Reduction of a High-Speed Boat (ACC) (2010)

He Shulong, Xing Shengde, Cheng Hongxia and Ni Qijun

China Ship Scientific Research Center (CSSRC)

Abstract

This paper analyzes the hydrodynamic characteristics of flow around a high-speed bubble boat (ACC). The complicated three-dimensional flow field is simplified into a two-dimensional one. Three flow patterns, named by the author as "no bubble," "closed bubble," and "half-open bubble," are conceived for the bubble boat flow and a similar principle is presented for every flow pattern. Through finite elements and the VOF method, a math model is constructed for the two-dimensional flow around the bubble. The relationship between the bubble and the variables, such as water speed, air pressure, and air volume speed, is studied by numerical testing. The results of this research can guide the design of air bubble boats.

4.7.1. Introduction

It is known that air bubbles will form around high-speed moving objects underwater. The bubbles are full of vapor and air. On the boundary of bubbles, the pressure is continuous, but the density changes from water to gas. The bubbles separate the object from the water, which causes the reduction of friction. So cavitation can be utilized to reduce the drag of vehicles.

Although there is no cavitation on the bottom of an ordinary boat, we can make bubbles, for example, by blowing air to the bottom to form a bubble overlaying the surface of the boat. The artificial bubble will move with and be stationary relative to the boat. Because of the high speed of the boat, there will be a distinct boundary between the air and the water. The pressure in the bubble will almost be a constant.

The main study of the bubble reduction determines the relationship of the bubble profile with relative variables and the effect on drag reduction. Once the profile of the bubble is determined, the surface covered by the bubble and the reduced friction can be estimated. The reduced friction should be larger than the added drag caused by introducing the bubble. This is achieved by choosing suitable parameters for the boat and the flow, including the pressure and volume speed of blowing air. The air volume consumed for maintaining the bubble and the power of the air jet can also be evaluated.

There is no difference between an artificial bubble and a cavitation bubble from the perspective of flow. So we can define the bubble number σ of the artificial bubble, just as for cavitation:

$$\sigma = \frac{P_0 - P_v}{\rho V_w^2/2} = \frac{P_0 - P}{\rho V_w^2/2} \tag{4.7.1}$$

where, P_0 is pressure at infinite forward or atmospheric pressure, P_v is saturated steam pressure, P is pressure in the bubble, V_w is water speed at infinite forward or boat speed, and ρ is water density. Saturated steam pressure is used to express the pressure in the bubble, which means the artificial bubble is similar to a cavitation bubble.

Bubble number is an important parameter, which has a decisive effect on the bubble flow. Reference pressure $P_0 = 0$, (in equation 4.7.1) can be simplified as:

$$\sigma = \frac{P_0 - P}{\rho V_w^2/2} = \frac{-P}{\rho V_w^2/2} \tag{4.7.2}$$

4.7.2. Math modeling for flow around bubble boat

1. Simplifying the three-dimensional flow into a two-dimensional flow
Considering the high speed of the boat, the transverse flow and viscous effect can be neglected when studying the flow around the bubble, so the complicated three-dimensional flow can be simplified into a two-dimensional non-viscous one. At the same time, because we are particularly interested in the flow around the bubble, we can further simplify the profile of the boat into a horizontal infinite wide plane with a wedge air jet, as shown in Fig. 4.7.1, where L is the length of the wedge air jet and L_a is the length of the bubble.

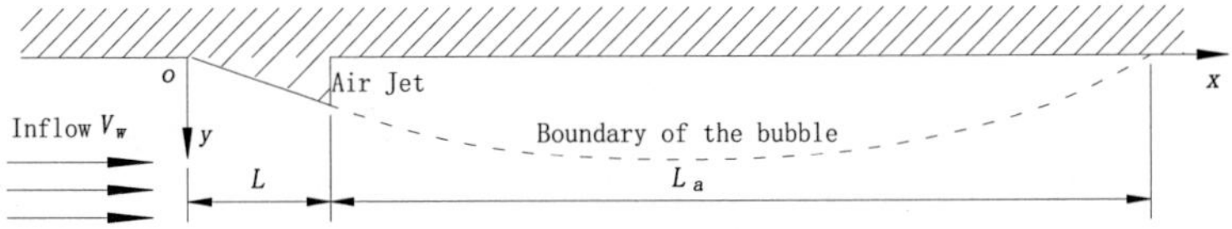

Fig. 4.7.1. Two dimensional flow around the artificial bubble

2. Analysis of the two-dimensional flow pattern

Actually, the flow shown in Fig. 4.7.1 is the closed-bubble flow pattern with a steady cavity. If the air jet pressure is large enough, or the air is jetted into the cavity continuously, then the cavity will stretch infinitely far along the X axis, which is called the half-open bubble flow pattern. On the contrary, if there is no air jetting into the flow, no bubble will be formed after the air jet (except the cavitation bubble caused by a high inflow speed), which is called the no-bubble flow pattern.

The three flow patterns have an internal relationship. There is a low pressure domain after the air jet in the no-bubble flow pattern. The pressure on the air jet is the minimum pressure for air jetting to form an artificial bubble. If the pressure at the air jet is large enough, the pressure will afterward drive the air flowing infinitely far, and the closed bubble flow pattern will be transformed into that of the half-open bubble.

3. Analysis of the similarity principle for two-dimensional flow

The similarity principle for different flow patterns is different. The simplest flow pattern is the no-bubble pattern. The independent variables for this pattern are water speed V_w and length of air jet L, so there is only one similarity number, Froude number F_w:

$$F_w = V_w / \sqrt{gL} \qquad (4.7.3)$$

There are three independent variables for the closed-bubble flow pattern, namely, water speed V_w, length of air jet L, and bubble pressure P. Two similarity numbers can be formed, namely Froude number F_w and bubble number σ (see equation 4.7.2 and equation 4.7.3).

It is reasonable to choose water speed V_w, length of air jet L, and jetting air-speed V_a as the independent variables for the half-open bubble flow pattern. So the similarity numbers for this pattern are Froude number F_w, based on water speed, and Froude number F_a, based on air speed. The definition of F_w is the same as equation 4.7.3; F_a can be defined as:

$$F_a = V_a / \sqrt{gL} \qquad (4.7.4)$$

4. Numerical simulation method for two-dimensional flow
The thin wing theory may be regard as a suitable method to solve the closed-bubble flow pattern. The difficulty of the method is the unknown boundary of the bubble beforehand. What's more, the singularity of the thin wing method may cause the results to be unsatisfactory, even giving an unrealistic flow that is contrary to fact. So the finite volume method is used in this paper to solve the closed-bubble flow pattern, with VOF to deal with two-phase flow. The numerical method is based on a non-viscous Euler equation, which can solve all the three flow patterns.

4.7.3. Analysis of the numerical results

Numerical model is shown in Fig. 4.7.1, where the length of the air jet is L = 2m and the height is H = 0.25m. The speed of the inflow is V_w = 4m/s, 7m/s, 10m/s, 13m/s.

1. Calculation results of no-bubble flow pattern
The minimum pressure after the air jet is listed in Table 4.7.1, which is in linear proportion to the square of the inflow speed.

Table 4.7.1. Minimum pressure at different inflow speeds

V_w/m/s	0	4	7	10	13
P_{min}/kPa	0	-5.9	-18.2	-37.4	-64.0

The coefficient C_p of pressure drag is defined as:

$$C_p = \frac{R_p}{\rho V_w^2 S/2} \tag{4.7.5}$$

where, R_p is the pressure drag acting on the wedge. R_p and C_p are listed in Table 4.7.2.

It can be seen that the coefficient C_p is almost a constant. Because there are differences between a wedge and the boat bottom, the drag acting on the wedge is different, with the added drag acting on the boat with the cavity and the bubble.

Table 4.7.2. Pressure drag acting on the wedge

V_w/m/s	0	4	7	10	13
R/N	0	721	2,208	4,506	7,615
F_w	0	0.903	1.580	2.258	2.934
$1{,}000C_p$	——	45.14	45.14	45.14	45.14

2. Calculation results of closed-bubble flow pattern

The length of the bubble at different air jets is shown in Fig. 4.7.2. It can be seen that the length of the bubbles converge to a constant at a low inflow speed, but the length of the bubble increases quickly with the increase of air jet pressure at a high inflow speed.

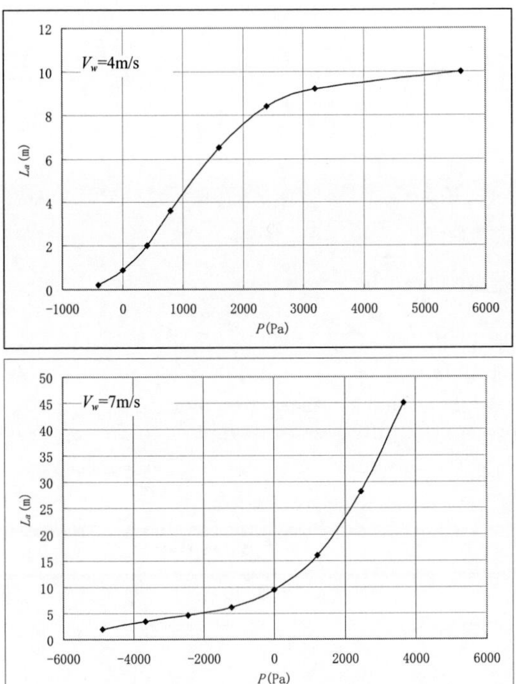

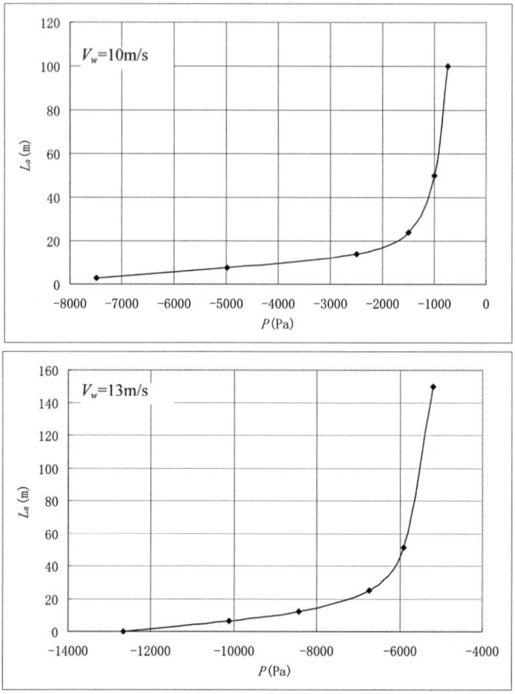

Fig. 4.7.2. Length of the bubble versus the air jet pressure

The profile of the bubble is shown in Fig. 4.7.3. It can be seen that the length of the bubble will increase with the increase of inflow speed at the same bubble number. That is to say, the high speed of the boat is in favor of the formation and stabilization of the bubble.

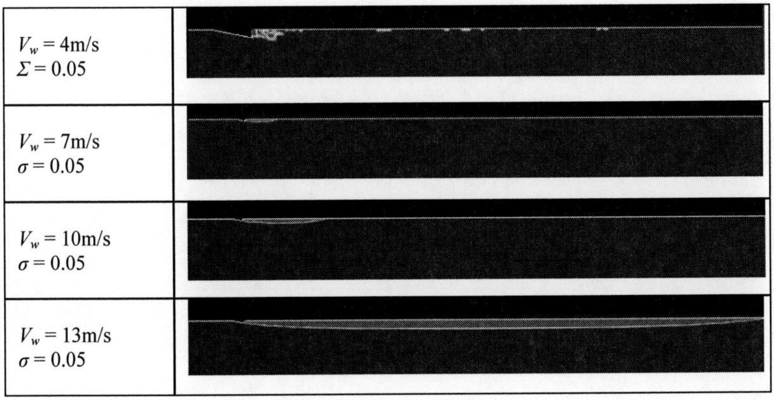

Fig. 4.7.3. Typical bubble profile

Pressure drag R_p acting on the wedge is shown in Fig. 4.7.4, where R_p will linearly decrease with the increase of the air jet pressure.

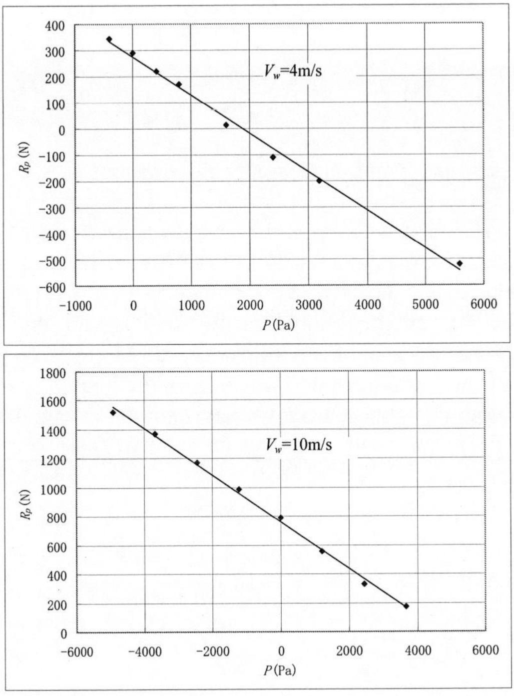

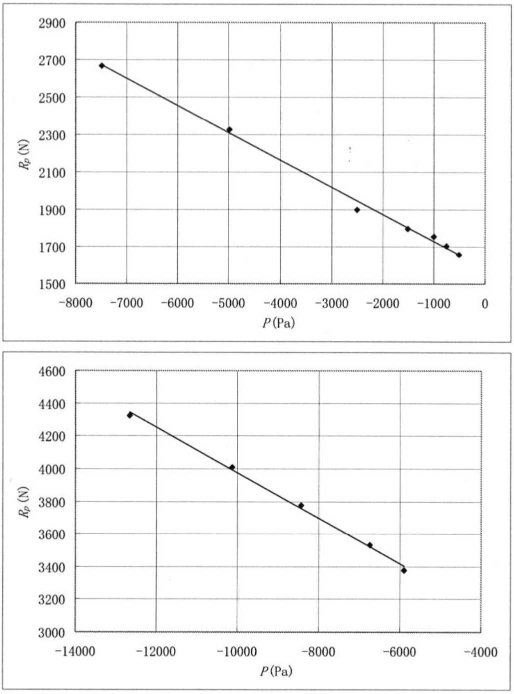

Fig. 4.7.4. Pressure drag acting on the wedge viz. air jet pressure

3. Calculation results of half-open bubble flow pattern

The inflow speed is still 2m/s, 5m/s, 10m/s, 20m/s. The typical profiles of the wave boundary of the bubble are shown in Fig. 4.7.5. It can be seen that the boundary of the bubble is fluctuating and the wave length increases with the increase of the inflow speed. At the same inflow speed, the speed of the air has almost no influence on the bubble.

Pressure drag R_p acting on the wedge is shown in Fig. 4.7.6. It can be seen that R_p decreases more with the increase of the air speed at a low inflow speed.

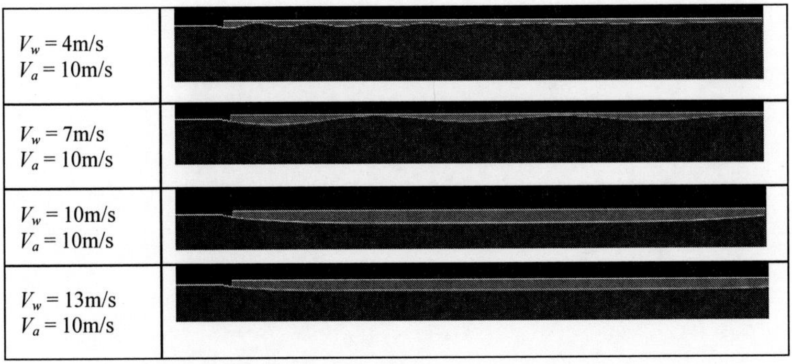

Fig. 4.7.5. Typical bubble profile

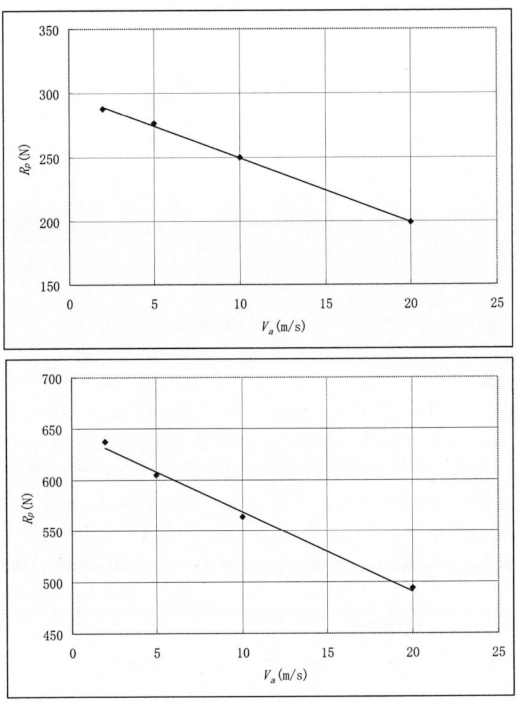

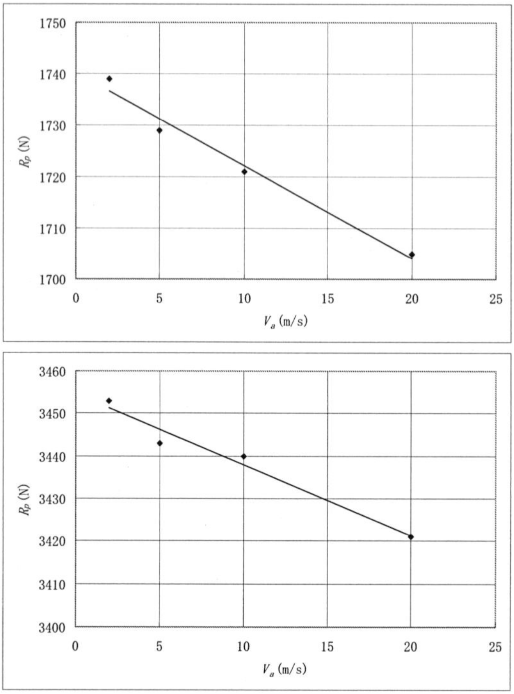

Fig. 4.7.6. Pressure drag acting on the wedge viz. inflow speed

It is also found that a larger air jet pressure is needed at a higher inflow speed when the air speed is the same. This is contrary to the closed bubble. This phenomenon can be explained from the viewpoint of energy.

When the bubble is closed, the bubble affects only the local flow. Because the air volume speed is 0, no energy is needed to support the bubble. If the bubble is half open, the effect of the bubble on the flow will extend infinitely. The continuous jetting of air will pull away the water. More energy is needed at a high inflow speed. This energy is provided by the air jet, which can be expressed as:

$$P_e = PQ \qquad (4.7.6)$$

where, P is the pressure at the air jet and Q is the air volume jetting into the bubble. Air volume is proportional to air speed. At the same air speed, if you want more energy to supply the bubble, you have to increase the air jet pressure. That is why a larger air jet pressure is observed at a higher inflow speed.

4.7.4. Concluding remarks

Calculating the results shows that the bubble will be longer and thinner at a high inflow speed. So the bubble may be more suitable to high-speed vehicles.

References

[4-7-1] C. S. Wu & S. L. He, "Numerical Simulation of Microbubble Flow around an Axisymmetric Body," Proceedings of 2nd PAAMES and AMEC 2006, pp. 552–58.

[4-7-2] C. S. Wu & S. L. He, "Numerical Simulation and Mechanical Analysis of Resistance Reduction by the Micro Air Bubble Flow on Crafts," *Ship Hydrodynamics* 9 (9): pp. 30–37, 2005 (in Chinese).

[4-7-3] Akihiro Kanai & Hideaki Miyata, "Direct Numerical Simulation of Wall Turbulent Flows with Micro-bubbles," *Int. J. Numer. Meth. Fluids*, 35: pp. 593–615, 2001.

[4-7-4] Yoshiaki Kodama, "Effect of Micro-bubble Distribution on Skin Friction Reduction," Proceedings of the International Symposium on Seawater Drag Reduction, 1998, Newport.

[4-7-5] S. Pal, C. L. Merkle, & S. Deutsch, "Bubble Characteristics and Trajectories in a Microbubble Boundary Layer," *Phys. Fluids*, 31 (4): pp. 744–51, 1988.

[4-7-6] N. K. Madavan, C. L. Merkle, & S. Deutsch, "Numerical Investigation into the Mechanisms of Micro-bubble Drag Reduction," *Journal of Fluids Engineering* 107: pp. 370–77, 1985.

[4-7-7] H. L. Hartmut, "A Simple Model for Gas Bubble Drag Reduction," *Phys. Fluids* 27 (12): pp. 2788–2790, 1984.

[4-7-8] V. G. Bogdevich, A. R. Evseev, & A. G. Malyuga, Gas Saturation Effect on Near-Wall Turbulence Characteristics," 2nd International Conference on Drag Reduction, Cambridge, BHRA, 27–37, 1977.

4.8

An Experimental Investigation to Reduce Drag with the Aid of Air Injection on the Bottom of Planing Hull Craft (2000)

Dong Wencai,[1] Guo Rixiu[1] and Liu Xiwu[2]

[1] Naval University of Engineering, Wuhan, China
[2] National Laboratory of Hydrodynamics, Wuxi, China

Abstract

This paper investigates experimentally the effect of air bubbles on the resistance reduction of stepped planing boats by several selected combinations with two longitudinal centers of gravity, two forms of step, two forms of spray strip, and five air flow rates. The total resistance is reduced by 25%, with air injected just ahead of the step. The main reason for the resistance reduction is the reduction of the wetted area by the formation of an air cavity, the alternation of the effective viscosity, and the density of the air–water mixed fluid.

4.8.1. Introduction

It has been reported that almost 15–30% of the total resistance of planing craft is reduced by injecting air between the bottom of the craft and its water boundary, and only 3% of the engine power is needed for the air injection.[4-8-1] The mechanism of the resistance reduction is the reduction of the wetted area due to the formation of the air cavity, and the alternation of flow characteristics of effective viscosity.[4-8-2–4] The step on the bottom of a planing craft is favorable to the formation of an air cavity, so the wetted area can be reduced remarkably and the efficiency of resistance reduction can be improved.

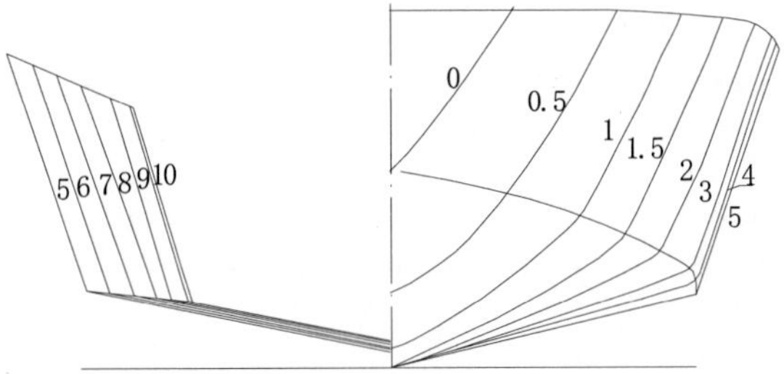

Fig. 4.8.1. Body plan of the experiment model

4.8.2. Model

The parent hull form selected in the present study is the American Series 62 model 4667-1. The dimensions of the model are of an overall length of 1673mm, a molded breadth of 485mm, a molded depth of 218.5mm, a projected hard chine length of 1590mm, a maximum breadth at the chines (excluding external spray strips) of 388mm, and a breadth over the chines at the transom (excluding external spray) of 253mm.

Two types of single step are attached to the bottom of the parent model respectively, that is, step 1 and step 2. Step 1 is a longitudinal step and located at the mid-ship; its horizontal projection is perpendicular to the longitudinal centerline.

Nomenclature

CG	center of gravity
$Draft_{10}$,$Draft_3$	draft at ordinate 10,3 respectively
F_∇	Froude number based on volume
Lp	projected chine length
LCG	location of longitudinal center of gravity
Q	air flow rate
Rt	total resistance
T	temperature of water
Trim	trim of craft
Zg	vertical location of CG
r_{zg}	rise of CG
τ_0	initial trim of model

Step 2 is the longitudinal diagonal step, the angle between its port (starboard) horizontal projection and the longitudinal centerline is 60°, and its endpoint is located at the intersection point of ordinate 5 and the keel. Two kinds of anti-strip rail are used: rail 1 and 2. Anti-spray rail 1 is a 15mm-wide horizontal anti-spray strip that lies along the chine from station 3 to 10. Rail 2 is an arc strip, its upper edge lies along the chine, from station 4.5 to 10, with a transition region from station 3 (horizontal strip) to 4.5 (arc strip). The arc length is 25mm, the chord length is 16mm, and the arc is wrapped down.

4.8.3. Experimental set-up

The tests had been carried out in the No. 1 towing tank at the China Ship Scientific Research Center. The models had been connected to the towing tank carriage in such a way that they were free to heave and pitch but restrained in all other modes of motion. The pivot of the construction was located at the intersection of the assumed centerline of the shaft and the cross section at the LCG. The resistance was measured by means of a strain gauge dynamometer UH53, pitch and heave had been measured by means of two wires over potentiometers on the stern and the bow of the model. The airflow rate was measured by a LZB-40 glass flow meter. The air pressure was measured by a manometer. During each run a photo had been taken through the bottom of the hull, to determine the wetted length and wetted area. The resistance, sinkage, and trim were measured in the speed range of 2–6.5m/s and in the air flow rate range of 0–8m^3/h. The selected combinations of three air injection forms, two types of step, two forms of spray strip, two longitudinal positions of CG, and five air-flow rates were investigated. The test schemes are shown in Table 4.8.1.

Table 4.8.1. Schemes of test

Test No.	W/kg	τ_0	LCG/mm	$Draft_{10}$ /mm	$Draft_3$ /mm	T/deg.	Towing angle	Step	Strip	Air injection form
1	19.9	0.41	-79.25	81	73	26.9	9.31	1	Without	Without
2	19.9	0.41	-79.25	81	73	26.9	9.31	1	Without	Double side pore
3	19.9	1.39	-158.5	90	63	26.9	10.71	1	Without	
4	19.9	1.39	-158.5	90	63	26.9	10.71	1	Without	Without
5	19.9	1.39	-158.5	90	63	26.6	10.71	1	2	Double side pore
6	19.9	1.39	-158.5	90	63	26.6	10.71	1	2	Double side slot
7	20.45	1.50	-158.5	92	63	26.2	10.71	2	2	Single side pore
8	20.45	1.50	-158.5	92	63	26.2	10.71	2	1	
9	20.45	1.50	-158.5	92	63	26.2	10.71	2	1	Double side pore
10	20.45	1.50	-158.5	92	63	26.2	10.71	2	Without	

4.8.4. Test results and discussion

1. Effect of air-flow rate

The test results of this experiment show that the variation of resistance due to the variation of air-flow rate is the same among the eight test schemes, as are the variations on rise of CG and trim. Thus, only the result of scheme 5 is shown in Figs 4.8.2–4, the percentages of R_t/W ,r_{zg} trim are given by (R_t/W)$_{(Q\neq0)}$/ (R_t/W)$_{(Q=0)}$, $(r_{zg})_{(Q\neq0)}$/ $(r_{zg})_{(Q=0)}$, $(\text{trim})_{(Q\neq0)}$/ $(\text{trim})_{(Q=0)}$ at the same velocity, respectively.

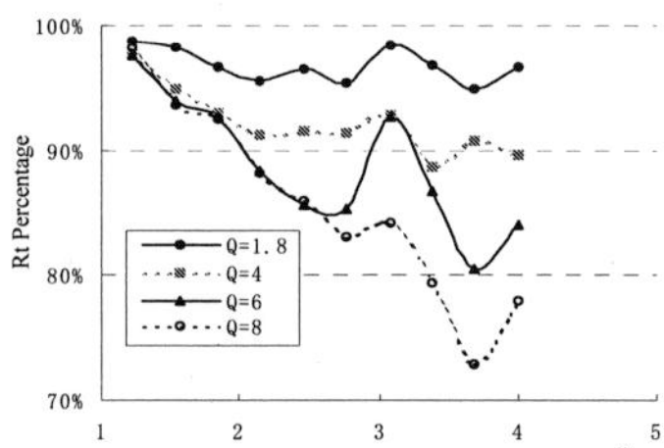

Fig.2 Effect of air flow rate on Rt/W

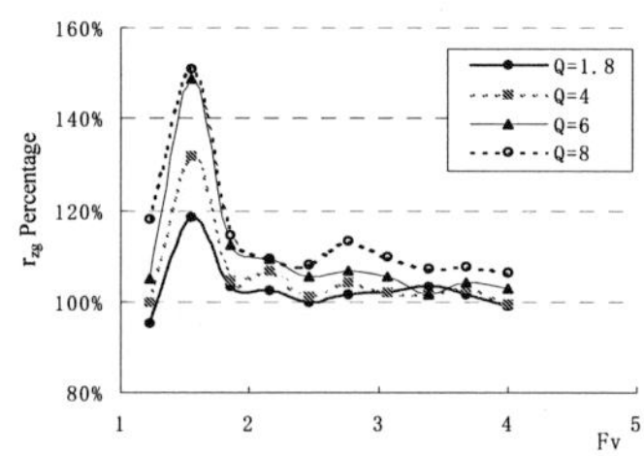

Fig.3 Effect of air flow rate on rise of zg

Fig. 4.8.2 (left). Fig. 4.8.3 (right)

As shown in Figs 4.8.2–4, as the air-flow rate increases, R_t/W decreases, r_{zg} increases, and the trim changes very little; however, with higher velocity (F_∇>2.46), the trim with higher air flow rate (Q=8m^3/h) is greater than that with other air flow rate (Q=0 ~ 6m^3/h).

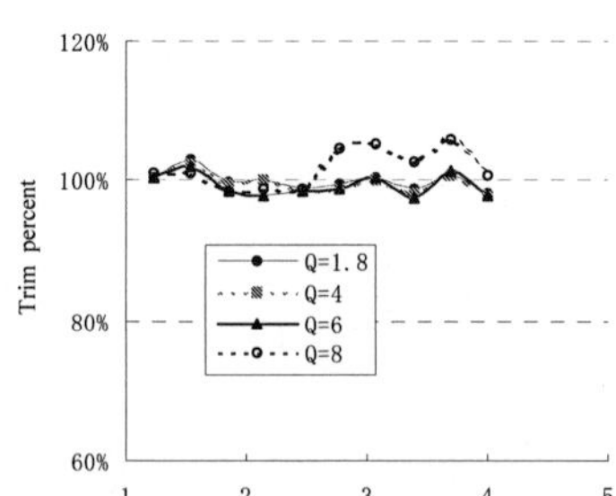

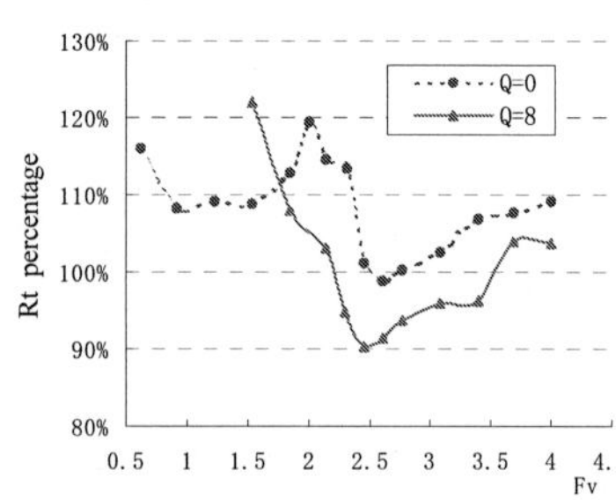

Fig. 4.8.4 (left). Fig. 4.8.5 (right)

1. Effect of the location of longitudinal center of gravity

The effects of the location of the longitudinal center of gravity on R_t/W, r_{zg} trim are shown in Figs 4.8.5–7. The percentages of R_t/W, r_{zg} trim are given by the R_t/W, r_{zg} trim at LCG = -5%Lp divided by those at LCG=-10%Lp, with the same velocity and the same air-flow rate, respectively.

As shown in Fig. 4.8.5, without air injection, R_t/W increases as the LCG shifts forward from –10%Lp to –5%Lp. Its average increment is about 10% within the velocity range of $1.0< F_\nabla<2.46$ and $F_\nabla \geq 3.08$. However, with air injection ($Q=8m^3/h$), its average decrement is about 5%, within the velocity range of $2.46< F_\nabla \leq 3.5$. This indicates that air injection can improve the performance of planing craft.

As shown in Figs 4.8.6–7, r_{zg} and trim decrease, as the LCG shifts move forward.

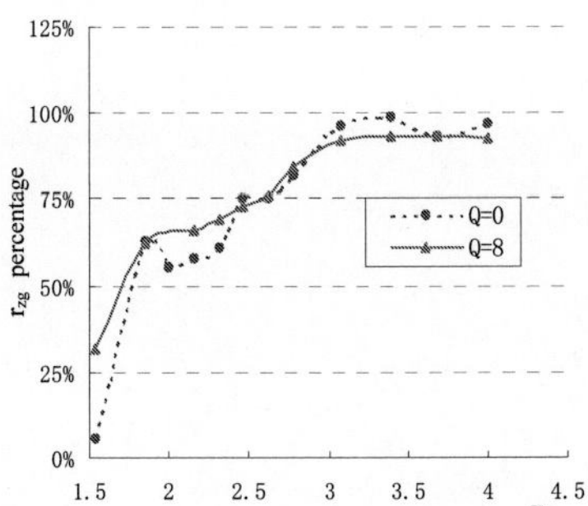

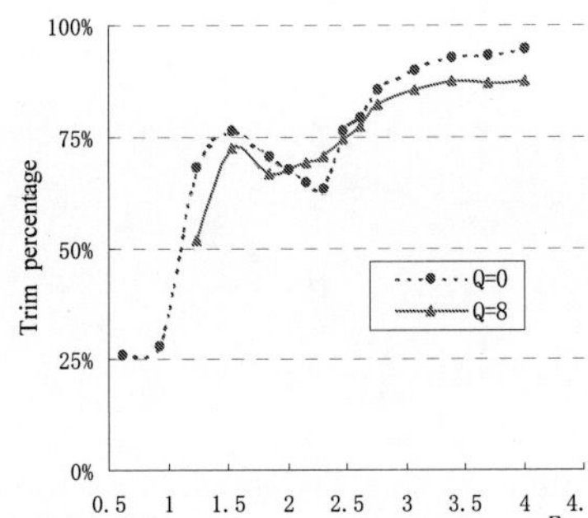

Fig. 4.8.6 (left). Fig. 4.8.7 (right)

3. Comparison between pore injection and slot injection

As shown in Table 4.8.1, scheme 5 and scheme 6 have the same step, the same anti-spray rail, and the same location of longitudinal center of gravity, but with a different air injection form. Pore injection was applied in scheme 5; slot injection was applied in scheme 6. Their different effects on the still water performance of a planing model are shown in Fig. 4.8.8.

The percentages of R_t/W, r_{zg} trim are given by the R_t/W, r_{zg} trim at slot air injection divided by those at the pore air injection, at the same velocity, respectively. The R_t/W of slot injection is greater than that of pore injection by 6–12.6% for all test velocity. The r_{zg} of slot injection is less than that for pore injection. When the velocity is within $F_\nabla \leq 2.77$, the trim of pore injection is less than that for slot injection. When the velocity is within $F_\nabla > 2.77$, the trim of pore injection is slightly greater than that for slot injection. Pore injection is more effective than slot injection on resistance reduction, probably due to the longitudinal location of air injection. The longitudinal location of pore injection is ahead of the step, for slot injection it is just at the step. Maybe air injection ahead of the step is favorable to the formation of an air cavity.

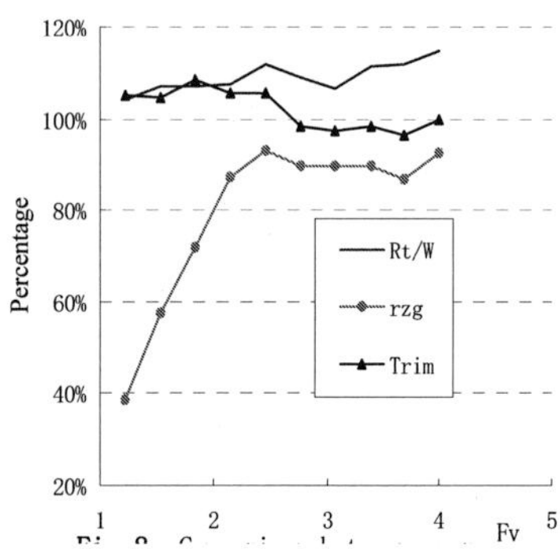

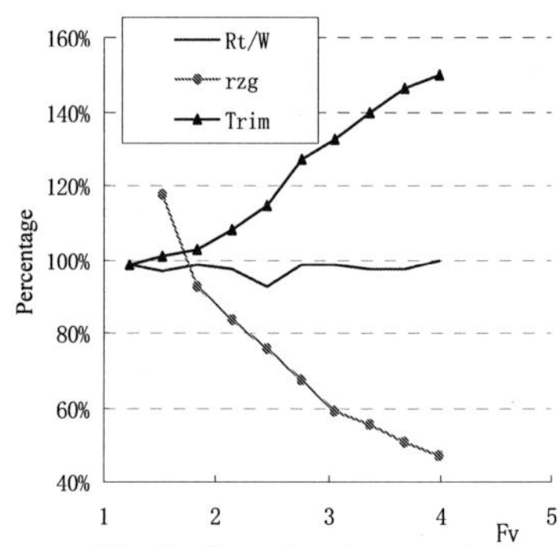

Fig. 4.8.8 (left). Fig. 4.8.9 (right)

4. Comparison between single and double-sided pore injection

As shown in Table 4.8.1, scheme 8 and scheme 9 have the same step, the same anti-spray rail, and the same longitudinal center of gravity, but have different injection forms. Single line pore injection was applied in scheme 8; double lines pore injection was applied in scheme 9. Their different effects on the results of the test are shown in Fig. 4.8.9. The percentages of R_t/W, r_{zg} trim is given by the R_t/W, r_{zg} trim with double-line pore injection divided by those with single-line pore injection, at the same velocity, respectively. The R_t/W of double-line pore injection is less than that for single-sided pore injection by 6–12.6%, within the velocity range of $F_\nabla > 1.84$. The r_{zg} of single-sided pore injection is greater than that of double-sided pore injection. The trim with single-sided injection is less than that for double-sided injection. It was also observed that the tail wake of the air–water mixed fluid at the transom for single or double line injection is approximately the same.

5. Effect of the step

As shown in Table 4.8.1, scheme 3 and scheme 10 have the same location of the longitudinal center of gravity, and the same air injection form, but with a different step. The longitudinal step with a straight transverse shape is used in scheme 3; the longitudinal diagonal step is used in scheme 10. The effect of the step on resistance is shown in Fig. 4.8.10. The percentages of R_t/W is given by the R_t/W with a longitudinal diagonal step divided by that with a straight transverse shape step, for the same velocity and air flow rate, respectively.

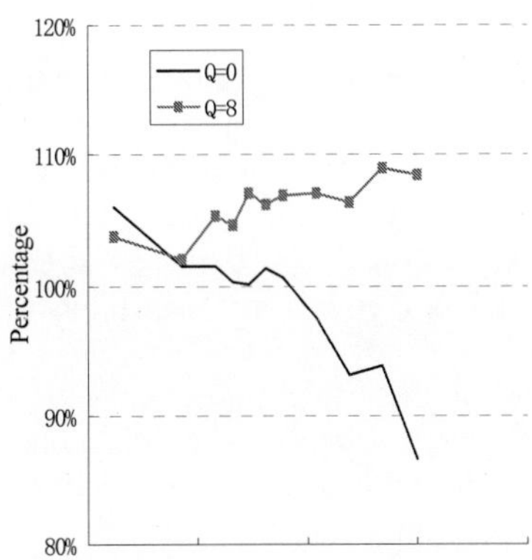

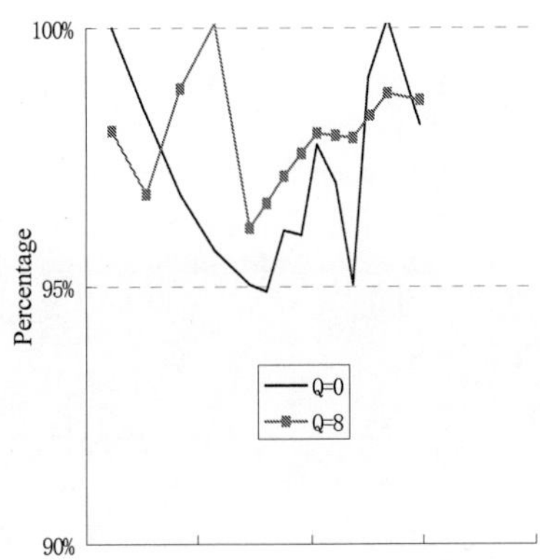

Fig. 4.8.10 (left). Fig. 4.8.11 (right)

As shown in Fig. 4.8.10, without air injection (Q=0m^3/h), the R_t/W with the longitudinal diagonal step is greater than that with another, for F_∇<2.84; however, the opposite result is obtained for F_∇>2.84. Concerning air-flow rate, the R_t/W with a longitudinal diagonal step is greater than that with a straight-shape transverse step by 7%, for all the test velocity. The different effect of the step on resistance is worth investigating further. Probably it is due to some leakage of air from the bottom of the bow of the model in the longitudinal diagonal step, owing to the pore injection location being too ahead of the midship.

6. Effect of an anti-spray rail

As shown in Table 4.8.1, scheme 7 and scheme 8 concern the same location of LCG, the same step, and the same air-injection form, but with a different spray rail. Where an arc-wrapped anti-spray rail is used in scheme 7, a horizontal anti-spray rail is used in scheme 8. The effect of the anti-spray rail on resistance is shown in Fig. 4.8.11. The percentages of R_t/W is given by the R_t/W with the arc-wrapped anti-spray rail divided by that with the horizontal anti-spray rail, for the same velocity and same air flow rate, respectively. Fig. 4.8.11 shows that the R_t/W with the arc-wrapped anti-spray rail is less than that with the horizontal spray strip by 3–4%.

4.8.5. Concluding remarks

1. Air injection just ahead of the step can effectively reduce resistance of a stepped planing craft, the main reason for the resistance reduction is the reduction of the wetted area by the formation of an air cavity, the alternation of the effective viscosity, and the density of the air-water mixed fluid.
2. As the air-flow rate increases, the increment in rise of the CG increases and the R_t/W decreases, but the trim changes very little.
3. As far as resistance reduction is considered, pore injection is obviously better than slot injection, double-line pore injection is somewhat better than single line pore injection.
4. As the LCG shifts forward, the R_t/W increases on the condition of a zero air-flow rate; however, with a suitable air-flow rate and a certain free stream velocity ($2.46 \le F_\nabla \le 3.5$), the R_t/W decreases.
5. Without air injection, the resistance of the model with longitudinal diagonal step is less than that of the model with a longitudinal step and a transverse straight step line; but with air injection, the opposite result is obtained.
6. The resistance of the model with a vertical arc-wrapped anti-spray rail along the chine is less than that of the model with a horizontal anti-spray rail along the chine; the R_t/W of the former is less than that of the latter by 3–4%.

References

[4-8-1] Dong Wencai & Guo Rixiu, "Research Progress of Air Injection Drag Reduction," *Journal of Ship Mechanics* 2 (5), pp. 73–78, 1998 (in Chinese).

[4-8-2] Dong Wencai & Guo Rixiu, "Experimental Study on the Skin Friction Reduction of Flat Plate by Formation of Air Cavity," International Workshop on Ship Hydrodynamics, IWSH '99, 1999, Wuhan, China.

[4-8-3] Seok-Ch'eon Go, Bam-Woo Han, Hee-Seok Yoo, & Hyochul Kim, "Formation of Air Cavity on the Bottom of Steeped Semi-Planing Boat and Its Effects on Resistance," China-Korea Marine Hydrodynamic Meeting, Shanghai China, 1997.

[4-8-4] Dong Wencai, Guo Rixiu, & Liu Xiwu, "Experimental Investigation on the Resistance of Stepped Planing Boat by Formation of Air Cavity," *Journal of Hydrodynamics*, forthcoming (in Chinese).

4.9

A Test Study on the Resistance Reduction of an Air Bubble Ship (2010)

Cheng Hong-xia, Ni Qijun, Yu Yangchun and Xing Shengde

China Ship Scientific Research Center, Wuxi, China

Abstract

Air is inducted to the bottom of a ship to form an air bubble around the bottom surface of the ship. The total resistance of the ship can be greatly decreased because the bottom flow type is changed, consisting mainly of air rather than water. The air bubble ship is a new type of ship that is more economical and efficient than conventional ships. Russia has successfully developed many bubble ships, while we still have little experience and stand at the initial stage of study of this ship type. CSSRC has tracked the development of the bubble ship for a long time, has developed a bubble ship, and has drawn up a set of experimental methods for this ship type, based on the former theoretical study results. A series of model tests for the target bubble ship and its original conventional chine ship were developed in a pool for resistance analysis. The ship model, test procedure, test parameters, conversion method, and results for the real ship are described in detail. This paper contrasts the results between the two real ships, and then presents the conclusion of the resistance decrease effect of the bubble ship. The results provide credible dependence for a performance study of the bubble ship, and have important reference value for the research and development of bubble ships.

4.9.1 Introduction

The bubble ship is a new type of ship that is more economical and efficient than conventional ships. It can induct air to the ship bottom and form an air bubble around the bottom surface. The total resistance of the ship can be greatly decreased because the bottom flow type is changed, consisting mainly of air rather than water. Besides, the seaworthiness can be improved since the air bubble in the ship bottom can let up its movement in a storm.

In contrast to a hydrofoil craft, a sidewall hovercraft, and other high-performance ships, high-speed bubble ships not only are economical but also have marked characteristics such as simple operating maintenance, berthing, and leaving dock facilities, a shallow draft, a need for a low guarantee request and capital cost since its structure is relatively simple compared with other ship types. Therefore, the ship has a wide range of market applications for both military and commercial use. Its potential community and economic benefits are remarkable, and the bubble ship is becoming more and more admired in international shipbuilding.

Until now, Russia has constructed more than 70 high-speed bubble ships, the tonnage of which ranges from 14t to 360t, and the ship speed between 30kn and 52kn. Such craft include *Linda*, *Lieda*, *Kitchen God*, *Diana*, *Chamois*, *Flat Deer*, *Water Star*, and other multi-type ships, which are used for military and commercial applications, frontier defense, and custom patrolling.

In China, people have expanded the embedded study of reducing resistance with micro-bubbles, but we still have little experience and stand at initial stage of the study of resistance reduction with artificial big bubbles.

CSSRC has tracked the development, application, and pulse of bubble ships at both home and abroad for a long time, and has developed abundant studies and argumentation. Now we have designed a bubble ship based on the former theoretical study results and have carried out resistance tests on the original conventional hard chine ship model for resistance-reduction effect analysis.

4.9.2. Test outline

The test objects are one conventional hard chine ship and one bubble ship that improves on the chine ship; the scale ratio of both is 5. The principal dimensions of the models and the real ships are listed in Table 4.9.1 and Table 4.9.2.

The bubble ship model is made with wholly transparent organic glass so that the formation of the bubble can easily be observed; the chine ship model is made with paulownia. The weight of the ship models was controlled strictly to satisfy the adjustment of the situation and the inertia of the gravity center. The model photos are shown in Figs 4.9.1 and 4.9.2.

Table 4.9.1. Key dimensions of the real bubble ship and its ship model

Item	Symbol	Real bubble ship	Bubble ship model
Length overall/m	L_{OA}	7.8	1.56
Waterline length/m	L_{WL}	7.18	1.436
Designed displacement/t	$\triangle$	4	0.032
Designed draft/m	T	0.525	0.105
Molded breath/m	B	2.48	0.496
Molded depth/m	H	1.55	0.31

Table 4.9.2. Key dimensions of the real hard chine ship and its ship model

Item	Symbol	Real chine ship	Chine ship model
Length overall/m	L_{OA}	7.8	1.56
Waterline length/m	L_{WL}	7.18	1.436
Designed displacement/t	$\triangle$	4.0	0.032
Designed draft/m	T	0.6	0.12
Molded breath/m	B	2.48	0.496
Molded depth/m	H	1.55	0.31

Fig. 4.9.1. Test model of the bubble ship

Fig. 4.9.2. Test model of the chine ship

4.9.3. Test method

Before tests, we established a feasible test plan and prepared a set of air fed fixtures and air ducts. A part of the test equipment is shown in Fig. 4.9.3.

We have expanded the tests according to the towing model method for the general planing boat and the main test equipment is an instrument for measuring resistance with four freedom degrees. The instrument is fixed at #7 of both ship models for the particularity of the bubble ship and the comparability between the two ship models. So the heaving of the towing point got from the test need to be converted to the heaving of the center of gravity.

Fig. 4.9.3. Semi plant—bobber flow meter and pressure reduction room

4.9.4. Test contents

During the tests, we first carried out flux debug tests for the bubble ship model to confirm a logical flux, and then we carried out resistance tests by adjusting the ship's running attitude continually to select a better state, where the angle of the trim is 0.941°. Under this state, we finished a resistance test with gas charging to get an integrate curve from low speed to high speed. In order to get the intuitional contrast, the resistance test was finished under the same state on the bubble ship model. The test data are model speed (V_m), model resistance (R_m), trim of model (θ_m), and heaving at center of gravity (Z_m); besides, we measured the pressure of five different locations in the air cavity. The temperature during the tests is 24 □, and the speed range is from 0.5m/s to 5.2m/s, which corresponds to the Froude number Fn=0.3–3.1, Fn is defined as $V/\sqrt{gL_{wl}}$ (V is the ship speed, L_{wl} is the waterline length).

4.9.5. Conversion method

The datum conversion method is adopted for the conventional ship, using two-dimensional Froude numbers to convert the test results to the real ship results, in which the frictional resistance coefficient uses ITTC formula from 1957, the compensate coefficient correlates to the model, and the real ship is 0.0004; thus, the conversion for resistance, angle of trim, and heaving of the center of gravity of the real ship is carried out with formulas as follows:

$$C_{Tm} = \frac{R_{Tm}}{\frac{1}{2}\rho_m S_m V_m^2} \quad (4.9.1)$$

where,

R_{Tm} Represents the total resistance of the ship model, the unit is N

C_{Tm} Total resistance coefficient of the ship model

ρ_m Water density (kgS^2/m^4), changed in terms of water temperature

V_m Ship model speed (m/s)

S_m Wetted surface of ship model (m^2)

$$C_{Rm} = C_{Tm} - C_{Fm} \quad (4.9.2)$$

while

$$C_{Rs} = C_{Rm} \quad (4.9.3)$$

where

C_{Rm} Residual resistance coefficient of the ship model

C_{Rs} Residual resistance coefficient of the real ship

C_{Fm} Frictional resistance coefficient of the ship model

$$C_F = \frac{0.075}{(\lg \mathrm{Re} - 2)^2} \quad (4.9.4)$$

Re is Reynolds number, which is $\mathrm{Re} = \frac{VL_{WL}}{\nu}$ (4.9.5)

where, V represents the ship speed of the model or the real ship (m/s),

L_{WL} Ship water line length (m)
ν Coefficient of viscosity of water (m^2/s), which may be changed in terms of the water temperature.
In that case, the total resistance coefficient of the real ship is:

$$C_{TS0} = C_{FS} + C_{Rs} + \Delta C_F \quad (4.9.6)$$

where,

ΔC_F is the conversion compensate coefficient, which we usually take as 0.4*10^{-3}.

After we get C_{TS0}, we will then calculate the total resistance for the real ship using the symbol of R_{TS} and effective power using the symbol of P_E:

$$R_{TS} = C_{TS0} \frac{1}{2} \rho_S S_s V_S^2 \quad (4.9.7)$$

$$P_E = R_{TS} V_S \text{ [kW]} \quad (4.9.8)$$

The trim of the real ship is equal to that of the ship model, and

$$\theta_s = \theta_m \quad (4.9.9)$$

The relation of heaving between the real ship and the ship model is

$$Z_s = \lambda Z_m \quad (4.9.10)$$

4.9.6. Results

After taking the conversions and dimensions, we contrast the ratio of resistance to displacement of the real ships, the angle of trim of the real ships, and the heaving of the center of gravity of the real ships, and we compare the effective power for the two real ships. The test results of the two real ships are shown in Table 4.9.3 and Figs 4.9.4–7.

Table 4.9.3 Test results of the real ships

Fn	$(R_S/\Delta_S)\times 10^5$			θ_S (°)		Z_S (mm)	
	Bubble ship	Chine ship	Resistance reduction percent of bubble ship	Bubble ship	Chine ship	Bubble ship	Chine ship
0.30	0.42	0.30	-37.71	-0.12	-0.05	1.66	-0.48
0.60	1.92	1.30	-47.45	-0.03	-0.12	-0.75	-1.98
0.90	5.75	4.77	-20.67	0.37	0.02	-2.85	-4.88
1.20	10.89	10.87	-0.12	2.72	2.77	-3.80	-7.59
1.49	12.52	13.52	7.41	3.61	3.92	2.83	-3.54
1.79	12.69	14.78	14.14	4.19	3.94	8.84	1.83
2.09	12.77	15.87	19.55	4.87	3.72	13.74	5.78
2.39	12.71	16.79	24.29	4.98	4.90	17.51	10.56
2.69	12.23	16.27	24.83	4.88	4.77	20.36	13.90
2.99	11.65	15.63	25.46	4.60	4.27	23.44	16.71
3.11	11.47	15.70	26.97	4.45	4.09	24.35	17.75

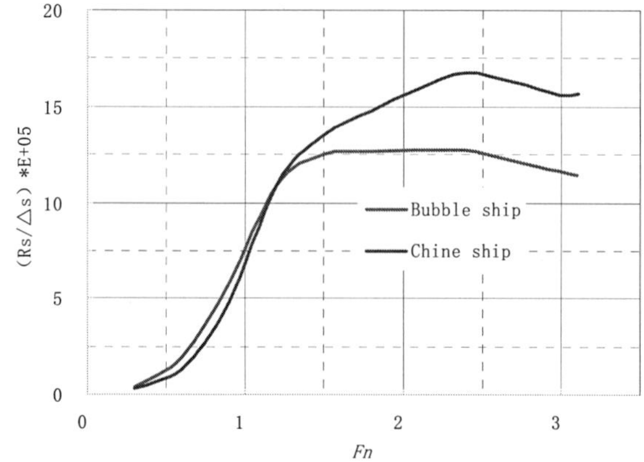

Fig. 4.9.4. Contrast curves of dimensionless resistance of the two real ships

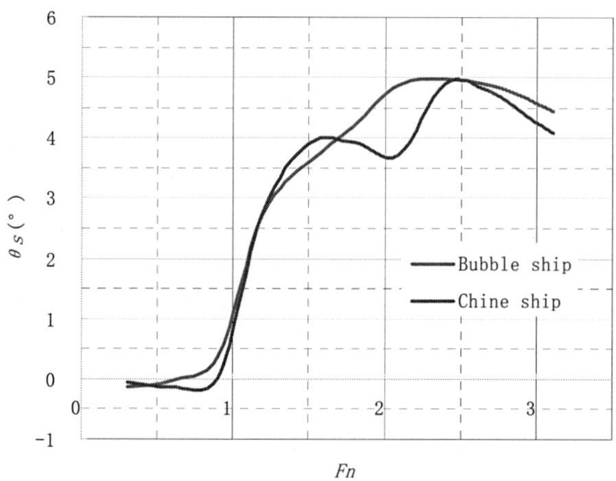

Fig. 4.9.5. Contrast curves of the trim of the two real ships

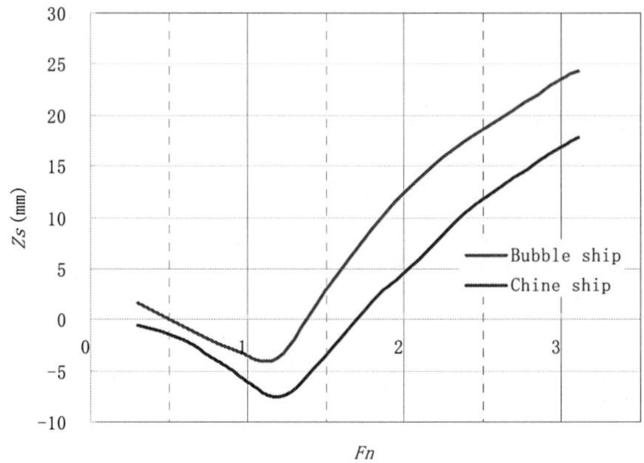

Fig. 4.9.6. Contrast curves of heaving of the two real ships

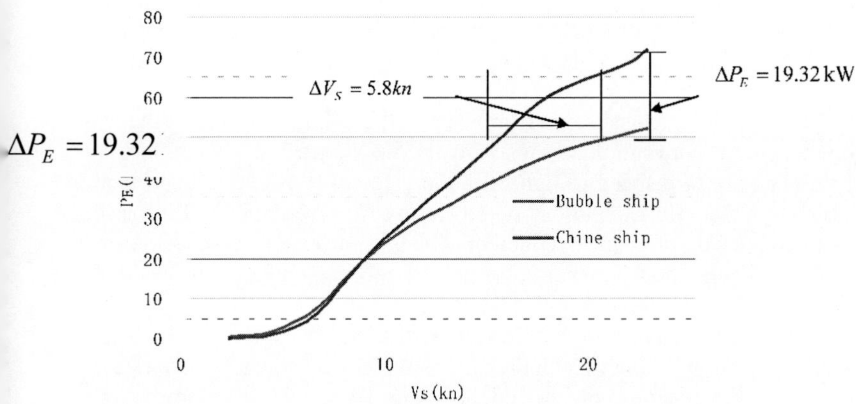

Fig. 4.9.7. Contrast curves of effective power for the two real ships

4.9.7. Conclusions

From Figs 4.9.4–7, it can be found that the calculated outcome between the bubble ship and the chine ship under a running attitude with 0.941 angle of trim is as follows:

1. When the speed is lower, the resistance of the hard chine ship is less than that of the bubble ship; with the speed increasing, the resistance of the bubble ship became decreasingly less than that of

the chine ship. When the *Fn* is 3.1, the devaluation percent of resistance is more than 27%. The reason for this is that when the ship speed is low, the length of the air bubble formed is short, the vaulted region formed by the air is not the same slickness as the bottom of the chine ship which induced the increase of resistance of the bubble ship; while, when the ship speed increases, the length of the formed bubble increases correspondingly, so the frictional resistance is reduced greatly which contributes to the reduction of the total resistance.

2. The trim of the bubble ship is larger than that of the chine ship during the whole speed range, but the difference is not that big, and the heaving of the bubble ship is still larger than that of the chine ship.
3. When the Froude number *Fn* is 3.1, the ship speed V_S is 22.6kn; the corresponding effective power for the bubble ship can be reduced by 27%, and when the effective power provided for the real ships is the same as 50kW, the bubble ship speed will increase by 5.2kn more than the chine ship—the increased percent is 38.1%.

To sum up, the resistance performance of the bubble ship excels that of commensurate conventional ships, and the resistance reduction effect is clear when the ship speed is high, with the value of resistance reduction at more than 27%. Thus, main engine power can be saved greatly. If we use the technique of resistance reduction with an artificial bubble added to high speed ships, the transportation will be more economic, and so it is important that the bubble ship is developed.

The performance of our bubble ship corresponds to that of the Russian bubble ship in public news. The results provide credible dependences for the performance study of the bubble ship type, and have important reference value for the bubble ship's research and development.

References

[4-9-1] "Resistance Test Method of Planing Ship Model" (CB/Z 244-1988).

[4-9-2] Shao Shiming, Zhao Lianen, & Zhu Nianchang, *Ship Resistance*, National Defense Industry Press, China, 1995 (in Chinese).

4.10

NUMERICAL EVALUATION OF THE RESISTANCE REDUCTION EFFECT FOR THREE-DIMENSIONAL HIGH-SPEED AIR CAVITY CRAFT (2013)

NI QIJUN,[1, 2] WANG LIYAN[1, 2] AND LI SHENGZHONG[1, 2]

[1] China Ship Scientific Research Center
[2] Jiangsu Key Laboratory of Green Ship Technology

Abstract

In this paper, the total resistance of high-speed air cavity crafts at different speeds are numerically simulated by solving RANS equations with $SST\ k-\omega$ turbulence models. The VOF method is used to deal with the free surface and gas-liquid flow field of the craft bottom. Computed results including free surface wave and pressure distribution are presented. Then the effect of the artificial cavity shape and air-jet velocity on resistance is analyzed. Drag reduction efficiency under different conditions is compared. The results are helpful for understanding the complex flow phenomenon of air cavity craft in order to optimize the artificial cavity geometry shape.

4.10.1. Introduction

Bubble drag reduction methods can reduce the total resistance by lowering fluid viscosity and changing the bottom flow patterns through injecting gas into the surface. The air cavity craft is a developing, new advanced hull form with higher hydrodynamic efficiency[4-10-1] that works by becoming an artificial cavity through injecting gas to the bottom of ships. This reduces the fluid density near the bottom of the ship, changing the structure of the boundary layer flow. Thus, it can reduce the total resistance of ships and also has an energy-saving purpose.

Russia is the most successful country in the research and application of high-speed bubble ships. It has accomplished more than 30 designs recently, finishing more than 70 high-speed air cavity craft, with a displacement of 14–360t and a speed of 30–52 kn. These craft include *Linda* and others. Many different products are widely used in the military, and by civil clients, border patrols, and customs, and so on. In China, the study of big bubble resistance reduction is still at the exploratory stage and some problems remain to be solved; nevertheless, some research has been carried out in various departments.

Research methods on high-speed planing air cavity craft mainly include theoretical analysis, model experiments, and numerical simulation. The model experiment method is reliable, but it takes a long time and many human work hours as well as significant material and financial resources. Computational fluid dynamics (CFD) technique as an effective complement to physical testing has been widely used in the field of ship research; it can speed up the prediction of ship hydrodynamic performance, and provide comprehensive, detailed flow field information, and the cost is low. For these reasons, it has become an indispensable and important means for scientific and engineering research. Therefore, this paper focuses on applying the numerical simulation method to glider and air cavity craft.

High-speed planing craft are different from low-speed ships as their draft is small. The weight of the craft is supported by hydrodynamic lift; in the process of high-speed movement, splashing waves are generated on both sides. Numerical simulation is more difficult for high-speed air cavity craft, because as there are two free liquid surface problems, it is difficult to tell how to fully consider the boundary conditions and simulate the real air cavity craft flow field. But in China, there has been less research on ACC craft because the air cavity craft hull is more complex. We generally use an unstructured grid, which greatly increases the dissipation data; but the complex flow phenomenon and splashing waves will affect the simulation, and numerical calculation precision is difficult to guarantee.

4.10.2. Numerical simulation

In this paper, a finite-volume method for solving the incompressible RANS equations is developed for the numerical simulation of a three-dimensional viscous flow with a free surface for high-speed air cavity craft with $SST\ k-\omega$ turbulence models. The VOF method is adopted to deal with the free-surface and gas–liquid interface. The method can track deformable complex free surfaces, such as happened in the free surface for turning, annexation, splash, and other complex free-surface phenomenon.

Because the planing craft is symmetrical, the half hull is considered. The computational domain is 9.8L ×1.2L × 0.7L (longitudinal × horizontal × vertical) and L is the length of the craft.

The detailed set-ups of the boundary condition are described as follows:

1. On the inlet boundary, the position of the free surface is prescribed by volume fractions of water and air, while using the ship model's speed as the inlet velocity.
2. On the outlet boundary, the flow is defined as a fully developed condition that the exit locates far from the ship model.
3. On the hull surface, a non-slip condition is applied, $u = v = w = 0$.
4. On the symmetry boundary, a symmetry condition is applied.

A planing craft shape is more complex, as it has a transverse oblique angle in the midship section and two angle lines on the craft side. It is very difficult to partition the grid. From the perspective of the related academic papers published, most authors generally use an unstructured grid,[4-10-1] which greatly increases the dissipation data. The complex flow phenomenon and splashing waves will be affected, and the numerical calculation precision is difficult to guarantee. In order to ensure accuracy of the numerical calculation, a partition technique is adopted, a boundary (interface grid) is used to solve the problem of the gas–liquid boundary conditions, and the whole domain is calculated for a structured grid.

The basic principle is that the appropriate encryption for the boat body and tail in the central grid must be used and the bow area in the air should have a relatively sparse grid; there should be mesh refinement near the surface of the model, and the first layer of the grid spacing should accord to y^+ (y^+ average about 50); near to the free surface grid should also be properly encrypted. The grid number is about 2.8 million on the whole calculation domain. The grid computing domain is shown in Fig. 4.10.1; the hull surface grid is shown in Fig. 4.10.2.

4.10.3 Numerical simulation of resistance and model experiment validation for the archetype (no cavity)

Using a numerical calculation method, the prototype of bubble boat (no air cavity) resistance is numerically simulated and compared with the model test results.

Due to the running attitude (trim and heave), the difference is bigger when the planing craft is in different Froude numbers ($F_{r\nabla}$); therefore, the model test results of trim and heave are adopted for numerical calculations at different Froude numbers. The comparisons of resistance coefficients between numerical calculation and model experiment results are presented in Fig. 4.10.3. The free surface wave is shown in Fig. 4.10.4 for the different Froude numbers.

Fig. 4.10.3 shows that the numerical calculation results of resistance change with the Froude number and that the model test results are basically identical. When $F_{r\nabla}$ =2.086, $F_{r\nabla}$ =2.607, the resistance value deviations are larger between numerical calculations and model testing (9.9% and 7.6%, respectively), and the numerical result is larger. When $F_{r\nabla} \geq$ 3.128, the numerical calculation results of resistance agree well with the model test results, the error being within 5%. As can be seen from the above comparison, resistance for planing craft is under the transition state and the numerical simulation precision is a bit poor; however, in the planing mode ($F_{r\nabla} \geq$ 3), the numerical simulation precision for the resistance is higher.

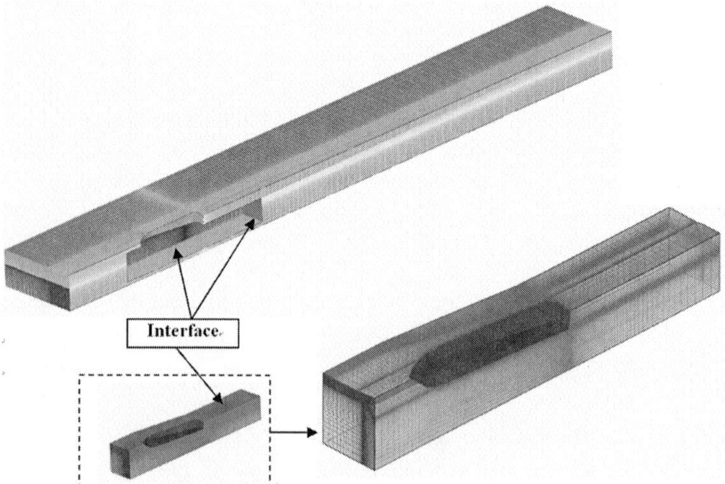

Fig. 4.10.1. Computation region

Fig. 4.10.2. Surface mesh

Fig. 4.10.4 shows that the free surface wave simulation is better, the computed results are in reasonable agreement with reality, and the free surface of the planing craft stern can be simulated very well. But the splashing wave is little different between the numerical simulation and the model experiment. From what has been discussed above, comparison results show that the numerical simulation has high accuracy on the planing craft resistance, which is a good foundation for the numerical simulation analysis of the air bubble drag reduction effect.

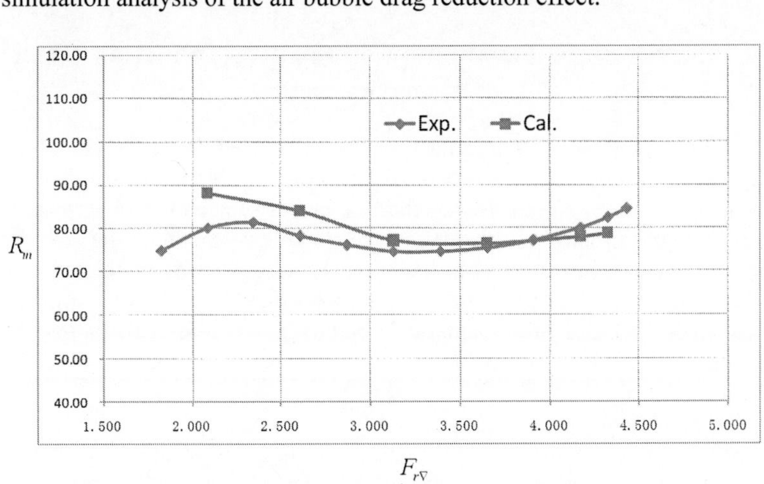

Fig. 4.10.3. Computational and experimental results of the total resistance for the planing craft

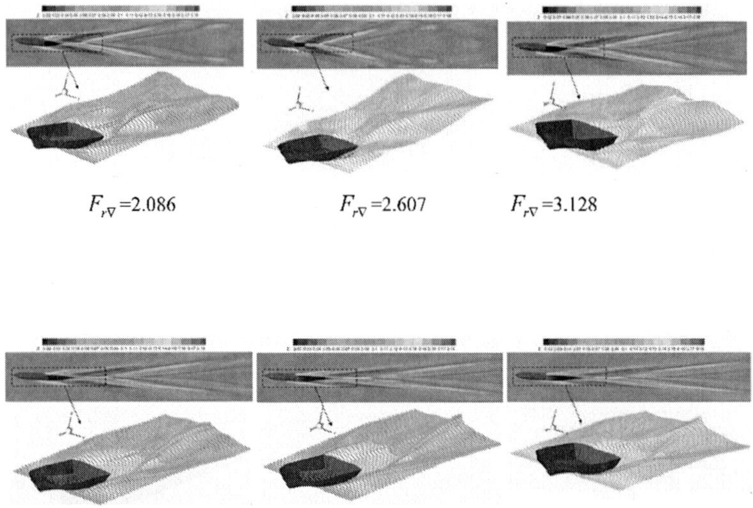

$F_{r\nabla}$=2.086 $F_{r\nabla}$=2.607 $F_{r\nabla}$=3.128

$F_{r\nabla}$=3.650 $F_{r\nabla}$=4.171 $F_{r\nabla}$=4.323

Fig. 4.10.4. Free surface wave contours for planing craft ($F_{r\nabla}$)

4.10.4. Numerical evaluation on the drag reduction effect of air cavity craft

The basic principle of drag reduction for high-speed air cavity craft is to reduce the fluid density near the bottom of the ship with an artificial cavitation, which can thus reduce the viscous resistance of ships. It usually sets a unique structure of the air cavity, owing to maintaining the stability of the air layer. There are two main forms: one is the internal cavity, the other is the outside baffle. In this paper, we focus on the first form, using two kinds of cases. The numerical method is used for calculating the effects.

1. Air cavity and gas injection scheme

On the basis of model test and numerical simulation results, two cavity cases were designed as shown in Fig. 4.10.5. Case 1 has one longitudinal step, setting a single injection, and Case 2 has two steps and two injectors with the same size.

To verify the drag reduction effect, and the relations between the drag reduction effect and the quantity of injecting, resistances are simulated

(Case 2's double injecting quantity is the same). The total injecting quantity of the two cases respectively is Q_0=0 m^3/h, Q_1=0.683m^3/h, Q_2=3.417 m^3/h, Q_3=6.834 m^3/h, Q_4=13.669 m^3/h, and Q_5=27.337 m^3/h.

The numerical simulation method of air cavity craft is similar to planing craft, with the same computational domain and the basic principles of meshing. The interface grid method is used to solve the set problem of the gas–liquid boundary condition. The total number of the grid is about 2.9 million.

Case 1 Case 2

Fig. 4.10.5. Diagrammatic sketch of the air cavity and injection

2. Numerical results and analysis

The resistance of air cavity craft at the design speed has been calculated. The drag reduction effect, the air-liquid distribution, surface pressure distribution, free surface wave, and gas path-line have all been analyzed in detail.

3. The drag reduction effect and the bubble covering

To compare the drag reduction effect in the different cases, the absolute drag reduction rate η_0 and the relative drag reduction rate η_1 are defined. The total resistance, frictional resistance, and residual resistance of the air cavity craft is R_{t0}, R_{f0} and R_{r0}, respectively. When not injecting, the total resistance, frictional resistance, and residual resistance of the air cavity craft are R_{t1}, R_{f1}, R_{r1}, respectively. When injecting, the total resistance, frictional resistance, and residual resistance are R_t, R_f, R_r, respectively. Their expression is as follows:

$$\eta_0 = \frac{R_t - R_{t0}}{R_{t0}} \times 100\% \quad \eta_1 = \frac{R_t - R_{t1}}{R_{t1}} \times 100\%$$

Table 4.10.1 shows the calculation results at the different gas flow rates; the analysis is as follows:

1. The resistance is increased (Case 2 increases resistance by 8.15%) without injection; bottom pressure distribution is discontinuous because of the air cavity. At the same time, pressure resistance and wetted surface area increasing owing to the cavity, which making frictional resistance increase.
2. The resistance is reduced with the increment of the air-flow rate for these two cases; the absolute maximum drag reduction rate can reach 14.23%, which can realize the purpose of drag reduction. By comparing the calculation results of the cases, the step and injector make the resistance increase at a lower air-flow rate.

In this paper, the air–liquid interface morphology is shown in Fig. 4.10.6. As can be seen from the figure:

1. The gas covering situation of the bottom improved by increasing the gas flow rate. Until a certain gas flow rate value, the covering does not change with the increment of the jet.
2. When the gas flow rate is bigger, the gas will escape from the cavity sides; the air covering improvement does not largely increase with the step and injector. The gas of injector 2 almost escapes from the injector to the stern.

Table 4.10.1 Resistance of cases with different gas flow rates

Parameter	Case 1						Case 2					
	Q_0	Q_1	Q_2	Q_3	Q_4	Q_5	Q_0	Q_1	Q_2	Q_3	Q_4	Q_5
R_r / R_{r0}	-0.23%	-4.20%	-1.91%	-2.58%	-3.26%	-2.26%	-0.62%	-6.75%	-7.08%	-8.25%	-9.45%	-9.34%
R_f / R_{f0}	11.86%	10.78%	-16.72%	-21.74%	-24.02%	-30.73%	20.38%	16.85%	5.03%	-19.71%	-20.90%	-19.47%
η_0	4.82%	2.05%	-8.09%	-10.58%	-11.93%	-14.15%	8.15%	3.10%	-2.02%	-13.03%	-14.23%	-13.57%
R_r / R_{r1}	—	-3.98%	-1.68%	-2.35%	-3.04%	-2.03%	—	-6.18%	-6.50%	-7.68%	-8.88%	-8.78%
R_f / R_{f1}	—	-0.97%	-25.55%	-30.04%	-32.07%	-38.07%	—	-2.93%	-12.75%	-33.30%	-34.29%	-33.10%
η_1	—	-2.64%	-12.32%	-14.69%	-15.98%	-18.09%	—	-4.67%	-9.40%	-19.59%	-20.69%	-20.08%

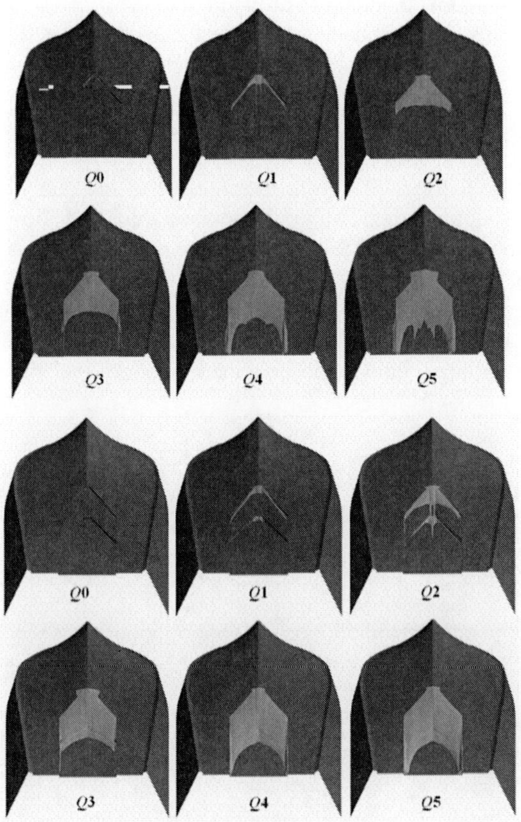

Case 1
Case 2

Fig. 4.10.6. The gas–liquid interfaces with different gas flow rates

3. When the air-flow rate is high enough, the triangle area in the stern has not been covered. Continuing to increase the air flow rate, the gas runs away from the sides to the stern, and a part of the gas escapes from two sides of the cavity directly. Because the trim angle is larger (2.376°) at the design speed, it leads higher to the tail of the static pressure. It is unable to break through the pressure difference owing to the small density of gas.

2. The flow phenomenon analysis around the injector

This part presents the flow phenomenon for the injector. The flow lines of the gas are shown in Figs 4.10.7 and 4.10.8, as can be seen from the figures:

1. When the gas flow rate is small (Q_1) in each case, the gas from the nozzle mainly flows along the edge of the cavity.
2. Gas flows downward from the injector to the rear after injecting; when it encounters water, there is a reversed flow around the injector. After that, the gas flows downward to the stern along the cavity sides.
3. With the increase of the gas flow rate, the gas coverage area gradually increases, until a certain gas flow rate. After that, the coverage area no longer increases; meanwhile, the gas escapes along the cavity sides.

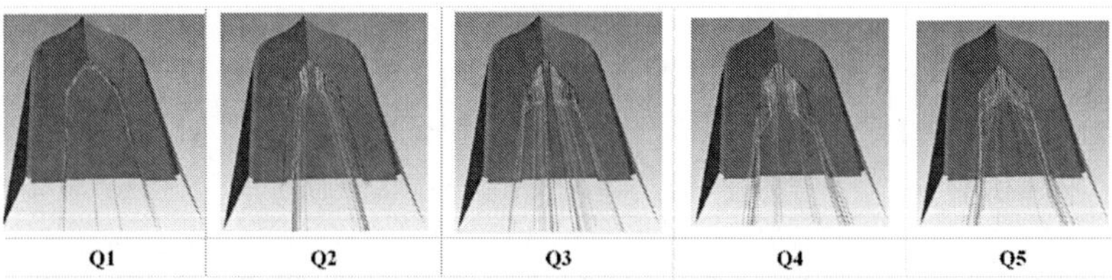

Fig. 4.10.7. Flow lines of the air injector in case 1

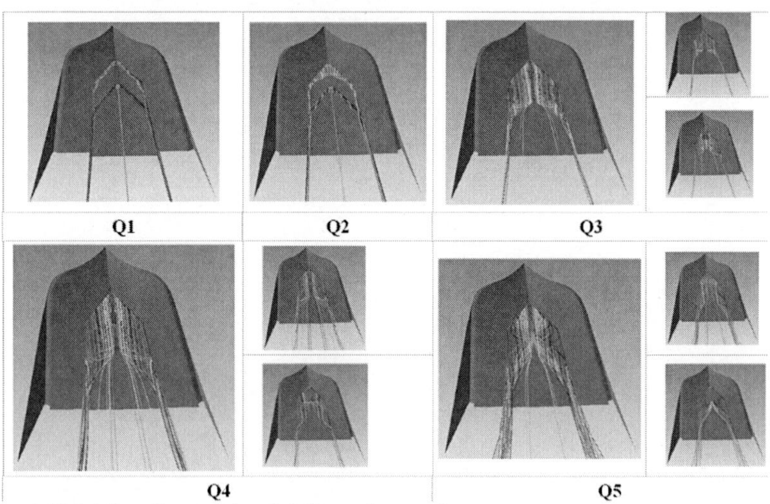

Fig. 4.10.8. Flow lines of the air injector in case 2

When the gas flow rate is very large, it will form a vortex behind the injector. The cavity internal flow phenomenon is very complex, it could also be the reason that the total resistance does not reduce with the gas flow rate increasing.

3. Surface pressure distribution
Fig. 4.10.9 shows the surface pressure distribution in two cases under different gas flow rates. From the figure, one can see that the crafts, both of which have steps, when they are not injecting have a very obvious low-pressure area. With the increment of the gas flow rate, the low-pressure area gradually disappears. When the gas flow rate is enough, the ship surface pressure distribution is similar to its original.

4. Free surface wave
Fig. 4.10.10 shows the free surface wave of Case 2 with its gas flow rate. It shows that the free surface wave shape is changed a little under other gas flow rates and injecting is little affected by the free surface wave shape (basically negligible).

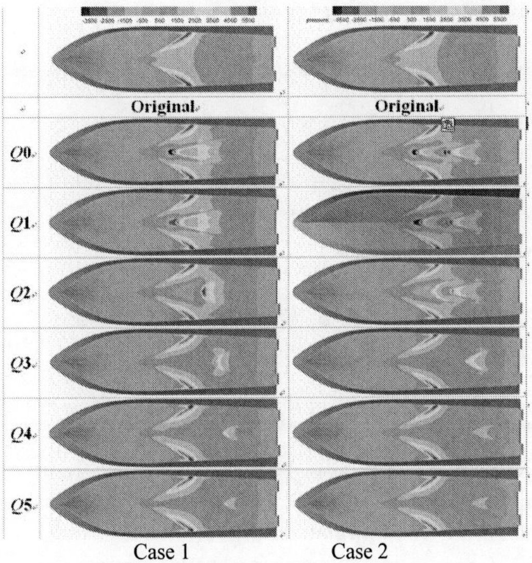

Fig. 4.10.9. Pressure distribution of Case 1 and Case 2

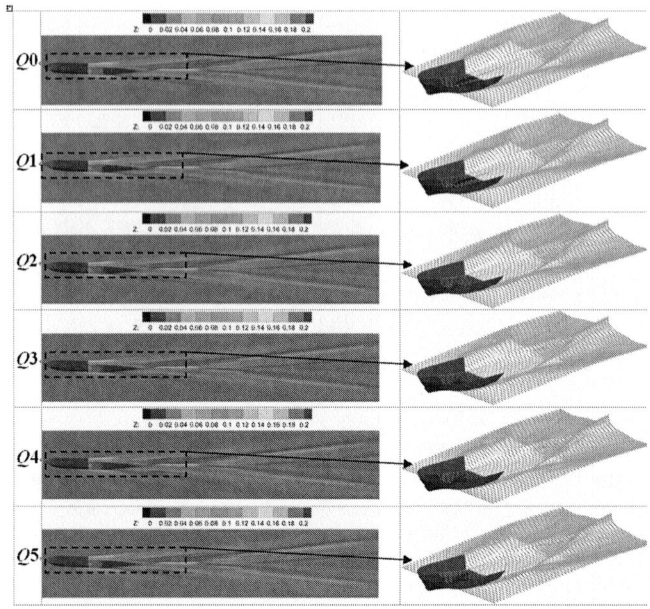

Fig. 4.10.10. Free surface wave of Case 2

4.10.5. Conclusion

The numerical simulation method is developed to solve the resistance of the high-speed air cavity craft to adopting a structured grid. The total resistances of high-speed air cavity craft at different speeds are numerically simulated by solving RANS equations with $SST\ k-\omega$ turbulence models. The VOF method is used to deal with the free surface and the gas–liquid flow field of the craft bottom. Then the effect of the artificial cavity shape and air-jet velocity on resistance are analyzed. This paper's main conclusions are as follows:

1. The numerical method for planing craft resistance has relatively high accuracy with model experimental validation.
2. The VOF method is adopted to simulate the gas–liquid interface of the air cavity. The numerical simulation methods are developed to analyze the drag reduction effect.
3. At the same flow velocity, the drag reduction effect increases with the gas flow rate. Meanwhile, there is an optimal gas flow rate. And if one continues to increase the jet volume, the drag reduction

effect is worse and the gas along the cavity will escape. So the gas flow rate should not be too high.

4. The free surface wave shape is changed little under other gas flow rates. This shows that injecting is little affected on the free surface wave (basically negligible).
5. In two cases, the gas flow downwards from the injector to the rear after injecting, when encountering water, led to a reversed flow around the injector. After that, the gas flows downward to the stern along the cavity sides.
6. On high-speed crafts, especially high-speed planing crafts, the static pressure is very high because the trim angle is great, and the gas is difficult to "launch" water. A triangle area is covered by water, and it limits the effect of drag reduction. So, it is estimated that running attitude is of great influence on the drag reduction effect.

References

[4-10-1] Cau Hongjian, *The Computation and Research on Resistance of Planing Craft Based on the Software Fluent*, Harbin: Harbin Engineering University, 2008 (in Chinese).

[4-10-2] Wang Liyan, *Research on a New Method of Air Lubrication Drag Reduction*, Wuxi: China Ship Scientific Research Center, 2012 (in Chinese).

[4-10-3] Ni Qijun & Zhong Zhende, "High Performance Ship Development and Application," 2nd Yangtze River Delta Industrial Development Forum, 2006 (in Chinese).

[4-10-4] Wendy Sanders, Jinhyun Cho, Eric Winkel, Elizabeth Ivy, & Robert Etter, "Bubble Drag Reduction at Large Scales and High Reynolds Numbers," 25th Symposium on Naval Hydrodynamics. 2004.

[4-10-5] He Shulong, Xing Shengde, Cheng Hongxia, & Ni Qijun, "Principle Study of Bubble Drag Reduction and Similarity Analysis for High Speed Vehicles," *Journal of Ship Mechanics* 14 (9): pp. 951–57, 2010 (in Chinese).

[4-10-6] He Shulong, Xing Shengde, Cheng Hongxia, & Ni Qijun "Numerical Simulation for 2 Dimensional Flow of Manmade Bubble," Proceedings of the 9th National Congress on Hydrodynamics and 22nd National Conference on Hydrodynamics, 2009 (in Chinese).

[4-10-7] He Shulong, Xing Shengde, & Cheng Hongxia, "Research on Artificial Bubble Regular of High Speed Air Cavity Craft," International Conference on High Performance Marine Vessels 2010

(HPMV CHINA 2010), 2010: C12.
[4-10-8] Yang Lei, "Experimental Reports of the High-Speed Craft," China Ship Scientific Research Center, scientific report, 2007 (in Chinese).

CHAPTER 5

WING IN GROUND EFFECT CRAFT (WIG)

5.1

Introduction to Performance, Technology, and Application of Wing in Ground Effect Craft (WIG)

Liang Yun

Marine Design and Research Institute of China

5.1.1. Basic principles of ground effect craft

What is a wing in ground effect craft? It is a craft running close to the ground or water screen while getting the ground effect. This effect is caused by the following physical phenomena, as shown in Fig. 5.1.1:

- Owing to the flow blockage between the underside of the wing and the ground, the pressure on the lower surface of the wing increases, so as to increase the lift. A comparison between a wing section operating in ground effect with the same section in free air is shown in Fig. 5.1.1.
- In such cases, the downwash velocities caused by wing tip vortices will be reduced, so as to reduce the induced drag caused by the induced velocity of the wing tip vortex.
- Therefore, the WIG gains inherited heaving stability; that is, the WIG is able to restore its original position, in case of the craft moving up or down vertically due to external disturbance.

Fig. 5.1.1 illustrates such phenomena. The upper figure shows the wing-up vortex without ground effect, while the lower figure shows that with ground effect.

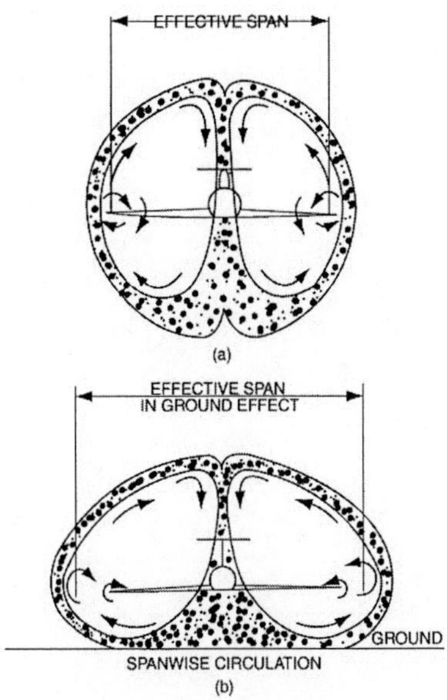

Fig. 5.1.1. Ground effect reduces the downwash and the lift induced drag

Fig. 5.1.2 shows the variation of the lift–drag ratio versus the aspect ratio of a wing, and the height of the wing tip above the sea or the land—that is,

$L/D = f(AR, h/c)$

where, L denotes lift; D, drag; h, flying height; c, chord length of the wing; and AR, aspect ratio of the wing. From the figure, one can see that the WIG is able to move up, when the craft is dropping down above the sea or ground, and vice versa, that means the WIG owns the inherited static heaving stability.

According to the attribution mentioned above, WIGs possess the following basic advantages:

- High speed
- Ability to operate on ground, water surface, ice, sand beach, swamp, and other surfaces where conventional boats and cars are unable to operate
- Amphibiousness
- Low fuel consumption, and high endurance
- Fine seaworthiness, in the case of crafts flying at high attitude

However, WIGs also have some disadvantages, such as:

- Difficulty in taking-off on water surfaces, particularly in rough seas
- Poor sea-keeping quality when operating in the zone of the low surface effect and in rough seas, due to the interference of waves

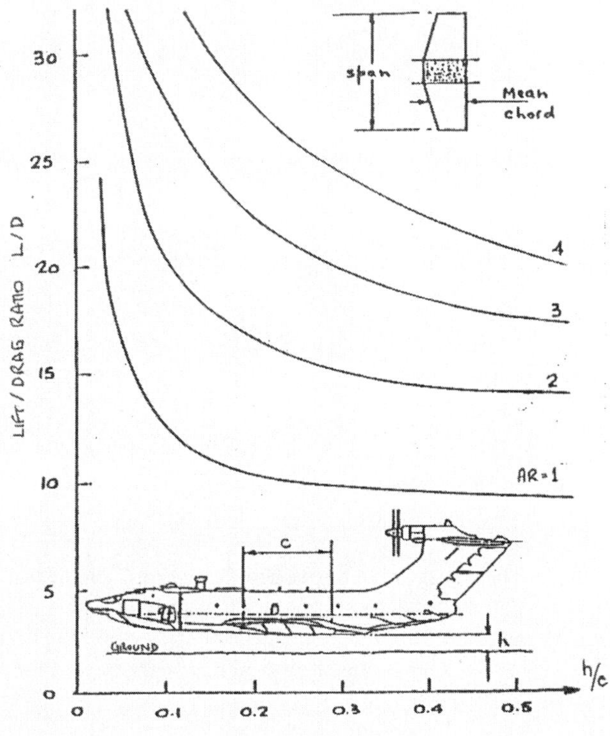

Fig. 5.1.2. The variation of the lift–drag ratio versus the aspect ratio of a wing

5.1.2. Some WIG technical terms

Since WIG is a special craft, a combination of ship, airplane, air cushion craft, and vehicle, with aero-hydrodynamics and air cushion dynamics applied, some technical terms have to be explained:

- Ground effect: The enhanced lift force acting on the wing running close to the ground or water surface; the height generally less than one chord of the wing. We call determine the range $0 < h/c < 1$ to be the ground effect zone, GEZ; $0.5 < h/c < 1$ is weak GEZ, and $0 < h/c < 0.1–0.15$ is strong GEZ. The enhanced lift is generated by the greater increase in pressure on the undersurface of the wing due to higher deceleration of the air trapped between the wing surface and the ground.
- Dynamic air cushion: This represents a high pressure region between the airfoil and the ground (or the water surface) in GEZ; it can be created by the following approaches: (1) air blown by the air propellers, or the air jets goes into the space under the wing at a much higher speed than the craft's forward speed, so as to enhance the lift; (2) the geometry of the craft hull can be properly shaped so that air entering in an opening can be retained, except for the release of air through a gap under the craft.
- Static air cushion: Just as in the ACV, the air cushion means a high pressure region between the airfoil and water surface, which is originated from the air jets and stays still or moves slowly before take-off.

5.1.3. Types of WIG and classification

There are many types of wing in ground effect craft with various air configurations developed over several decades; however, the generic designation covering all such vehicles is WIG. Other names used are the Ekranoplan (a Russian name that means a lifting surface close to a solid screen), flaircraft, wingship, hoverwing, ground effect machine (GEM), wing in surface effect ship (WISES), power-augmented-ram wing in ground effect craft (PARWIG), dynamic air cushion craft (DACC), and dynamic air cushion wing in ground effect craft (DACWIG). According to the air-hydrodynamical characteristics and configurations, the main types of WIG can be outlined as classic WIG, PARWIG, DACC, DACWIG, and so on.

- A classic WIG is characterized by a reverse delta main wing with an anhedral configuration, tapered chord, and forward sweep, and by winglets with up dihedral angles for keeping good longitudinal and transverse stabilities. The propulsion system can be mounted above the hull. Perhaps this is the most simple, efficient air configuration; it was developed by Alexander Lippisch in Germany during the 1960s and 1970s. Some typical crafts of this type are shown in the following figures.

Fig. 5.1.3(a). Lippisch X-113 (lateral view)

Fig. 5.1.3 (b) Lippisch X114 (front view);

- A PARWIG is characterized by a pair of ducted air propellers (turbofans) mounted in front of the leading edge of the main wing to provide pressurized air for creating a dynamic air cushion under the main wing and improving the take-off performance. The sea-keeping quality below the take-off speed is also improved by creating a high lift force at a low speed, and the slamming of the hull in waves is reduced, as shown in Fig. 5.1.4. The PARWIG can fly in GEZ, and fly even out of GEZ for a short time but without amphibious ability, as in the classic WIG.

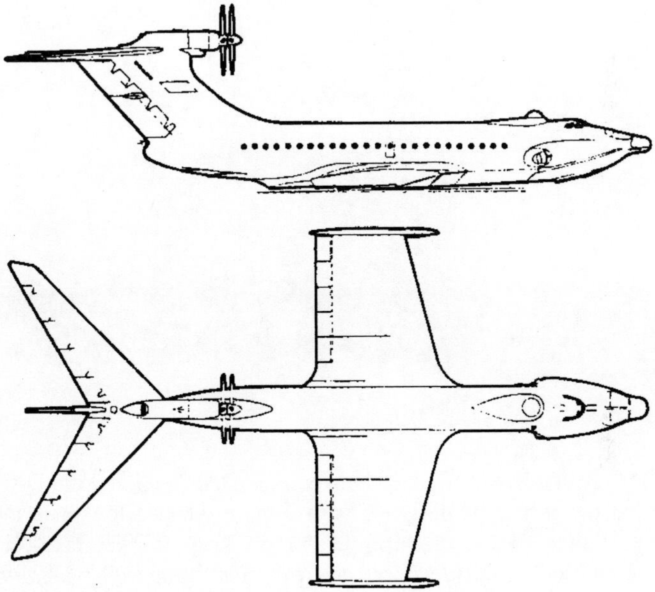

Fig. 5.1.4 General Profile of a PARWIG, *Orlyonok*, Russia

- The DACC (dynamic air cushion craft) is characterized by its main wing with low aspect ratio, deepened side buoys on both sides of the wing, and flaps at the main wing tailing edge, the configuration of which is similar to the ram air cushion supporting craft, with the added ability to take off and turn into surface effect flight. Forward-mounted ducted air propellers blow pressured air into the cushion, which enables the craft to hover on water or/and ground surface statically. So the air configuration and forward-mounted propellers of the craft ensure it has both static and dynamic air cushions, and has the amphibious ability to operate on the water surface, ground, ice, sand beach, and swamp, and so on. However, the aspect ratio of the main wing is so low that the craft cannot operate at high speed with a high altitude, and has poor sea-keeping quality. Fig. 5.1.5 shows the Russian DACC type *Volga-2*.

Fig. 5.1.5. Russian DACC type *Volga-2*

- DACWIG. The difference between the DACWIG and the DACC is that the former is designed with a main wing, a low aspect ratio, and forwarded ducted propellers, to keep a nice take-off and amphibious ability, as well as with side composed wings with a higher aspect ratio to enable the craft to fly at both strong and weak SEZ. Thus, the dynamic air cushion wing in ground effect craft, DACWIG, is a hybrid type of craft developed from DACC and PARWIG, as shown in the figures of Section 5.6.

According to the operational characteristics of various WIG, and the requirements of the International Maritime Organization (IMO), WIG can be classified as follows:

- "A" Craft, incapable of operation without ground effect: includes DACCs and some DACWIGs

- "B" Craft, capable of increasing its altitude outside the influence of GE: includes Classic WIGs, PARWIGs, and DACWIGs; the maximum of such "over flight" should be less than the minimum safe altitude of an aircraft prescribed by the International Civil Aviation Organization, ICAO
- "C" Craft, capable of taking off from the ground or water surface and to cruise at an altitude that exceeds the minimum safe altitude of an aircraft prescribed by ICAO.

5.1.4. Development of WIG

In the mid-1930s, Kaario in Finland started to build and test a craft that operated in a very strong ground effect zone—that is, at a very low altitude, but with poor longitudinal stability, which had been solved by Dr Alexander Lippisch in Germany. At beginning of the 1960s, Dr Lippisch developed the X series of WIGs, using the reverse delta main wing with an anhedral, tapered chord and a forward sweep configuration, and out winglets with an up dihedral angle. This solved both the longitudinal and transverse stability problems perfectly and enabled the WIG not only to fly in both strong and weak GEZ but also to fly out of GEZ for a short time. Such types of WIG have been developed worldwide, as a classic WIG (Fig. 5.1.3) in Germany, in the United States as X series crafts, in China as XTW series WIGs, and in Australia.

At the beginning of 1960s, Dr R. Y. Alexeev, began to develop the Ekranoplan in the Former USSR, which would lift the craft completely from the water surface and glide just above it, so as to overcome the "wave-making barrier" for conventional ships and the "cavitation barrier" for hydrofoil crafts at high speed. He developed a series of PARWIGs (Fig. 5.1.4), especially for military application, as landing crafts and guided missile attack craft, owing to their fine aerodynamic properties, stability, and maneuverability, and to their ability to operate at various altitudes, even flying out of GEZ for a short time. However, such crafts have some disadvantages, particularly their high cost of building and maintenance, which makes them unsuitable to be developed for civilian applications. In the late 1970s, Alexeev began to develop the DACC type (Fig 5.1.5), with low speed requirements (100–130km/s) and seaworthiness for civil application, which could operate in river mouths and inland rivers, on ice and swamps, and so on.

In the 1990s, researchers in both Russia and China, merging the advantages of both PARWIGs and DACCs, begin to develop the a WIG type of DACWIG, which was able to operate at high speed, had fine sea-keeping quality and maneuverability, and could fly on seas and rivers and on swamps, earth, ice, and sand, with amphibiousness and landing ability.

This chapter presents several papers concerned with the performance, technology, and application of WIGs, all of which were presented at the International Annual Shanghai HPMV Conference over the past dozen years.

1. First, we would like to introduce the research, design, performance, and application of the PARWIG in sections 5.2 and 5.3.
2. Then we will introduce the DACWIG in sections 5.4 and 5.5.
3. A new concept of WIG is presented in sections 5.6 and 5.7.
4. Last, sections 5.8–10 introduce various theoretical hydrodynamic calculations for new airfoil (S profile) and the longitudinal stability of WIG, as well as a study of the hydrodynamic parameters of WIG with an automatic control system for improving longitudinal stability.

5.2

A WIG Craft as a Taiwan Straits Passenger Liner (2010)

Shi YaJun, Zhang Wei and Ye XiuNai

China Ship Scientific Research Center, Jiangsu Wuxi, China

Abstract

This article analyzes the characteristics and the advantages of WIG craft and the hydro-meteorological situations around the sea routes of the Taiwan Straits. It is demonstrated that a WIG craft is feasible for straits passenger services. A proposal for a 50t ferry craft for straits passenger is provided, while the economy of the project is analyzed.

5.2.1 Introduction

At 8:00 on April 22, 2009, the ro-ro ship *Taihua* carrying 909 passengers starts the first direct cross-straits transport of passengers by sea. Although the conventional cross-straits passenger ship has the advantage of being low cost, the shortcomings are a long sailing time and poor comfort, which will restrict the development of cross-straits transport of passengers by sea. Thus, it is necessary to develop a fast, comfortable ship to carry passengers across the Taiwan Straits.

A WIG craft has a speed as high as an aircraft and very good comfort. In the process of cruising, the craft can fly over the surface of the water, and passengers would not suffer from wave disturbance.

We try to introduce a proposal for WIG craft from the aspects of technical feasibility, convenience of use and management, security, economy, and so on, which will be helpful to straits transportation.

5.2.2. Characteristics and advantages of WIG craft

1. Principles and characteristics
A WIG is a multimodal craft that, in its main operational mode, flies above the water or some other surface by using surface effect, without constant contact with such a surface, and is supported in the air mainly by aerodynamic lift generated by a wing (wings), hull, or their parts, which are intended to utilize the surface effect.

Compared with conventional ships, WIG speed (100–300kn) is much faster, and they are not affected by waves when cruising, and even have the capability to move vertically.

WIG craft have a lot of the same characteristics as helicopters, such as high speed, long range, good safety, and security. But WIG can also take off and land on water safely and can stay on the surface of the sea for a long time.

Compared with seaplanes, WIG craft can fly over the sea at very low altitudes, with remarkable endurance, security, and convenience.

2. Advantages in the application to the straits passenger ferry
(1) High speed
We know that the speed of conventional ships is 20–35km/h and the speed of hovercraft and hydrofoil boats are up to 50–80km/h. However, WIG craft can fly completely over the ground or the water, and resistance is reduced greatly; moreover, their speed usually ranges between 120–600km/h and is thus much faster than conventional vessels. Such a high speed would make the straits passenger service more convenient, and it would thus be possible to build the "West of Taiwan Straits" one-hour economic circle.

(2) High safety
WIG craft can fly around the ground-effect area, which means flying close to the surface safely. When a WIG craft is flying at an altitude of 1–5m, in case of emergency situations or troubles preventing flying, the craft can descend safely to land on the surface, or slide at a low speed, or park to wait for rescue; the disastrous consequence of a crash do not occur.

(3) Good economy
WIG craft fly over the water and their operation and parking don't rely so much on airports; thus, the requirements for station equipment are relatively low. In addition, they are easy to drive and maintain. With low fuel consumption, small levels of pollution, and highly effective load

levels, their achievements are large. Nowadays, the global shortage of energy is becoming more and more serious; however, WIG crafts can save a lot of operating costs.

(4) Comfort
During a WIG craft's cruising, there is no high noise emitted by either the craft nor the swing and bounce caused by a boat rocking and bumping, and the noise level in the cabin is equal to that in an airplane, so it is very comfortable for passengers.

(5) Convenience of management and security
According to IMO regulations, if a WIG craft works under the height of 150m, it classes as a ship. Without an open case of low altitude in the country, the use of small airplanes needs approval in advance. Both the domestic design and the use of WIG craft are taken over by the Ship Inspection Bureau and Maritime Safety Administration. Therefore, they can be sailed without being reported to air traffic control.

(6) Good time efficiency
Although the speed of a WIG craft is lower than that of a jet aircraft, taking the total time into consideration the WIG craft's time efficiency will be better than civil aviation. The reasons are as follows: passenger planes need to make many preparations before take-off, such as waiting for a runway, whereas WIG craft can take off on a wide water surface; carrying commuter passengers on a WIG craft between port cities will shorten the time of the commute; moreover, the distance from the port to the urban area in general is far less than the distance from the airport to the urban area, and thereby it makes the total travel time much shorter. These factors give the WIG craft significant advantages on a shorter route.

5.2.3. A brief introduction to the Taiwan Straits route

According to the book *Northwest Pacific Ocean Wave Statistics Collection*, the Taiwan Straits is divided into two blocks: E5 and S1. The annual wave statistics of the two blocks can be seen in Table 5.2.1.

Table 5.2.1. Wave scale of the Taiwan Straits

Sea state	E5	S1	Average
1		1%	0.5%
2	10%	10.4%	10.2%
3	34.3%	38.5%	36.4%
4	70.7%	78.1%	74.4%
5	91.4%	95.2%	93.3%

From Table 5.2.1 we can see that the seaworthiness of the straits ferry should meet the requirement of sea state 4. According to the current level of technology, a 50t WIG craft can meet these operational requirements. The distance between Xiamen and west Taiwan is 160 miles, and the sailing duration is less than 1 hour. For example, if the traveling time is 1 hour, the preparation time is half an hour, and the working time is from 8:00 am to 5:00 pm, then we can figure out that a single boat could travel three round-voyages every day. According to this requirement, the speed should not be less than 300km/h.

5.2.4. A WIG project for a 50t straits ferry

In this paper, the straits passenger ferry WIG craft project concerns a composite wing layout and two turbofan DV-2C engines in the fore. And an R&R MK21 turboprop engine is high-placed in the rear as the cruise propulsion. Its hull structure is built with aluminum alloy and composite materials. The main technical parameters of the project are as follows:

Length overall	36.8m
Width overall	28m
Height overall	10.9m
Normal gross weight at take-off	45,000kg
Max weight at take off	50,000kg
Number of passengers	100
Fuel	10t
Cruise speed	360km/h
Max speed	420km/h
Cruise altitude	2–5m
Max altitude	≥50m
Sea-keeping	Sea State 4
Flight sea-keeping	Sea State 6
Max. range	1,500 miles

Fig 5.2.1. The profile of the 50t straits ferry

5.2.5. Economic analysis of operation

1. During the daytime, the craft can make three round trips every day. The proposal shows that the cruising speed is 360km/h, the maximum flight speed is 420 km/h, and the distance from Kaohsiung port to Xiamen port is 180 nautical miles; thus, the flying time is about 55 minutes. Assuming that it would take 5 minutes to take off and land, respectively, for the WIG, then the time for the single trip is 1 hour and 5 minutes.

The proposal shows that the maximum range is 1,500 miles, and it can travel four round trips from Xiamen to Kaohsiung, as shown in Table 5.2.2; refueling is only needed once. Therefore, the time for landing is much shorter than before. Supposing that the time for taking off is 30 minutes, then the WIG can travel three round trips between 8:00 am and 5:00 pm. The time for taking off and landing in each voyage can be seen in Table 5.2.2.

Table 5.2.2. The WIG craft's schedule from Xiamen to Kaohsiung

Xiamen		Kaohsiung	
To	From	To	From
	8:00	9:05	
10:40			9:35
	11:10	12:15	
13:50			12:45
	14:20	15:25	
17:00			15:55

2. The cost of small-scale production is 80 million yuan per ship, supposing that the annual interest rate of the bank loans is 8% with a 10-year loan period.

3. The supposed life cycle of the hull is 15 years, and the engine's TBO is 1,600h.

4. It is supposed that the management expenses (taxes, insurance, port charges, waterway management fees, etc.) account for 35% of the total income.

5. Oil and fuel expenses
Under sea state 4 conditions, the average annual expense is 74.4%. Under these conditions, the sailing rate should be under 74.4%. When WIG craft take off and land in a sheltered harbor, the sailing rate should be at 70%.

WIG craft will consume about 1.2 tons of fuel in a single voyage. As the kerosene price is 3,900 RMB/ton, it will cost 4,680 yuan for a single voyage, and the annual expenses of the fuel is 52,000 yuan.

6. Maintenance funds: 15% of the cost per year (engine overhaul costs are included).

7. Wages and benefits
It is assumed there will be 25 people working for the WIG craft operation and that each person will be paid 20,000 yuan per month, with 5% increments every year.

8. Tickets
The WIG craft will carry 107,000 passengers per year. If the number of commuters is three per day, the ratio of carrying passengers is 70%, the ratio of the voyage is 70%, and the average price of the ticket is 750 yuan per person, which is in accordance with 2% increments each year. If so, the first year would generate 15.5 million yuan, and a total of about 354.66 million yuan would be generated in 15 years.

5.3

DEVELOPMENT, OPERATION, AND EXPERIENCE ON THE PRACTICAL WIG CRAFT *TIANYI-1* (2003)

WANG EN GUANG, CHU LIN TANG AND SONG MING DE

Aviation Industry Corporation of China, Special Vehicle Research Institute

Abstract

This paper details the following aspects of China's first practical small-size WIG-effect Aircraft *Tianyi-1*, which was developed by the Aviation Industry Corporation of China Special Vehicle Research Institute cooperating with the China Academy of Science and Technology Development and other units, and has been employed in sightseeing, tourism, cargo, and passenger transportation. The paper further contains a brief description of its development, its overall configuration features, the situation after it was put into commercial service, and experiences and lessons extracted from its service, and so on. The paper also briefly describes the problems encountered in operating the aircraft *Tianyi-1* and the technical improvements achieved. In addition, the Aviation Industry Corporation of China Special Vehicle Research Institute, as a unit which has successfully developed and utilized WIG-effect craft in China, presents its viewpoint and position concerning the development and application prospect of WIG-effect aircraft.

Finally, the Aviation Industry Corporation of China Special Vehicle Research Institute is described in terms of its superiority and actual strength in scientific research, and how its products in special flying vehicles (seaplanes) filled the gap in China since it was founded in 1962.

5.3.1. Development of *Tianyi-1* (DXF-100)

Tianyi-1 (DXF-100), mainly designed by the Aviation Industry Corporation of China Special Vehicle Research Institute in cooperation with the China Academy of Science & Technology Development, and others, is the first small practical WIG-effect aircraft used for passenger/cargo transportation, tourism, and sightseeing. The development of *Tianyi-1* was started in May 1995. It was manufactured by Jingmen Hongtu Manufactory of the Aviation Industry of China. In September 1998, *Tianyi-1* made its first flight successfully in a seaplane trial-flight base in Jingmen City, Hubei province. Upon the approval of the China Ship Classification Society, *Tianyi-1* has obtained a permit for operation on water issued by the Chinese Ministry of Communication, and a training permit and piloting license issued by Zhejiang Bureau of Maritime Affairs for pilot and crew in 1999. It was put into business operation from the flying base in Huzhou in the same year.

Tianyi-1 has the overall configuration of an airplane, including the main parts of fuselage, wing, tail wing, and end plate, and so on. The whole aircraft body is made of aluminum alloy, with various technical treatments such as anti-corrosion against seawater, salt, and fog. In order to improve the T-O hover lift and the performance of T-O running on water, reduce T-O speed and running distance on water, and decrease hydrodynamic impacting loading during T-O and landing on water, a power-augmented lift system with high effect for supporting 85% of the aircraft's weight was adopted on the craft.

The main technique parameters and performance indexes of *Tianyi-1* WIG aircraft are as follows:

Overall length	16m
Overall width	11m
Overall altitude	4.9m
Number of passengers	15 (It can hold 13 persons after being equipped with air conditioner and internal decoration)
Full load range	200km (max. fuel range: 360km)
Max. take-off weight	4800kg
Max. speed	180 km/h (cruising speed: 145km/h)
Cruising flight altitude	0.6–1.2m
Take-off/landing anti-wave alt.	0.6–0.9m
Cruising flight anti-wave alt.	1.2m
Power plant	Equipped with two IO-540 aviation engines made by Lycoming Company, USA, each rated at 300 hp.

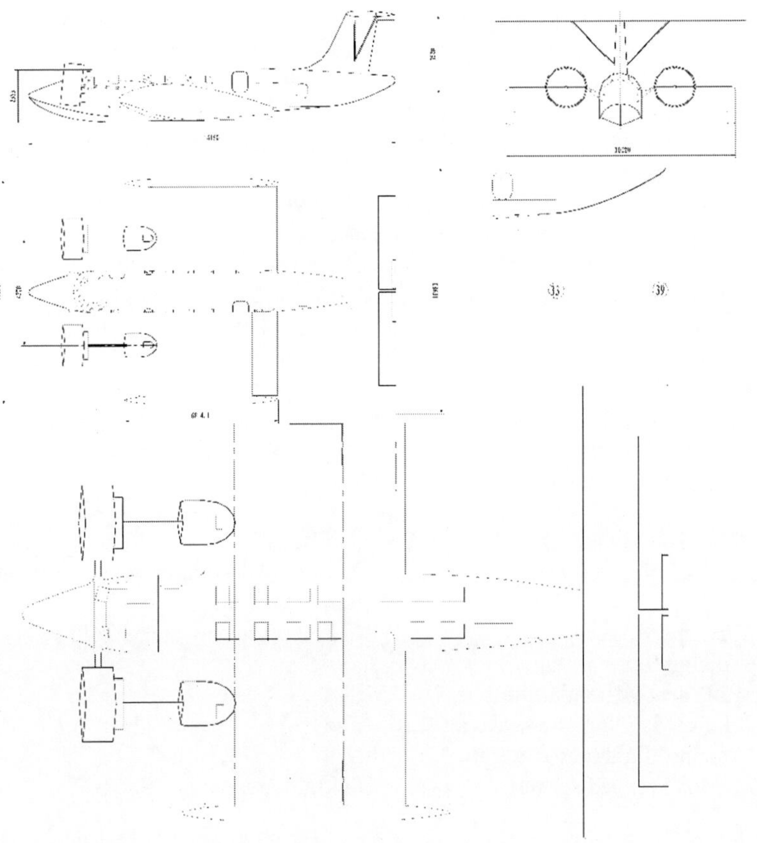

Fig. 5.3.1. Three-view drawing and general arrangement of WIG "Skywing-1" (DXF100)

5.3.2. Operational situation of *Tianyi-1*

The *Tianyi-1* WIG aircraft is the first practical WIG-effect aircraft in China to have obtained the certificate of the China Ship Classification Society. Two *Tianyi-1* WIGs were made as a prototype for flight tests. It has flown 1,700 sorties and carried 18,000 passengers in the last four years since it was put into business operation in Taihu base in June 1999. During business operation, it has received a large number of visiting groups from state government leaders and officials, and it has been investigated by groups of international experts who have been to China. It has aroused general attention and won high praise from personages of various circles.

The successful business operation of the *Tianyi-1* WIG has generated good social effects and economic benefits. Undoubtedly, this is a significant trial that has ended the long-term backwardness in high-speed water transportation. This kind of trial will actively influence the development of WIG craft.

Business operation over the past four years has proved that *Tianyi-1* has good flight ability on water surface and air cruising flight stability, with good handling quality, high anti-wave and wind capability at take-off and landing, and good safety, reliability, and flight comfort. A flight accident has never happened in service. The design principle and technology performance index of the prototype airplane have been verified and checked in its practical operation. Except for the max indexes, other performance indexes meet the design requirements, and the T-O anti-wave and steady flight altitude even exceeds the design requirement.

Main leading particulars are as follows:

M (weight at take-off)	4,800kg
Number of passengers	12, equipped with air-conditioner and internal decoration
Max. speed	165km/h (cruising speed 140km/h)
Cruising flight altitude	0.6–1.2m
Max. steady flight altitude	1.8m
Take-off/landing anti-wave alt.	1.0m
Cruising flight anti-wave alt.	1.2m
Full load range (normal)	200km, max. range 360km

The power-to-weight ratio of *Tianyi-1* is 0.125hp/kg, which is smaller than other WIGs with the same tonnage. However, various performance indexes for *Tianyi-1* still come up to the above-mentioned standard. It is shown that the design of *Tianyi-1* is successful.

5.3.3. Development and improvement of *Tianyi-1* in service

Any new products are not perfect in the service, and the new type WIG-effect aircraft is no exception; thus, only two WIG-effect aircraft of the *Tianyi-1*-type were manufactured as prototypes.

There were some problems with the design and manufacture of *Tianyi-1* in service. The problems in its design concerned maintenance in service and the system reliability; the problems of its manufacture concerned the quality of the aircraft body and technology. In view of this situation, our technicians in the design and manufacture departments designed a revision program, and modified and tested on board. They have solved some difficult technical problems to improve maintainability in service and system operation reliability to ensure the safety of the business operation and the flight rate. Technicians and crews have solved some technical problems that affected normal operation of *Tianyi-1* in service over the past four years. Some of the measures taken are as follows:

Causes of starter motors damage were analyzed and found; the line system was replaced and strictly checked; the start switch was repaired, and a breakdown of the starter motors was fixed.

The problem of positive yaw existed in flight, when *Tianyi-1* No. 01 was put into operation in the early stage. The technicians made a survey of the causes, and improved the implementation of the program, thus solving the problem.

To eliminate frequent leakage on the end plates of both sides of the wings, the technicians analyzed the trouble on the spot and changed the structure of the end plates at the hull bottom and the forms of sealing. Through ground testing and trial flights, these measures have proved successful.

The support air conditioner driving system was modified to improve the supporting rigidity, and the accessories for the auto air conditioner was equipped to solve problems in the effect of the air conditioning, the belt looseness in the driving system, and the problem that the air conditioner could not normally work for long periods.

At the beginning of *Tianyi-1*'s service, the temperature of the engine lubricant was high, affecting the normal operation of the WIG-effect vehicle. The radiator fairing of the lubricant was modified to improve the ventilation condition of the lubricant radiator, and the high temperature of the lubricant was reduced.

Another problem related to fire prevention of the fuel tank. There was a danger of fire affecting the fuel tanks installed on both side wings. Furthermore, the smell of petrol spread easily to the compartment, affecting the life and working environment of the crews, because the insulation between the fuel tank and the other systems was not considered earlier in the structural design process. For these reasons, the fuel structure and the surrounding structure were modified and service maintainability of the fuel tank was improved.

High flickering frequency of the windshield wiper in front of the cockpit affected the view of the pilot and his comfort. It was finally modified on the vehicle, after much work had been done in research, analysis, and testing.

5.3.4. Experience of the *Tianyi-1*'s development and service

There is no experience of predecessors to be used for reference regarding WIG-effect vehicle development and service in China. The finished product must be repaired at its manufacturing base in case of faults, but in accordance with aviation specification that may not be appropriate. It would take few flights a month to train pilots under Taihu lake water conditions if we persisted in aviation standard and flight clearance conditions. We boldly determined flight clearance conditions and operation procedures, and prepared documents and manuals for the WIG-effect vehicle service and maintenance based on scientific analysis, safety considerations, and actual conditions, as well as paying attention to daily checks and the regular maintenance of the structure and system to ensure flight safety and operation rate. *Tianyi-1* has made 1,700 flights without serious accident for four years. This proves that the ways and methods of exploring are correct.

WIG craft for business flights is just starting in China and in a few other countries in the world. How can one correctly handle the relations between service and crew training? Training is regarded in the service through careful arrangement and good coordination "to assure service and also quick training of its own pilots and ground crews." Service experience for the development of WIG craft in China has accumulated, and large numbers of WIG craft crews have been trained.

It is unusual in the research history of vehicles in China that *Tianyi-1* was successfully developed over three years and conducted business flights. The successful accomplishment of this system engineering shows that the development period can be effectively shortened and the benefit of the product can be achieved as quickly as possible with lateral cooperation of tackling key problems in project operation. The successful practice of *Tianyi-1* in the above aspects provides a very convincing example.

At present, there is no published standard and specification for WIG-craft design in China or the world. The Aviation Industry Corporation of China Special Vehicle Research Institute started the research on a design specification appropriate to this model design to guide the whole model design. A design specification appropriate to the *Tianyi-1* WIG-effect vehicle has been prepared, and the systematic design criterion and technical requirements is in an advanced stage of development, with technical achievements and experience spanning over 30 years. WIG-craft technical features must be paid more regard and should be consulted concerning specifications for upgrades and the manufacture of high-speed ships, which are based on the full digestion of materials relating to WIG craft both domestic and foreign.

5.3.5. Development and application prospects for WIG crafts

Because of their unique operation mode and potential purposes, WIG craft have many of the advantages of both ships and aircraft: they can operate in displacement mode on a water surface like a ship, but can also take-off and land on water and fly in sky like an airplane. Meanwhile, WIG craft can fly steadily close to the water surface for a long time using ground effect like a seaplane. WIG craft have the safety and reliability of a ship and the high speed and comfort of an aircraft. Compared with ordinary aircraft, WIG craft can take-off and land on water without a well-equipped, expensive, and special airport. They are able to fly at extremely low altitudes. All countries in the world pay much attention to WIG craft due to these advantages and features. Many countries dedicate labor forces and financial resources to develop WIG crafts. China has conducted theoretical and engineering application research in this technical field for over 30 years and has made gratifying achievements. China is not far behind other countries, and is at the forefront among countries of the world in civil business application.

The successful business flight of *Tianyi-1* encourages us to use WIG crafts for high-speed water transportation and develop a WIG-craft industry. China has the vastness of its seas and oceans with 18,000km of coastline and numerous coastal ports and docks. Many economic and practical transportation platforms with high speed and safety are needed for passengers and cargo transportation, marine resources exploitation, and marine activities. WIG craft are undoubtedly an ideal transportation platform. On the other hand, there are many rivers, lakes, and reservoirs throughout the country. The situation is favorable to the development of small-scale WIG craft.

We believe that the application and development of WIG craft will bring change to water transportation, and improve low-speed water transportation in our country. WIG craft will surely form an industry to promote the rapid development of water transportation and will bring along change and development to related fields.

5.3.6. Development situation of special vehicles at the Aviation Industry Corporation of China Special Vehicle Research Institute

The Aviation Industry Corporation of China Special Vehicle Research Institute is a unique center for developing seaplanes and other special vehicles. Over 40 years since the foundation of the institute, it has developed the first-generation of large seaplane—SH5—the first large forest fire-fighting aircraft—SH5 B seaplane—the first ultra-light seaplane—A2C—the first large helium manned airship—FK4—the first practical wing in ground effect aircraft—DXF100, *Tianyi-1*, and so on. These craft fill the gap in aviation technology in our country. The high-speed aviation towing tank possessed by the Aviation Industry Corporation of China Special Vehicle Research Institute may be "the first in Asia" and in "the front row in the world" in its construction scale and the performance indexes of the main equipment.

In addition, some experts researching the WIG craft in this institute had been to Russia to participate in the demonstration of the overall technology program for large or medium civil WIG craft. They made full preparations for the development of WIG craft.

The background of experiences encompassing the development of seaplanes and WIG craft, the special technology team, the test conditions and the concerned design teams make the Aviation Industry Corporation of China Special Vehicle Research Institute a powerful technical force in the development of WIG-effect aircraft. It intends to cooperate extensively at home and abroad. The institute will make a contribution to the development and industrialization of WIG craft.

The types of special vehicles successfully developed by the Aviation Industry Corporation of China Special Vehicle Research Institute are:

1. *Tianyi-1* WIG-effect aircraft (see Fig. 5.3.2)
2. SH5 seaplane (see Fig. 5.3.3)
3. SH5 B forest fire-fighting aircraft (see Fig. 5.3.4)
4. A2C ultra-light seaplane (see Fig. 5.3.5)
5. High-speed aviation towing tank (see Fig. 5.3.6)

Fig. 5.3.2. DXF100 wing in ground effect aircraft

The main particulars:

Max. flight speed	180km/h
Cruising speed	145km/h
Stable flying height	≤1.5m
Cruising height	0.6–1.2m
Max. wave height, take-off/landing	0.6–0.9m
Max. wave height, in flight	1.2m
Max. range	360km
Max. take-off weight	4,800kg
Payload	1,150kg (12–15 persons)

Fig. 5.3.3

Fig. 5.3.4

The main particulars:

Length overall	38.9m
Wing span	36.0m
Height overall	9.8m
Max. take-off weight	45,000kg
Max. fuel capacity	16,500kg
Max. level speed	555km/h
Cruising speed	390km/h
Service attitude	7,000m
Max. service range	4,600km

Fig. 5.3.5. Seagull 200 (A2C) Ultra-light Amphibian

The main particulars:

Length overall	6.641m
Wing span	10.91m
Height overall	3.198m
Max. take-off weight	520kg
Max. payload	252kg
Max. level speed	110km/h
Cruising speed	90km/h
Flight range	200km

Fig. 5.3.6. High-speed aviation towing tank

The main particulars:

Tank length	510m
Tank depth	6.5m
Tank width	6.8m
Water depth	5.0m
Speed of trailer	0.1–22m/s
Capability of wave making machine	Wave length 2–12 m
Wave height	0.05–0.3 m

5.4

DESIGN FEATURES OF WIG CRAFT EL-7 *IVOLGA* AND TEST RESULTS (2006)

V. V. KOLGANOV & V. G. SERGEEV

Moscow, Russia

Abstract

In this paper, the features and flight-test results of an experimental WIG craft are presented. A WIG craft based on an aero-hydrodynamic design is known as "a composite wing" with a T-form twin tail and a turn-over ring propeller (TRP) disposed ahead. The centerplane is a small finite wing with a large central chord, and with an opposite angle of a crossing "V" and a front swept on the back edge with simple shields. On the sides of the centerplane are installed thc displaced floats with a plane bottom ski. The consoles are connected with the centerplane under positive angle "V" and are equipped with a drive for maneuvering and mooring WIG craft in the mode of navigation. The performance of contours without acting elements has ensured low aerodynamic resistance and high aerodynamic quality to WIGE craft at small heights of flying. A WIGE craft control system in flight is like an airplane system, with a rudder for flying height control, and a perpendicular rudder for direction control and ailerons. The control of air flowing on the bottom at movement is provided with a synchronous and independent turning of both TRP and the section profile of the centerplane. A catamaran scheme ensures stability, and the division of the floats into waterproof compartments ensures the flood-ability of WIGE craft.

Flight tests have shown that WIGE craft can be operated in hull-borne mode, with movement on a dynamic air cushion and in flight. Meanwhile, the WIGE craft has good stability of movement while in wing in ground effect at a flying altitude of 0.2–3 m, and has good control response, even up to more than 15m, and is simple in handling. The power of the main propulsion is sufficient for continuation of movement even with the failure of one engine. The low flying height provides safety during an emergency landing. When flying at a height of more than 15m, the WIGE craft has the characteristics of stability and control response, and is like an easy hydroplane. Therefore, when at a height outside the wing in ground effect zone, the WIGE craft can easily and safely return to the wing in ground effect zone, without the use of any additional devices or automatic control system.

V. V. Kolganov
—Technical Director of the Industrial Complex composed by Hong Kong Engain Science and Technology Ltd, Joint Stock Company "Research & Production Complex 'TREK,'" Russia, and Ukraine Feodosia Shipbuilding Company "MORYE"
—Chief designer of WIG craft,
—General director of JSC RPC "TREK"
—Academician of the Russian Academy of Natural Sciences

V. G. Sergeev
—Deputy Chief Designer of WIG craft,
—Deputy General Director of JSC RPC "TREK"
—Corresponding member of the Russian Academy of Natural Sciences

5.4.1. Introduction

The following three companies composed an industrial complex at the beginning of 2002 to develop HPMV, such as WIG, ACC, ACV, HYF, and so on:

- Hong Kong Engain Science and Technology Ltd
- Joint Stock Company "Research & Production Complex 'TREK,'" Russia
- Feodosia Shipbuilding Company "MORYE," Ukraine

As a unique general agency of the industrial complex mentioned above for the China and Southeast Asia District, Hong Kong Engain Science and Technology Ltd is responsible for all the technical inquiry, design, production, manufacture, sales, after-sales service, and so on of such products in these districts.

5.4.2. Aero-hydrodynamic features of configuration

WIG craft EL-7 *Ivolga* is a high-speed, amphibian, all-weather, highly economic and reliable vehicle with non-contact movement near water/ground surface and a flying height of up to 2–3m for universal purposes. WIG craft EL-7 *Ivolga* is intended for operation on inland water areas, including areas covered by snow and ice, coastal sea zones, flat plots of land, and waterlogged and flooded districts. WIG craft can be based on land and water with an independent exit without application of hydro-slopes on both coastal and onboard devices. The conditions of operation were determined by a choice of a range of speeds and modes of movement, ensuring the amphibiousness and simplicity of service. The basic characteristics of WIG craft are submitted below; a general view is shown in Fig. 5.4.1.

Table 5.4.1. Basic flying-technical characteristics of WIG craft EL-7

Weight/kg	3,700
Crew (passengers)	1 (10)
Commercial loading/kg	1,100
Engine type	BMW-S38
Quantity, power of engine/hp	2,286
Fuel capacity/kg	200
Range of flight/km	
at altitude 0.3 m	1,520
at altitude 0.8 m	1,210
Speed of flight/ km/h	
cruising	180
maximal	210
Draft at hull borne/m	0.45
Sea state	3
Draft in planing mode/m	up to 0.1
Surface of operation	water, snow, ice, ground, bog

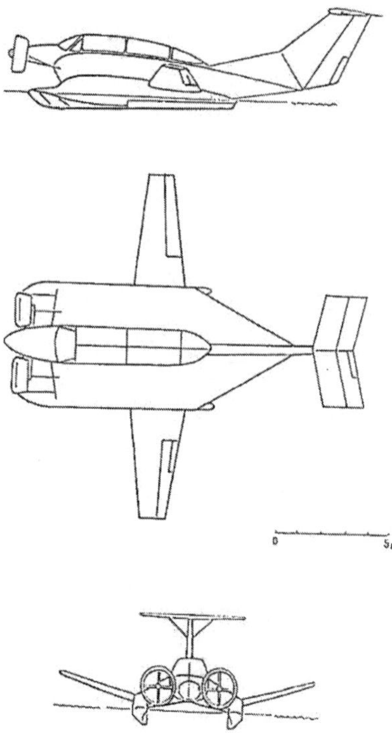

Fig. 5.4.1. General view of experimental WIG craft EL-7 *Ivolga*

The WIG craft is composed in the compound wing arrangement with a single-keel T-form tail-wing unit. The compound wing consists of a hull central plane (the wing with a large chord and small scope), to which two composite wings are connected, with a small chord and large span—that is, with a high aspect ratio. The comparatively large size of the centerplane has aerodynamically allowed not only leaving for theoretical contours, in which the engines and all units of power installation and equipment are placed, but also having the cabin of the driver within the passenger-cargo interior, which is closed from above by the streamlined lantern. This reduces the frontal air resistance of the WIG craft and has considerably raised the aerodynamic quality, especially at low heights.

On the basis of the WIG craft, the development is focused on the application of the concept of an "air cushion" formed under the centerplane in movement without contact with a surface. At a low speed of movement (planing mode) or take-off, the cushion is organized for blowing under the center-plane bottom, when air jets from the air screws located before the center-plane enter into the ring shrouds. At a cruising speed of movement, the wing in ground effect shown as the "dynamic air cushion" under the centerplane is supported by the approaching air flow near to the surface ("wing in ground effect"), and the craft goes up to a height of 2m, owing to which the lifting force is considerably increased and the resistance decreased. Therefore, the aerodynamic quality and the range of flight are essentially increased. The dependence of the technical flight range of WIG craft EL-7 on the height of flight due to the wing in ground effect is presented in Fig. 5.4.2.

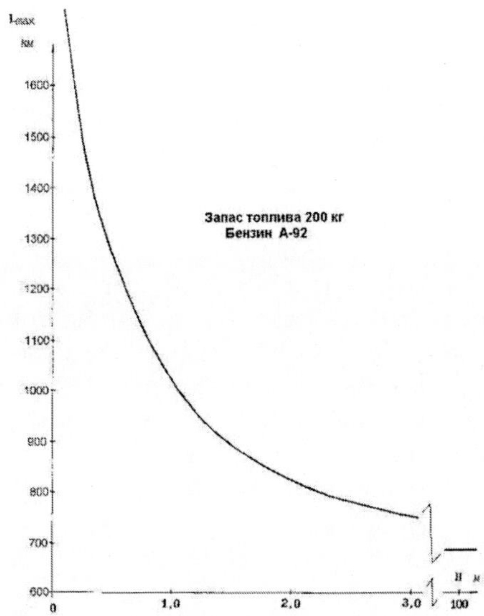

Fig. 5.4.2. Maximum flight range versus flight altitude

For the most full realization of an "air cushion," specially profiling ski-floats located by the sides of the centerplane are allowed, working as the aerodynamic boundary plates, while the deflected flaps are at the rear edge of the centerplane, and the steering flaps stick out under the hinge edge of the centerplane. In order to increase the efficiency of braking, the reverse-brake chamber is organized with lower and upper flaps in a tail part of the centerplane.

Low thrust is required at cruising due to the high aerodynamic quality of the wing in ground effect and the application of the blowing cushion at take-off. An essential measure excluded hydrodynamic resistance to the movement has ensured the reduction of the thrust-to-weight ratio. The blowing cushion ensures all-weather and all-surface non-contact operations in various conditions of spreading surfaces (water, ice, snow, ground) and exiting on water and flat coast with steering but without additional coastal and onboard devices.

The blowing of air jets under the centerplane and the ensuring of thrust at cruising is carried out with the unified engines of the WIG craft by the synchronous turn of air screws connected to the WIG craft engines with cardan shafts. The air propellers are fixed with ring shrouds, which increase the propeller thrust by up to 30%, considerably reduce the noise level, and ensure the safety of the environment at rotation.

The dimension and weight of WIG craft together with its high speed allow it to keep an average height of flight in rough seas during its movement with inertia; “rolling motion” is practically excluded. Because of the super low height of motion and the uniformity of temperature on a spreading surface (wing in ground effect) the vertical impulses of air are excluded, and air “turbulence” is absent in this case. During cruising the WIG’s movements are quiet and comfortable.

The catamaran arrangement of the WIG craft’s hull, formed by the centerplane and two ski-floats, with their organization of hermetic and watertight compartments, ensures high seaworthiness and flood-ability in modes of navigation and drift on water in heavy seas. A combination of the catamaran hull with high tail wing and consoles and the presence of blowing from screws under the centerplane can reduce the trim angle by 1.5–2 times in comparison with usual boats and catamarans with the same dimensions.

The increase in seaworthiness is also promoted by the wings, which fold vertically at a steep angle above the centerplane. The resulting aerodynamic force is directed downwards, because the catamaran scheme increases stability, especially in a strong lateral wind and heavy sea, in comparison with craft with horizontal wings fixed at the consoles. The simplified folding wings raise safety when mooring to another floating vehicle, essentially reducing the dimensions of the WIG craft, and expanding its opportunities for movement on narrow rivers, closed reservoirs, and water areas occupied by floating vehicles (Fig. 5.4.3).

Fig. 5.4.3. Profile of WIG with folded wings

The large sweepback on the rear edge of the centerplane is placed near to the wing consoles and is situated high on the horizontal tail wing, which are combined into a significant crossing V-form with moderate sweep of the keel. The selection of areas of rudders based both on their weight and their aerodynamic balancing, and the installation of aerodynamic trimmers and other layout features have ensured the stability and simplicity of control in all movement modes of WIG craft. WIG craft have lower requirements for mass preparations for the crew.

Two automobile engines with a capacity of more than 200kW and the turn-over ring propellers are installed on WIG craft EL-7, which, in the case of the failure of one engine, allows it to continue wing in ground effect flight or to changeover to movement in the mode of navigation (winter, in the aero-sledge mode) with one working engine. Therefore, the efficiency of rudders in flight provides safe compensation of moments arising from the dead engine.

It is necessary to note that the characteristics of stability and the control response of the WIG craft at a height higher than the one having an influence on the wing in ground effect will correspond to the norms of airworthiness of light aircraft. Here the point of view of the developers is reflected: a light WIG craft, flying above out of wing in ground effect and being under the influences and impulses of wind, heavy seas, and mistakes of control, should easily and safely come back to the wing in ground effect mode of movement without application of additional devices such as automatic devices for improvement of stability, and so on, due to the aerodynamic stability and control response incorporated in it.

5.4.3. Test results of the experimental WIG craft EL-7 *Ivolga*

The tests began in Moscow in September 1998 and since January 1999 have been carried out on Lake Baikal and the Angara River located in the city of Irkutsk, with material and financial support by the Verhne-Lenskiy River Shipping Company, which is ready to operate these WIG craft in East Siberia.

The first flight with the use of blowing was carried out on the Irkutsk Reservoir in winter conditions on February 16, 1999. The first flight in wing in ground effect mode and with cruise configuration (shields were removed, and the beam with screws located at the cruise rule position) was carried out at a speed of 80–110km/h on February 20, 1999.

On August 20, 1999, the craft demonstrated going from coast to water in blowing state, and flew with wing in ground effect in cruise mode, and carried on to the coast in blowing state (Fig. 5.4.2). Until December 1999 the flights were carried out in wing in ground effect mode, and the WIG craft was driven by V. V. Kolganov in flight.

Fig. 5.4.4

Since December 1999, the flying of WIG craft has been mastered by D. G. Schebliakov. On December 10, 1999, a flight in wing in ground effect mode was performed at a height up to 4m with maneuvering at the track (Fig. 5.4.5). On December 15, flight and maneuvering outside wing in ground effect (more than 15m) was demonstrated. On February 10, 2000, the flight on Baikal was executed and the craft went back.

Fig. 5.4.5

Fig. 5.4.6 shows that the WIG craft flew above snow and above water (Angara does not freeze in a range of 10–12km from Baikal), and above ice on Baikal in both wing in ground and aircraft modes. On October 14, 2000, the experimental WIG craft EL-7 with a weight of 3,700kg performed taking off, flying with wing in ground effect at heights up to 4m and landing on the water with the height of waves at more than 1m, corresponding to sea state 3.

Fig. 5.4.6

Test results of the experimental prototype were received, which confirmed that the working capacity of WIG craft EL-7 *Ivolga* incorporated scientific-technical and designing ideas. It is determined that the WIG craft has good longitudinal and lateral stability at all ranges of flight height, is simple to control, including at heights with a faint display of wing in ground effect (5–10 m); in “set free” control conditions, a stable mode of flight still continues. During tests, the control of the WIG craft and the flying technique are fulfilled for maneuvering the track, speed, and height at a non-contact movement state based on blowing and flight in the wing in ground effect mode. The “airplane” mode of flight was tested. Turning tests in the operative range of wing in ground effect were carried out at flight height up to 1.5m and with a roll up to 15° (Fig. 5.4.7); emergency turning tests were also carried out at a flight height of 3–5m and with a roll 30°–50°. The capacity of the power plant with the BMW S38 engines is sufficient for continuation of the wing in ground-effect flight even with the failure of one engine. When moving in the operative range of the wing in ground effect, the experimental sample of the WIG craft EL-7 *Ivolga* has a much higher aerodynamic quality ($K \geq 25$) in comparison with “airplane” modes (K=11-12) at the same flight speed, which causes substantial growth of the flight range at small heights at identical take-off weight and fuel capacity (Fig. 5.4.7). At cruising movement with a variable profile of flight and at maneuvering speed, track and height, the average fuel consumption was 25–30 liters of petrol Ai-95 for a route of 100 km with a take-off weight of 3,700 kg and a speed of 150–180 km/hr; nevertheless, it was 75–90 liters of petrol Ai-95 for a route of 100 km with movement in the "airplane" mode.

The tests have confirmed the safe properties of the WIG craft in the environment: the ground and grass are practically unbroken at base, and wave formation is absent during movement; in addition, noise and toxicity are comparable with an automobile.

There are the remarks of test-pilots and passengers: absence of rolling and turbulence, low noise in cabins and in the environment, good survey, and comfortable and pleasant flight.

Now the research and production complex "TREK" together with Verhne-Lenskiy River Shipping Company is conducting works on certification and preparation for the all-year-round operation of WIG craft EL-7 on Lake Baikal and the River Lena.

Fig. 5.4.7(a). WIG flying and maneuvering

Fig. 5.4.7(b). WIG maneuvering

5.5

DEVELOPMENT OF AIR CUSHION TECHNOLOGY—FROM STATIC AIR CUSHION (SAC) TO DYNAMIC AIR CUSHION WING IN GROUND EFFECT TECHNOLOGY (DACWIG) (2003)

LIANG YUN

Consultant of Ship Design Academic Committee, CSNAME Technical Director of Flying Dragon (& Engain) Science & Technology Ltd (Hong Kong)

Abstract

Developments in air cushion technology over the past four decades are described briefly in this paper. Owing to their fine amphibious ability and speed performance, ACV craft have been developed greatly in both civil and military applications; for example, both the United States and Russia have established the largest air cushion landing fleet in the World. Nevertheless, due to the static air cushion principal, the lift power can't be converted to propulsion power; in addition, the hull does not have a wing and its profile is not wing shaped, and thus the aerodynamic property of the ACV is unsatisfied, particularly at super high speed.

This paper introduces amphibious wing in ground effect craft (AWIG) (or DACWIG), which merge surface effect technology and aviation technology into air cushion technology. This therefore provides AWIG with both static and dynamic air cushion effects, providing super high-speed, good maneuverability and sea-keeping quality, low specific fuel consumption, a large navigation range, and fine amphibious behavior. It is suitable not only for civil applications (water transportation for passengers at short range, touring, salvage, survey, etc.), but also for military applications.

5.5.1 The evolution of ACV over the past 40 years[5-5-1], [5-5-3–5]

Sir Christopher Cockerell started to study peripheral jet-type hovercraft in 1956. The first operation of such a craft, named *SRN1*, to cross the English Channel was carried out successfully on July 25, 1959. China also worked on the research of plenum-type hovercraft—that is, with a solid sidewall at both sides and air cushion seals at both bow and stern. Since there are two cushion seals at bow and stern, the cushion will not be blown out from both sides if the seals are pulled up and when the craft operates on the water surface, which saves flow rate and lift power. Thus, the side wall hovercraft or surface effect ship (SES) was formed. The lift power is reduced 1–2 times, compared with amphibious hovercraft.

The flying height of the hovercraft was lower at the earlier stage of development, only about 10–20mm for the plenum type and 200mm for the peripheral jet type, due to the momentum transfer of the jet flow (*SRN1*). However, it is still unsatisfactory for amphibiousness, and only provides small abilities on land—that is, the craft needs to land on prepared cement ground or a smooth sand beach. This is due to its lower cushion pressure, only about 100–500kg/m^2, compared with the footprint pressure of a human, which is about 600kg/m^2.

The amphibiousness of the hovercraft has been solved smoothly by using a skirt. Since the skirt is composed of flexible neoprene-coated cloth, a series of advantages and functions can be listed as follows:

- Being able to land on a sandy beach and operate on a water surface, ice, swamp, grassland, and so on
- Being able to land on a steep pier and get through the ground barrier with a height of 0.6–0.7 skirt height, and ditches with a width of about 1.4L, where L is the length of the craft
- Improvement of sea-keeping quality, and decrease of the drag of the craft in waves
- Lowered cost and maintenance fee for the skirt, because of the detachability of the skirt fingers
- Decrease of the drag of the craft at high speed, and improvement in take-off ability

Therefore, the application of the skirt should be considered as a revolutionary measure for the hovercraft to generate the ACV from the laboratory to practical application, leading to a golden age for the hovercraft (in the 1960s–70s).

From the perspective of hydrodynamics, the plenum chamber principle had been widely used on ACV/SES instead of the momentum transfer principle of jet flow on *SRN1* at the earlier development stage of hovercraft.

However, ACV also has a series of disadvantages, as follows:

- It is difficult to develop into a large craft due to the use of air propellers
- It has high lift power, which is unable to be transferred to propulsion power at high speed
- Low range
- High cost
- Serious corrosion of propellers under the action of water/sand spray as well as foreign objects

To sum up, ACV/SES underwent considerable development and application over the past 30 years. For example, ACVs have been used as landing craft, car-passenger ferry ships, transport craft between islands, on inland rivers and swamps, in shallow water and rapid streams, as touring craft, as salvage and flood protection craft, as heavy-load transport craft, as oil exploration platforms, and so on, due to their amphibiousness, wide deck and cabin area, high speed, and smaller physical field under water. Actually, both the United States and Russia have a large ACV landing craft fleet.

Fig. 5.5.1 shows the United States' air cushion landing craft (LCAC), weighing 150t and with one main battle tank on board. The craft can be accommodated in a dock landing ship for transportation across the ocean, which ensures it is possible to land on more shores in the world (increased from 17% to 70%).

Fig. 5.5.1. LCAC of US

Fig. 5.5.2 shows the former USSR's landing ship *Zybr*, which is the biggest air cushion landing craft in the world to date, with an AUW of 550t, a payload of 150t, a max speed of 60km/h, and a large range; the craft can carry out direct shore-to-shore landings.

Fig. 5.5.2. Russian *Zybr*

Fig. 5.5.3 shows the largest civil ACV, SRN4 MK3, from the United Kingdom, which is able to accommodate 420 passengers and 55 cars, with a max speed of 65kn; it is used for crossing the English Channel.

Fig. 5.5.3. The UK's SRN4 MK3

SES are supported mostly by air cushions and are lifted greatly off the water surface; they are equipped with water propulsion to reduce water drag significantly and get a very high speed (92.3 km/h for the United States' SES100B). In addition, lift power decreases greatly due to use of side-walls (only one third of that of a ACV), and it is able to use water propulsion. Therefore, SES can be used as a passenger or passenger/car ferry, mine-sweeper, missile ship, and so on.

Fig. 5.5.4 shows the Russian patrol frigate SES *Bora*, which has a weight of 1,000t, a length of 65m, and a dash speed of 55kn. Fig. 5.5.5 shows a Chinese passenger transport SES; a Chinese ACV driven by a 31m gas turbine is shown in Fig. 5.5.6.

Fig. 5.5.4 Russian Missile SES *Bora*

Fig. 5.5.5. Chinese passenger transport SES

Fig. 5.5.6. Chinese gas turbine driving ACV landing test craft

5.5.2. The challenge to ACV/SES for competition in the commercial market[5-5-1] [5-5-6] [5-5-14]

Since the end of the 1980s, with the appearance of high-speed catamaran (HSCAT) and wave-piercing craft (WPC), civil ACV/SES have lost their competitive ability, as shown in Table 5.5.2. The reasons for this are as follows:

Table 5.5.2. The orders and deliveries of various high speed crafts in the 1980s and 1990s

Year	87	88	89	90	91	92	93	94	99	87	88	89	90	91	92	93	94	99
Catamaran	29	38	25	26	30	27	39	37	23	31	27	31	34	23	20	33	36	36
CAT with hydrofoils	0	0	0	0	0	1	6	7	0	0	0	0	2	3	9	7	3	0
Monohull	12	16	7	12	5	5	12	10	6	13	5	11	10	8	12	11	10	17
ACV	2	1	1	3	2	2	0	0	0	0	2	6	4	2	0	0	1	0
SES	0	7	5	7	3	3	1	4	1	11	17	9	10	7	4	10	2	1
HYC	6	5	11	9	16	6	4	5	2	11	14	16	16	8	9	7	2	0
WPC	1	3	5	4	1	3	3	2	2	9	7	7	4	5	3	3	3	3
SWATH	0	0	2	1	0	0	0	0	1	0	3	2	0	0	2	1	1	0
Total	50	70	56	62	57	47	66	65	35	75	75	82	80	56	59	72	58	57

1. High cost (for ACV)
At the early stage of development, most engines and equipment for ACV are of aviation types; therefore, the cost of craft is high. After some technical improvements, the total power is reduced drastically, from nearly 100kw/t to 30–50kw/t (the state of the art). Thus a lift engine of a diesel type can be used on ACV (AP1.88, UK) instead of aviation types, and then the cost of the craft can be reduced greatly. However, it still can't compete with CAT and other similar craft.

2. Low life expectancy and high maintenance fee for the skirt
Although there are a lot of advantages of a skirt, in our experience, the maintenance of the skirt was a troublesome problem for users, causing high costs and maintenance fees, and a short inspection and overhaul period as well as length of life (only 250–500h for the finger and about 2,000h for the bag).

In addition, inspection and overhaul have to be carried out on the ground or pier, thus some extra equipment and payments as well as labor power are needed by users. Possibly, this is an important cause limiting the application of both ACV and SES.

Due to the factors mentioned above, the civil market of ACV/SES has dropped greatly, as listed in Table 5.5.2. However, owing to its special function—that is, its amphibiousness and land-ability—the military application of ACV craft has been highly valued in the world.

3. Low speed, low total hydrodynamic property, and low efficiency for the power plant
People always pursue transportation tools with high speed. A speed of 30–40km/h was not bad for water transportation in the last century; however, in the twenty-first century, and especially in comparison with airplanes, it sounds too slow. Therefore, on the one hand, ship designers try to keep to 30–45km for HSCAT and WPC with simple equipment and structure as well as a large displacement, so as to keep a reasonable Froude number ($F_{rL} \cong 1$); on the other hand, they seek more high-speed craft. But pursuing super high speed for ACV craft—from 40–50km/h to 70–150km/h—has many probable difficulties and disadvantages, as follows:

1. At such high speed, because an ACV craft does not take advantage of wings or a winged hull, the hull only generates resistance but has no contribution to lift, which drops its aerodynamic property.
2. The lift power on an ACV can only generate lift by a static air cushion, but this can't be converted to propulsion power at high

speed. However, in the case of high speed, craft with a winged hull (or with wings) will generate enough dynamic lift. The lift power will doubtless be a large burden for such craft.

To sum up, the aerodynamic property of ACV is rather unsatisfactory, compared with craft with amphibious wing in ground effect, as shown in Table 5.5.3.

Table 5.5.3. Specific fuel consumption for various high speed craft (kg/person km)

Craft type	ACV	SES	HYC	Helicopter	AWIG (second generation in Russia) *Ivolga*
Specific fuel consumption (Kg/person km)	0.06–0.1	0.05–0.08	≈ 0.06	0.12–0.19	≅ 0.02

4. Unsatisfactory sea-keeping quality (especially for ACV)

In order to save lift power, the hovering height for ACV craft is small; an ACV is basically planing on the water surface. Therefore, the interference of waves is extremely serious for the following reasons:

(1) Speed loss

Due to wave disturbance, the leakage area of the air cushion of an ACV/SES operating in waves will be changed; thus, the cushion pressure and motion of the craft (heave, heel, and pitch, etc.) will be changed. Although both the bow and stern skirt are flexible, the wave drag still increases drastically due to the large frontal area of the skirt. Consequently, the speed will drop down greatly. In the case of using water jet propulsion, the speed might drop down even more owing to the air ingestion from the cushion into the water inlet of the waterjet system.

Fig. 5.5.7 denotes a comparison of speed loss between various HPMV. It is found that the speed loss of ACV is most serious due to the ratio of cushion length to beam, which is unable to be higher due to transverse stability. However, SES craft might extend the ratio of length to beam to improve speed loss in waves.

The speed loss of both ACV and SES craft can be improved by using responsive skirts (particularly bow skirts). Fig. 5.5.8 shows the brief configuration of the responsive skirt. It is denoted that the skirt bag might be raised up, in the case of a wave impacting on it, so as to reduce the skirt drag. However, the responsiveness couldn't be too high, as otherwise, serious tuck of the skirt might occur, which will result in a plough-in of the crafts.

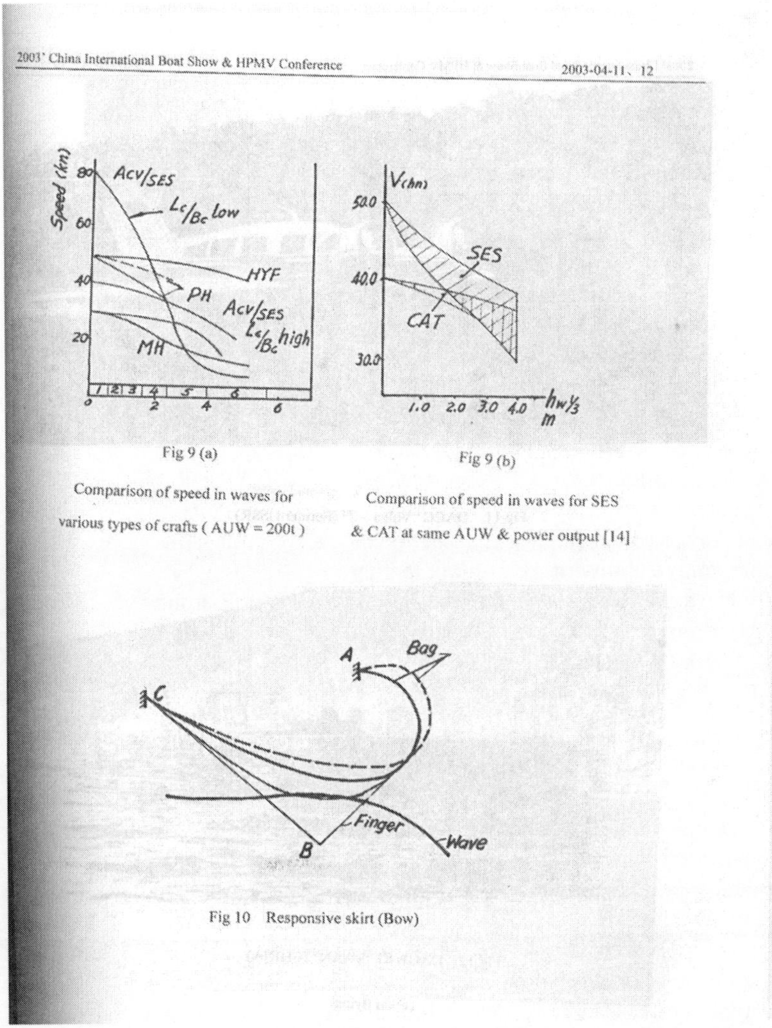
2003' China International Boat Show & HPMV Conference 2003-04-11、12

Fig 9 (a)

Comparison of speed in waves for various types of crafts (AUW = 200t)

Fig 9 (b)

Comparison of speed in waves for SES & CAT at same AUW & power output [14]

Fig 10 Responsive skirt (Bow)

Fig. 5.5.7 (up) shows a comparison of speed loss in waves between various HPMV

Fig. 5.5.8 (low) shows the deformation of the responsive skirt on ACV under the action of a heading wave.

(2) Vertical acceleration

Even in waves of small height or in short waves, a certain vertical acceleration will affect ACV/SES craft due to the so-called cobble stone effect—that is, the motion of a hovercraft is like a car running on a road paved with cobble stones. The cushion flow rate of crafts running on the waves will be changed, because of the change in wave profile, even with the same air leakage area. Thus, the cushion pressure will be changed, and the so-called wave-pumping effect will be generated in practice.

Besides, in the case of ACV/SES craft running on short waves, even with a short wave length, the air compression effect also creates a heaving motion and vertical acceleration. In the case of craft running on long waves, serious motion will be generated due to the change in leakage area of air in the cushion.

All the issues mentioned above make for significant speed loss and vertical acceleration, and result in discomfort for passengers and crews; thus, ACV craft lost their competitive ability in the civil commercial market, particularly after the appearance of HSCAT and WPC of a large size and speeds of 30–40kn. However, owing to their amphibious behavior and lower underwater physical field, ACV craft still have definite competitive ability in the military market.

Nevertheless, the speed of all these HPMV is still around 30–50km. In order to overcome this speed barrier and design one that goes faster than 100km/h, people have to look to WIG craft.

5.5.3. The features of amphibious wing in ground effect craft (AWIG)[5-5-7–11] [5-5-13]

As detailed in the references above, the WIG family includes power-augmented-ram wing in ground effect craft (PARWIG), dynamic air cushion craft (DACC), and dynamic air cushion wing in ground effect craft (DACWIG).

A PARWIG craft is characterized by high speed, high altitude, and fine sea-keeping quality after takeoff, and DACC is characterized by fine amphibious capabilities, lower altitude, and poor sea-keeping quality.

However, the DACWIG combined with the advantages of both PARWIG and DACC not only has high speed and altitude (even flying off the surface effect zone, SEZ), but also has fine amphibious performance. Amphibious performance is particularly important for WIG craft, especially of small and medium size, for the reasons discussed below.

- For civil applications, it is difficult for the current WIG to compete with jumbo jets owing to poor economy.[5-5-10] Therefore WIGs are limited to short range routes, especially routes between islands, where no pier, airport, or similar is available. A WIG with its amphibious performance is able to land/launch directly on the sea or beach.
- In general, a WIG with amphibious performance will improve take-off ability in both calm water and waves.
- The Slamming load of the craft is improved during taking off and touching down.

Fig. 5.5.9 and Fig. 5.5.10 show a Chinese DACWIG, named *Swan*, weighing 8.1t, taking 20 passengers, with a speed of 140 km/h, and an altitude of 1m. Here, we call it a DACWIG, with the dual meanings described as follows:

- The power mounted in the front creates the static air cushion under the main wings
- The dynamic air cushion should be formed under the main wing during high-speed flight of craft in the SEZ

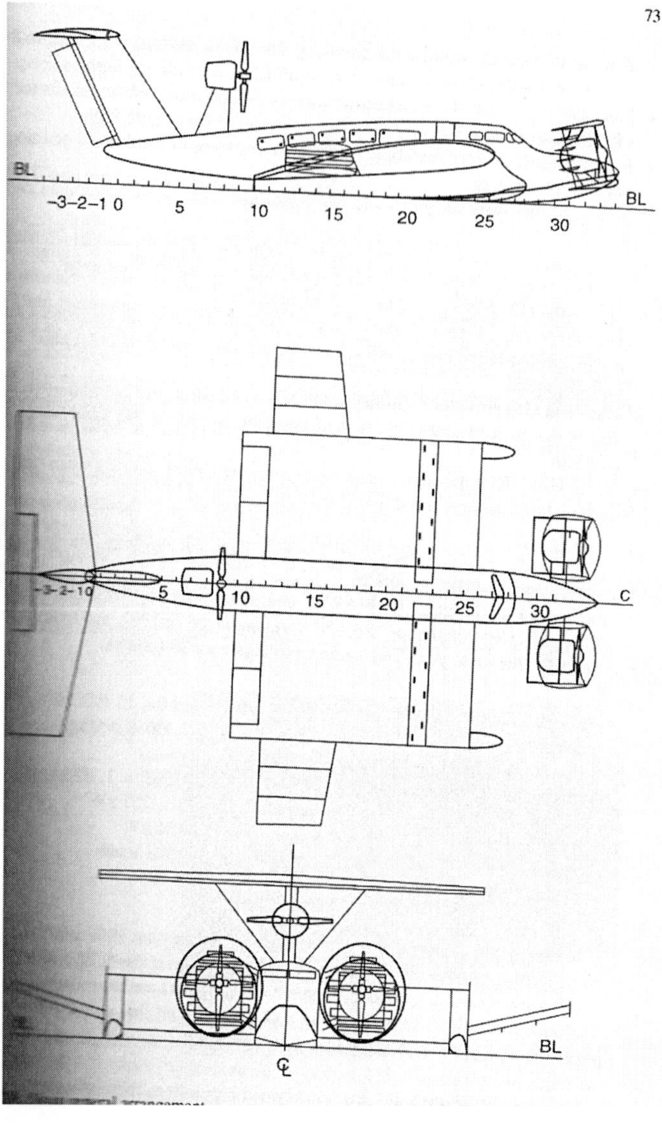

Fig. 5.5.9. Aerodynamic configuration of DACWIG type *Swan*

Fig. 5.5.10. *Swan* in flight

Fig. 5.5.9 expresses the general and aerodynamic arrangement of *Swan*. It shows the main wings with a low aspect ratio, together with side buoys, flaps, and mounted ducted air propellers; this creates the static hovering and amphibious performance. The side composite wing with high AR improves the aerodynamic property of the craft at high speed, and the high stern wing extended backward keeps satisfactory longitudinal stability. Fig. 5.5.10 shows *Swan* flying.

Therefore, the features of this craft are as follows:

1. The long main wing (or central wing) makes for a large air cushion and a small aspect ratio, which is favorable for the static air cushion. The air jet blows from the ducted air propeller mounted fore into the air cushion, the aerodynamic mechanism of which is just like that of the ACV with peripheral jets—that is, transforming the direction of air momentum to make the static air cushion, with higher efficiency.
2. When the guide vanes, located after the fore mounted air propellers, turn up, and the flap after the main wings opens, the dynamic air cushion of the crafts will be formed, with higher flying speed and fine aerodynamic property.
3. The protruding composite wings on both sides enlarge the aspect ratio of the wings and bring about fine aerodynamic property

during high speed.

4. In the case of high craft speed with the guide vane up, the function of the fore air propeller will be transformed from providing lift to propulsion with no bad efficiency of propulsion.

At present, there are some methods to make static lift (or amphibious) for WIG; these are listed as follows:

1. Use the air jet, which is blown from the fore-mounted ducted air propeller, and guided into the air cushion with the help of the guide vane, as in the Russian *Volga-2* (Fig. 5.1.5) and the Chinese *Swan*.
2. Use the air jet, blown from the fore-mounted ducted air propeller (the disc plane of which can be rotated) into the air cushion, to get fine aerodynamic property craft during both take-off and flying operation mode, as in the Russian *Ivolga* (Fig. 5.4.1).
3. Use the air jet, blown from the fore-mounted ducted air propeller (without a guide vane) directly into the air cushion, as in the Chinese 750 type of craft (Fig. 5.5.11).
4. Use the air jet, blown from a thruster-air propeller, via the air duct and into the air cushion, with bow retractable skirt applied to seal the leakage of air during the stage of static hovering and before take-off, as in the German *Hover-wing* type (Fig. 5.5.12).

Fig. 5.5.11. Chinese 750 type launching in water.

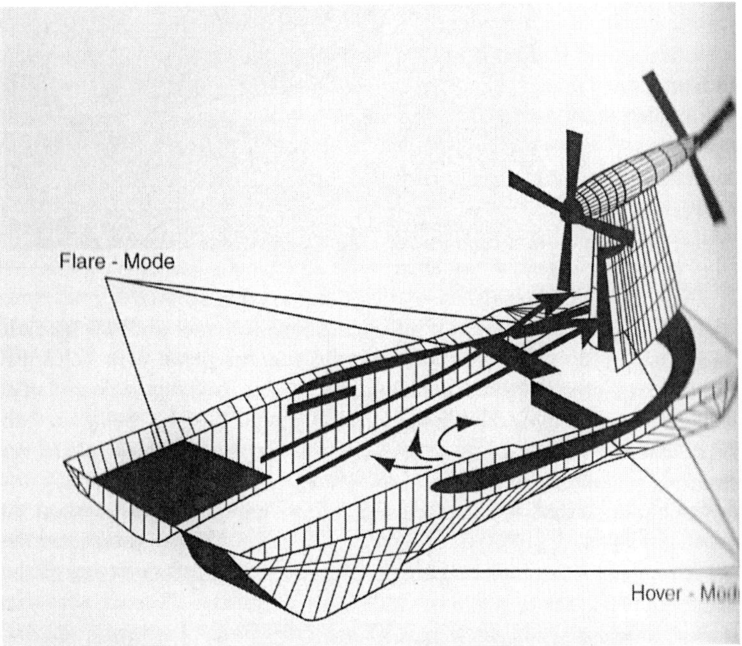

Fig. 5.5.12(a). Aerodynamic configuration of the German *Hoverwing* type

Fig. 5.5.12(b). *Hoverwing* in flight

1. Static hovering performance
A comparison of static hovering performance between AWIG and ACV is listed in Table 5.5.4.

Table 5.5.4

Craft name	SRN1	"33"	Ap1.-88	*Swan*	*Ivolga*
Country	UK	China	UK	China	Russia
Year of built	1959	1959	1982–2000	2000	2002
Craft type	Peripheral jet ACV without skirt	Plenum ACV	Modern ACV with skirt	AWIG	AWIG
Total weight, W, /t	3.4	4.0	29	8.0	3.6
Lift power, N_L /kW	224	176	750	514	430
Total power, N_t /kW	319	294	1500	724	420
N_L/W	65..9	44.1	25.9	64.25	116.7
N_t/W	93.9	73.5	51.8	90.5	116.7
Speed on calm water /km/h	24	50	55	140	200
Static hovering height/ mm	≈ 200	10–20		10–20	

Form Table 5.5.4, one can see that SRN1 is characterized by a high static hovering height (about 200mm), good land-ability, and a certain amphibious performance. Following the development of the skirt, the specific lift power has been dropped drastically, as in the AP1-88, having only half the power of SRN1, or even lower, as needed.

At present, the static hovering performance of AWIG is just close to that found in the early development stage of ACV (the precise specific lift power for *Ivolga* is unclear), with a low hovering height. However, it should be noted that the hovering height will be raised quickly (even with a lower speed) due to the dynamic lift contributed by the main wing, composite wing, and tail wing, and so on.

Fig. 5.5.13(a) shows the change in the static and dynamic lift, L_c, L_d, versus the relative flying height h/c (where h is flying height and c is wing chord). Fig. 5.5.13(b) shows the change of the lift coefficient C_y with the relative altitude and angle of attack. It is noted that the dynamic lift is drastically greater than the static lift, and C_y will be raised greatly with h dropped down. In addition, the hovering height will be increased by adjusting the trimming angle and inclination angle of the fore ducted air propeller, even at lower speed.

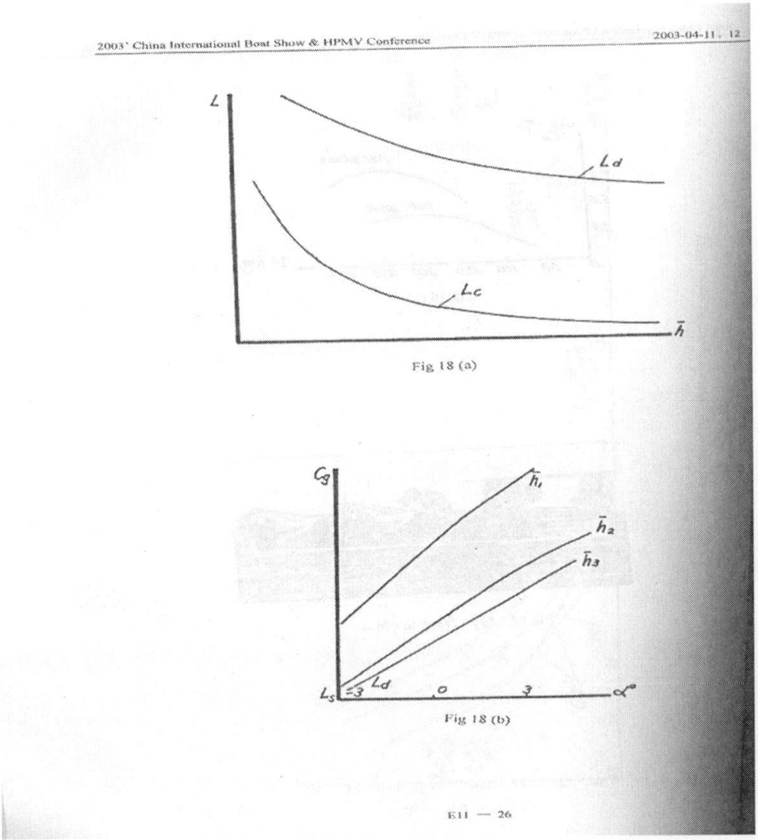

Fig. 5.5.13(a). L_c, L_d versus relative flying height h/c
Fig. 5.5.13(b). C_y versus relative altitude and angle of attack of main wing

The Russian *Ivolga* is able to fly on ground with an unevenness of 0.2–0.5cm, and *Swan* is able to land on a slope of 1/10. In civil applications, the land-ability of such crafts might be satisfactory; however, for military applications, landing pads for such craft should be improved. At present, the landing pad might be soft (*Volga-2*) or hard (*Swan*, *Ivolga*).

2. Comparison of other performances of AWIG with other high-speed crafts.[5-5-10] [5-5-11]
Here we take the Russian AWIG type *Ivolga* (second generation WIG) as an example, and compare it with other HPMVs.

(1) Hydro-aerodynamic performance at high-speed flying
As described above, thanks to the rotatable ducted air propellers and composite wing, the craft possesses not only amphibiousness but also high speed and altitude and can avoid operating on the air-water interference surface; this creates serious disadvantages in terms of aero-hydrodynamics. In addition, it does not need special lift power as does the PARWIG; thus, the lift power can be converted to propulsion power by rotating the duct air propeller, meaning that the propulsion efficiency is higher.

Helicopters also have high speed, fine maneuverability, vertical take-off and landing, static suspension, and good operation performance at lower altitude. However, due to the lack of lift increment for the air wing and because of the special profile of helicopters (with retractable landing gear, the huge boss for the rotated wing, the stern propeller at a high position, some protruding parts), the aerodynamic property (i.e., $K_p = \frac{C_y}{C_x}\eta_p$) used is two times lower than for conventional airplanes. Therefore, the fuel consumption of helicopters is much high, with k_η lower and the cost more expensive.

Fig. 5.5.14 shows the hydrodynamic coefficient versus F_{rD} for various HPMV, a WIG, and a helicopter, where $k_\eta = Wv/102P = \frac{\frac{W}{R} \bullet vR}{102P} = k\frac{C_y}{C_x} \bullet \eta_p$; W, overall weight of craft (kg); v, craft speed (m/s); P, total power (kw); k, coefficient. $F_{rD} = \frac{v}{\sqrt{gD^{1/3}}}$, where D denotes the volume of displacement (m^3).

In this figure, HYF denotes Hydrofoil craft; CAT, Catamaran; MH, Monohull craft; ACC, high-speed air cavity craft; and the symbol "○" denotes SES. In addition, all of them have the same tendency for the hydrodynamic coefficient versus Froude number. Also plotted in the figure is Swan, which is a type of Amphibious WIG; ⊗, which is the AWIG type *Ivolga*; and ⊕, which is the EH101 helicopter, made by both the UK and Italy.

Although there are few data on the figure, a general tendency can be found on it: conventional HPMV will be suitable for crafts only in the lower relative speed range (Frd < 4.5); otherwise, the hydrodynamic coefficient will drop down drastically. However, WIGs are suitable for higher speeds, with Frd as high as 9–15, or even higher, where hydrodynamic coefficient is high ($k_\eta \cong 4-5$). Helicopters are also suitable for such high speed but with lower k_η (–2.2).

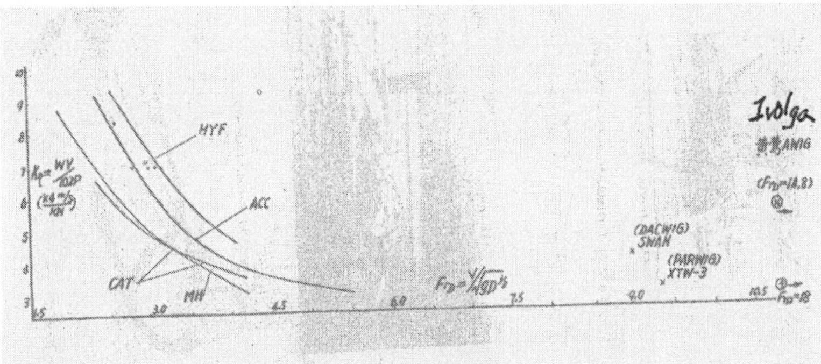

Fig. 5.5.14

(2) Fine sea-keeping quality
Owing to the high altitude and dynamic air cushion, the sea-keeping quality of the craft will be fine; it is mostly not limited to waves after take-off, and it has better take-off ability in waves, as shown in Table 5.5.5.

(3) Good maneuverability
Since the flight altitude of such a craft is high, and there is an aileron on the composite wing, the craft is able to perform banked turns with a large heeling angle in order to get a small turning radius. A heeling angle of 15° can be made in the case of an altitude of 1m, and 25–30° in the case of an altitude of 2m or over (Fig. 5.4.7). Table 5.5.5 shows a comparison between various performances between a WIG, an HPMV, and a helicopter.

Table 5.5.5

Transportation Tool	Overall weight l/kg	Cruising speed /km/h	Price (reference for 2002 (x1,000 USD)	Fuel consu-mption /kg/km	Amphibi-ousness	Sea-keeping quality $h_{w3\%}$,m	$h_{w3\%}/W^{1/3}$	Passengers
Airplane *AN-2* (Russia)	5,500	180	800	0.78				9–12
Helicopter *MI-2* (Russia)	3,550	170	1,500	1.38				6-8
HYF *Nefka* (Russua)	3,000	70	200	0.62	No			10
HYF *Raketa* (Russia)	27,000	58			No	0.8	0.27	58
HYF *Kometa* (Russia)	60,000	58			No	1.5	0.38	
ACV *716II* (CHN)	20,000	72			Yes	1.0	0.36	36
SES *719II* (CHN)	123,000	51			No	1.5	0.30	257
WIG *Orlyonok* (Russia)	140,000	400				1.5 *	0.29	150
AWIG *Volga-2* (Russia)	2,700	120			Yes	0.5	0.36	10
AWIG *Ivolga* second generation Russian	3,600	180	500–700	0.2	Yes	1.25 *	0.82	10

* These data denote the upper limit of the wave height to the craft during take-off; however, the wave height is unlimited after take off, since the altitude of such crafts is as high as 30m.

To sum up, the advantages of an AWIG of the second-generation-type *Ivolga* can be listed as follows:

1. Good land-ability and amphibious performance, being able to land or launch from an unprepared shore, seashore, ice and snow covered surface, swamp, and so on, and running on uneven ground with raise of 0.2–0.7m
2. High speed, up to 200km/h or higher
3. Multiple operation modes—that is, hull-borne, cushion-borne, planing, and flying
4. Being able to jump as high as 30m with banked turn at high altitude, or maneuver at medium speed in cushion-borne operation
5. Thanks to the surface effect and dynamic cushion function, the lift power can be used as both lift power at low speed and propulsion power at high speed; therefore, the specific fuel consumption is lower, with higher aerodynamic property and a larger range compared with other HPMV and helicopters
6. Fine sea-keeping quality and seaworthiness
7. Because it uses car engines, the price and maintenance fee are lower, which improves its competitiveness in the market significantly

Fig. 5.5.15[5-5-2] shows the WIG landing craft, developed by Lockheed-Martin, USA. From the figure, one can see that the stealthy superstructure and wing against the radar wave from the air on the hostile side are considered carefully. In addition, since the AWIG can be operated close to water and the ground surface, the threat from the ground and ships should be small.

According to news reports from both the *Washington Post* (October 7, 2002), and *Aviation Week & Space Journal* (October 15, 2002), Boeing has been developing a huge WIG named *Peculiar* with a wing span of 152m/500' (195' for Boeing 747-400), payload 1272t (40.2 t for 747-400), AUW 2727 t (395.5t for 747-400), which is able to accommodate 17 main tanks M1, and fly in SEZ (6.2m), or beyond SEZ (6000m), with a range of 10,000 miles. It is planned to develop the huge airplane for transporting divisions from the US mainland to any corner of the world in a short period of time.

5.5.4. Summary

1. The appearance of ACVs 40 years ago made a great stir around the world owing to its high speed and fine amphibious performance. It has been developed in not only civil applications but also military ones. Both the United States and Russia have established a large landing ACV fleet. However, since the basic function of an ACV is static hovering, the unwinged hull profile and lack of ability to convert the lift power to propulsion power make the aerodynamic property of the ACV at high speed unsatisfactory.
2. By merging air cushion technology and surface effect technology into the aviation technology, AWIG not only has high speed, ability to fly at various altitudes, and fine maneuverability and sea-keeping, but also has fine amphibious performance, less specific fuel consumption, and a large range.
3. According to the advantages mentioned above, it is believed that there is a bright future for the AWIG.

References

[5-5-1] L. Yun & A. Bliault, "Theory and Design of Air Cushion Craft," London: Arnold, 2000.

[5-5-2] "Caspian Sea Monsters to Marauders?" *Hovercraft Bulletin*, October–December 2001.

[5-5-3] *Jane's Yearbook* (1990–2000).

[5-5-4] ЦМКБ "Almaz" 50 set ,*Судостроение* 4,1999 (in Russian).

[5-5-5] D. R. Havis, "Hovercraft Development," HPMV '92 Proceedings, Arlington, VA, USA.

[5-5-6] L. Yun & Peng Guihua, "The Evolution of Chinese ACV/SES."

[5-5-7] L. Yun, C. H. Wu, & Y. N. Xie, "Design Feature and Evolution of Dynamic Air Cushion Wing-in-Ground Effect Craft," Proceedings of 3rd International HPMV Conference, Shanghai, China, April, 1999.

[5-5-8] L. Yun, C. H. Wu, & Y. N. Xie, "Fluid-Aerodynamics," Performance Investigation of Dynamic Air Cushion Wing-in-Ground Effect Craft, Proceedings of 3rd International HPMV Conference, Shanghai, China, April, 1999.

[5-5-9] L. Yun, Y. N. Xie, & G. H. Peng, "On the Design Consideration of the Aerodynamic Configuration on the Amphibious Wing-in-Ground Type 'SWAN' During its Convention," 6th International Boat Show & HPMV Conference, April, 2001, Shanghai, China.

[5-5-10] K. P. Wong & L. Yun, "Development of Russian WIG,"

Proceedings of 7th International Boat Show & HPMV Conference, April, 2002, Shanghai, China.

[5-5-11] V. V. Kolganov & V. G. Sergeev, "Design Deatures of WIG Craft EL-7 'Ivolga' and Its Test Results," Proceedings of 7th International Boat Show & HPMV Conference, April, 2002, Shanghai, China.

[5-5-12] V. I. Zukov, "Aerodynamic Characteristics, Stability, and Controllability of WIG Craft," 1997, Moscow (in Russian).

[5-5-13] L. Yun & G. H. Peng, "Development Tendency of Fast Water Transportation in 21st-Century Dynamic Air Cushion Wing-in-Ground Effect Craft," *Shipbuilding Engineering*, 1996 (in Chinese).

[5-5-14] L. Yun, "Comparison of Performance of Various HPMV," Lectures at Seoul University, South Korea.

5.6

A New Concept in Wing in Surface Effect Ships (2003)

Hiromichi Akimoto, Syozo Kubo and Takahiro Taketsume

Department of Applied Mathematics & Physics, Faculty of Engineering, Tottori University (Japan)

Abstract

A wing in surface effect ship (WISES) is a high-speed marine vehicle that uses aerodynamic wings enhanced by the ground effect on the water surface. Stable and economical operation of a WISES depends on lowering its take-off and landing speeds with high lift mechanisms. Power-augmented ram (PAR) is one of the popular approaches to lift enhancement of WISES. However, PAR depends greatly on the calmness of the water surface under the main wing; in addition, complexity in structure of the ship is required.

The present authors propose a new concept in WISES. This is a canard type of WISES with propellers mounted on its canard wing. The ship can simply make a higher angle of attack position to gain a higher lift force, because the hull is short after the step for planing. Accelerated streams from forward-mounted propellers prevent the main wing from stalling even in a high-lift condition. To demonstrate the concept, the present authors built a radio controlled self-propulsion model with a length of 1.8m. This paper describes the problems related to the commercialization of WISES, the new concept of the canard type of WISES, and some results of the ongoing experiments on the self-propulsion model.

5.6.1. Introduction

When an aerodynamic wing comes close to the ground or water surface, a considerable increase in its lift–drag ratio is shown. This effect is often called the "ground effect," since it is mainly discussed applied to aircraft on runways. However, the present authors prefer the term "surface effect" as it is simpler to apply to marine vehicles.

A wing in surface effect ship (WISES) is a high-speed marine vehicle which utilizes aerodynamic wings enhanced by the surface effect. It is also a synonym for a wing in ground (WIG) ship, a ground effect machine (GEM), Ekranoplan, and 飞船 (fěi-chuán, in Chinese).

Many types of WISESs have been proposed and built since the 1960s. In the beginning, there was an optimistic view that the commercialization of WISESs would happen immediately and that many WISESs would soon be cruising around the world.

Unfortunately, no WISES has commercially flourished until now. Although, there are many WISES around the world, as shown in *JANE's High-Speed Marine Transportation*,[5-6-1] and others might be under development, only a few vehicles are in practical service.

In Japan, the major problem in developing WISES concerns its seaworthiness. Ship builders and owners of marine liners are suspicious of the safety of this new type of ship in the rough sea states that sometimes occur around the country. We do not discuss the use of a WISES on a large river or lake due to the geographical features of Japan. Therefore, we need a new concept of WISES which can be operated on the rough seas around Japan.

This paper describes problems related to the commercialization of WISESs, the new concept of WISES, and some experimental results of the concept.

5.6.2. Problems in commercialization

Japan is a country of mountainous islands; compared to those in other countries, the rivers and lakes are small. Therefore, we must consider the sea-keeping capability of WISESs at the very early stages of development, as the primary concern of ship builders and owners for an experimental WISES is its safety and endurance on waves.

The Kubo and Kobe ship yard of Mitsubishi Heavy Industry built *μ-Sky 1* and *μ-Sky 2* in 1990,[5-6-4] which are one- and two-seated preproduction models of WISES, respectively. In some sea trials of *μ-Sky* on the inland sea of Setonaikai, it easily cruised in a rough sea state after take-off from a harbor, while an accompanying displacement type of boat with a length of 8m could not go forward due to wave resistance. Although verified data were not recorded, it is obvious that after a WISES takes off from a harbor properly protected from rough ocean waves, its sea keeping capability is better than any other conventional ships of the same size.

Since regular services of sea transportation are usually from one harbor to another, a WISES does not need to be able to take off and land on rough waves, except for specially designed rescue ships. Even when a WISES has to alight in a severe sea state in an emergency or in cases of breakdown, the requirement for the ship is to remain afloat in the water, waiting for the arrival of rescue vessels, or to move slowly to the nearest harbor in displacement mode. Of course, commercial WISESs should not have excessively high performances for their economy. Therefore, seaworthiness on the take-off/landing operation in protected waters should be considered separately from the performance on the ocean.

To leave from the water, WISES should accelerate itself to the take-off speed. In waves, this procedure requires additional power for wave resistance. Thus, there is a marginal wave height for take-off. When the wave height in the departure zone exceeds the limitation, services are delayed or canceled.

To reduce the cancellation rate of a WISES, the take-off speed must be reduced by high-lift devices. The popular technique of power-augmented ram (PAR) utilizes the high-speed air flow from forward-mounted thrusters ramming into the chamber-like space between the water and the main wing. With an elaborately designed PAR mechanism, WISES supports almost all its weight even when its speed is zero (i.e., very low-altitude hovering). However, PAR requires additional complexity of ship structure and raises the construction cost. Besides, PAR is not fully effective in rough sea states, because air leakage through the gaps between the side walls and the wavy free surface diminishes the pressure under the wing.

An all-wing type (or flying-wing type) of WISES is one of the solutions to this problem. For example, the *Sky* series with a light wing load enables them to take off at the speed of 60km/h without a PAR system. However, the light wing load also means that the area of the wing is excessive in cruising conditions. Therefore, the approach of an all-wing type is not suitable for large sized WISES intended for commercial transportation.

5.6.3. WISES with a canard configuration

To use a WISES in regular transportation services in Japan, it should have cost-effective high lift mechanisms that are not dependent on PAR effect.

When an aircraft takes off or lands, it unfolds the high lift devices, such as a multi-element flap, and takes a high angle of attack position to exert sufficient lift force with the low velocity at that moment. However, flaps are not effective when the clearance between the trailing edge and the water surface is very small. The flow at the trailing edge easily peels off from the flap and goes along the water surface. And the large long tail structure, required for the stability of WISES, limits the angle of attack of the ship.

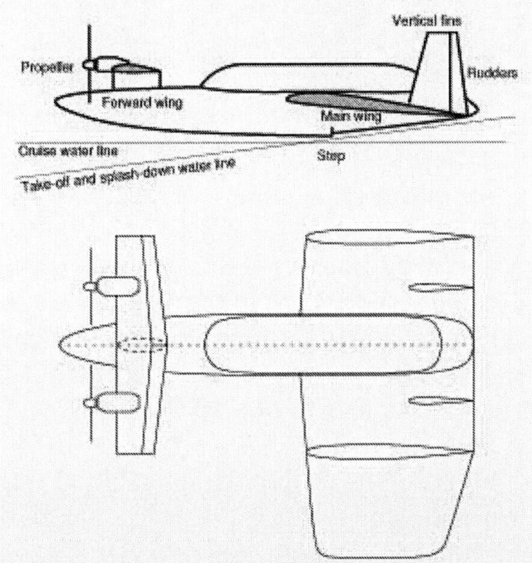

Figure 1: Canard type WISES

Fig. 5.6.1

The Present authors propose a new type of WISES with better performance in waves. It is a canard type of WISES with forward-mounted propellers (see Figs 5.6.1 and 5.6.2).

The features of this concept are summarized as follows:

1. A short tail due to the canard type of horizontal stabilizer makes it possible to take a high angle of attack position at the instant of a take-off/landing maneuver.
2. High-speed wakes from the propellers wash the canard and the main wing, so that these wings will not stall even in high lift conditions.
3. At a high angle of attack position, the canard generates positive lift force for the pitch-up operation of the ship, while the horizontal stabilizer of the conventional concept generates negative lift for the same operation.

Fig. 5.6.2. A computer graphic image of the canard type WISES

4. Propellers and power plants are mounted in a spray-free region. Especially during take-off and landing, power units are lifted high from the water by the pitch-up operation. This is quite favorable for the durability and maintenance of the propulsion system.
5. Rudders on two vertical fins and the elevator on the canard are located in the slip stream of the propeller. This ensures the controllability of the ship during take-off and landing and even in low-speed conditions.
6. A long fore-hull in this concept is favorable for high performance in waves when the ship is running under the speed of take-off.

In this concept, the combination of the canard and the forward-mounted propeller is essential. A problem in this concept is the heavy weight of the power plants near the bow. However, turbo prop. engines of high power/weight ratio in the market will provide sufficient performance for this purpose. On the other hand, it is not easy to build a small-sized WISES of this type, because small turbo prop. engines are not easily available in the market, and other types of engines are relatively heavy for a small ship.

Although the concept is simple in its structure, there is complicated aerodynamic interaction between the canard, the main wing, and the propeller's slip stream. Therefore, it is favorable to demonstrate the concept by operating self-propulsion models, along with analytical approaches. This will be described in section 5.6.5.

5.6.4. Stability consideration of the canard concept

One distinctive feature of WISES is the interaction between the water plane and the dynamics of the vehicle. When a wing comes close to the water surface, the center of lift shifts rearward and causes a large pitch-down moment. Therefore no fixed aerodynamic center of pitching moment exists when the WISES is in a vertical motion.

For this reason, up to now many proposed WISESs have a large horizontal stabilizer mounted on a long tail structure. Mathematical explanations of the stability criteria are found in Staufenbiel[5-6-2] and Rozhdestvensky.[5-6-3] It is said that two separate aerodynamic centers of the height and the attack angle are defined locally around the concerning cruising condition, and that for the height stability the aerodynamic center of the height $x_{ac,h}$ should be placed in front of the aerodynamic center of the attack angle $x_{ac,\alpha}$

$x_{ac,h} > x_{ac,\alpha}$ (x-axis is in the forward direction)

To move $x_{ac,\alpha}$ rearward in the design procedure of a WISES, it is necessary to enlarge the area of the stabilizer and/or to lengthen the distance from the gravity center to the stabilizer.

In the canard configuration, a horizontal stabilizer is mounted in front of both the gravity center and the main wing. Therefore, we should reconsider the height stability of a WISES with canard configuration. The equation of motion of a WISES with a thruster is as follows:

$$\frac{dp}{dt} = M(\alpha, h) + r^t \times F^t, \quad (5.6.1)$$

$$m\frac{d^2x}{dt^2} = F(\alpha, h) + F^t + mg, \quad (5.6.2)$$

where p is the angular momentum vector to the gravitational center of the ship, x is the position of the gravitational center, and $M(\alpha; h)$ and $F(\alpha; h)$ are the moment vector and the force vector of the wing system, which includes the main wing and stabilizers. These two vectors depend on the attack angle of the main wing, α, and the clearance between the water surface and the trailing edge of the main wing, h. F^t , g, r^t, and m are the thrust, acceleration of gravity, relative position of the thruster to the gravitational center, and the mass of the ship, respectively.

By considering the perturbation of the pitch angle θ (bow-up is positive) and the displacement from the cruising altitude z, equations 5.6.1 and 5.6.2 are reduced to:

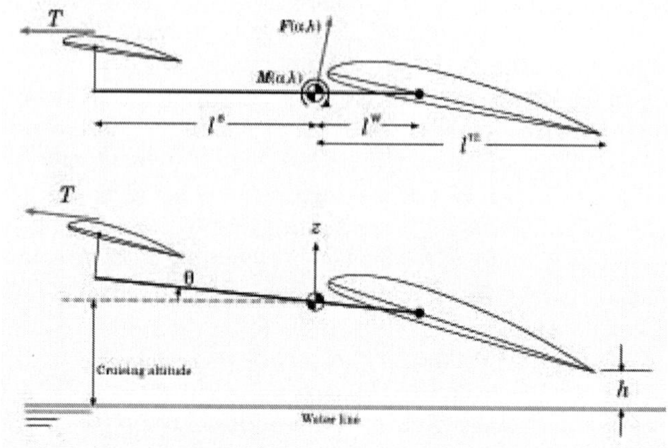

Fig. 5.6.3. A simplified canard model

$$I\ddot{\theta} = M_\alpha \theta + M_h (h_\theta \theta + z), \quad (5.6.3)$$

$$m\ddot{z} = L_\alpha \theta + L_h (h_\theta \theta + z) + T\theta \quad (5.6.4)$$

where I is the inertial moment around the pitching axis; M_α, M_h, L_α and L_h are the first derivatives of $M(\alpha; h)$ and $L(\alpha; h)$, respectively; and T is the thrust. Due to the horizontal distance from the gravity center to the trailing edge of the main wing, h depends not only on z but also θ, and then the term h_θ is not zero. For simplicity, the velocity of the ship is assumed to be constant and damping terms related to $\dot{\theta}$ and $\dot{z}$ are omitted.

For the consideration of the static stability of height, we assume that the perturbation of the pitching angle balances in a shorter time scale than that of the perturbation of vertical motion. Then θ is expressed as:

$$\theta = -Mhz/(m_\alpha \theta + M_h h_\theta) \quad (5.6.5)$$

By applying 5.6.5, 5.6.4 can be rewritten as,

$$m\ddot{z} = \left\{ \frac{L_\alpha + L_h h_\theta + T}{M_\alpha + M_h h_\theta} (-M_h) + L_h \right\} z \quad (5.6.6)$$

The static stability of height requires $\{\cdots\} < 0$. In the surface effect, naturally $L_h < 0$. And for the canard configuration, M_h is positive, because the main wing subjected to the surface effect is at rear of the gravity center.

Fig. 5.6.3 shows a simplified configuration of the concept. l^{TE} is the lever length from the gravitational center to the trailing edge, hence $h_\theta = -l^{TE}$.

Then, the static stability condition becomes an inequality:

$$\frac{M_h}{L_h} > \frac{M_\alpha - l^{TE} M_h}{L_\alpha - l^{TE} L_h + T} \qquad (5.6.7)$$

$$(\mathrm{x}_{\mathrm{ac,h}} > x^*_{ac,\alpha})$$

The right-hand side of 5.6.7 is the modified aerodynamic center of attack angle $x^*_{ac,\alpha}$. If we neglect the contribution of the inclined thruster and the lever l^{TE}, then $x^*_{ac,\alpha}$ becomes the usual aerodynamic center of attack angle $x_{\mathrm{ac},\alpha}$ (= M_α = L_α).

In conventional configurations, a large horizontal tail mounted some distance from the gravity center assures $x_{\mathrm{ac},h} > x_{\mathrm{ac},\alpha}$. However, this does not work in the canard type concept. In $x^*_{ac,\alpha}$ of 5.6.7, all three terms in the denominator, as well as M_α and M_h in the numerator, are all positive. Therefore, the static stability condition $x_{\mathrm{ac},h} > x^*_{ac,\alpha}$ largely depends on -l^{TE} in the numerator. In the canard configuration, this can be satisfied through a proper distribution of the total weight between the canard and the main wing.

We do not yet discuss the dynamic stability of the new concept. However, it should be noted that, usually, stabilization of a vehicle is not so difficult with a simple automatic control mechanism, and self-stability is not always a requirement of economical transportation. Downsizing stabilizers through the use of an automatic control often contributes to the total efficiency of a vehicle.

5.6.5. Self-propulsion model

For a new concept vehicle, its verification by a self-propulsion model (or a flying test bed) is an indispensable step in its development. Although the unconditional environment of test fields causes the ambiguity in quantitative evaluations, a test flight simply demonstrates the rationality of the new concept.

Table 5.6.1 shows the specification of the model *Kaien-1*[1] which is now subjected to ongoing field tests.[5-6-5] Fig. 5.6.4 shows photographs from the field test. The test of the model is still in progress, and a detailed report on it will appear in the near future.

---

[1] *Kaien* is written 海燕 in Kanji (Chinese characters). It means “sea petrel.”

At present, the test shows that the canard-type concept works properly in surface effect, and the simple bow-up operation exerts sufficient lift, as in the design, even in waves. The maximum wave height at take-off is 20cm (22% chord length of the main wing) in the actual results.

Table 5.6.1. Specifications of *Kaien-1*

Total length/mm	1,900
Width/mm	1,240
Height/mm	560
Total weight/g*	4,210
Chord length (main wing)/mm	900
Propulsion	2 propellers (30cm in diameter)
Take-off speed/m/s	6–7

* A GPS receiver and an onboard camera are included in the total weight.

Fig. 5.6.4. *Kaien-1* in field tests

5.6.6. Concluding remarks

The present authors propose the new concept of a wing in surface effect ship. It is a combination of the canard configuration and forward-mounted propellers. This concept has favorable features due to the interaction between the canard system and the slip stream of the propeller; it is expected to have better performance than similar conventional crafts, especially in severe sea states. Although the concept is totally different from the conventional approach of PAR and it has a large tail volume, it is possible to ensure the static height stability of the ship. A primary validation of the concept by means of a small self-propulsion model is in progress. At the moment, the test shows the effectiveness of the concept.

The model test will be continued along with improvement of the system and possible scaled-up models. Beside these, there are some projects including numerical simulations and experiments in a towing tank and a wind tunnel.

Acknowledgements

This research was partially supported by the Ministry of Education, Science, Sports and Culture of Japan, Grant-in-Aid for Scientific Research (A), 14205145, 2002.

References

[5-6-1] S. J. Phillips, ed., *JANE's High-Speed Marine Transportation 2001–2002*, JANE's Information Group, 2001.

[5-6-2] R. W. Staufenbiel, "On the Design of Stable Ram Wing Vehicles," Proceedings of Symposium "Ram Wing and Ground Effect Craft," Royal Aeronautical Society, London, May 19, 1987, pp. 110–36, 1987.

[5-6-3] Kirill V. Rozhdestvensky, *Aerodynamics of a Lifting System in Extreme Ground Effect*, Springer, 2000.

[5-6-4] Syozo Kubo, "A Production Model of WIG as a High Speed Marine Craft, 'Marine Slider -Sky 2,'" Proceedings of FAST '91, Trondheim, Norway, pp. 607–22, 1991.

[5-6-5] Takahiro Taketsume, "Field Tests of a Small Self-Propulsion Model of Wing-In-Surface-Effect-Ship with the Canard Configuration," Graduate thesis, Department of Applied Mathematics and Physics, Faculty of Engineering, Tottori University, February 2003 (in Japanese).

5.7

Study on Stability of WIG Craft during Take-off and Landing (2006)

Shi Yajun, Yuan Changhua and Xu Xinfa

China Ship Scientific Research Center, Jiangsu, Wuxi

Abstract

A WIG craft is a new type of ship that flies stably in the zone using ground effect. Only the flying stability was focused on in the early years, but the stability of the WIG craft during take-off and landing is also important. During take-off and landing, WIG craft rely on hydrodynamic and aerodynamic forces. In this paper, characteristics of take-off and landing of WIG craft are analyzed, and suitable motion equations are achieved where space motions of WIG craft on take-off and landing are divided into individual longitudinal motion and lateral motion. Lateral stability and longitudinal stability are studied using small disturbance and linear analysis, respectively. The effects of wind and waves are also considered.

5.7.1. Introduction

Motion modes of WIG craft include moving in waves and flying in the air. The flying stability of WIG craft has been studied for many years, but the stability of motion in waves, especially stability during take-off and landing, has seldom been studied.

Aerodynamic and hydrodynamic forces are coupled in WIG craft during take-off or landing. Aerodynamic force cannot be neglected and actually plays a key role in the take-off and landing processes. So the role and effect of the aerodynamic layout should be fully considered in the problem of take-off and landing stability of WIG craft. Hydrodynamic force is also considered.

In terms of symmetry, this stability problem can be studied with the corresponding lateral stability and longitudinal stability.

The stability of the craft motion in waves or in lateral sliding mode at low velocity (where the aerodynamic force does not play a main role) is not studied in this paper, as it also refers to conventional ships. The lateral sliding state is now studied in the velocity range $0.6V_t \sim V_t$, and $0.6V_L \sim V_L$, where V_t and V_L are takeoff velocity and landing velocity, respectively.

5.7.2. Motion Equations

Coordinate system

Coordinate system O-XYZ is defined as follows: original point O is the CG of the craft, and the OX axis is the symmetric plane (longitudinal plane) with a positive forward, and the OY axis is perpendicular to the OX axis with a positive upward.

The coordinate system O-XYZ is fixed on a WIG craft, and moves together with the craft, so it is called a coordinate onboard system. The OX axis is called the longitudinal axis; it is in the symmetric plane and is parallel to the basic line of the hull (or the chord line of the main wing) with a positive forward. OY is called the perpendicular axis, and is also in the symmetric plane with a positive upward. The OZ axis is called the transverse axis, which is perpendicular to the symmetric plane with a positive rightward. The symmetric plane can be expressed with XOY.

The thrust of the propeller can be expressed directly in the coordinate on board system. Hydrodynamic moment, including controlling moment caused by controlling plane can also be expressed directly in the coordinate onboard system. Gravity and other hydrodynamic moments can be given on other coordinates, and expressed on the board coordinate system via the coordinate transfer system. The moment of inertia of the WIG craft to the CG in the board coordinate system does not include the consumption of fuel, so they are constant. Therefore, it is more convenient to investigate the motion of the WIG by means of the board coordinate system.

Nomenclature

V_t	take-off speed
V_L	touchdown speed
V_0	initial speed

V_x speed on X direction

V_y speed on Y direction

V_z speed on Z direction

m_x added mass in X direction

m_y added mass in Y direction

m_z added mass in Z direction

ω_x angular velocity about the longitudinal axis

ω_y angular velocity about the vertical axis

ω_z angular velocity about the transverse axis

J_x added moment of inertia about the longitudinal axis

J_y added moment of inertia about the vertical axis

J_z added moment of inertia about the transverse axis

J_{xy} added moment of inertia about the longitudinal and vertical axis

X drag, including aerodynamic and hydrodynamic forces

Y drag, including the projection of aero-hydrodynamic forces on the vertical axis (lift)

Z lateral force, including the projection of aero-hydrodynamic forces on the transverse axis

M_x heeling moment, with respect to the external forces acting on the longitudinal axis

M_y deviation moment, with respect to the external forces acting on the vertical axis

M_z trimming moment, with respect to the external forces acting on the transverse axis

ϕ heeling angle

ϕ_0 initial heeling angle

ϑ pitching angle

ϑ_0 initial pitching angle

β side slip angle

C_{za} lateral aerodynamic force coefficient

C_{zw} lateral hydrodynamic force coefficient

m_{xa} aerodynamic heeling moment coefficient

m_{ya} aerodynamic deviation moment coefficient

m_{xw} hydrodynamic heeling moment coefficient

m_{yw} hydrodynamic deviation moment coefficient

$\overline{\omega}_y$ non-dimensional angular velocity about vertical axis

V_{wind1} force component on longitudinal axis, caused by gust

V_{wind2} force component on transverse axis, caused by gust

m_H added mass in Y direction (vertical direction)

P thrust of propeller

y_P thrust level

$\dot{\vartheta}$ pitching angular velocity

$\ddot{\vartheta}$ pitching angular acceleration

$\overline{\ddot{H}}$ non-dimensional vertical acceleration

$\overline{\dot{H}}$ non-dimensional vertical velocity

ρ air density

S wing area

$\overline{H} = H / \overline{C}$ relative flying height of CG of the WIG craft from the supported surface

H flying height of CG of the craft from the supported surface

C_y^{ϑ} first derivative of the aerodynamic lift coefficient with respect to the pitching angle, when $\overline{H}$ keeps constant

$C_y^{\overline{H}}$ first derivative of the aerodynamic lift coefficient with respect to the relative flying height, when ϑ keeps constant

$C_y^{\overline{\dot{H}}}$ first derivative of the aerodynamic lift coefficient with respect to the non-dimensional vertical speed, when ϑ keeps constant

$C_y^{\overline{\dot{\vartheta}}}$ first derivative of the aerodynamic lift coefficient with respect to the non-dimensional pitching angular velocity, when $\overline{H}$ keeps constant

$\Delta m_{z\vartheta}$ perturbation moment coefficient

ξ_H heaving attenuation coefficient caused by the hydrodynamic load

ξ'_H heaving attenuation coefficient including the influence of aerodynamic forces

ξ_ϑ pitching attenuation coefficient caused by hydrodynamic forces

ξ'_ϑ pitching attenuation coefficient including the influence of aerodynamic forces

ω_H inherited heaving frequency caused by the hydrodynamic load

ω'_H inherited heaving frequency, including the effect of the aerodynamic force

ω_ϑ inherited pitching frequency caused by the hydrodynamic load

ω'_ϑ inherited pitching frequency, including the effect of the aerodynamic force

It should be noted that the motion equations of the WIG craft at take-off and landing differ from those of a normal ship or airplane: here the aerodynamic force should be considered as part of the ship's motion equations and the added water mass should be considered as part of the airplane's motion equations.

Suppose that the WIG craft is rigid and has a symmetry profile, as well as the center of the added mass lies on the CG; in that case, the motion equations in the body axis coordinate system O-XYZ can be expressed as follows:

$$(m+m_x)\frac{dV_x}{dt}+(m+m_z)\omega_y V_z-(m+m_y)\omega_z V_y=X$$

$$(m+m_y)\frac{dV_y}{dt}+(m+m_x)\omega_z V_x-(m+m_z)\omega_x V_z=Y$$

$$(m+m_z)\frac{dV_z}{dt}+(m+m_y)\omega_x V_y-(m+m_x)\omega_y V_x=Z$$

$$(I_x+J_x)\frac{d\omega_x}{dt}+(I_z-I_y+J_z-J_y)\omega_y\omega_z+(I_{xy}+J_{xy})\left(\omega_x\omega_z-\frac{d\omega_y}{dt}\right)+(m_z-m_y)V_yV_z=M_x$$

$$(I_z+J_z)\frac{d\omega_z}{dt}+(I_y-I_x+J_y-J_x)\omega_x\omega_y+(I_{xy}+J_{xy})\left(\omega_y^2-\omega_x^2\right)+(m_y-m_x)V_xV_y=M_z$$

$$(I_y + J_y)\frac{d\omega_y}{dt} + (I_x - I_z + J_x - J_z)\omega_x\omega_z - (I_{xy} + J_{xy})\left(\omega_y\omega_z + \frac{d\omega_x}{dt}\right) + (m_x - m_z)V_xV_z = M_y$$

(5.7.1)

Where m_x, m_y, m_z are the added masses in the directions of *X*, *Y*, *Z*, respectively; J_x, J_y, J_z, J_{xy} are the added inertia moments, respectively, and the inertia product of the corresponding axis, *X*, *Y*, *Z*, are the projected values of the external force, respectively, including the aerodynamic and hydrodynamic forces. M_x, M_y, M_z are the moments to the corresponding axis component of the external force, respectively, including the aerodynamic force and hydrodynamic forces. $\vec{V}_c = V_x\vec{i} + V_y\vec{j} + V_z\vec{k}$, $\vec{\omega} = \omega_x\vec{i} + \omega_y\vec{j} + \omega_z\vec{k}$. And the body axis coordinate system *O-XYZ* is defined as follows: the point *O* lies at the CG, while *OX* axis points to the fore of the craft; the *OY* axis points upward, which is within the symmetry section (longitudinal mid-ship section) and perpendicular to the *OX* axis; the *OZ* axis is perpendicular to symmetry (*OXY*) and points to the starboard.

The difference between the above equations group and the motion equations group of the WIG craft flying in the air are that there are additional added mass and inertial moments, and there are linear velocity product terms in the moment equation and aerodynamic force and hydrodynamic force in the external force and moment. Thus, $m_x \neq m_y \neq m_z$, $J_x \neq J_y \neq J_z$ in the normal condition, and the motion equations group is more complicated.

How we can solve the problem of motion stability of the WIG craft at take-off and landing by the motion equations group will be introduced in the following sections.

5.7.3. Course Stability of WIG craft during take-off and landing

We now turn to studying the course stability that keeps the WIG craft navigation in line. It is supposed that $V_x = V_{x0}$, $V_y = V_{y0} = 0$, $V_{z0} = 0$, $\omega_{x0} = \omega_{y0} = \omega_{z0} = 0$, and $\Delta\omega_z = 0$; therefore, the above movement equations group can be simplified as follows:

$$\begin{cases}(m+m_z)\dfrac{d\Delta V_z}{dt}-(m+m_x)V_{x0}\Delta\omega_y=\dfrac{\partial z}{\partial V_z}\Delta V_z+\dfrac{\partial z}{\partial\phi}\Delta\phi+\dfrac{\partial z}{\partial\omega_x}\Delta\omega_x+\dfrac{\partial z}{\partial\omega_y}\Delta\omega_y\\(I_x+J_x)\dfrac{d\Delta\omega_x}{dt}-(I_{xy}+J_{xy})\dfrac{d\Delta\omega_y}{dt}=\dfrac{\partial M_x}{\partial V_z}\Delta V_z+\dfrac{\partial M_x}{\partial\phi}\Delta\phi+\dfrac{\partial M_x}{\partial\omega_x}\Delta\omega_x+\dfrac{\partial M_x}{\partial\omega_y}\Delta\omega_y\\(I_y+J_y)\dfrac{d\Delta\omega_y}{dt}-(I_{xy}+J_{xy})\dfrac{d\Delta\omega_x}{dt}+(m_x-m_z)V_{x0}\Delta V_z=\dfrac{\partial M_y}{\partial V_z}\Delta V_z+\dfrac{\partial M_y}{\partial\phi}\Delta\phi+\\\qquad\dfrac{\partial M_y}{\partial\omega_x}\Delta\omega_x+\dfrac{\partial M_y}{\partial\omega_y}\Delta\omega_y\end{cases}$$

(5.7.2)

There are 3 linear ordinary differential equations with four variables $\Delta V_z, \Delta\phi, \Delta\omega_x, \Delta\omega_y$ in the equations group, so it is unclosed. One more equation should be added to make the equations group closed. According to the kinematic relation the following equation is introduced into the equations group:

$$\frac{d\Delta\phi}{dt}=\Delta\omega_x-tg\vartheta_0\Delta\omega_y \qquad (5.7.3)$$

Equation 5.7.2 together with Equation 5.7.3 forms a closed system. The motion equations can be expressed in dimensionless form:

$$\begin{cases}2\dfrac{d\beta}{d\bar{t}}-\left(aC_{za}^{\phi}+C_{zw}^{\phi}\right)\Delta\phi-\left(aC_{za}^{\beta}+C_{zw}^{\beta}\right)\Delta\beta-\left(aC_{za}^{\bar{\omega}_x}+C_{zw}^{\bar{\omega}_x}\right)\Delta\bar{\omega}_x\\-\left(aC_{za}^{\bar{\omega}_y}+C_{zw}^{\bar{\omega}_y}+2\mu_x\right)\Delta\bar{\omega}_y=0\\\dfrac{d\Delta\bar{\omega}_x}{d\bar{t}}-A\dfrac{d\Delta\bar{\omega}_y}{d\bar{t}}-\left(a'\bar{m}_{xa}^{\phi}+\bar{m}_{xw}^{\phi}\right)\Delta\phi-\left(a'\bar{m}_{xa}^{\beta}+\bar{m}_{xw}^{\beta}\right)\Delta\beta\\-\left(a'\bar{m}_{xa}^{\bar{\omega}_x}+\bar{m}_{xw}^{\bar{\omega}_x}\right)\Delta\bar{\omega}_x-\left(a'\bar{m}_{xa}^{\bar{\omega}_y}+\bar{m}_{xw}^{\bar{\omega}_y}\right)\Delta\bar{\omega}_y=0\\\dfrac{d\Delta\bar{\omega}_y}{d\bar{t}}-B\dfrac{d\Delta\bar{\omega}_x}{d\bar{t}}-\left(a'\bar{m}_{ya}^{\phi}+\bar{m}_{yw}^{\phi}\right)\Delta\phi-\left[a'\bar{m}_{ya}^{\beta}+\bar{m}_{yw}^{\beta}+\dfrac{(\mu_z-\mu_x)}{i'_y}\right]\Delta\beta\\-\left(a'\bar{m}_{ya}^{\bar{\omega}_x}+\bar{m}_{yw}^{\bar{\omega}_x}\right)\Delta\bar{\omega}_x-\left(a'\bar{m}_{ya}^{\bar{\omega}_y}+\bar{m}_{yw}^{\bar{\omega}_y}\right)\Delta\bar{\omega}_y=0\\\dfrac{d\Delta\phi}{d\bar{t}}=\mu_z\Delta\bar{\omega}_x\end{cases} \qquad (5.7.4)$$

Where C_{za}, m_{xa}, m_{ya} are the coefficients of aerodynamic force and moment, while C_{zw}, m_{xw}, m_{yw} are the coefficients of hydrodynamic force and moment.

By utilizing equation 5.7.4, the course stability of the WIG craft during take-off and landing states is analyzed with the well-known method.[5-7-2]

The generalized condition of the course motion stability should be studied first to derive the course in a standard and simple way. And then the relationship of the course stability to the original state and overall dimensions coefficient is analyzed to guide the design and driving.

1. General condition of course movement stability

The characteristic equation of the system of equation 5.7.4 is:

$$A\lambda^4 + B\lambda^3 + C\lambda^2 + D\lambda + E = 0 \qquad (5.7.5)$$

For $A = 1 - A'B' = 1 - \frac{i'_{xy}}{i'_x} \cdot \frac{i'_{xy}}{i'_y} \approx 1$, Equation 5.7.5 can be simplified as:

$$\lambda^4 + B\lambda^3 + C\lambda^2 + D\lambda + E = 0 \qquad (3.7.6)$$

The necessary and sufficient conditions of lateral movement stability are each of the four roots of equation 5.7.6 being non-zero and having a negative real part. The necessary conditions of the negative real part of all roots of equation 5.7.6 are the following equations being founded simultaneously:

$$\begin{cases} B > 0,\ C > 0,\ D > 0,\ E > 0, \\ (BC - D)D - B^2 E > 0. \end{cases} \qquad (5.7.7)$$

Because the values of coefficient *B*, *C*, *D*, and *E* are connected with their original state, it is difficult for them to be stable for a real WIG craft in all the original states. It is one of the design aims for them to be stable for a WIG craft in the practical design working range.

Analysis for the different original states is as follows.

2 Analysis of the lateral stability of WIG craft

First, we studied the coefficients of equation 5.7.5. Usually, $B>0$, $C>0$, $D>0$ are always founded.

With respect to the necessary and sufficient condition of lateral stability shown in equation 5.7.7, the two conditions $E>0$ and $(BC - D)D - B^2E > 0$ are uniquely left for checking. $E>0$ is an aperiodic stability condition, and $(BC - D)D - B^2E > 0$ is an oscillation stability condition.

Therefore, for the course stability study, it is only necessary to check whether the condition $E>0$ is satisfied. As the value *E* is related to motion, it is different in its different original states, which are the combinations of

ϕ_0, V_0, ϑ_0. Generally, the original heeling angle is considered to be zero, which means $\phi_0 = 0$. The route speed V_0 is related to the lateral sliding state and the proportion of aerodynamic force to the hydrodynamic one. The trim angle (pitch angle) will change the value and location of the hydrodynamic force. Therefore the different combination of V_0 and ϑ_0 affects the value of *E* strongly. The WIG craft's course stability in surface take-off and landing should be studied under different original states (V_0, ϑ_0). Then every factor in the expression of *E* should be analyzed in detail. Now a more detailed expression of *E* can be written as follows:

$$E = \frac{1}{2}\mu_z \Big\{ \left(aC^{\beta}_{za} + C^{\beta}_{zw}\right)\Big[\left(a\bar{m}^{\bar{\omega}_y}_{xa} + \bar{m}^{\bar{\omega}_y}_{xw}\right)\left(a\bar{m}^{\phi}_{ya} + \bar{m}^{\phi}_{yw}\right) - \left(a\bar{m}^{\phi}_{xa} + \bar{m}^{\phi}_{xw}\right)\left(a\bar{m}^{\bar{\omega}_y}_{ya} + \bar{m}^{\bar{\omega}_y}_{yw}\right)\Big] +$$
$$\left(a\bar{m}^{\beta}_{xa} + \bar{m}^{\beta}_{xw}\right)\Big[\left(aC^{\phi}_{za} + C^{\phi}_{zw}\right)\left(a\bar{m}^{\bar{\omega}_y}_{ya} + \bar{m}^{\bar{\omega}_y}_{yw}\right) - \left(aC^{\bar{\omega}_y}_{za} + C^{\bar{\omega}_y}_{zw}\right)\left(a\bar{m}^{\phi}_{ya} + \bar{m}^{\phi}_{yw}\right)\Big] +$$
$$\left[a\bar{m}^{\beta}_{ya} + \bar{m}^{\beta}_{yw} + \frac{(\mu_z - \mu_x)}{i'_y}\right]\Big[\left(aC^{\bar{\omega}_y}_{za} + C^{\bar{\omega}_y}_{zw}\right)\left(a\bar{m}^{\phi}_{xa} + \bar{m}^{\phi}_{xw}\right) - \left(aC^{\phi}_{za} + C^{\phi}_{zw}\right)\left(a\bar{m}^{\bar{\omega}_y}_{xa} + \bar{m}^{\bar{\omega}_y}_{xw}\right)\Big]\Big\}$$

(5.7.8)

For $E = E(V_0, \vartheta_0)$, it should be discussed with respect to the different states.

After omitting the secondary item in equation 5.7.8, the following equation is got:

$$E \cong \frac{1}{2}\mu_z \Big\{ C^{\beta}_{zw}\left(\bar{m}^{\bar{\omega}_y}_{xw}\bar{m}^{\phi}_{yw} - \bar{m}^{\phi}_{xw}\bar{m}^{\bar{\omega}_y}_{yw}\right) + \bar{m}^{\beta}_{xw}\left(C^{\phi}_{zw}\bar{m}^{\bar{\omega}_y}_{yw} - C^{\bar{\omega}_y}_{zw}\bar{m}^{\phi}_{yw}\right)$$
$$+\left(\bar{m}^{\beta}_{yw} + \frac{\mu_z - \mu_x}{i'_y}\right)\left(C^{\bar{\omega}_y}_{zw}\bar{m}^{\phi}_{xw} - C^{\phi}_{zw}\bar{m}^{\bar{\omega}_y}_{xw}\right)\Big\}$$

(5.7.9)

At least for the sliding state with $V_0 \le 0.8V_t$, it is enough to judge the course stability by the value of *E* in equation 5.7.9. Through more analysis of every item in equation 5.7.3.8, it is known that C^{ϕ}_{zw} is equal to zero, and $C^{\bar{\omega}_y}_{zw}$ is related to ϑ_0. When $\vartheta_0 < \vartheta_c$, there is $C^{\bar{\omega}_y}_{zw} > 0$. When $\vartheta_0 \ge \vartheta_c$, there is $C^{\bar{\omega}_y}_{zw} \le 0$, $C^{\beta}_{zw} < 0$, $\bar{m}^{\phi}_{xw} < 0$, $\left|\bar{m}^{\phi}_{xw}\right| > \left|\bar{m}^{\phi}_{yw}\right|$, $\bar{m}^{\bar{\omega}_y}_{xw} < 0$, and $\bar{m}^{\bar{\omega}_y}_{yw} < 0$. When $\vartheta_0 < \vartheta_c$, there is $\bar{m}^{\beta}_{yw} > 0$. When

$\vartheta_0 \geq \vartheta_c$, there is $\overline{m}_{yw}^{\beta} < 0$.

Assuming $C_{zw}^{\phi} = 0$, there is:

$$E = \frac{1}{2}\mu_z \left\{ C_{zw}^{\beta} \left(\overline{m}_{xw}^{\overline{\omega}_y} \overline{m}_{yw}^{\phi} - \overline{m}_{xw}^{\phi} \overline{m}_{yw}^{\overline{\omega}_y} \right) - \overline{m}_{xw}^{\beta} C_{zw}^{\overline{\omega}_y} \overline{m}_{yw}^{\phi} + \left(\overline{m}_{yw}^{\beta} + \frac{\mu_z - \mu_x}{i_y'} \right) C_{zw}^{\overline{\omega}_y} \overline{m}_{xw}^{\phi} \right\}$$

(5.7.10)

After studying the above items, it is known that when $\vartheta_0 \leq \vartheta_c$, there is *E*<0, which means the course is not stable, and when $\vartheta_0 > \vartheta_c$, there is *E*>0, which means the course is stable.

3. Gust disturbance

Horizontal gusts are one of the outside disturbances for the WIG craft. When studying the effect of wind, it is divided into two components V_{wind1} and V_{wind2}, which are along the *OX* axis and *OZ* axis, respectively.

The component V_{wind1} along the *OX* axis will affect the take-off speed of the WIG craft (the speed over the ground). The component V_{wind2} along the *OZ* axis will affect the course stability of the WIG craft, only when the aerodynamic force plays a leading role ($V_0 \geq 0.9V_t$ or in landing process). In this case, equation 5.7.2 is changed to the following form:

$$\begin{cases} (m+m_z)\dfrac{d(\Delta V_z + V_{\text{wind2}})}{dt} - (m+m_x)V_{x0}\Delta\omega_y \\ = \dfrac{\partial z}{\partial V_z}(\Delta V_z + V_{\text{wind2}}) + \dfrac{\partial z}{\partial \phi}\Delta\phi + \dfrac{\partial z}{\partial \omega_x}\Delta\omega_x + \dfrac{\partial z}{\partial \omega_y}\Delta\omega_y \\ (I_x + J_x)\dfrac{d\Delta\omega_x}{dt} - (I_{xy} + J_{xy})\dfrac{d\Delta\omega_y}{dt} \\ = \dfrac{\partial M_x}{\partial V_z}(\Delta V_z + V_{\text{wind2}}) + \dfrac{\partial M_x}{\partial \phi}\Delta\phi + \dfrac{\partial M_x}{\partial \omega_x}\Delta\omega_x + \dfrac{\partial M_x}{\partial \omega_y}\Delta\omega_y \\ (I_y + J_y)\dfrac{d\Delta\omega_y}{dt} - (I_{xy} + J_{xy})\dfrac{d\Delta\omega_x}{dt} + (m_x - m_z)V_{x0}(\Delta V_z + V_{\text{wind2}}) \\ = \dfrac{\partial M_y}{\partial V_z}(\Delta V_z + V_{\text{wind2}}) + \dfrac{\partial M_y}{\partial \phi}\Delta\phi + \dfrac{\partial M_y}{\partial \omega_x}\Delta\omega_x + \dfrac{\partial M_y}{\partial \omega_y}\Delta\omega_y \end{cases} \quad (5.7.11)$$

Referring to 1, it can be analyzed, if remarking $\Delta\beta = \dfrac{\Delta V_z + V_{\text{wind2}}}{V_0}$.

5.7.4. Longitudinal Stability of WIG Crafts during take-off and landing

1. General study of longitudinal stability

When studying the longitudinal motion stability, there are $V_x = V_0$, $V_z = 0$, $\omega_x = 0$, $\omega_y = 0$. While studying the stability of heave and pitch motion (porpoise motion), using the earth's axes system is more convenient:

$$\begin{cases} (m+m_H)\ddot{H} = P\sin(\vartheta+\varphi_P) + Y\cos\Theta - X\sin\Theta - G \\ (I_z+J_z)\ddot{\vartheta} = M_z P y_P \\ \dot{H} = V\sin\Theta \end{cases} \quad (5.7.12)$$

where, m_H is the added mass in direction *Y* (vertical direction). When $\vartheta = 0$, $m_H = m_y$.

J_z is the added moment of inertia around axis *Z*, P is engine thrust, and y_P is thrust level.

X and *Y* are the components of drag and lift, respectively, in the craft axis coordinate system, and $X = X_a + X_w$, $Y = Y_a + Y_w$, which is the sum of the aerodynamic and hydrodynamic forces.

By the linearization of the small disturbance, equation 5.7.12 can be expressed as:

$$\begin{cases} \dfrac{2C}{V_0}\tau\Delta\ddot{\overline{H}} + \left(\dfrac{C}{V_0}C_x - C_y^{\dot{\overline{H}}}\right)\Delta\dot{\overline{H}} - C_y^{\overline{H}}\Delta\overline{H} - \left(C_x + C_y^{\vartheta}\right)\Delta\vartheta = 0 \\ \mu'\Delta\ddot{\vartheta} - m_z^{\dot{\vartheta}}\Delta\dot{\vartheta} - m_z^{\vartheta}\Delta\vartheta - m_z^{\dot{\overline{H}}}\Delta\dot{\overline{H}} - m_z^{\overline{H}}\Delta\overline{H} = 0 \end{cases} \quad (5.7.13)$$

Where $\tau = \dfrac{m+m_H}{\rho V_0 S}$, $\mu' = \dfrac{2(I_z+J_z)}{\rho V_0^2 SC}$, $\Delta\overline{H} = \dfrac{\Delta H}{C}$.

The corresponding characteristic equation is as follows:

$$A\lambda^4 + B\lambda^3 + C\lambda^2 + D\lambda + E = 0,$$

$$\begin{cases} A = \dfrac{2C}{V_0}\tau\mu' \\ B = \left(\dfrac{C}{V_0}C_x - C_y^{\dot{\bar{H}}}\right)\mu' - \dfrac{2C\tau}{V_0}m_z^{\dot{\vartheta}} \\ C = -\mu' C_y^{\bar{H}} - \dfrac{2C\tau}{V_0}m_z^{\vartheta} - \left(\dfrac{C}{V_0}C_x - C_y^{\dot{\bar{H}}}\right)m_z^{\dot{\vartheta}} \\ D = C_y^{\bar{H}} m_z^{\dot{\vartheta}} - \left(\dfrac{C}{V_0}C_x - C_y^{\dot{\bar{H}}}\right)m_z^{\vartheta} - \left(C_x + C_y^{\vartheta}\right)m_z^{\dot{\bar{H}}} \\ E = C_y^{\bar{H}} m_z^{\vartheta} - \left(C_x + C_y^{\vartheta}\right)m_z^{\bar{H}} \end{cases} \quad (5.7.14)$$

Then the necessary and sufficient conditions of the longitudinal stability of the WIG on take-off and landing are:

$$\begin{aligned} &A > 0,\ B > 0,\ C > 0,\ D > 0,\ E > 0 \\ &R = D(BC - AD) - B^2 E > 0 \end{aligned} \quad (5.7.15)$$

They should be right simultaneously.

If *A*, *B*, *C*, *D*, and *E* are obtained through equation 5.7.14, the longitudinal motion in the take-off and landing state can be judged to be stable or divergent with the condition 5.7.4.4. Thus, the problem of longitudinal stability is already solved really.

Under the condition that lateral sliding speed is not near the take-off speed—that is, $V_0 < 0.8\ V_t$—the aerodynamic force and moment can be omitted, and then the stability of the longitudinal movement (porpoise motion) can be analyzed completely with hydrodynamic characteristics as analyzed for the planing craft. Thus the fluid dynamic coefficients in equation 5.7.14 are directly written as the hydrodynamic coefficients, such as $C_y \to C_{yw}$.

Thus, the coefficients of $C_{xw}, C_{yw}^{\vartheta}, C_{yw}^{\bar{H}}, C_{yw}^{\dot{\bar{H}}}, m_{zw}^{\vartheta}\ m_{zw}^{\dot{\vartheta}}, m_{zw}^{\bar{H}}, m_{zw}^{\dot{\bar{H}}}$ are the key in the analysis of longitudinal stability and the hydrodynamic design in the lateral sliding state.

2. Natural oscillation frequency and attenuation coefficient of longitudinal motion in take-off and landing

To obtain the natural oscillation frequency and attenuation coefficient, the cross-coupling of heave and pitch motion can be decoupled artificially, which means that:

$$\begin{cases} \dfrac{2C_w}{V_0}\tau\Delta\ddot{\overline{H}} + \left(\dfrac{C_w}{V_0}C_{xw} - C_{yw}^{\dot{\overline{H}}}\right)\Delta\dot{\overline{H}} - C_{yw}^{\overline{H}}\Delta\overline{H} = \Delta f_H \\ \mu'\Delta\ddot{\vartheta} - m_{zw}^{\dot{\vartheta}}\Delta\dot{\vartheta} - m_{zw}^{\vartheta}\Delta\vartheta = \Delta m_{z\vartheta} \end{cases} \quad (5.7.16)$$

where, Δf_H and $\Delta m_{z\vartheta}$ are the coefficients of disturbing force and moment, respectively. There is:

$$\Delta\overline{H}(P) = \frac{1}{\dfrac{2C_w}{V_0}\tau P^2 + \left(\dfrac{C_w}{V_0}C_{xw} - C_{yw}^{\dot{\overline{H}}}\right)P - C_{yw}^{\overline{H}}}\Delta f_H(P)$$

$$= \frac{1}{(P^2 + 2\xi_H P + \omega_H^2)}\left(\frac{V_0}{2C_w\tau}\Delta f_H(P)\right)$$

where,

$$\begin{cases} \omega_H = \sqrt{\dfrac{(-C_{yw}^{\overline{H}})\rho_w S_w V_0^2}{2C_w(m+m_H)}} \\ \xi_H = \dfrac{\rho_w S_w V_0^2}{4C_w(m+m_H)}\left(\dfrac{C_w}{V_0}C_{xw} - C_{yw}^{\dot{\overline{H}}}\right) \\ \omega_\vartheta = \sqrt{\dfrac{(-m_{zw}^{\vartheta})C_w\rho_w S_w V_0^2}{2(I_z + J_z)}} \\ \xi_\vartheta = \dfrac{-m_{zw}^{\dot{\vartheta}}C_w\rho_w S_w V_0^2}{4(I_z + J_z)} \end{cases} \quad (5.7.17)$$

Obviously, if $\xi_H > 0,\ \xi_\vartheta > 0$, both heave and pitch motions are damped, which means that they are stable. On the contrary, if $\xi_H < 0,\ \xi_\vartheta < 0$, both heave and pitch motions are diffused, which means that they are unstable.

According to the characteristics of the WIG craft motion, the factor $S_w V_0^2$ reduces rapidly with the increase of V_0. The tendency of variation can be shown as in Fig. 5.7.1.

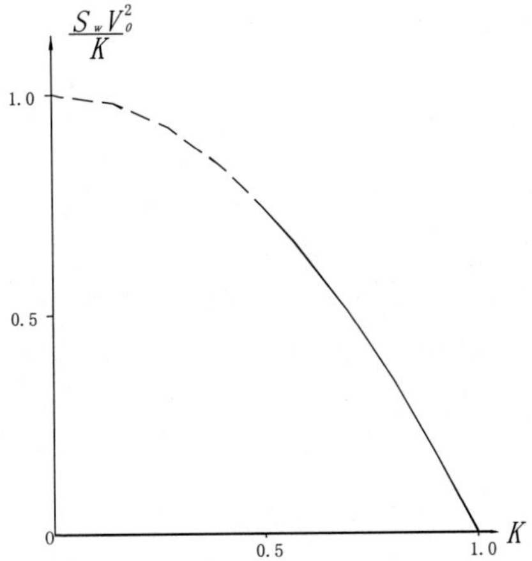

Fig. 5.7.1. The variation tendency of $S_w V_0^2$

According to Fig. 5.7.1, $S_w V_0^2$ reduces rapidly with the increase of V_0. Natural frequency and attenuation coefficients induced by hydrodynamic load will reduce rapidly; likewise ξ_H will reduce more rapidly too, as shown in Fig. 5.7.2.

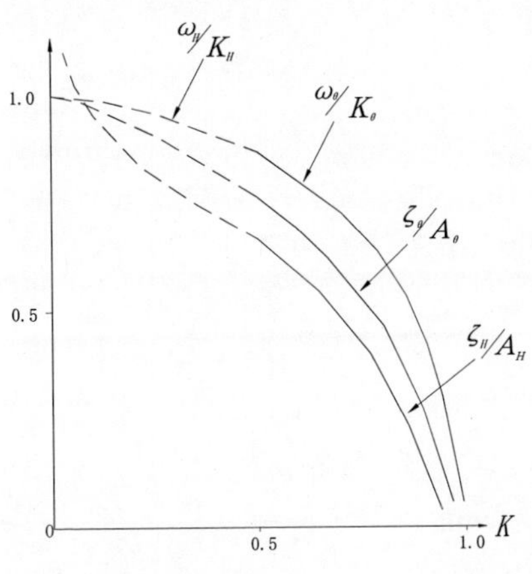

Fig. 5.7.2. The variation tendency of ω_H、ω_ϑ、ξ_H and ξ_ϑ

When the lateral sliding speed V_0 is gradually close to the take-off speed, the coefficients of natural frequency and attenuation induced by hydrodynamic load reduce rapidly, so that the original damping oscillation becomes a non-periodic divergent motion step by step. But this case does not occur on the WIG craft, since the expressions of $\omega_H, \omega_\vartheta, \xi_H, \xi_\vartheta$ should include the effect of the aerodynamic coefficient; thus, equation 5.7.4.17 should be replaced by the following formula:

$$\begin{cases} \omega'_H = \sqrt{\dfrac{-\rho_w S_w V_0^2 \left(aC_{ya}^{\overline{H}} + C_{yw}^{\overline{H}}\right)}{2C_w(m+m_H)}}, \\ \xi'_H = \dfrac{\rho_w S_w V_0^2}{4C_w(m+m_H)}\left[\dfrac{C_w}{V_0}(C_{xw} + aC_{xa}) - \left(aC_{ya}^{\dot{\overline{H}}} + C_{yw}^{\dot{\overline{H}}}\right)\right], \\ \omega'_\vartheta = \sqrt{\dfrac{-C_w \rho_w S_w V_0^2 \left(a'm_{za}^{\vartheta} + m_{zw}^{\vartheta}\right)}{2(I_z + J_z)}}, \\ \xi'_\vartheta = -\dfrac{C_w \rho_w S_w V_0^2 \left(a'm_{za}^{\dot{\vartheta}} + m_{zw}^{\dot{\vartheta}}\right)}{4(I_z + J_z)} \end{cases} \tag{5.7.18}$$

Here, the symbol $'$ expresses the corresponding values induced by aerodynamic forces.

$\omega'_H \xrightarrow[k\to 1]{} \sqrt{\dfrac{g(-C_{ya}^{\overline{H}})}{C_w C_{yt}}}$, when the lateral sliding speed $V_0 \to V_t$.

Attenuation coefficient ξ'_H can't be reduced unlimitedly, and will tend to be a fixed value $-\dfrac{C_{ya}^{\dot{\overline{H}}} g}{2C_w C_{yt}}$. It is clearly shown that the aerodynamic load increases the longitudinal motion stability of the WIG craft in the take-off and landing state. Thus, it is entirely possible that the "porpoise motion" does not happen during the take-off and landing process.

The natural frequency of pitch tends to a fixed value, while $V_0 \to V_t$, which means that $\omega'_\vartheta \xrightarrow[k\to 1]{} \sqrt{\dfrac{(-m_{za}^{\vartheta})g}{i_z C_{yt} C_a}}$, and the coefficient of longitudinal wave attenuation $\xi'_\vartheta \xrightarrow[k\to 1]{} \dfrac{g}{2i_z C_a C_{yt}}(-m_{za}^{\dot{\vartheta}})$.

It can be concluded that the contribution of the aerodynamic will raise the stability of pitch and heave motions.

It is indicated that owing to the aerodynamic contribution, the longitudinal motion of the WIG craft always tends toward being stable in the take-off and landing state or when nearing take-off and landing (suppose that the aerodynamic layout has proven the flying stability). So the key is to analyze the motion with the planing velocity not being so high; thus: $V_0 < 0.8V_t$.

If this is so, the hull design of the WIG craft can refer back to some basic rules of planing craft in the stability design.

3. Wave disturbance

The above discussion on longitudinal stability is focused on the stability of the craft itself without wave disturbance. If the craft itself is unstable, it will be unstable with any disturbance, no matter what the external disturbance characteristic may be. The so-called porpoise motion means the motion is unstable after the disturbance disappears. If the wave disturbance is continual, the main characteristic is the disturbance frequency—that is, the encountering frequency $\omega = 2\pi \frac{V_0}{\lambda}$, where λ is the wave length (suppose the wave is steady). When the disturbance frequency is close or even equal to the natural frequency ω_H or ω_ϑ, the amplitude of the motion will increase significantly. When the disturbance frequency is far from the natural frequency, the amplitude of the motion will be reduced greatly, and the wave has little effect on the motion.

5.7.5. Conclusions

The course stability of WIG craft during take-off and landing depends on the hydrodynamic characteristics at the condition $V_0 < 0.8V_t$ (or $0.8V_L$). The course stability can be assured, so long as the pitch angle is greater than the critical angle, $\vartheta_0 > \vartheta_c$. When $\vartheta_0 < \vartheta_c$, an unstable course can easily occur, which will lead to a sudden dangerous turn. When $V_0 \geq 0.9V_t$ (or $0.9V_L$), the course stability of the WIG craft depends mainly on the aerodynamic characteristics and not on the hydrodynamic characteristics. Since the fore part of the craft used to be raised in the take-off state, it is most possible that it will happen in the landing state: $\vartheta_0 < \vartheta_c$. It should be regarded as significant by the pilot.

The longitudinal motion stability of the WIG craft during take-off and landing depends on its hydrodynamic characteristics, aerodynamic characteristics, and the characteristics of dynamical augmentation of lift. The hydrodynamic force often increases with the increase of the planing speed, and debases the stability, so as to lead to the porpoise motion. Aerodynamic characteristics take active action on the longitudinal stability (only if the aerodynamic layout is stable). The aerodynamic characteristics will be of greater help when the planing velocity is closer to the take-off speed. Dynamical augmentation always enhances the take-off and landing stability of WIG craft, especially in a low-speed planing state.

5.8

Study on Optimization of Automatic Control Parameters for WIG Craft (Ekranoplan) (2006)

Zhou Qian, Xu Xingfa
and Yuan Changhua

China Ship Scientific Research Center, Wuxi, China

Abstract

In this paper, a method based on the space motion equation of ekranoplan for optimization of automatic control parameters is introduced.

First, the space motion equation of "WIG + Automatic Control System" is established. Then MATLAB is chosen as the main software support for establishing the mathematical model of the optimization system with the help of its non-linear multivariable function fmincon (). Finally, the optimum procedure for the automatic control parameters is programmed on the platform MATLAB/GUIDE, and the optimization calculation of automatic control parameters can be carried out.

The results show that, through the optimized calculation of automatic control parameters of WIG, optimal results may be achieved, and for solving the questions of the dynamic stability and maneuverability of WIG, it is unnecessary purely to depend only on aerodynamic design.

5.8.1. Introduction

Usually the questions about the motion stability and maneuverability of WIG are solved using the aerodynamic design, but there exist certain limitations: under some situations it is often difficult to satisfy the purpose, and when an operating region is quite wide, the limitation is more obvious. The aerodynamic properties of WIG without an automatic control system are completely determined by the aerodynamic design and the configuration parameters of WIG, and it is extremely difficult to satisfy the prescriptive technical specifications through modifying the aerodynamic design, because the aerodynamic design often has been modified many times and certain sacrifices (for example aerodynamic quality or structural weight and so on) must be made for design, even though under some conditions it is impossible to modify the aerodynamic design. Therefore, a WIG needs to be equipped with an automatic control system, and it is possible to satisfy all kinds of technical specifications for the motion stability of WIG through reasonable selection of the automatic control parameters—for instance, a period of vibration, the maximum amplitude of vibration, the time of attenuation at half amplitude, and so on. Compared with modifying the aerodynamic design, this should be much easier.

Therefore, using the space motion equation with control parameters, the optimized computation of automatic control parameters for WIG may be carried out by the design variable, the objective function, and the constraint condition, respectively, until the optimal automatic control parameters satisfying the purpose are obtained.

5.8.2. The space motion equation of WIG with automatic control parameters

The coordinate system and nomenclature can be taken from section 5.7.2. Let $X_1 = \Delta\overline{V}$, $X_2 = \Delta\alpha$, $X_3 = \Delta\overline{H}$, $X_4 = \Delta\vartheta$, $X_5 = \Delta\beta$, $X_6 = \Delta\phi$, $X_7 = \Delta\overline{\omega}_x$, $X_8 = \Delta\overline{\omega}_y$, $X_9 = \Delta\overline{\omega}_z$, $X_{10} = \Delta\psi$, $X_{11} = \Delta\delta_{ta}$, $X_{12} = \Delta\delta_{ea}$, $X_{13} = \Delta\delta_{fa}$, $X_{14} = \Delta\delta_{aa}$, $X_{15} = \Delta\delta_{ra}$.

And $K_{\dot{V}} = k_1$, $K_V = k_2$, $K_{\dot{\vartheta}} = k_3$, $K_{\vartheta} = k_4$, $K_{\dot{H}} = k_5$, $K_H = k_6$, $K_{\dot{\phi}} = k_7$, $K_{\phi} = k_8$, $K_{\dot{\psi}} = k_9$, $K_{\psi} = k_{10}$. $a_{11,1}^0 = \dfrac{k_1}{\tau}$, $a_{11,1} = k_2$, $a_{12,4}^0 = -\dfrac{k_3}{\tau}$, $a_{12,4} = -k_4$, $a_{13,3}^0 = \dfrac{k_5}{\tau}$, $a_{13,3} = k_6$, $a_{14,6}^0 = -\dfrac{k_7}{\tau}$, $a_{14,6} = -k_8$, $a_{15,10}^0 = -\dfrac{k_9}{\tau}$, $a_{15,10} = -k_{10}$.

The coordinate system and nomenclature has been introduced in the previous section (section 5.7.2).

The linear and dimensionless space motion equation of WIG with automatic control parameters is educed as follows:

$$\begin{cases}
a_{11}^{0}\dfrac{dX_1}{d\bar{t}}+a_{11}X_1+a_{12}^{0}\dfrac{dX_2}{d\bar{t}}+a_{12}X_2+a_{13}^{0}\dfrac{dX_3}{d\bar{t}}+a_{13}X_3+a_{14}^{0}\dfrac{dX_4}{d\bar{t}}+a_{14}X_4+a_{15}^{0}\dfrac{dX_5}{d\bar{t}}+\\
+a_{15}X_5+a_{16}X_6+a_{17}X_7+a_{18}X_8+a_{19}X_9+a_{1,11}X_{11}+a_{1,12}X_{12}+a_{1,13}X_{13}+a_{1,15}X_{15}=\\
=-a_{1,11}\Delta\delta_{tc}-a_{1,12}\Delta\delta_{ec}-a_{1,13}\Delta\delta_{fc}-a_{1,15}\Delta\delta_{rc}+\Delta C_{xp}\\
a_{21}^{0}\dfrac{dX_1}{d\bar{t}}+a_{21}X_1+a_{22}^{0}\dfrac{dX_2}{d\bar{t}}+a_{22}X_2+a_{23}^{0}\dfrac{dX_3}{d\bar{t}}+a_{23}X_3+a_{24}^{0}\dfrac{dX_4}{d\bar{t}}+a_{24}X_4+a_{25}^{0}\dfrac{dX_5}{d\bar{t}}+\\
+a_{25}X_5+a_{26}X_6+a_{27}X_7+a_{28}X_8+a_{29}X_9+a_{2,11}X_{11}+a_{2,12}X_{12}+a_{2,13}X_{13}+a_{2,15}X_{15}=\\
=-a_{2,11}\Delta\delta_{tc}-a_{2,12}\Delta\delta_{ec}-a_{2,13}\Delta\delta_{fc}-a_{2,15}\Delta\delta_{rc}-\Delta C_{yp}\\
a_{31}^{0}\dfrac{dX_1}{d\bar{t}}+a_{31}X_1+a_{33}^{0}\dfrac{dX_3}{d\bar{t}}+a_{33}X_3+a_{34}^{0}\dfrac{dX_4}{d\bar{t}}+a_{34}X_4+a_{35}^{0}\dfrac{dX_5}{d\bar{t}}+\\
+a_{35}X_5+a_{36}X_6+a_{37}X_7+a_{38}X_8+a_{3,12}X_{12}+a_{3,13}X_{13}+a_{3,15}X_{15}=\\
=-a_{3,12}\Delta\delta_{ec}-a_{3,13}\Delta\delta_{fc}-a_{3,15}\Delta\delta_{rc}+\Delta C_{zp}\\
a_{41}X_1+a_{43}X_3+a_{44}X_4+a_{45}X_5+a_{46}X_6+\dfrac{dX_7}{d\bar{t}}+a_{47}X_7+a_{48}^{0}\dfrac{dX_8}{d\bar{t}}+a_{48}X_8+\\
+a_{4,14}X_{14}+a_{4,15}X_{15}=-a_{4,14}\Delta\delta_{ac}-a_{4,15}\Delta\delta_{rc}+\Delta\overline{m}_{xp}\\
a_{51}X_1+a_{53}X_3+a_{54}X_4+a_{55}X_5+a_{56}X_6+a_{57}^{0}\dfrac{dX_7}{d\bar{t}}+a_{57}X_7+\dfrac{dX_8}{d\bar{t}}+a_{58}X_8+\\
+a_{5,14}X_{14}+a_{5,15}X_{15}=-a_{5,14}\Delta\delta_{ac}-a_{5,15}\Delta\delta_{rc}+\Delta\overline{m}_{yp}\\
a_{61}X_1+a_{63}^{00}\dfrac{d^2X_3}{d\bar{t}^2}+a_{63}^{0}\dfrac{dX_3}{d\bar{t}}+a_{63}X_3+a_{64}^{0}\dfrac{dX_4}{d\bar{t}}+a_{64}X_4+a_{65}X_5+a_{66}X_6+\dfrac{dX_9}{d\bar{t}}+\\
+a_{6,11}X_{11}+a_{6,12}X_{12}+a_{6,13}X_{13}=-a_{6,11}\Delta\delta_{tc}-a_{6,12}\Delta\delta_{ec}-a_{6,13}\Delta\delta_{fc}+\Delta\overline{m}_{zp}
\end{cases}$$

$$
\begin{cases}
a_{71}X_1 + a_{72}X_2 + \dfrac{dX_3}{d\bar{t}} + a_{74}X_4 + a_{75}X_5 + a_{76}X_6 = 0 \\
\dfrac{dX_4}{d\bar{t}} + a_{88}X_8 + a_{89}X_9 = 0 \\
\dfrac{dX_6}{d\bar{t}} + a_{97}X_7 + a_{98}X_8 + a_{99}X_9 = 0 \\
\dfrac{dX_{10}}{d\bar{t}} + a_{10,8}X_8 + a_{10,9}X_9 = 0 \\
a_{11,1}^0 \dfrac{dX_1}{d\bar{t}} + a_{11,1}X_1 + a_{11,11}^0 \dfrac{dX_{11}}{d\bar{t}} + X_{11} = 0 \\
a_{12,4}^0 \dfrac{dX_4}{d\bar{t}} + a_{12,4}X_4 + a_{12,12}^0 \dfrac{dX_{12}}{d\bar{t}} + X_{12} = 0 \\
a_{13,3}^0 \dfrac{dX_3}{d\bar{t}} + a_{13,3}X_3 + a_{13,13}^0 \dfrac{dX_{13}}{d\bar{t}} + X_{13} = 0 \\
a_{14,6}^0 \dfrac{dX_6}{d\bar{t}} + a_{14,6}X_6 + a_{14,14}^0 \dfrac{dX_{14}}{d\bar{t}} + X_{14} = 0 \\
a_{15,10}^0 \dfrac{dX_{10}}{d\bar{t}} + a_{15,10}X_{10} + a_{15,15}^0 \dfrac{dX_{15}}{d\bar{t}} + X_{15} = 0
\end{cases}
$$

Where $V_0 = \sqrt{\dfrac{2G}{\rho S C_{y_0} \cos\phi_0}}$, $\tau = \dfrac{2G}{g\rho V_0 S}$, $\mu_c = \dfrac{2G}{g\rho S\overline{C}}$,
$\mu_b = \dfrac{2G}{g\rho Sb}$, $\Delta\overline{V} = \dfrac{\Delta V}{V_0}$, $\Delta\overline{H} = \dfrac{\Delta H}{\overline{C}}$, $\Delta\overline{\omega}_x = \dfrac{b}{2V_0}\Delta\omega_x$,
$\Delta\overline{\omega}_y = \dfrac{b}{2V_0}\Delta\omega_y, \Delta\overline{\omega}_z = \dfrac{\overline{C}}{V_0}\Delta\omega_z$.

5.8.3. Mathematical model of optimized systems

1. Design variables

In this paper, the automatic control parameters in the space motion equation of WIG are taken as the design variable. These automatic control parameters are: automatic control parameter K_V and $K_{\dot{V}}$, corresponding with the velocity; automatic control parameter K_ϑ and $K_{\dot{\bar{\vartheta}}}$, corresponding with the longitudinal motion pitch; automatic control parameter $K_{\overline{H}}$ and $K_{\dot{\overline{H}}}$, corresponding with the longitudinal motion flying altitude; automatic control parameter K_{ϕ} and $K_{\dot{\phi}}$, corresponding with the transversal motion heel; automatic control parameter K_{ψ} and $K_{\dot{\psi}}$, corresponding with the transversal motion yaw.

2. Constraint conditions

In this paper, the constraint conditions are the boundary conditions of the automatic control parameters; the lower and upper bounds of these constraint conditions, which are based on our experience, are given below: $K_{\dot{V}} = 0 \sim 0.5$, $K_{V} = 0 \sim 0.8$, $K_{\dot{\overline{H}}} = 0 \sim 1.1$, $K_{\overline{H}} = 0 \sim 0.6$, $K_{\dot{\bar{\vartheta}}} = 0 \sim 1.3$, $K_{\vartheta} = 0 \sim 1.8$, $K_{\dot{\phi}} = 0 \sim 1.2$, $K_{\phi} = 0 \sim 1$, $K_{\dot{\psi}} = 0 \sim 5$, $K_{\psi} = 0 \sim 1.5$.

lb= [0 0 0 0 0 0 0 0 0 0]

ub=[0.5 0.8 1.1 0.6 1.3 1.8 1.2 1 5 1.5]

Here, lb is the lower bounds of the automatic control parameters, and ub is the upper bounds of the automatic control parameters.

3. Objective function

In this paper, the square of the attenuated value for the initial amplitude of the motion parameters is defined as the objective function:

$$J = \sum_{i}^{m} \left(f_i(X_i) - f_{0i}(X_i) \right)^2$$

Where $f_{0i}(X_i)$ is the initial amplitude of the motion parameter under the condition of the initial automatic control parameters; $f_i(X_i)$ is the initial amplitude of the motion parameter under the condition of the optimized automatic control parameters; X_i is the motion parameter of WIG. Here $i = 1$ represents the velocity V , $i = 3$ represents the flight height $\dot{H}$, $i = 4$ represents the pitch angle ϑ, $i = 5$ represents the sideslip angle β, $i = 6$ represents the bank angle ϕ, $i = 10$ represents the yaw angle ψ. Our aim is to find the optimal values of the design variables and the minimum value of the objective function J.

5.8.4. Optimization calculation

In this paper the nonlinear multivariable function fmincon (), which is provided by the MATLAB optimization toolbox, has been selected, and the optimization calculation of automatic control parameters for WIG may be carried out by fmincon ().
General format: x = fmincon (fun,x0,A,b,Aeq,beq,lb,ub)
where x, b, beq, lb and ub are vectors, and A and Aeq are matrices, respectively.

On the basis of the nonlinear multivariable function fmincon () and the solver of ordinary differential equation ode45, the optimum procedure of automatic control parameters for WIG may be programmed with the help of MATLAB/GUIDE.

Thus the practical format will be:
k=fmincon(@ObjS_Equation,k0,[],[],[],[],lb,ub,[],[])
where, k0 is the initial design variable, lb and ub are lower and upper bounds of the design variables, the objective function name is ObjS_Equation, and k is the optimal solution of the automatic control parameters.

5.8.5. Example

In this paper, we take a WIG as an example to demonstrate the optimization calculation.

Fig. 5.8.1 illustrates the convergence of curve along with time for the primary motion parameters. The initial condition is $\vartheta_0 = 1^0, \overline{H}_0 = 0.3$ and the automatic control parameters are zero—that is,
k_0=[$K_{\dot{V}}, K_V, K_{\dot{\overline{H}}}, K_{\overline{H}}, K_{\dot{\overline{\vartheta}}}, K_\vartheta, K_{\dot{\phi}}, K_\phi, K_{\dot{\psi}}, K_\psi$] = [0, 0, 0, 0, 0, 0, 0, 0, 0, 0].

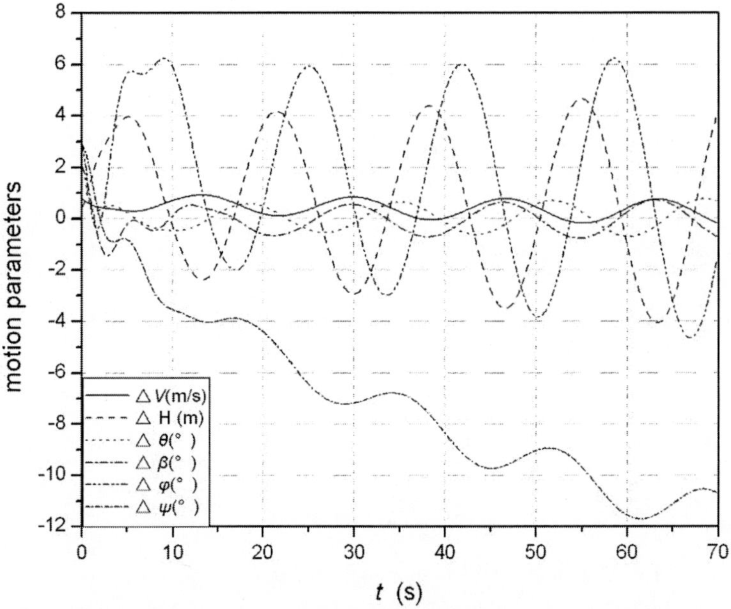

Fig. 5.8.1. The curve of the primary motion parameters before optimization

Suppose that the initial values of the design variables are k_0, the attenuated multiple of the initial amplitude for the primary motion parameters is 0.2, and the iterative number of the optimization calculation is 100, then the optimal results of the design variables are obtained, and the optimized automatic control parameters of WIG are $\boldsymbol{k}$—that is,

$\boldsymbol{k} = [K_{\dot{V}}, K_V, K_{\dot{\bar{H}}}, K_{\bar{H}}, K_{\dot{\bar{\vartheta}}}, K_{\vartheta}, K_{\dot{\phi}}, K_{\phi}, K_{\dot{\psi}}, K_{\psi}]$ = [0.0019, 0.7796, 1.1000, 0.1896, 1.2716, 1.7969, 1.0270, 0.0225, 4.9846, 0.7446]

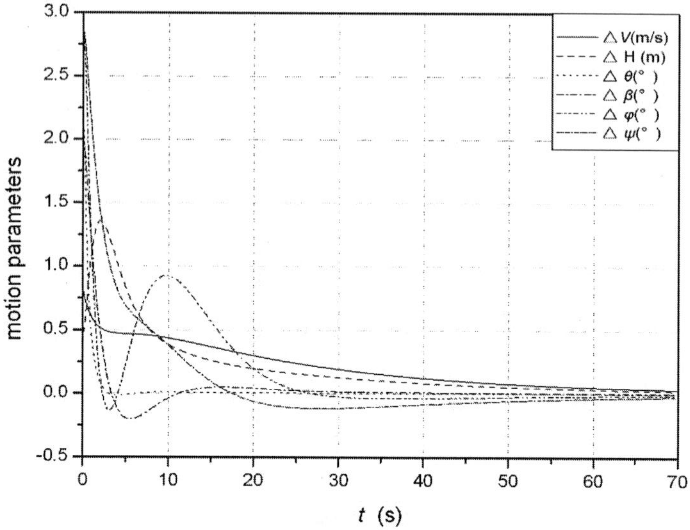

Fig. 5.8.2. The curve of the primary motion parameters after optimization

Fig. 5.8.2 illustrates that after optimization the initial amplitudes of all motion parameters have been largely attenuated, and before optimization the flight height and the bank angle are two maximal initial amplitudes; however, after optimization the initial amplitude of the flight height has been attenuated from about 3.8m to about 1.25m, and the initial amplitude of the bank angle has been attenuated from about 5° to about 1°.

5.8.6. Conclusions

Large quantities of results obtained from other optimization calculations and the above example show that after the aerodynamic characteristics of a WIG have been confirmed and the performance index of a WIG has been given in advance, it is unnecessary to only depend upon the aerodynamic design to solve questions about the motion stability and maneuverability of a WIG. Through the optimization calculation of automatic control parameters, the prescriptive performance index may always be achieved by means of the space motion equation of "WIG + Automatic Control System" and the optimum procedure.

References

[5-8-1] Yuan Changhua & Ye Yonglin, "Study on the Space Motion of Wing-In-Ground Effect Craft," FAST 2003.

[5-8-2] Huang Jianghua, "Practical Computer Simulation of Chemical Processes—MATLAB's Application in Chemical Engineering," Chemical Industry Press, China, 2004 (in Chinese).